Algebra 2

An Incremental Development

Second Edition

Algebra 2

An Incremental Development

Second Edition

JOHN H. SAXON, JR.

SAXON PUBLISHERS, INC.

Algebra 2: An Incremental Development
Second Edition

Copyright © 1997 by Saxon Publishers, Inc.

Printed in the United States of America

ISBN: 0-939798-62-X

Editor: Nancy Warren
Production Supervisors: Joan Coleman and David Pond
Graphic Artists: Scott Kirby, John Chitwood, and David Pond

Eleventh printing: June 2001

Printed on recycled paper

┌─ *Reaching us via the Internet* ─┐
WWW: www.saxonpub.com
E-mail: info@saxonpub.com
└────────────────────────────────┘

Saxon Publishers, Inc.
2450 John Saxon Blvd.
Norman, OK 73071

Contents

Preface

This is the second edition of the second book in an integrated three-book series designed to prepare students for calculus. In this book we continue the study of topics from algebra and geometry and begin our study of trigonometry. Mathematics is an abstract study of the behavior and interrelationships of numbers. In Algebra 1, we found that algebra is not difficult—it is just different. Concepts that were confusing when first encountered became familiar concepts after they had been practiced for a period of weeks or months—until finally they were understood. Then further study of the same concepts caused additional understanding as totally unexpected ramifications appeared. And, as we mastered these new abstractions, our understanding of seemingly unrelated concepts became clearer.

Thus, mathematics does not consist of unconnected topics that can be filed in separate compartments, studied once, mastered, and then neglected. Mathematics is like a big ball made of pieces of string that have been tied together. Many pieces touch directly, but the other pieces are all an integral part of the ball, and all must be rolled along together if understanding is to be achieved.

A total assimilation of the fundamentals of mathematics is the key that will unlock the doors of higher mathematics and the doors to chemistry, physics, engineering, and other mathematically based disciplines. In addition, it will also unlock the doors to the understanding of psychology, sociology, and other nonmathematical disciplines in which research depends heavily on mathematical statistics. Thus, we see that mathematical ability is necessary in almost any field of endeavor.

One must be able to apply the fundamental concepts of mathematics automatically if these fundamentals are to be useful. There is insufficient time to relearn basics every time a basic principle must be applied, and familiarity or a slight acquaintance with a basic principle does not suffice for its use. Testing has indicated that many students have only a tenuous grasp of basics at this point even though some topics have been practiced for almost a full year.

Thus, in this book we go back to the beginning—to signed numbers—and then quickly review all of the topics of Algebra 1 and practice these topics as we weave in more advanced concepts. We will also practice the skills that are necessary to apply the concepts. The applicability of some of these skills, such as completing the square, deriving the quadratic formula, simplification of radicals, and complex numbers, might not be apparent at this time, but the benefits of having mastered these skills will become evident as your education continues.

We will continue our study of geometry in this book. Lessons on geometry appear at regular intervals, and one or two geometry problems appear in every homework problem set. We begin our study of trigonometry in Lesson 43 when we introduce the fundamental trigonometric ratios—the sine, cosine, and tangent. We will practice the use of these ratios in every problem set for the rest of the book. The long-term practice of the fundamental concepts of algebra, geometry, and trigonometry will make these concepts familiar concepts and will enable an in-depth understanding of their use in the next book in this series, a pre-calculus book entitled *Advanced Mathematics*.

Problems have been selected in various skill areas, and these problems will be practiced again and again in the problem sets. **It is wise to strive for speed and accuracy when working these review problems.** If you feel that you have mastered a type of problem, don't skip it when it appears again. If you have really mastered the concept, the problem should not be troublesome; you should be able to do the problem quickly and accurately. If you have not mastered the concept, you need the practice that working the problem will provide. **You must work every problem in every problem set to get the full benefit of the structure of this book**. Master musicians practice fundamental musical skills every day. All experts practice fundamentals as often as possible. To attain and maintain proficiency in mathematics, it is necessary to practice fundamental mathematical skills constantly as new concepts are being investigated. And, as in the last book, you are encouraged to be diligent and to work at developing defense mechanisms whose use will protect you against every humans' seemingly uncanny ability to invent ways to make mistakes.

One last word. There is no requirement that you like mathematics. I am not especially fond of mathematics—and I wrote the book—but I do love the ability to pass through doors that knowledge of mathematics has unlocked for me. I did not know what was behind the doors when I began. Some things I found there were not appealing while others were fascinating. For example, I enjoyed being an Air Force test pilot. A degree in engineering was a requirement to be admitted to test pilot school. My knowledge of mathematics enabled me to obtain this degree. At the time I began my study of mathematics, I had no idea that I would want to be a test pilot or would ever need to use mathematics in any way.

I thank Tom Brodsky for his help in selecting geometry problems for the problem sets. I thank Joan Coleman and David Pond for supervising the preparation of the manuscript. I thank Margaret Heisserer, Scott Kirby, John Chitwood, Julie Webster, Smith Richardson, Tony Carl, Gary Skidmore, Tim Maltz, Jonathan Maltz, and Kevin McKeown for creating the artwork, typesetting, and proofreading.

I again thank Frank Wang for his valuable help in getting the first edition of this book finalized and publisher Bob Worth for his help in getting the first edition published.

John Saxon
Norman, Oklahoma

Basic

Course

LESSON A *Geometry review • Angles • Review of absolute value • Properties and definitions*

A.A
geometry review

Some fundamental mathematical terms are impossible to define exactly. We call these terms **primitive terms** or **undefined terms**. We define these terms as best we can and then use them to define other terms. The words **point**, **curve**, **line**, and **plane** are primitive terms.

A **point** is a location. When we put a dot on a piece of paper to mark a location, the dot is not the point because a mathematical point has no size and the dot does have size. We say that the dot is the **graph** of the mathematical point and marks the location of the point. A **curve** is an unbroken connection of points. Since points have no size, they cannot really be connected. Thus, we prefer to say that a curve defines the path traveled by a moving point. We can use a pencil to graph a curve. These figures are curves.

A mathematical **line** is a straight curve that has no ends. **Only one mathematical line can be drawn that passes through two designated points.** Since a line defines the path of a moving point that has no width, a line has width. The pencil line that we draw marks the location of the mathematical line. When we use a pencil to draw the graph of a mathematical line, we often put arrowheads on the ends of the pencil line to emphasize that the mathematical line has no ends.

We can name a line by naming any two points on the line in any order. The line above can be called line *AB*, line *BA*, line *AC*, line *CA*, line *BC*, or line *CB*. Instead of writing the word "line," we can put a bar with two arrowheads above the letters, as we show here.

$$\overleftrightarrow{AB} \quad \overleftrightarrow{BA} \quad \overleftrightarrow{AC} \quad \overleftrightarrow{CA} \quad \overleftrightarrow{BC} \quad \overleftrightarrow{CB}$$

These notations are read as "line *AB*," "line *BA*," etc. We remember that a part of a line is called a **line segment** or just a **segment.** A segment has two endpoints. A segment can be named by naming the two endpoints in any order. The following segment can be called segment *AB* or segment *BA*.

A ●————————————● B

Instead of writing the word "segment," we can draw a bar that has no arrowheads above the letters. Segment *AB* and segment *BA* can be written as

$$\overline{AB} \qquad \text{and} \qquad \overline{BA}$$

If we write the letters without using the bar, we are designating the length of the segment. If segment *AB* has a length of 2 centimeters, we could write either

$$AB = 2 \text{ cm} \qquad \text{or} \qquad BA = 2 \text{ cm}$$

A **ray** is sometimes called a **half line.** A ray has one endpoint, the beginning point, called the **origin.** The ray shown here begins at point A, goes through points B and C, and continues without end.

When we name a ray, we must name the origin first and then name any other point on the ray. We can name a ray by using a line segment with one arrowhead. The ray shown above can be named by writing either

$$\overrightarrow{AB} \qquad \text{or} \qquad \overrightarrow{AC}$$

These notations are read by saying "ray AB" and "ray AC."

A **plane** is a flat surface that has no boundaries and no thickness. Two lines in the same plane either **intersect** (cross) or do not intersect. Lines in the same plane that do not intersect are called **parallel lines.** All points that lie on either of two intersecting lines are in the plane that contains the lines. We say that these intersecting lines determine the plane. Since three points that are not on the same line determine two intersecting lines, we see that three points that are not on the same line also determine a plane.

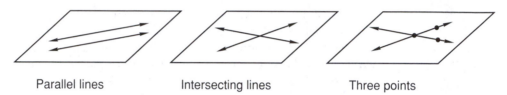

Parallel lines   Intersecting lines   Three points

A.B
angles

The word **angle** comes from the Greek word *ankulos*, meaning "crooked" or "bent." An angle is formed by two rays that have a common endpoint. If the rays point in opposite directions, we say that the angle formed is a **straight angle.** If the rays make a square corner, we say that the rays are **perpendicular** and that the angle formed is a **right angle.** We often use a small square, as in the following figure, to designate a right angle. If the angle is smaller than a right angle, it is an **acute angle.** An angle greater than a right angle but less than a straight angle is called an **obtuse angle.** An angle greater than a straight angle but less than two straight angles is called a **reflex angle.**

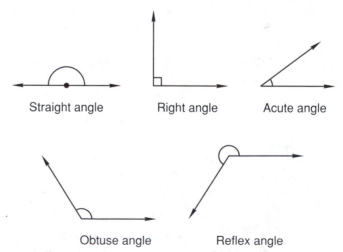

Straight angle   Right angle   Acute angle

Obtuse angle   Reflex angle

If a right angle is divided into 90 parts, we say that each part has a measure of 1 degree. Thus, a right angle is a **90-degree angle.** Two right angles make a straight angle, so a **straight angle is a 180-degree angle.** Four right angles form a **360-degree angle.** Thus, the measure of a circle is 360 degrees. We use a small raised circle to denote degrees. Thus, we can write 90 degrees, 180 degrees, and 360 degrees as 90°, 180°, and 360°.

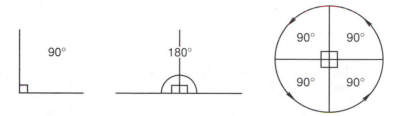

European authors tend to define an angle to be the **opening** between two rays. Authors of U.S. geometry books tend to define the angle to be the **set of points** determined by the two rays.

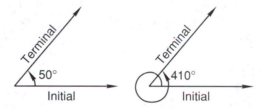

Opening   Set of points

Authors of trigonometry books prefer to define an angle to be a **rotation** of a ray about its endpoint from an **initial position** to a final position called the **terminal position.** We see that the rotation definition permits us to distinguish between a 50° angle and a 410° angle even though the initial and terminal positions are the same.

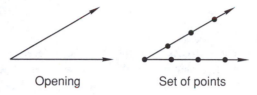

Some angles can be named by using a single letter preceded by the symbol ∠ for angle. The notation ∠A is read as "angle A." Some angles require that we use three letters to name the angle. The notation ∠BAD is read as "angle BAD." When we use three letters, the middle letter names the **vertex** of the angle, which is the point where the two rays of the angle intersect. The other two letters name a point on one ray and a point on the other ray.

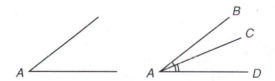

The angle on the left is ∠A. The figure on the right has three angles. The big angle is ∠BAD. Angle BAC and angle CAD are called **adjacent angles** because they have the same vertex, share a common side, and do not overlap (i.e., do not have any common interior points).

If the sum of the measures of two angles is 90°, the angles are called **complementary angles.** If the sum of the measures of two angles is 180°, the angles are called **supplementary angles.**

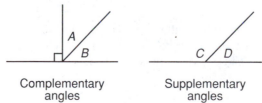

Complementary   Supplementary
angles   angles

In the figures in this book, lines that appear to be straight are straight. Two intersecting straight lines (all lines are straight lines) form four angles. The angles that are opposite each other are called **vertical angles.** Vertical angles are equal angles.

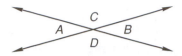

In this figure, angle *A* has the same measure as angle *B*, and angle *C* has the same measure as angle *D*.

It is important to remember that only numbers can be equal. **If we say that two angles are equal, we mean that the number that describes the measure of one angle is equal to the number that describes the measure of the other angle. If we say that two line segments are equal, we mean that the numbers that describe the lengths of the segments are equal.** Both of the following notations tell us that the measure of angle *A* equals the measure of angle *B*.

$$\angle A = \angle B \qquad m\angle A = m\angle B$$

Because excessive attention to the difference between *equal* and *equal measure* tends to be counterproductive, in this book we will sometimes say that angles are equal or that line segments are equal because this phrasing is easily understood. **However, we must remember that when we use the words *equal angles* or *equal segments*, we are describing angles whose measures are equal and segments whose lengths are equal.**

example A.1   Find *x* and *y*.

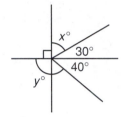

solution   The 30° angle and angle *x* form a right angle, so *x* equals **60.** Thus, angle *x* and the 30° angle are **complementary** angles. The 40° angle and angle *y* form a straight angle. Straight angles are 180° angles, so *y* equals **140.** Thus, angle *y* and the 40° angle are **supplementary** angles.

example A.2   Find *x*, *y*, and *p*.

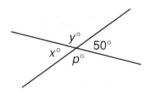

solution   Angle *y* and the 50° angle form a 180° angle. Thus, *y* equals **130.** Because vertical angles are equal angles, *x* equals **50** and *p* equals **130.**

example A.3   Find *x*, *y*, and *p*.

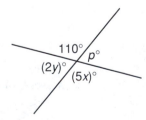

solution   This problem allows us to use the fact that if two angles form a straight angle the sum of their measures is 180°. We see that angle 2*y* and 110° form a straight angle. Also, 5*x* must equal 110 because vertical angles are equal.

STRAIGHT ANGLE	VERTICAL ANGLE
$2y + 110 = 180$	$5x = 110$
$2y = 70$	$x = 22$
$y = 35$	

Since *y* is 35, 2*y* is 70. Thus, **p = 70** because vertical angles are equal.

A.C
review of absolute value

A number is an idea. A numerical expression is often called a numeral and is a single symbol or a collection of symbols that designates a particular number. We say that the number designated is the value of the expression. All of the following numerical expressions designate the number positive three, and we say that each of these expressions has a value of positive three.

$$3 \qquad \frac{7 + 8}{5} \qquad 2 + 1 \qquad \frac{12}{4} \qquad \frac{75}{25} \qquad \frac{16}{2} - 5$$

We have agreed that a positive number can be designated by a numeral preceded by a plus sign or by a numeral without a sign. Thus, we can designate positive three by writing either

$$+3 \qquad \text{or} \qquad 3$$

The number zero is neither positive nor negative and can be designated with the single symbol

$$0$$

Every other real number is either positive or negative and can be thought of as having two qualities or parts. One of the parts is designated by the plus sign or the minus sign. The other part is designated by the numerical part of the numeral. The two numerals

$$+3 \qquad \text{and} \qquad -3$$

designate a positive number and a negative number. The signs of the numerals are different, but the numerical part of each is

$$3$$

We say that this part of the numeral designates the **absolute value** of the number. It is difficult to find a definition of absolute value that is acceptable to everyone. Some people object to saying that the absolute value is the same thing as the "bigness" of a number because "bigness" might be confused with the concept of "greater than" which is used to order numbers. Some explain absolute value by saying that all nonzero real numbers can be paired, each with its opposite, and that the absolute value of either is the positive member of the pair. Thus,

$$+3 \qquad \text{and} \qquad -3$$

are a pair of opposites, and both have an absolute value of 3. Other people prefer to define the absolute value of a number as the number that describes the distance of the graph of the number from the origin. If we use this definition, we see that the graphs of $+3$ and -3 are both 3 units from the origin, and thus both numbers have an absolute value of 3.

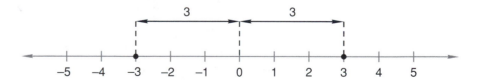

Some people feel that words should not be used to define absolute value because absolute value can be defined exactly by using only symbols and using two vertical lines to indicate absolute value. This definition is in three parts. Unfortunately, the third part can be confusing.

$$\text{(a)} \quad \text{If } x > 0 \qquad |x| = x$$

$$\text{(b)} \quad \text{If } x = 0 \qquad |x| = x$$

$$\text{(c)} \quad \text{If } x < 0 \qquad |x| = -x$$

Part (c) does not say that the absolute value of x is a negative number. It says that if x is a negative number (all numbers less than zero are negative), the absolute value of x is the opposite of x.

Since –15 is a negative number, its absolute value is its opposite, which is +15.

$$|-15| = -(-15) = 15$$

In the same way, if we designate the absolute value of an algebraic expression such as

$$|x + 2|$$

and x has a value such that $x + 2$ is a negative number, then the absolute value of the expression will be the negative of the expression.

If $x + 2 < 0$, $|x + 2| = -(x + 2)$

To demonstrate, we give x a value of -5, and then we will have

$$|-5 + 2| = |-3| = -(-3) = +3$$

No matter how we think of absolute value, we must remember that the absolute value of zero is zero and that the absolute value of every other real number is a positive number.

$$|0| = 0 \qquad |-5| = 5 \qquad |5| = 5 \qquad |-2.5| = 2.5$$

In this book, we will sometimes use the word "number" when the word "numeral" would be more accurate. We do this because overemphasizing the distinction between the two words can be counterproductive.

example A.4   Simplify: $-|-4| - 2 + |-5|$

solution   We will simplify in two steps.

$$-4 - 2 + 5 \qquad \text{simplified}$$

$$-1 \qquad \text{added algebraically}$$

A.D
properties and definitions

Understanding algebra is easier if we make an effort to remember the difference between properties and definitions. A **property** describes the way something is. We can't change properties. We are stuck with properties because they are what they are. For instance,

$$3 + 2 = 5 \qquad \text{and} \qquad 2 + 3 = 5$$

The order of addition of two real numbers does not change the answer. We can understand this property better if we use dots rather than numerals.

Here we have represented the number 5 with 5 dots. Now, on the left below we separate the dots to show what we mean by $3 + 2$, and on the right we show $2 + 3$. The answer is 5 in both cases because there are a total of 5 dots regardless of the way in which they are arranged. We call this property the **commutative property of real numbers in addition.**

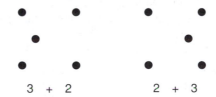

$$3 \;+\; 2 \qquad\qquad 2 \;+\; 3$$

Definitions are different because they are things that we have agreed on. For instance,

$$3^2 \qquad \text{means} \qquad 3 \text{ times } 3$$

It didn't have to mean that. We could have used 3^2 to mean 3 times 2, but we didn't . We note that the order of operations is also a definition. When we write

$$3 + 4 \cdot 5$$

we could mean to multiply first or to add first. Since we cannot have two different answers to the same problem, it is necessary to agree on the meaning of the notation. We have agreed to do multiplication before algebraic addition, and so this expression represents the number +23.

Also, when we wish to write the negative of 3^2, we write

$$-3^2$$

When we wish to indicate that the quantity -3 is to be squared, we write

$$(-3)^2$$

These are definitions of what we mean when we write

$$-3^2 \quad \text{and} \quad (-3)^2$$

and there is nothing to understand. We have defined these notations to have the meanings shown.

The first problem set contains review problems that require us to simplify expressions that contain signed numbers. When these expressions are simplified, try to remember which steps can be justified by properties and which steps can be justified by definitions.

example A.5   Simplify: $(-2)^3 - 2^2 - (-2)^2$

solution   First we simplify each expression and then work the problem.

$$-8 - 4 - 4 \quad \text{simplified}$$

$$\mathbf{-16} \quad \text{added algebraically}$$

problem set A

1. Find y.

2. Find x.

3. Find x, y, and p.

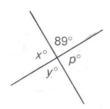

4. Find x, y, and z.

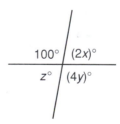

5. The supplement of an angle is $40°$. What is the angle?

6. The complement of an angle is $40°$. What is the angle?

The following problems review operations with signed numbers. Remember that $(-2)^2$ means $(-2)(-2)$, which equals $+4$, and that -2^2 means $-(2)(2)$, which equals -4.
Simplify:

7. $-2 - (-2)$

8. $-3 - [-(-2)]$

9. $-2 - 3(-2 - 2) - 5(-5 + 7)$

10. $-[-2(-5 + 2) - (-2 - 3)]$

11. $-2 + (-2)^3$

12. $-3^2 - 3 - (-3)^2$

13. $-3(-2 - 3 + 6) - [-5(-2) + 3(-2 - 4)]$

14. $-2 - 2^2 - 2^3 - 2^4$

15. $|-2| - |-4 - 2| + |8|$

16. $-|-3(2) - 3| - 2^2$

17. $-2^2 - 2^3 - |-2| - 2$

18. $-3[-1 - 2(-1 - 1)][-3(-2) - 1]$

19. $-3[-3(-4 - 1) - (-3 - 4)]$

20. $-2[(-3 + 1) - (-2 - 2)(-1 + 3)]$

21. $-2[-2(-4) - 2^3](-|2|)$

22. $-8 - 3^2 - (-2)^2 - 3(-2) + 2$

23. $-\{-[-5(-3 + 2)7]\}$

24. $-5 - |-3 - 4| - (3)^2 - 3$

25. $3(-2 + 5) - 2^2(2 - 3) - |-2|$

26. $\dfrac{-5 - (-2) + 8 - 4(5)}{6 - 4(-3)}$

27. $(-2)[|-3 - 4 - 5| - 2^3 - (-1)]$

28. $\dfrac{-3 - (-2) + 9 - (-5)}{7\left(|-3 + 4|\right)}$

29. $4(-2)[-(7 - 3)(5 - 2)2]$

30. $4 - (-4) - 5(3 - 1) + 3(4)(-2)^3$

LESSON B *Perimeter • Area • Volume • Surface area • Sectors of circles*

B.A

perimeter

The **perimeter** of a closed, planar geometric figure is the distance around the figure. The perimeter of this figure is 12.5 units.

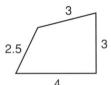

Perimeter = 2.5 + 3 + 3 + 4 = 12.5

We call the perimeter of a circle the **circumference** of the circle. The **radius** of a circle is the distance from the center of the circle to any point on the circle. A **chord** of a circle is a line segment whose endpoints are on the circle. A **diameter** is a chord that passes through the center of the circle. The length of a diameter is twice the length of a radius.

Radius

Chord

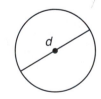
Diameter

It takes about 3.14 diameters to go all the way around a circle. The exact number is a number we call *pi*. We use the symbol π to represent this number. It takes π diameters to equal the circumference, and it takes 2π radii to equal the circumference. When we use 3.14 as an approximation for π, we use the symbol $\approx$, which means "approximately equal to."

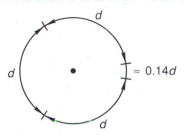

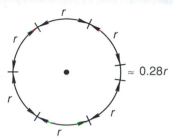

Circumference $= \pi d$
Circumference $\approx 3.14d$

Circumference $= 2\pi r$
Circumference $\approx 6.28r$

From this we see that the number π is the number we get when we divide the circumference of any circle by its diameter.

For any circle $\qquad \dfrac{\text{Circumference}}{\text{Diameter}} = \pi$

We can use this relationship to find the circumference of a circle if we know the diameter and to find the diameter if we know the circumference.

$\pi \times$ diameter $=$ circumference $\qquad\qquad$ Diameter $= \dfrac{\text{circumference}}{\pi}$

To write π as a decimal number would require an infinite number of digits because π is an **irrational number.** A calculator gives a decimal approximation of π as

$$\pi \approx 3.141592654$$

In our calculations that involve the number π we can use 3.14 as an approximation or use the π key on a calculator, which gives a more accurate approximation.

B.B
area

The area of a flat, closed geometric figure is a number that tells us how many squares of a certain size it will take to completely cover the figure. On the left we show a rectangle that is 2 centimeters high and 4 centimeters long.

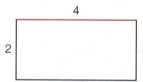

On the right we see that the rectangle can be divided into 2×4, or 8, squares whose sides are 1 centimeter long. The area of each square is 1 square centimeter. From this we see that the area of a rectangle equals the length times the width. Thus the area of the rectangle is 8 square centimeters, which we write as 8 cm^2.

The area of a circle is greater than the area of three squares whose sides have the same length as the radius of the circle, and the area of the circle is less than the area of four of these squares.

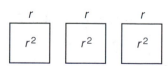

The area of a circle is greater than $3r^2$ and less than $4r^2$. Mathematicians have proved that the area of a circle equals πr^2. Thus, if we know the radius of a circle we can find the circumference of the circle and the area of the circle. Both of these relationships contain the number π.

$$\text{Circumference of circle} = 2\pi r$$

$$\text{Area of circle} = \pi r^2$$

The area of a triangle equals one-half the product of the base and the altitude. The altitude is the perpendicular distance from the base to the opposite vertex. The altitude of a triangle is also called the height of the triangle. The words "height" and "altitude" are used interchangeably.

$$\text{Area of a triangle} = \frac{\text{base} \times \text{height}}{2}$$

The area of each of the triangles shown here is 6 square units, because each triangle has a base of 4 and an altitude of 3. The small squares indicate perpendicular lines.

  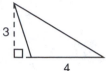

$$\text{Area} = \frac{4 \times 3}{2} = 6 \text{ square units}$$

example B.1   Find the perimeter and the area of this figure. Lines that look parallel are parallel. Dimensions are in meters.

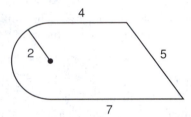

solution   We can subdivide the figure into a semicircle, a rectangle, and a triangle.

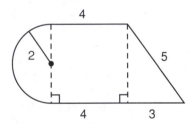

The perimeter is the distance around the figure, which equals the sum of the lengths of the straight sides and half the perimeter of the circle whose radius is 2 m.

$$\text{Perimeter} = 4 + 3 + 5 + 4 + \frac{2\pi(2)}{2}$$

$$= (16 + 2\pi) \text{ m} \approx \textbf{22.28 m}$$

The area of the figure equals the sum of the areas of the semicircle, the rectangle, and the right triangle.

$$\text{Area} = \text{◖} + \square + \triangle$$

$$= \frac{\pi(2)^2}{2} + (4 \times 4) + \frac{3 \times 4}{2} \approx \textbf{28.28 m}^2$$

example B.2   Find the area of the shaded portion of this figure. Dimensions are in meters.

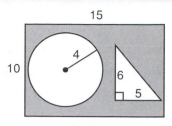

solution   The area of the shaded portion equals the area of the rectangle minus the areas of the circle and the triangle.

$$\text{Area} = 10 \times 15 - \pi(4)^2 - \frac{6 \times 5}{2}$$

$$= 150 - 16\pi - 15 \approx \textbf{84.76 m}^2$$

B.C
volume

A **cube** is a three-dimensional geometric figure whose six sides are identical squares.

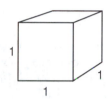

A cube whose edges are all 1 centimeter long has a volume of 1 cubic centimeter. A cube whose edges are all 1 inch long has a volume of 1 cubic inch. A cube whose edges are all 1 kilometer long has a volume of 1 cubic kilometer.

A geometric figure that has three dimensions and thus takes up space is called a **geometric solid.** The volume of a geometric solid is a number that tells us how many cubes of a certain size it will take to fill up the solid. It is helpful to think of sugar cubes when discussing volume. If we have a rectangle that measures 2 cm by 3 cm, it has an area of 6 cm². If we put 1 sugar cube on each square, we see that one layer of cubes has a volume of 6 cubic centimeters. If we stack the cubes 3 deep, we see that we have 3 times 6, or 18 cubes. Thus, the volume of the figure on the right is 18 cubic centimeters. Geometric solids whose sides are perpendicular to the base are called **right geometric solids,** or just **right solids.**

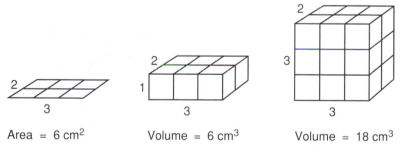

Area = 6 cm²	Volume = 6 cm³	Volume = 18 cm³

The figure on the left below is the figure from example B.1. It has an area of approximately 28.28 square meters. If we use this figure as a base and build sides that are perpendicular to the base and 1 meter high, the container formed will hold 28.28 crushed 1-m sugar cubes.

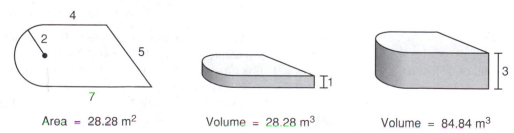

Area = 28.28 m²	Volume = 28.28 m³	Volume = 84.84 m³

If we make the sides 3 meters high, the volume would be 3 times 28.28 m³, or 84.84 m³. We can extend this process to determine the volume of any right solid. We see that **the volume of a right solid equals the area of the base times the height**.

A right solid can also be called a **right cylinder.** If the sides of the base are line segments (the base is a polygon), the cylinder can also be called a **right prism.**

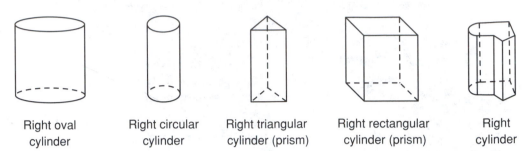

| Right oval cylinder | Right circular cylinder | Right triangular cylinder (prism) | Right rectangular cylinder (prism) | Right cylinder |

The right cylinder on the left is a right **oval** cylinder because the top and the bottom are identical ovals whose surfaces are parallel. The next cylinder is a right **circular** cylinder. The bases of this cylinder are identical circles whose surfaces are parallel. The third cylinder is a right **triangular** cylinder with identical bases that are triangles whose surfaces are parallel. This cylinder can be called a **prism** because its base is a shape whose sides are line segments. The **rectangular** cylinder is also a prism. A cylinder is formed by moving a **line segment** called an **element** around a closed, flat geometric figure. The element is always parallel to a given line.

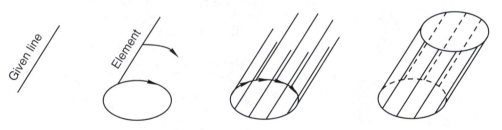

The volume of a cylinder or a prism equals the area of the base times the perpendicular distance between the bases, whether or not the cylinder is a right cylinder.

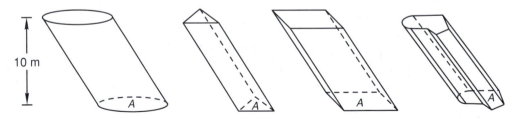

If the perpendicular distance between the bases of each cylinder shown above is 10 m and if the area of each base is 10 m², the volume of each cylinder is 100 m³.

Volume of cylinder (prism) = area of base × height

$$= 10 \text{ m}^2 \times 10 \text{ m} = 100 \text{ m}^3$$

A **cone** is defined as a solid bounded by a closed, flat base and the surface formed by line segments which join all the points on the boundary of the base to a fixed point not in the plane of the base. This point is called the **vertex.** The line from the center of the base to the vertex is called the **axis** of the cone. In a right circular cone the axis is perpendicular to the base. A cone whose base is a polygon is called a **pyramid.** Thus, a pyramid is a cone whose base has straight sides.

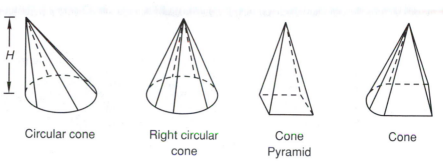

| Circular cone | Right circular cone | Cone Pyramid | Cone |

The altitude of a cone or pyramid is the perpendicular distance from the base to the vertex. **The volume of a cone (pyramid) is exactly one-third the volume of the cylinder (prism) that has the same base and the same height.**

A **sphere** is a perfectly round, three-dimensional shape. Every point on the surface of a sphere is the same distance from the center. This distance is the radius of the sphere.

Sphere

The volume of a sphere is exactly two-thirds the volume of the right circular cylinder into which the sphere fits. The radius of the cylinder equals the radius of the sphere, and the height of the cylinder is twice the radius of the sphere.

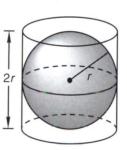

Volume of the sphere equals $\frac{2}{3}$ the volume of the cylinder

Close your eyes and try to remember this diagram. It will help you remember the formula for the volume of a sphere. The first proof of this method of finding the volume of a sphere is attributed to the Greek philosopher Archimedes (287–212 B.C.). There is a formula for the volume of a sphere. See if you can use the figure above to find the formula.

example B.3   Find the volume of this cylinder. The area of the base is 242 m².

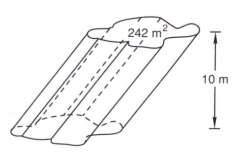

solution   The volume of any cylinder equals the area of the base times the altitude.

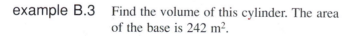

Volume = 242 m² × 10 m = **2420 m³**

example B.4   Find the volume of this cone. The area of the base is 242 m².

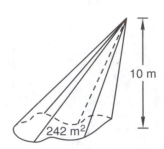

solution   The volume of a cone equals one-third the volume of the cylinder that has the same base and altitude.

$$\text{Volume} = \frac{1}{3}(242 \text{ m}^2)(10 \text{ m}) = \frac{2420}{3} \text{ m}^3 \approx \mathbf{806.67 \text{ m}^3}$$

example B.5   Find the volume of a sphere whose radius is 3 centimeters.

solution   The volume of a sphere is exactly two-thirds the volume of the right circular cylinder that will contain it.

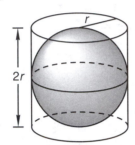

Area of base $= \pi(3)^2 = 9\pi$ cm²

Height $= 2r = 6$ cm

Volume of cylinder $= 6(9\pi) = 54\pi$ cm³

The sphere takes up two-thirds of this volume.

Volume of sphere $= \frac{2}{3}(54\pi)$ cm³ $= \mathbf{36\pi \text{ cm}^3}$

B.D
surface area

The surface area of a geometric solid is the sum of the areas of the faces of the solid. In this book we will restrict our investigation of the surface areas of cylinders to the surface areas of right cylinders.

example B.6   Find the surface area of this right prism. All dimensions are in centimeters.

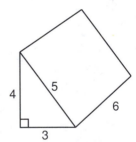

solution   The prism has two ends that are triangles. It has three faces that are rectangles.

$$\text{Area of one end} = \frac{4 \text{ cm} \times 3 \text{ cm}}{2} = 6 \text{ cm}^2$$

$$\text{Area of other end} = \frac{4 \text{ cm} \times 3 \text{ cm}}{2} = 6 \text{ cm}^2$$

Area of bottom $= 3 \text{ cm} \times 6 \text{ cm} = 18 \text{ cm}^2$
Area of back $= 4 \text{ cm} \times 6 \text{ cm} = 24 \text{ cm}^2$
Area of front $= 5 \text{ cm} \times 6 \text{ cm} = \underline{30 \text{ cm}^2}$
Surface area $=$ total $= \mathbf{84 \text{ cm}^2}$

example B.7   Find the surface area of the right circular cylinder shown. Dimensions are in meters.

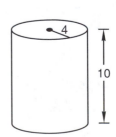

solution   The cylinder has two ends that are circles. The area of one end is πr^2, so the area of both ends is

$$\pi r^2 + \pi r^2 \approx (3.14)(4 \text{ m})^2 + (3.14)(4 \text{ m})^2$$
$$\approx 100.48 \text{ m}^2$$

We can easily calculate the lateral surface area if we think of the cylinder as a tin can which we can cut down the dotted line and then press flat.

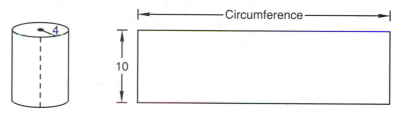

We note that the height of the rectangle is 10 meters and that the length is the circumference of the circle, which is π times the diameter. The radius of this cylinder is 4 meters, so the diameter is 8 meters.

$$\text{Circumference} = \pi d \approx (3.14)(8) \approx 25.12 \text{ m}$$

Thus,

$$\text{Lateral surface area} \approx 10 \text{ m} \times 25.12 \text{ m} \approx 251.2 \text{ m}^2$$

Thus,

$$\text{Total surface area} \approx 100.48 \text{ m}^2 + 251.2 \text{ m}^2 \approx \textbf{351.68 m}^2$$

example B.8   The base of a right cylinder 10 feet high is shown. Find the surface area of the solid. Dimensions are in feet.

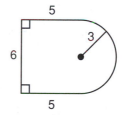 

solution   The surface area equals the sum of the areas of the two equal bases and the lateral surface area.

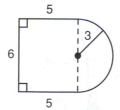

$$\text{Area of one base} = \;\square\; + \;D$$

$$= (\ell \times w) + \frac{\pi r^2}{2}$$

$$= (5 \times 6) + \frac{\pi(3)^2}{2}$$

$$= 30 + \frac{9\pi}{2} \approx 44.13 \text{ ft}^2$$

The lateral surface area of any right solid equals the perimeter times the height. We can see this if we cut our solid and mash it flat. We get

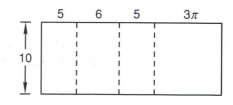

The length of the curved side equals the circumference of a whole circle divided by 2.

$$\text{Length of curve} = \frac{\pi d}{2} = \frac{\pi(2r)}{2} = \pi r = 3\pi$$

The perimeter of the figure is

$$\text{Perimeter} = 5 + 6 + 5 + 3\pi \approx 25.42 \text{ ft}$$

Thus, the lateral surface area is the area of the rectangle.

$$\text{Lateral surface area} \approx (10 \text{ ft})(25.42 \text{ ft}) \approx 254.2 \text{ ft}^2$$

We add this to the surface area of both bases to get the total surface area.

Base area	44.13 ft²
Base area	44.13 ft²
Lateral surface area	254.20 ft²
Total surface area	**342.46 ft²**

There is an easy way to remember how to find the surface area of a sphere. The surface area of a sphere equals the combined areas of four circles, each of which has a radius the same length as the radius of the sphere. A picture to aid your memory shows a grapefruit and four halves of the grapefruit.

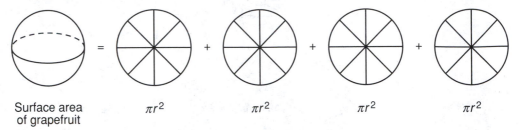

Surface area of grapefruit πr^2 πr^2 πr^2 πr^2

Close your eyes and try to place the diagram above in your memory. If you do, it will be easier to remember that the surface area of a sphere equals the sum of the areas of four circles, each with a radius whose length equals the length of the radius of the sphere.

$$\textbf{Surface area of a sphere} = 4\pi r^2$$

There are formulas for the lateral surface areas of many types of cones, and they are all different. The lateral surface area of a right circular cone equals πrs, where r is the radius of the base and s is the distance from any point on the perimeter of the base to the vertex. This distance is called the **slant height**. The lateral surface area of a pyramid is the sum of the areas of the faces of the pyramid.

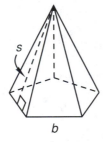

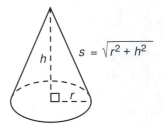

Pyramid

Area of one face $= \dfrac{1}{2}bs$

Right circular cone

Lateral surface area $= \pi rs$

B.E

sectors of circles

A part of a circle is called an **arc.** If we draw two radii to connect the endpoints of the arc to the center of the circle, the area enclosed is called a **sector** of the circle.

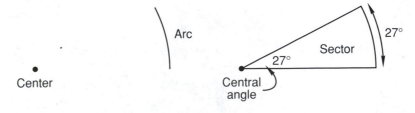

Arc   27°   Sector

Center   Central angle   27°

We note that the degree measure of the arc is the same as the degree measure of the **central angle** formed by the two radii. There are 360 degrees in a circle. One degree of arc is $\frac{1}{360}$ of a circle. Twenty-seven degrees of arc is $\frac{27}{360}$ of a circle. The sector designated by 27° of arc is $\frac{27}{360}$ of the area of the circle. The length of a 27° arc is $\frac{27}{360}$ of the circumference of the circle. If the radius of the circle is 6 m, the area of the circle is $\pi(6)^2$ m² and the circumference of the circle is $2\pi(6)$ m.

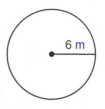

Area $= \pi r^2 = \pi(6)^2$ m² $\approx$ **113 m²**

Circumference $= 2\pi r = 2\pi(6)$ m $\approx$ **37.7 m**

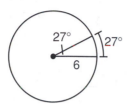

Area of 27° sector $\approx \dfrac{27}{360} \times 113$ m² $\approx$ **8.48 m²**

Length of 27° arc $\approx \dfrac{27}{360} \times 37.7$ m $\approx$ **2.83 m**

example B.9　The radius of this circle is 10 cm. Find the length of a 45° arc and the area of a 45° sector.

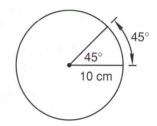

solution　First we find the area of the circle and the circumference of the circle.

$$\text{Area} = \pi r^2 = \pi(10)^2 \text{ cm}^2 \approx 314 \text{ cm}^2$$

$$\text{Circumference} = 2\pi r = 2\pi(10) \text{ cm} \approx \textbf{62.8 cm}$$

The area of the 45° sector is $\frac{45}{360}$ times the area of the whole circle. The length of a 45° arc is $\frac{45}{360}$ times the circumference of the whole circle.

$$\text{Area of 45° sector} \approx \dfrac{45}{360}(314) \text{ cm}^2 \approx \textbf{39.3 cm}^2$$

$$\text{Length of 45° arc} \approx \dfrac{45}{360}(62.8 \text{ cm}) \approx \textbf{7.85 cm}$$

problem set B

1.　The area of the shaded region is the area of the square minus the area of the circle. Find the area of the shaded region. Dimensions are in meters.

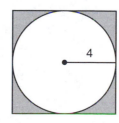

2.　Find the area of the shaded region of this figure by subtracting the areas of the rectangle and the two small triangles from the area of the big triangle. Dimensions are in meters.

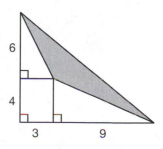

3. The base of the triangle is a diameter of the circle. The altitude of the triangle is a radius of the circle. Find the area of the shaded region. Dimensions are in centimeters.

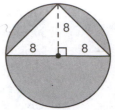

4. Find the perimeter of this figure. Dimensions are in meters.

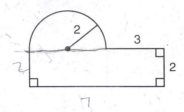

5. Find the area of the 40° sector of the circle. Dimensions are in meters.

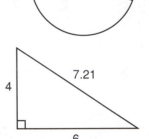

6. The base of a pyramid 10 cm high is the triangle shown. Find the volume of the pyramid. Dimensions are in centimeters.

7. Find the area of the figure. Dimensions are in meters. Then find the volume of a cylinder 8 meters high that has this figure as its base. Corners that look square are square.

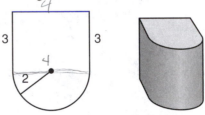

8. Find the volume and the surface area of a sphere whose radius is 6 centimeters.

9. Find the area of a 72° sector of a circle whose radius is 10 cm.

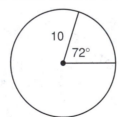

10. Find the perimeter of this figure. Dimensions are in yards. Corners that look square are square.

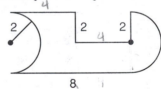

11. Find x, y, and z.

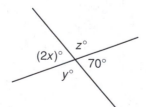

12. Find A.

13. Find B.

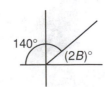

14. The complement of an angle is 10°. What is the measure of the angle?

15. The supplement of an angle is 60°. What is the measure of the angle?

Simplify: -4-8-4-2

16. $-2^2 - 2^3 - (-2)^2 - 2$

17. $-2^2 - |-4| + |4|$

18. $-|-3| - 3 - 3^2$

19. $-4 - (-3)^3 - 2^2 + |-4|$

20. $-3^2 - 2(-4 + 6)$

21. $-4(-2^2 - 3) - 5 + |-3|$

22. $-2[-1 - (-5)] - [-6(-2) + 3]$

23. $-2^2 - 2^3 - 2 - |-2|$

24. $-2 - |-3 - 4 + 8| - 2^2$

25. $-|-2 - 3 - 4| - |-2|$

26. $\dfrac{-5 - (-2) + 8 - 4(5) - 3}{6 - 4(-3)}$

27. $(-2)\big[|-3 + 4 - 5| - 2^3 - (-1)\big]$

28. $\dfrac{-|-5| - (-2) + 6 - 4(3 - |6 - 9|)}{5 - |(4)(-3)|}$

29. $\dfrac{-2 - (-3 - 2) - (-2 + 5)}{-4(2^2 - 3)(-2)}$

30. $-2(-3 + 4 - 6) - 2^2(-2) - 3(-2) - |-5|$

LESSON 1   *Polygons • Triangles • Transversals • Proportional segments*

1.A
polygons

Definitions often change. The definition of a polygon is a good example. The word is formed from the Greek roots *poly*, which means "more than one" or "many," and *gonon*, which means "angle." Thus, polygon literally means "more than one angle." In 1571 Diggs said that "Polygona are such figures that haue moe than foure sides." In 1656 Blount said that a polygon was a geometrical figure that "hath many corners."

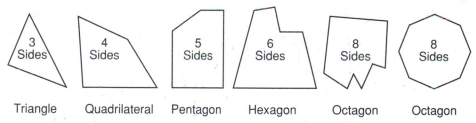

| Triangle | Quadrilateral | Pentagon | Hexagon | Octagon | Octagon |

All of the figures shown here fit Blount's definition of a polygon, but the two on the left do not have enough sides for Diggs's definition. Modern authors tend to define polygons as simple, closed, flat geometric figures whose sides are straight lines. The figures below are not polygons.

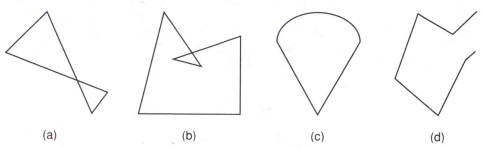

(a)   (b)   (c)   (d)

The sides of the figures (a) and (b) cross, so these are not simple, closed geometric figures. One "side" of (c) is not a straight line, and figure (d) is not a closed figure. The five figures shown below are all polygons. **Note that in each figure the number of vertices (corners) is the same as the number of sides.**

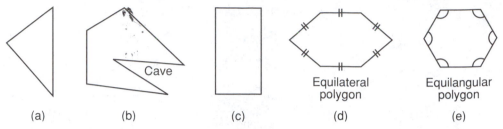

(a)   (b)   (c)   (d)   (e)

Figure (b) has an indentation that we think of as a cave, and this polygon is called a **concave polygon.** Any polygon that does not have a cave is a **convex polygon.** Any two points in the interior of a convex polygon can be connected with a line segment that does not cut a side of the polygon.

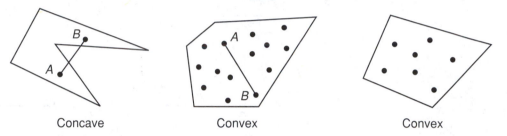

Concave   Convex   Convex

If all the angles of a polygon have equal measures and all sides have equal lengths, the polygon is called a **regular polygon.** Polygons that are not regular polygons are called **irregular polygons.**

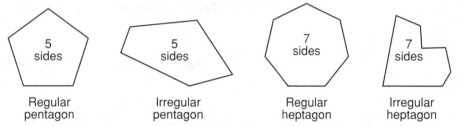

Regular   Irregular   Regular   Irregular
pentagon   pentagon   heptagon   heptagon

The name of a polygon tells the number of sides the polygon has.

NUMBER OF SIDES	NAME	NUMBER OF SIDES	NAME
3	Triangle	9	Nonagon
4	Quadrilateral	10	Decagon
5	Pentagon	11	Undecagon
6	Hexagon	12	Dodecagon
7	Heptagon	n	n-gon
8	Octagon		

Although these names are useful, we will not concentrate on memorizing them. Some polygons of more than 12 sides have special names, but these names are not used often. Instead, we use the word **polygon** and tell the number of sides or use the number of sides with the suffix **-gon**. Thus, if a polygon has 143 sides, we would call it a polygon with 143 sides or a 143-gon. The endpoints of one side of a polygon are called **consecutive vertices,** and two adjacent sides are called **consecutive sides.** A **diagonal** of a polygon is a line segment that connects two nonconsecutive vertices. In the figures below, the dashed lines represent diagonals.

Diagonals   Diagonals

1.B
triangles

If a triangle has a right angle, the triangle is a **right triangle**. If one angle is greater than 90°, the triangle is an **obtuse triangle**. If all angles are less than 90°, the triangle is an **acute triangle.**

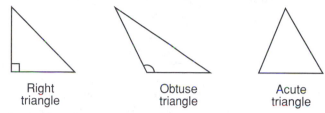

| Right triangle | Obtuse triangle | Acute triangle |

The sum of the measures of the three angles of any triangle is 180°. The greatest angle is opposite the longest side, and the smallest angle is opposite the shortest side. In the triangle on the left below, we know that the length of side C is greater than the length of side B because 80° is greater than 70°.

In the triangle on the right, we know that C is the smallest angle because it is opposite the shortest side. Angle A is the largest angle because it is opposite the longest side.

In a triangle, the angles opposite sides of equal lengths have equal measures. The sides opposite angles of equal measures have equal lengths.

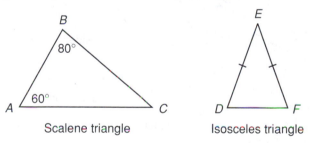

Scalene triangle   Isosceles triangle

Angle C in the figure on the left must be a 40° angle, because the sum of all three angles must be 180°. **In the same figure all the angles have different measures, so all the sides must have different lengths.** If all the sides of a triangle have different lengths, the triangle is called a **scalene triangle.** The identical tick marks on two sides of the triangle on the right above tell us that these two sides have equal lengths. Thus, angle D and angle F must have equal measures. A triangle that has at least two equal sides (and two equal angles) is called an **isosceles triangle.** The word **isosceles** derives from the Greek prefix *iso-*, meaning "equal," and the Greek word *skelos,* meaning "leg."

The triangle shown below has three sides whose lengths are equal, so we call this triangle an **equilateral triangle,** from the Latin *equi-* meaning "equal" and *latus* meaning "side."

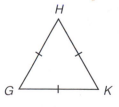

Equilateral triangle

The identical tick marks indicate that the lengths of the sides are equal. Since angles opposite equal sides have equal measures, all three angles in this triangle must have equal measures.

Each angle must have a measure of 60° because 3 × 60° equals 180°. Since an equilateral triangle has at least two sides whose lengths are equal, an equilateral triangle is also an isosceles triangle. We summarize these very important properties of triangles in the following boxes.

> If two sides of a triangle have equal lengths, the angles opposite these sides have equal measures. If two angles of a triangle have equal measures, the sides opposite these angles have equal lengths.

> When the three sides of a triangle have equal lengths, all three angles are 60° angles. If the three angles of a triangle are equal, they must be 60° angles and the three sides must have equal lengths.

example 1.1   Find x and y.

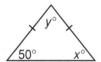

solution   **In any triangle the angles opposite equal sides are equal angles.** Thus x is 50 and angle x is a **50° angle.** The sum of three angles in a triangle is 180°, so y must be 80 and angle y is an **80° angle.**

example 1.2   Find x and y.

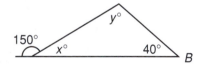

solution   The 150° angle and angle x form a 180° angle. Thus, angle x is a **30° angle.** Since angle B is a 40° angle, angle y must be a **110° angle** so that the sum of the angles will be 180°. We check by adding all three angles.

$$30 + 40 + 110 = 180 \qquad \text{check}$$

example 1.3   This triangle is an equilateral triangle. Find D.

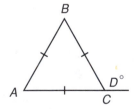

solution   If the triangle is an equilateral triangle, all three angles are equal and each angle is a 60° angle. Angle D and one of the 60° angles form a straight angle. Thus, angle D is a 120° angle.

$$m\angle D = \mathbf{120°}$$

1.C

transversals   A **transversal** is a line that cuts or intersects two or more other lines. **If a transversal intersects two or more lines that are parallel and if the transversal is perpendicular to one of the parallel lines, it is perpendicular to all the parallel lines.** We use the symbol ∥ to mean parallel and ∦ to mean not parallel.

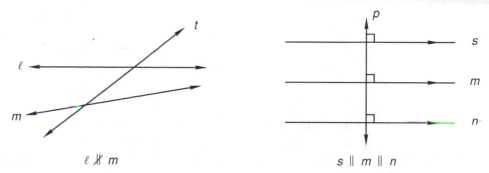

$\ell \not\parallel m$ $s \parallel m \parallel n$

In the left-hand figure, line *t* is a transversal because it intersects both line *m* and line *ℓ*. In the right-hand figure, line *p* is a transversal because it intersects lines *s*, *m*, and *n*. These lines are parallel lines, as indicated by the arrowheads that are not on the ends of the segments. Because the transversal *p* is perpendicular to one of the parallel lines, it is perpendicular to all the parallel lines. We omit the arrowheads on the ends of these lines because the arrowheads would clutter the diagram.

If the transversal is not perpendicular to the lines, two groups of equal angles are formed. **Half the angles are "large angles" that are equal angles and are greater than 90°. Half the angles are "small angles" that are equal angles and are less than 90°.**

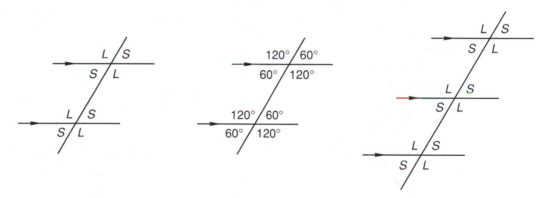

On the left we use the letters *L* and *S* to mean "large" and "small." We note that together a large angle and a small angle form a straight angle (180°), so the large angles and the small angles are supplementary angles. In the center figure, the large angles are 120° angles and the small angles are 60° angles. Thus, the sum of any large angle and any small angle is 180°. In the figure on the right, we see that when a transversal cuts three parallel lines, six equal large angles are formed and six equal small angles are formed.

example 1.4   Find *A*, *B*, and *C*.

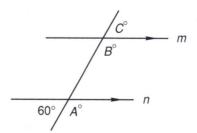

solution   Angle *A* and the 60° angle form a straight angle, which measures 180°. Thus, *A* is **120.** Lines *m* and *n* are parallel, so all the small angles are equal and all the large angles are equal. Thus, *C* equals **60** and *B* equals **120.**

1.D

proportional segments   When three or more parallel lines are cut by two transversals, the lengths of the corresponding segments of the transversals are proportional. This means that the lengths of the segments of one transversal are related to the lengths of the corresponding segments of the other transversal by a number called the **scale factor.**

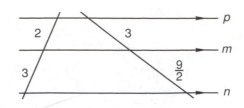

In this figure we use arrowheads that are not on the ends of the segments to indicate that lines p, m, and n are parallel. The left-to-right scale factor for this figure is $\frac{3}{2}$. The arrowhead tells us that the scale factor is from left to right. This means that $\frac{3}{2}$ times the length of any segment on the left equals the length of the corresponding segment on the right.

Segment Length	Times Scale Factor	Corresponding Segment Length
2	$2 \times \overrightarrow{\frac{3}{2}}$ =	3
3	$3 \times \overrightarrow{\frac{3}{2}}$ =	$\frac{9}{2}$

In the same figure the scale factor from right to left is $\frac{2}{3}$. This number is the reciprocal of the scale factor from left to right.

Segment Length	Times Scale Factor	Corresponding Segment Length
3	$3 \times \overleftarrow{\frac{2}{3}}$ =	2
$\frac{9}{2}$	$\frac{9}{2} \times \overleftarrow{\frac{2}{3}}$ =	3

example 1.5 The arrowheads indicate that the lines are parallel. Find x.

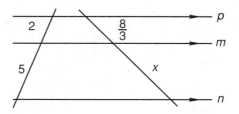

solution The segments whose lengths are 2 and $\frac{8}{3}$ are corresponding segments. Thus, 2 times the left-to-right scale factor equals $\frac{8}{3}$.

$$2\overrightarrow{SF} = \frac{8}{3}$$

To solve for $\overrightarrow{SF}$, we multiply both sides by $\frac{1}{2}$ and simplify.

$$\left(\frac{1}{2}\right)2\overrightarrow{SF} = \frac{1}{2}\left(\frac{8}{3}\right) \quad \text{multiplied by } \frac{1}{2}$$

$$\overrightarrow{SF} = \frac{4}{3} \quad \text{simplified}$$

Now 5 times the left-to-right scale factor equals x.

$$5\overrightarrow{SF} = x \quad \text{equation}$$

$$5\left(\frac{4}{3}\right) = x \quad \text{used } \frac{4}{3} \text{ for } \overrightarrow{SF}$$

$$\frac{20}{3} = x \quad \text{multiplied}$$

practice   **a.** $m\angle A = 35°$. Find $m\angle C$ and $m\angle B$.

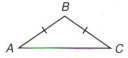

b. Find x and y.

c. Find A, B, and C.

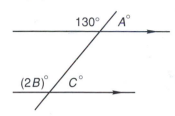

d. Find x.

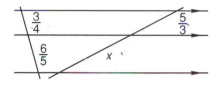

problem set 1

1. Find x and y.

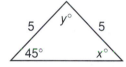

2. Find x and y.

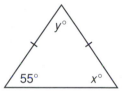

3. Find A, B, and C.

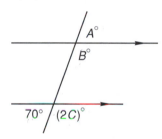

4. Find x.

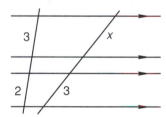

5. The diameter of the circle is 8 cm, as shown. The sum of the top two shaded areas is the area of half the circle minus the area of the triangle. The shaded area below is a 60° sector of the circle. Find the sum of the three shaded areas.

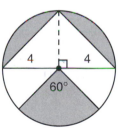

6. Find the area of the shaded portion of this figure. Dimensions are in centimeters.

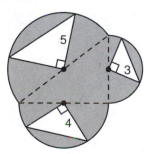

7. Find the perimeter of the figure. All angles that look like right angles are right angles. Dimensions are in feet.

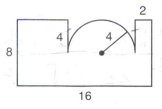

8. Find the volume of a right circular cylinder whose radius is 4 ft and whose height is 8 ft. Find the volume of a sphere whose radius is 4 ft.

9. Find x, y, and p.

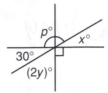

10. Find x, y, and p.

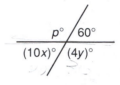

11. The complement of an angle is 17°. What is the measure of the angle?

12. The figure below is the base of a cone whose altitude is 7 meters. Find the volume of the cone.

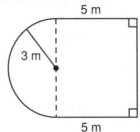

13. The radius of each circle is r ft. What is the area of the square?

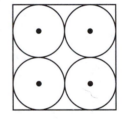

Simplify:

14. $-[-2(-3 - 2) - (-2 - 3)]$

15. $-2[-2 - 3(-2 - 2)][-2(-4) - 3]$

16. $-2^2 - 2^3[-2 + 3(-2)] - |-2^3|$

17. $-3 - 2^3 - 4^2 - |-2 - 3(2)|$

18. $-\{-2[(-3 + 7) - (-2)][-3(-2 + 1)]\}$

19. $3^2 - 3^3 + 3^4 - (-3)^3 - 3$

20. $-(-4)^2 - 4|-2| - 2^3 + |-11 - 4|$

21. $6 - \{[3^2 - 8 + (-2)][-(4 - 6)(-3)^2 + 2]2\}$

22. $-[-(-2)] - |-4 - 3|2^2 - 4$

23. $(-|-3|)[(2 - 7)(-3 - 2) + (-2)^2]$

24. $\dfrac{-|-4| - (-3) + 7 - 6(4 - |7 - 11|)}{7 - |(3)(-2)|}$

25. $-(-3 - 2)(-7 - |-3 - 2|) - (-3)^2$

26. $(-3)[|-2 - 7 - 2| - (-3)^2 - (-2)]$

27. $\dfrac{-4 - (-3) + 7 - 6(2)}{7 - (3)(-2)}$

28. $3 - 5 - 2^2 - 4^2(-1)(-3 - |-2 - 5| - 3)$

29. $-8 + (-3)(-2)^2 + (-7) - 2(-4 - 2)$

30. $6(-3)[-(5 - 4)(6 - 2)3]$

LESSON 2   *Negative exponents • Product and power theorems for exponents • Circle relationships*

2.A
negative exponents

Negative exponents cannot be "understood" because they are the result of a definition, and thus there is nothing to understand. We define 2 to the third power as follows:

$$2^3 = 2 \cdot 2 \cdot 2$$

We have agreed that 2^3 means 2 times 2 times 2. In a similar fashion, we define 2 to the negative third power to mean 1 over 2 to the third power.

$$2^{-3} = \frac{1}{2^3}$$

Thus, we have two ways to write the same thing. We give the formal definition of negative exponents as follows:

DEFINITION OF x^{-n}

If n is any real number and x is any real number that is not zero,

$$x^{-n} = \frac{1}{x^n}$$

This definition tells us that when we write an exponential expression in reciprocal form, the sign of the exponent must be changed. If the exponent is negative, it is positive in reciprocal form; and if it is positive, it is negative in reciprocal form. In the definition we say that x cannot be zero because division by zero is undefined.

example 2.1   Simplify: (a) $\dfrac{1}{3^{-2}}$ (b) 3^{-3} (c) -3^{-2} (d) $(-3)^{-2}$ (e) $-(-3)^{-3}$

solution   (a) $\dfrac{1}{3^{-2}} = 3^2 = \mathbf{9}$ (b) $3^{-3} = \dfrac{1}{3^3} = \dfrac{1}{27}$

(c)   Negative signs and negative exponents in the same expression can lead to confusion. If the negative sign is not "protected" by parentheses, a good ploy is to cover the negative sign with a finger. Then simplify the resulting expression and remove the finger as the last step.

$$-3^{-2} \qquad \text{problem}$$

$$ 3^{-2} \qquad \text{covered minus sign}$$

$$ \frac{1}{3^2} \qquad \text{equivalent expression}$$

$$ \frac{1}{9} \qquad \text{simplified}$$

$$-\frac{1}{9} \qquad \text{removed finger}$$

(d)   When we try to slide our finger over the minus sign in (d), we find that we cannot because the minus sign is "protected" by the parentheses.

$$(-3)^{-2} \qquad \text{problem}$$

$$\overrightarrow{\parallel\!D}\,(-3)^{-2}\qquad\text{``protected''}$$

$$\frac{1}{(-3)^2}\qquad\text{equivalent expression}$$

$$\frac{1}{9}\qquad\text{simplified}$$

(e)   One of the minus signs is "unprotected."

$$-(-3)^{-3}\qquad\text{problem}$$

$$\overrightarrow{\parallel\!D}\,(-3)^{-3}\qquad\text{covered minus sign}$$

$$\overrightarrow{\parallel\!D}\,\frac{1}{(-3)^3}\qquad\text{equivalent expression}$$

$$\overrightarrow{\parallel\!D}\,-\frac{1}{27}\qquad\text{simplified}$$

$$-\left(-\frac{1}{27}\right)=\frac{1}{27}\qquad\text{removed finger}$$

2.B
product theorem for exponents

We remember that x^2 means x times x

$$x^2 = x \cdot x$$

and x^3 means x times x times x

$$x^3 = x \cdot x \cdot x$$

Using these definitions, we can find an expression whose value equals to x^2 times x^3.

$$x^2 \cdot x^3 \quad\text{means}\quad x \cdot x \quad\text{times}\quad x \cdot x \cdot x \quad\text{which equals } x^5$$

This demonstrates the product theorem for exponents, which we state formally in the following box.

PRODUCT THEOREM FOR EXPONENTS

If m and n and x are real numbers and $x \neq 0$,

$$x^m \cdot x^n = x^{m+n}$$

This theorem holds for all real number exponents.

example 2.2   Simplify: $x^2yx^{-5}y^{-4}x^5x^0$

solution   We simplify by adding the exponents of like bases and get

$$x^2y^{-3}$$

example 2.3   Simplify: $\dfrac{yy^{-3}x^4y^5x^{-10}}{y^{-6}x^{-3}y^{10}x^2}$

solution   First we simplify the numerator and the denominator. Then we decide to write the answer with all factors in the numerator.

$$\frac{y^3x^{-6}}{y^4x^{-1}} = y^{-1}x^{-5}$$

2.C
power theorem
for exponents

We can use the product theorem to expand $(x^2)^3$ as

$$(x^2)^3 = x^2 \cdot x^2 \cdot x^2 = x^6$$

This procedure generalizes to the power theorem for exponents.

POWER THEOREM FOR EXPONENTS

If m and n and x are real numbers,

$$(x^m)^n = x^{mn}$$

This theorem can be extended to any number of exponential factors.

EXTENSION OF THE POWER THEOREM

If the variables are real numbers,

$$(x^m y^a z^b k^c \ldots)^n = x^{mn} y^{an} z^{bn} k^{cn} \ldots$$

---

example 2.4   Simplify: $\dfrac{x(x^{-3})^2 y(xy^{-2})^{-3}}{(y^2)^3 y^{-3}(x^2)^3}$

solution   First we will use the power theorem in both the numerator and the denominator and get

$$\frac{xx^{-6}yx^{-3}y^6}{y^6 y^{-3} x^6}$$

Now we simplify both the numerator and the denominator, and as the last step, we decide to write all exponential expressions with positive exponents.

$$\frac{x^{-8}y^7}{y^3 x^6} = \frac{y^4}{x^{14}}$$

2.D
circle
relationships

If we know the area of a circle, we can find the diameter of the circle and can find the radius of the circle. If we know the circumference of a circle, we can also find the diameter and the radius of the circle.

example 2.5   The area of a circle is 12.2 m². What is the approximate circumference of the circle?

solution   First we find the radius.

$$\pi r^2 = \text{area} \qquad \text{equation}$$

$$\pi r^2 = 12.2 \qquad \text{substituted}$$

$$r^2 = \frac{12.2}{\pi} \qquad \text{divided by } \pi$$

$$r = \sqrt{\frac{12.2}{\pi}} \qquad \text{square root of both sides}$$

$$r \approx 1.97 \text{ m} \qquad \text{simplified}$$

We used a calculator and rounded the answer to two decimal places, so the answer is not exact. We indicate that the answer is not exact by using the symbol ≈ for "approximately equal to." The circumference equals $2\pi r$, so now we can find the circumference.

$$\begin{aligned}
\text{Circumference} &= 2\pi r & \text{equation} \\
&\approx 2\pi(1.97) & \text{substituted} \\
&\approx \textbf{12.38 m} & \text{simplified}
\end{aligned}$$

example 2.6   The circumference of a circle is 8π cm. What is the area of the circle?

solution   First we find the radius.

$$\begin{aligned}
\text{Circumference} &= 2\pi r & \text{equation} \\
8\pi &= 2\pi r & \text{substituted} \\
\frac{8\pi}{2\pi} &= r & \text{divided by } 2\pi \\
4 \text{ cm} &= r & \text{simplified}
\end{aligned}$$

Now we can use 4 cm for r to find the area.

$$\begin{aligned}
\text{Area} &= \pi r^2 & \text{equation} \\
&= \pi(4 \text{ cm})^2 & \text{substituted} \\
&= \textbf{16}\pi \textbf{ cm}^2 & \text{simplified}
\end{aligned}$$

practice   Simplify:

a. -4^{-2}

b. $-(-4)^{-2}$

c. $\dfrac{(x^2 y^{-2})^0 (x^{-3} y)^{-2}}{y^{-8} x^4 y^2 x^3}$

d. The area of a circle is 49π cm². What is the circumference of the circle?

problem set 2

1. Find x.

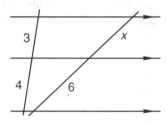

2. Find x and y.

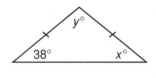

3. The base of a cylinder is a right triangle topped by a 60° sector of a circle, as shown. If the dimensions are in meters and the height of the cylinder is 8 meters, what is the volume of the cylinder?

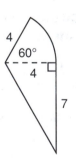

4. Find A, B, and C.

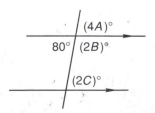

5. Find A, B, and C.

6. The area of the square is 16 cm². What is the length of one side? The circles inside the square are all the same size. What is a radius of one circle? What is the area of one circle?

7. The volume of this circular cylinder is 250π cm³. What is the height of the cylinder? Dimensions are in centimeters.

8. The figure shown is the base of a cone whose altitude is 4 meters. What is the volume of the cone? Dimensions are in meters.

Simplify. Write answers with all exponential expressions in the numerator.

9. $\dfrac{xx^2(x^0y^{-1})^2}{x^2x^{-5}(y^2)^5}$

10. $\dfrac{m^2p^0(m^{-2}p)^2}{m^{-2}p^{-1}(m^{-3}p^2)^3}$

11. $\dfrac{(x^2y)^0xy}{x^2(y^{-2})^3}$

12. $\dfrac{(a^2b^0)^2ab^{-2}}{a^2b^{-2}(ab^{-3})^2}$

Simplify. Write answers with positive exponents.

13. $\dfrac{(xm^{-1})^{-3}x^2m^2}{(x^0y^2)^{-2}xy}$

14. $\dfrac{(c^2d)^{-3}c^{-5}}{(c^2d^0)^{-2}d^3}$

15. $\dfrac{(m^2n^{-5})^{-2}m(n^0)^2}{(m^2n^{-2})^{-3}m^2}$

16. $\dfrac{(x^{-2}y^5)^3(x^2)^0y}{xy^{-3}x^{-2}}$

17. $\dfrac{(b^2c^{-2})^{-3}c^{-3}}{(b^2c^0b^{-2})^4}$

Simplify. Write answers with negative exponents.

18. $\dfrac{(abc)^{-3}c^2b}{a^{-4}bc^2a}$

19. $\dfrac{kL^2k^{-2}}{(k^0L)^2L^{-3}k}$

20. $\dfrac{s^2ym^{-3}}{(s^0t^2)^{-3}m^{-3}st}$

21. $\dfrac{(x^{-3}yz^{-3})^2xy^0}{(xy^0z^{-2})^{-3}xy}$

22. $\dfrac{x^{-3}y^2xy^4}{(x^{-2}y)^3y^{-3}x}$

Simplify:

23. -3^{-2}

24. $\dfrac{1}{-2^{-3}}$

25. $-3^2 - [-2^0 - (3-2) - 2]$

26. $-2\{[-3 - 2(-2)][-2 - 3(-2)]\}$

27. $2\{-3^0[(-5-2)(-3) - 2]\}$

28. $-3[4^0 - 7(2-3) - 2^2]$

29. $-|-2-3| - (-5) - 3^3$

30. $-|-3^2 - 2| - 2^0 - (-3)$

LESSON 3 *Evaluation of expressions • Adding like terms*

3.A
evaluation of expressions

We remember that a numerical expression is a meaningful arrangement of numerals and symbols that designate operations. Thus, each of the following can be called a numerical expression.

$$4 \qquad 2 + 2 \qquad \frac{7 + 3 + 14}{6} \qquad \sqrt{16} \qquad 48 \div 12$$

Every numerical expression represents a single number. We say that this number is the **value** of the expression. The value of each of the above expressions is 4, for each expression is a different way to designate the number 4.

Every numerical expression is also an **algebraic expression.** An algebraic expression can contain letters that represent unspecified numbers. We call the letters **variables.** The value of the algebraic expression

$$x + 4$$

depends on the number we use as a replacement for x. If we replace x with -32, the expression will have a value of -28.

$$(-32) + 4 = -28$$

When we replace the variables in an expression with selected numbers and simplify, we say that we have **evaluated** the expression.

example 3.1 Evaluate: $x^2y - y$ if $x = -2$ and $y = -4$

solution We replace x with -2 and y with -4.

$$(-2)^2(-4) - (-4) = (+4)(-4) + 4 = -16 + 4 = -12$$

example 3.2 Evaluate: $a(-b - a) - ab$ if $a = -2$ and $b = 4$

solution We replace a with -2 and b with 4.

$$-2[-(4) - (-2)] - (-2)(4) = -2(-2) - (-8) = 4 + 8 = 12$$

3.B
adding like terms

Like terms are terms whose literal components represent the same number regardless of the numbers used to replace the variables. Thus,

$$3xyz \qquad \text{and} \qquad -2zyx$$

are like terms because xyz and zyx have the same value regardless of the replacement values of the variables. We demonstrate this by replacing x with 2, y with 3, and z with 4.

xyz	zyx
$(2)(3)(4)$	$(4)(3)(2)$
$(6)(4)$	$(12)(2)$
24	**24**

We add like terms by adding the coefficients of the terms, as shown in the following examples.

example 3.3 Simplify by adding like terms: $3xy - 2x + 4 - 6yx + 3x$

solution We add like terms and get

$$-3yx + x + 4$$

example 3.4   Simplify by adding like terms:

$$\frac{3a^{-2}b}{c} - \frac{4b}{a^2c} + 7a^{-2}c^{-1}b - \frac{4ba^2}{c}$$

solution   If we write the terms in the same form, we can see which terms are like terms. This time we choose to write the terms with all exponents positive, and we get

$$\frac{3b}{a^2c} - \frac{4b}{a^2c} + \frac{7b}{a^2c} - \frac{4ba^2}{c}$$

We see that the first three terms are like terms and can be added by adding the numerical coefficients. We do this and get

$$\frac{6b}{a^2c} - \frac{4ba^2}{c}$$

practice
a.   Evaluate: $ab^2 - b$ if $a = 2$ and $b = -3$

b.   Evaluate: $xy - (-xy + y)$ if $x = 2$ and $y = -3$

c.   Simplify by adding like terms: $\dfrac{2a^{-3}x}{m} - \dfrac{5x}{a^3m} + \dfrac{a^3}{m^{-1}x}$

problem set 3

1.   Find x, y, and z.

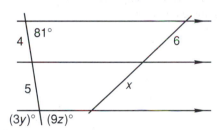

2.   The volume of this circular cone is 48π m^3. What is the circumference of the base?

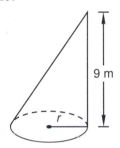

3.   Find A and B.

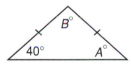

4.   This figure is the base of a cone that is 10 cm tall. What is the volume of the cone? Dimensions are in centimeters.

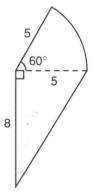

5.   In this figure two circles have a radius of 8 cm. One circle has a radius of 6 cm, and one circle has a radius of 5 cm. If the pairs of circles are tangent (touching at a single point) as shown, what is the perimeter of the quadrilateral $PQRS$?

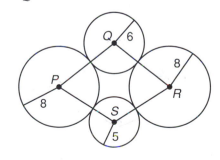

Evaluate:

6. $x - |x|y^2 - xy$ if $x = -2$ and $y = -3$

7. $(a - b) - a(-b)$ if $a = -5$ and $b = 3$

8. $-a(a - ax)(x - a)$ if $a = -2$ and $x = 4$

9. $a^2 - y^3(a - y^2)y$ if $a = -2$ and $y = -3$

10. $-p^2 - p(a - p^2)$ if $a = 4$ and $p = -3$

11. $a^2 - y^3(a - y^2)y^2$ if $a = -2$ and $y = 3$

12. $a^2 - a(x - ax)$ if $a = 1$ and $x = 2$

Simplify by adding like terms:

13. $\dfrac{p^2 x^4}{m^5} - \dfrac{2p^4 m^5}{x^{-4}} - \dfrac{3p^2 x^2}{x^{-2} m^5} + \dfrac{7p^2 m^{-5} m^{10} x^4}{p^{-2}}$

14. $-\dfrac{m^4 x^5}{k^5} + \dfrac{2m^2 x^5}{k^5 m^{-2}} - \dfrac{3m^3 x^2 k}{m^{-4} x^3 k^4}$

15. $-2x^5 y^4 + \dfrac{3xy^3}{x^{-4} y^{-1}} + \dfrac{4x^3 y^2}{x^2 y}$

16. $2xy^2 m - \dfrac{3x^2 y^2 m^4}{m^3 x} + \dfrac{2x^2 ym^3}{mx^2}$

Simplify:

17. $\dfrac{(x^0 y^2)^{-3} y^{-2} p^0}{(x^2)^{-4}(y^2)^0 (p^3)^{-2}}$

18. $\dfrac{(mxy^2 p)^2 p^{-2} x^2}{(p^0 xmy^3)^{-2} xp^{-2}}$

19. $\dfrac{xx^2(x^{-2})^{-2}(mpx^2)^{-4}}{(x^0)^2(x^2)^0(x^2 mp^{-2})^3}$

20. $\dfrac{p^2 x^{-4} k^5 (p^2 k)^{-2}}{(p^2 x^{-3})^{-2}}$

21. $\dfrac{x^2 xx^0 (x^{-2})^2}{xx^3 x^{-14}(x^{-2})^{-3}}$

22. $\dfrac{(x^2 y^{-2} p^0)^{-3} p^2}{x^2 (x^{-4})^0 (p^{-2} y^5)^{-2}}$

23. $\dfrac{(x^{-2} y^2)^5 (x^0 y^{-2})^{-4}}{(x^{-4} yy^2)^2 (p^{-4})^0}$

Simplify:

24. $-3^{-2} + \dfrac{1}{2^{-3}}$

25. $-(-2)^3 - \dfrac{1}{(-3)^{-2}}$

26. $-3[-2 - 2 - (-3)](-2 - 3)$

27. $-2(-3 + 7^0) - |(-2 - 3)|$

28. $|-2| - 3^2 - (-3)^3 - 2$

29. $-2\{[(-3 - 2)(-2)](-2 - 3)\}$

30. $-[(-3)(-2) - (-3)(-2 + 4)]$

LESSON 4   *Distributive property* • *Solution of equations* • *Change sides–change signs*

4.A

distributive property

The distributive property is a property of real numbers that permits two approaches to the simplification of expressions such as

$$3(4 + 5)$$

We find that we can get the same answer with either approach.

ADD FIRST	MULTIPLY FIRST
3(4 + 5)	3(4 + 5)
3(9)	$3 \cdot 4 + 3 \cdot 5$
27	12 + 15
	27

On the left, we added 4 and 5 to get 9 and then multiplied by 3. On the right, we multiplied first and then added. The answer was the same in both cases. We say that on the right we *distributed* the multiplication over the addition.

example 4.1   Expand: $\dfrac{4a^2}{b}\left(\dfrac{b^{-2}}{a^4} - \dfrac{3ba}{a^2}\right)$

solution   **Since letters stand for unspecified numbers, all the rules for numbers also hold for letters.** Two multiplications are indicated. We do them both and then simplify.

$$\frac{4a^2 b^{-2}}{ba^4} - \frac{12a^2 ba}{ba^2} = \frac{4}{b^3 a^2} - 12a$$

example 4.2   Expand: $\dfrac{a^{-3}b^0}{c}\left(\dfrac{a^2 bc}{c^2} - \dfrac{3a^{-2}}{b^{-2}}\right)$

solution   The distributive property permits two multiplications. Then we simplify.

$$\frac{a^{-3}a^2 bc}{c^3} - \frac{3a^{-3}a^{-2}}{cb^{-2}} = \frac{a^{-1}b}{c^2} - \frac{3a^{-5}}{cb^{-2}}$$

4.B
solutions of equations

We remember from algebra 1 that the two rules for solving equations are the addition rule and the multiplication/division rule. These rules are extensions of the additive property of equality and the multiplicative property of equality, and these rules apply to both true equations and false equations.

(a) 4 + 3 = 7   true   (b) 4 + 3 = 5   false

We can add the same number to both sides of an equation without changing the truth or falsity of the equation. We will demonstrate this by adding -5 to both sides of equations (a) and (b).

(a)	4 + 3 = 7	true	(b)	4 + 3 = 5	false
	4 + 3 − 5 = 7 − 5	added −5		4 + 3 − 5 = 5 − 5	added −5
	2 = 2	still true		2 = 0	still false

We can use the additive property of equality to prove that the same number can be added to both sides of an equation without changing the solution set of the equation. To demonstrate, we will use the equation

$$x + 4 = 6$$

The number 2 is the solution to this equation. If we add -5 to both sides of the equation, we get

$$x + 4 - 5 = 6 - 5$$

or $$x - 1 = 1$$

We did not change the solution by adding -5 to both sides, as 2 is also the solution to the new equation. **Equations that have the same solution sets are called *equivalent equations*.** Thus, the new equation and the original equation are equivalent equations.

A similar explanation could be used for the multiplication/division rule. We will forgo this explanation and state the two rules as follows.

> ### ADDITION RULE FOR EQUATIONS
>
> The same quantity can be added to both sides of an equation without changing the solution set of the equation.

> ### MULTIPLICATION/DIVISION RULE FOR EQUATIONS
>
> Every term on both sides of an equation can be multiplied (or divided) by the same nonzero quantity without changing the solution set of the equation.

We remember that we always use the addition rule before we use the multiplication/division rule. This is because the solution of an equation undoes a normal order of operations problem. To demonstrate, we will begin with 4, then multiply by 3, and then add -2 to get 10.

$$3(4) - 2 = 10$$

Now, to undo what we have done and get back to 4, we must undo the addition of -2 first and then undo the multiplication. To demonstrate this procedure, we replace 4 with x and get the equation

$$3x - 2 = 10$$

Now we solve to find that x equals 4.

$$
\begin{array}{ll}
3x - 2 = 10 & \text{equation} \\
\underline{+2 \quad +2} & \text{add } +2 \text{ to both sides} \\
3x = 12 &
\end{array}
$$

$$
\begin{array}{ll}
\dfrac{3x}{3} = \dfrac{12}{3} & \text{divided by 3}
\end{array}
$$

$$x = 4$$

We remember from algebra 1 that the five steps for solving simple equations with one variable are:

1. Eliminate parentheses.
2. Add like terms on both sides.
3. Eliminate the variable on one side or the other.
4. Eliminate the constant term on the side with the variable.
5. Eliminate the coefficient of the variable.

We will use these steps to solve the equations in the next two examples.

example 4.3   Solve: $12 - (2x + 5) = -2 + (x - 3)$

solution   As the first step, we eliminate the parentheses, remembering that if the parentheses are preceded by a minus sign, we must change all signs therein.

$$12 - 2x - 5 = -2 + x - 3$$

Now we simplify on both sides of the equation.

$$7 - 2x = x - 5$$

Next we eliminate the x term on the left side by adding $+2x$ to both sides.

$$
\begin{array}{l}
7 - 2x = x - 5 \\
\underline{+2x \quad +2x} \\
7 = 3x - 5
\end{array}
$$

Now we eliminate the -5 on the right by adding $+5$ to both sides.

$$\begin{array}{rcl} 7 &=& 3x - 5 \\ +5 && +\ 5 \\ \hline 12 &=& 3x \end{array}$$

Then we complete the solution by dividing both sides by 3.

$$\mathbf{4 = x} \qquad \text{divided by } 3$$

The same procedure is used when the numbers in the equation are fractions or mixed numbers.

example 4.4   Solve: $3\left(\dfrac{5}{6} - \dfrac{5}{3}x\right) = -\left(-\dfrac{1}{2} + x\right)$

solution   As the first step, we eliminate the parentheses. Then we solve.

$$\begin{array}{rcll} \dfrac{5}{2} - 5x &=& \dfrac{1}{2} - x & \text{multiplied} \\[2mm] +\ 5x & & +\ 5x & \text{add } 5x \text{ to both sides} \\[1mm] \hline \dfrac{5}{2} &=& \dfrac{1}{2} + 4x & \\[2mm] -\dfrac{1}{2} & & -\dfrac{1}{2} & \text{add } -\dfrac{1}{2} \text{ to both sides} \\[1mm] \hline 2 &=& 4x & \end{array}$$

$$\dfrac{1}{2} = x \qquad \text{divided both sides by } 4$$

4.C

change sides– change signs

It is important to understand why we do things in algebra, but it is also important not to let the emphasis on understanding interfere with our ability to do. The use of the addition rule for equations is a case in point. We can use this rule to eliminate a term from one side of an equation by adding the opposite of the term to both sides of the equation. For example, if we wish to solve the equation

$$y + 2x = 4$$

for y, we add $-2x$ to both sides of the equation.

$$\begin{array}{rcll} y + 2x &=& 4 & \text{equation} \\ -\ 2x && -\ 2x & \text{add } -2x \text{ to both sides} \\ \hline y && = 4 - 2x & \end{array}$$

We were able to eliminate the $2x$ term from the left-hand side of the equation, but when we did, the same term appeared on the right-hand side of the equation with its sign changed. **This happens every time we use the addition rule. The term will disappear on one side of the equation and will appear on the other side with its sign changed.** Many people use this thought process. Rather than mentally adding the same quantity to both sides, they simply pick up a term, carry it across the equals sign, and change the sign of the term. This leads to the adage

Change sides—change signs

Authors of algebra books published in the late 1800s called this process **transposition.** If we use transposition to solve the preceding equation for y, we transpose the $+2x$ to the right-hand side, where it becomes $-2x$.

$$y\ \boxed{+\ 2x} = 4 \quad \longrightarrow \quad y = 4 - 2x$$

example 4.5   Use the rule change sides–change signs to solve for x: $x - 2 = 7$

solution   We move the -2 from the left side to the right side and change its sign.

$$x\left(-\,2\right)= 7 \quad \longrightarrow \quad x = 7 + 2 \quad \longrightarrow \quad x = 9$$

example 4.6   Use transposition to solve for p: $p - 3x + 4 = 7y$

solution   We move $-3x + 4$ from the left side to the right side, where it becomes $+3x - 4$.

$$p\left(-\,3x + 4\right)= 7y \quad \longrightarrow \quad p = 7y + 3x - 4$$

example 4.7   Solve for y: $3y - 2x + 5 = 0$

solution   We move $-2x + 5$ to the right side and change both signs. Then we complete the solution by dividing by 3.

$$3y - 2x + 5 = 0 \qquad \text{equation}$$
$$3y = 2x - 5 \qquad \text{changed sides and changed signs}$$
$$y = \frac{2}{3}x - \frac{5}{3} \qquad \text{divided}$$

practice   **a.**   Expand: $\dfrac{2a^{-4}b^{0}}{c}\left(\dfrac{2ab^{2}c}{x^{2}} - \dfrac{a^{-2}}{b^{-4}}\right)$

b.   Solve: $2\left(\dfrac{1}{8} - \dfrac{3}{2}x\right) = -\left(-\dfrac{1}{4}x + 2\right)$

problem set 4

1.   If the radius of a circle is r cm, what is the area of the circle? If the radius is increased to $2r$ cm, what is the area of the circle?

2.   Find x, y, and z.

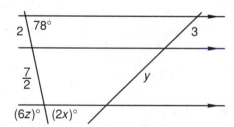

3.   The measure of angle BCD is $100°$, as shown. The tick marks on the figure indicate that $BC = CD$ and that $AB = BD$. First find x and y. Then find P. Then find R and Q.

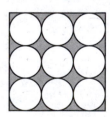

4.   The length of each side of the square is 6 cm. What is the area of the square? The circles have equal areas. What is the radius of a circle? What is the area of the shaded portion of the figure?

5. The radius of the circle is 1 cm. The base of the triangle is 3 cm. The area of the circle equals the area of the triangle. What is the height of the triangle?

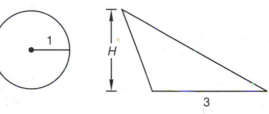

Solve:

6. $15(4 - 5b) = 16(4 - 6b) + 10$ **7.** $3\frac{1}{3}x - \frac{5}{6} = -\frac{2}{3}$

8. $3(-2x - 3) - 2^2 = -(-3x - 5) - 2$ **9.** $-2(2x - 3) - 2^3 - 3 = -x - (-4)$

10. $4\frac{1}{3}x - \frac{1}{2} = 3\frac{2}{5}$ **11.** $-\frac{3}{5}x + \frac{2}{7} = 4\frac{3}{8}$

Expand:

12. $\dfrac{xy^2}{x^0 x^{-3}}\left(\dfrac{xy^{-2}}{x(y^2)^0} - \dfrac{3y^{-2}}{x^4}\right)$ **13.** $\dfrac{ay^{-4}}{p}\left(\dfrac{p^{-2}}{ay^2} - \dfrac{3a^{-1}y}{p^{-2}}\right)$

Simplify. Write all exponential expressions in the numerator.

14. $\dfrac{(3x^2)^{-2}y^0 y^5}{(9y)^{-2}yy^2 x^{-3}}$ **15.** $\dfrac{(2yx^{-2})^{-2}yx^2}{(x^2)^0 y^{-3}x^2}$ **16.** $\dfrac{2(x^{-2})^{-2}yx^2 y^{-3}}{x^0 xx^2 x^{-5}(x^2)^3}$

17. $\dfrac{(x^2 y2x)^{-2}y}{(x^{-4})^0 xxy^2}$ **18.** $\dfrac{3x^2 xy^2 x^{-4}}{(x^2 y)^{-2}(-2)^{-2}}$

Simplify by adding like terms:

19. $\dfrac{2x^2 xyx}{x^2 y^{-1}} - \dfrac{3x^2 y^4}{yy} + \dfrac{7xx^{-3}y^{-2}}{x^{-4}y^{-4}}$ **20.** $\dfrac{3x}{y} - 7x^2 x^{-1}y^{-1} + 2y^2 y^{-1}x^{-1}$

21. $\dfrac{2ay^2}{x} + \dfrac{5a^2 x^{-1}}{ay^{-2}} + \dfrac{2xy^2}{ay}$

Evaluate:

22. $a^2(a - ab)$ if $a = -2$ and $b = 3$

23. $x^0 yx(xy - x^2)$ if $x = -3$ and $y = -1$

24. $a^{-2}b(a - b)(b - a)$ if $a = -1$ and $b = -2$

25. $ab(a^2 - b)a - b$ if $a = -3$ and $b = -1$

Simplify:

26. $-3^0 - 2^0 - 2^0(-2 - 3^2) - (-2 + 7) - |-2 - 3|$

27. $-2\{2[(-3 - 2^2) - (-3 + 7)] - 2\}$

28. $-3^0 - (-2 - 3 - 2^0)(-3) - (-2 - 4) + (-6)$

29. $-2^{-2}(-16)$ **30.** $-(-2^{-3}) - \dfrac{1}{(-2)^{-2}}$

LESSON 5   *Word problems • Fractional parts of a number*

5.A

word problems

Word problems that contain one statement of equality can usually be solved by writing one equation and using one unknown (variable). Word problems that contain two statements of equality can usually be solved by writing two equations and using two unknowns. Three statements of equality require three equations and three unknowns, etc. **In general, to obtain a unique solution, the number of equations must equal or exceed the number of unknowns. If the number of equations exceeds the number of unknowns, at least one of the equations is redundant.**

We will begin with problems that can be solved by writing one equation in one unknown. **If the equation tells us how much two quantities differ, then one of the quantities must be increased or decreased as required so that a statement of equality can be written.**

example 5.1   Twice a number is decreased by 7, and this quantity is multiplied by 3. The result is 9 less than 10 times the number. What is the number?

solution   **In this kind of problem, we can prevent the most common mistake if we begin by writing an equation that we know is untrue.**

$$(2N - 7)3 = 10N \qquad \text{untrue}$$

We were told that the left side is 9 less than the right side. We can make the sides equal by adding 9 to the left side or by adding -9 to the right side. We choose the second option and get

$$(2N - 7)3 = 10N - 9 \qquad \text{added} -9 \text{ to the right side}$$

Now we solve to find that the number is -3.

$$6N - 21 = 10N - 9 \qquad \text{multiplied}$$
$$-12 = 4N \qquad \text{added} -6N + 9 \text{ to both sides}$$
$$\mathbf{-3 = N} \qquad \text{divided by 4}$$

example 5.2   The number of ducks on the pond was doubled when the new flock landed. Then, 7 more ducks came. The resulting number of ducks was 13 less than 3 times the original number. How many ducks were there to begin with?

solution   Again we begin with an equation that is untrue.

$$2N_D + 7 = 3N_D \qquad \text{untrue}$$

We can make this a true equation by adding $+13$ to the left side or by adding -13 to the right side. We decide to add $+13$ to the left side. Then we solve

$$2N_D + 7 + 13 = 3N_D \qquad \text{added 13 to the left side}$$
$$2N_D + 20 = 3N_D \qquad \text{simplified}$$
$$\mathbf{20 = N_D} \qquad \text{added} -2N_D \text{ to both sides}$$

example 5.3   The sum of -7 and 6 times a number is multiplied by 5. The result is 332 less than 3 times the number. What is the number?

solution   Again we begin by writing an equation that is untrue.

$$(6N - 7)5 = 3N \qquad \text{untrue}$$

We were told that the left side is 332 less than the right side. Thus we add 332 to the left side to make the sides equal.

$$(6N - 7)5 + 332 = 3N$$

Now we solve to find that N equals -11.

$$30N - 35 + 332 = 3N \qquad \text{multiplied}$$
$$30N + 297 = 3N \qquad \text{added}$$
$$27N = -297 \qquad \text{rearranged}$$
$$\mathbf{N = -11} \qquad \text{divided}$$

5.B
fractional parts of a number

When we multiply a number by a fraction, we say that we have taken a fractional part of the number. For instance, if we multiply $\frac{3}{8}$ by 40, we get 15.

$$\frac{3}{8} \times 40 = 15$$

We say this with words by saying that three-eighths of 40 is 15. We see that 40 associates with the word *of* and 15 associates with the word *is*. Thus, the general form of the equation is

$$F \times of = is$$

example 5.4   One-fifth of the clowns had red noses. If 30 clowns had red noses, how many clowns were there in all?

solution   We can write the statement of the problem as

$$\frac{1}{5} \text{ of the clowns is } 30$$

Now we write the equation

$$F \times of = is$$

and replace F with $\frac{1}{5}$, *of* with C, and *is* with 30. Then we solve.

$$\frac{1}{5} \cdot C = 30 \qquad \text{equation}$$

$$\frac{5}{1} \cdot \frac{1}{5} \cdot C = \frac{5}{1} \cdot 30 \qquad \text{multiplied both sides by } \frac{5}{1}$$

$$\mathbf{C = 150} \qquad \text{solution}$$

example 5.5   Seven-eighths of the Tartar horde rode horses. If 140,000 were in the horde, how many did not ride horses?

solution   If seven-eighths rode, then one-eighth did not ride.

$$F \times of = is$$

We have

$$\frac{1}{8} \times 140{,}000 = N_R$$

$$\mathbf{17{,}500 = N_R}$$

practice   **a.**   The sum of -4 and 5 times a number is multiplied by 3. The result is 212 less than 7 times the number. What is the number?

b.   Three-eighths of the pieces unloaded were Victorian. If 1624 pieces were unloaded, how many were not Victorian?

problem set 5

1. Twice a number is decreased by 9, and this sum is multiplied by 4. The result is 8 less than 10 times the number. What is the number?

2. The number of ducks on the pond tripled when the new flock landed. Next 11 more ducks came. The resulting number of ducks was 13 less than 4 times the original number. How many ducks were there to begin with?

3. The sum of -8 and 5 times a number is multiplied by 4. The result is 116 less than 6 times the number. What is the number?

4. One-eighth of the clowns had red noses. If 12 clowns had red noses, how many clowns were there in all?

5. Five-sevenths of the Tartar horde rode horses. If 140,000 were in the horde, how many did not ride horses?

6. The area of the big triangle, PQR, is 27 in.^{2}. First find H. Then find AQ. Then find the area of the small triangle, QAB.

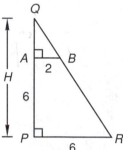

7. $\overline{OP}$ and $\overline{OQ}$ are radii of circle O. Find y. Then find x.

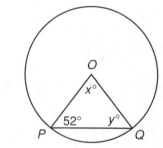

8. In the figure shown, the arrowheads tell us that the pairs of opposite sides are parallel. Find K. Then find P, Q, and C. Then find D and x.

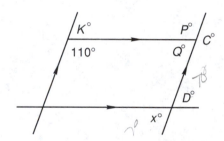

9. The circumference of a circle is 16π inches. Find the radius of the circle and the area of the circle. Then find the volume of a circular cylinder 5 inches tall that has this circle as its base.

Solve:

10. $-3x^0(2x - 3) - (-2^0) - 2 = 5(x - 3^0)2$

11. $-2^2(-2 - x) - x^0(3 - 2) = -2(x + 3)$

12. $3\frac{1}{2}x + 2\frac{1}{4} = -\frac{1}{8}$

13. $\frac{1}{2}(6 - 8x) + \frac{3}{4}(8x - 12) = 4x + 6$

14. $-3 - 3^0 - 3^2(2x - 5) - (-2x - 3) = -x^0(x - 3)$

15. $-2^3 - \frac{1}{-2^{-2}}(x + 2) - 3x = -2^0(-2x^0 - 4)$

16. $-3[x - 2 - 3(2)] + 2[x - 3(x - 2)] = 7(x - 5)$

Expand:

17. $\dfrac{2ab}{c^2}\left(\dfrac{c^2a^{-1}}{b} - \dfrac{3ac}{b}\right)$

18. $-\dfrac{ax^2}{b}\left(\dfrac{bax^3}{a^2} - 3ax\right)$

Simplify:

19. $\dfrac{(xm^{-2})^0\,x^0\,m^0}{xx^2\,m^0\,(2x)^{-2}}$

20. $\dfrac{4c^2\,dc^{-3}(2cd^{-2})^{-2}}{c^0c^{-3}(c^{-2}d)^2}$

21. $\dfrac{p^2m^5(p^{-3})(2p)^{-3}}{m^6(m^{-2})^2\,mp^3}$

Simplify by adding like terms:

22. $\dfrac{x^2xy}{y^{-2}} - \dfrac{3x^5}{xxy^{-3}} + \dfrac{7x^7}{y^3x^4}$

23. $-\dfrac{3a^2x^4}{x} + \dfrac{2aax^2}{x} - \dfrac{5x^3}{a^{-2}}$

Evaluate:

24. $mx - m(m - mx^2)$ if $m = -2$ and $x = -1$

25. $a^2 - b(a - b)$ if $a = -\dfrac{1}{2}$ and $b = \dfrac{1}{4}$

26. $a - ba(a^2 - b)$ if $a = -\dfrac{1}{2}$ and $b = -\dfrac{1}{4}$

Simplify:

27. $-2(-2 - 3^2) - 2[-2(-3)]$

28. $-3^2 - (-3)^3 - \dfrac{1}{-2^2}$

29. $-3^0[-2^0 - 2^2 - 2^3(-2 - 3)]$

30. $-3\big[(-2^0 + 5) - (-3 + 7) - |-2|\big]$

LESSON 6   *Equations with decimal numbers • Consecutive integer word problems*

6.A
equations with decimal numbers

When an equation contains decimal numbers, it is sometimes helpful to multiply every term in the equation by a power of 10 that will turn all the numbers into integers.

example 6.1   Solve: $0.003x + 0.4 = 2.05$

solution   If we begin by multiplying every term by 1000, we get

$$3x + 400 = 2050$$

which is an equation that contains only the variable and integers. Then we solve the equation.

$$3x = 1650 \qquad \text{added} -400 \text{ to both sides}$$

$$x = \mathbf{550} \qquad \text{divided both sides by 3}$$

Many people use the words **decimal fraction** to describe a number that has an internal decimal point. This is because numbers such as

$$2.0413$$

can be written in fractional form. We can write 2.0413 as a fraction as

$$\frac{20{,}413}{10{,}000}$$

Thus, the general equation for a fractional part of a number can also be used for problems that involve a decimal part of a number.

example 6.2   The students found that 0.015 of the teachers were either brave or completely fearless. If 300 teachers fell into one of these categories, how many teachers were there in all?

solution   We use the equation for a fractional part of a number and replace F with WD for "what decimal."

$$WD \times of = is$$

We replace WD with 0.015, of with T, and is with 300.

$$0.015T = 300$$

We finish by dividing both sides by 0.015.

$$T = 20{,}000$$

example 6.3   An analysis of the old woman's utterances showed that 0.932 were vaticinal. If she spoke 2000 times during the period in question, how many utterances were not vaticinal?

solution   If 0.932 were vaticinal, the decimal fraction that was not vaticinal was

$$1 - 0.932 = 0.068$$

So we can write

$$WD \times of = is$$

$$(0.068)(2000) = NV$$

$$136 = NV$$

So 136 utterances were not vaticinal.

6.B
consecutive integer word problems

In algebra, we study problems whose mastery will provide the skills necessary to solve problems that will be encountered in higher mathematics and in mathematically based disciplines such as chemistry or physics. Problems about consecutive integers are of this type. They help us remember which numbers are integers and allow us to practice our word problem skills.

We remember that we designate an unspecified integer with the letter N and greater consecutive integers with $N + 1$, $N + 2$, etc.

Consecutive integers $N, N + 1, N + 2$, etc.

Consecutive odd integers are 2 units apart, and consecutive even integers are also 2 units apart. Thus, we can designate both of them with the same notation.

Consecutive odd integers $N, N + 2, N + 4, N + 6$, etc.

Consecutive even integers $N, N + 2, N + 4, N + 6$, etc.

example 6.4   Find three consecutive even integers such that 5 times the sum of the first and the third is 16 greater than 9 times the second.

solution   We designate the consecutive even integers as

$$N \qquad N + 2 \qquad N + 4$$

and write the necessary equation and solve.

$$5(N + N + 4) - 16 = 9(N + 2) \qquad \text{equation}$$

$$10N + 20 - 16 = 9N + 18 \qquad \text{multiplied}$$

$$10N + 4 = 9N + 18 \qquad \text{simplified}$$

$$N = 14 \qquad \text{added} -9N - 4$$

So the desired integers are **14, 16,** and **18.**

example 6.5 Find four consecutive integers such that 5 times the sum of the first and the fourth is 1 greater than 8 times the third.

solution We designate the consecutive integers as

$$N \quad N+1 \quad N+2 \quad N+3$$

Now we write the equation and solve

$$5(N + N + 3) - 1 = 8(N + 2) \qquad \text{equation}$$
$$10N + 15 - 1 = 8N + 16 \qquad \text{multiplied}$$
$$10N + 14 = 8N + 16 \qquad \text{simplified}$$
$$2N = 2 \qquad \text{added } -8N - 14$$
$$N = 1 \qquad \text{divided by 2}$$

Thus, the desired integers are **1, 2, 3**, and **4**.

practice

a. The astronomers found that 0.017 of the stars examined were red dwarfs. If 29,000 stars were examined, how many were not red dwarfs?

b. Find three consecutive even integers such that 3 times the sum of the first and the third is 84 less than 12 times the second.

problem set 6

1. The students found that 0.016 of the teachers were either brave or completely fearless. If 480 teachers fell into one of these categories, how many teachers were there in all?

2. An analysis of the old man's statements showed that 0.653 were prophetic. If he spoke 3000 times during the period in question, how many statements were not prophetic?

3. A number is multiplied by -3 and then this product is decreased by 7. The result is 4 less than twice the opposite of the number. What is the number?

4. When Cleopatra called for barge workers, $2\frac{1}{2}$ times the number needed showed up. If 175 showed up, how many barge workers did she need?

5. Find three consecutive odd integers such that 6 times the sum of the first and the third is 28 greater than 8 times the second.

6. Find four consecutive integers such that 4 times the sum of the first and the fourth is 24 greater than 6 times the third.

7. The surface area of a sphere is 46π cm^2. What is the radius of the sphere?

8. In the figure shown, sides of equal lengths are indicated by equal tick marks. First find A and B. Then find K. Then find M.

9. Find x, A, and B.

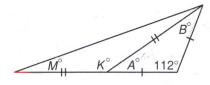

10. If A is a measure of an angle, $180 - A$ is the measure of the supplement of the angle. If the measure of an angle equals twice the measure of its supplement, what is the measure of the angle?

Solve:

11. $0.005x + 0.6 = 2.05$

12. $3\frac{2}{5}x + 1\frac{1}{4} = 7\frac{1}{3}$

13. $-3(x - 2 + 1) - (-2)^2 - 3(x - 2) = 5x^0(2 - x) - 2x$

14. $-3 - 2^2 - 2(x - 3) = 2[(x - 5)(2 - 5)]$

15. $4(x + 3) - 2^0(-x - 3) = 2x - 4(x^0 - x) - 3^2$

Expand:

16. $\dfrac{xy}{p}\left(\dfrac{-3p^{-1}}{xy} + \dfrac{2p}{x^{-1}y}\right)$

17. $-\dfrac{x^0k}{p}\left(\dfrac{k^0p}{x} - 2p\right)$

Simplify:

18. $\dfrac{(2x^{-2}y^0)^{-2}yx^{-2}}{xxxy^2(y^{-2})^2}$

19. $\dfrac{a^0bc^0(a^{-1}b^{-1})^2}{ab(ab^0)abc}$

20. $\dfrac{(2x^2)^{-3}(xy^0)^{-2}}{2xx^0x^1xxy^2}$

Simplify by adding like terms:

21. $-2xy + \dfrac{5x^0xy^{-1}}{y^{-2}} - \dfrac{5xx^{-1}x^2}{(x^{-1})^{-1}}$

22. $-\dfrac{3x^2xy^2}{y^4} + \dfrac{2xxx}{y^{-2}} - \dfrac{3xy}{x^{-2}y^{-1}}$

Evaluate:

23. $xy - x^2y - y$ if $x = -2$ and $y = -4$

24. $a^{-2}b - a(a - b)$ if $a = -\dfrac{1}{2}$ and $b = \dfrac{1}{4}$

25. $m^2p(mp - p^2)$ if $m = -\dfrac{1}{4}$ and $p = \dfrac{1}{5}$

Simplify:

26. $-3^0[-3^2 - 2(-2 - 3)][-2^0]$

27. $-3 - (-3)^2 + (-3)(-6)$

28. $-3^2 + (-3)^2 - 4^2 - |-2 - 2|$

29. $-3^{-2} - \dfrac{2}{-2^{-3}} - 2^0$

30. $-(-2)^{-3} - 3^{-2} - 3$

LESSON 7   *Percent • Equations from geometry*

7.A

percent   The Latin word for "by" is *per* and the Latin word for "hundred" is *centum*. Thus, the word **percent** literally means "by the hundred." The percent equation (b) is exactly the same equation as the fractional-part-of-a-number equation (a) except that the denominator of the fraction is 100.

$$\text{(a)} \quad WF \times of = is \qquad \text{(b)} \quad \frac{P}{100} \times of = is$$

There are two other forms of the percent equation that are often used.

$$\text{(c)} \quad \frac{P}{100} = \frac{is}{of} \qquad \text{and} \qquad \text{(d)} \quad \text{Rate} \times of = is$$

We call (c) the **ratio form** of the percent equation. In form (d) the rate is the percent divided by 100. If the percent was 20 percent, then the rate would be 0.2, which is 20 divided by 100. **Any of the three percent equations can be used. They are not different equations but are three different forms of the same equation.**

There are two types of percent problems. In one type, the original quantity is divided into two parts and the final percent is less than 100. In the second type, the original quantity increases and the final percent is greater than 100. It is helpful to be able to draw diagrams that give us a picture of the problem.

example 7.1   Eighteen is 20 percent of what number? Work the problem and then draw the completed diagram.

solution   We will use the fractional form of the percent equation.

$$\frac{P}{100} \times of = is \quad \longrightarrow \quad \frac{20}{100} \times WN = 18$$

Now we multiply both sides by $\frac{100}{20}$ to solve.

$$\frac{100}{20} \cdot \frac{20}{100} WN = \frac{100}{20} \cdot 18 \quad \longrightarrow \quad WN = 90$$

If one part of 90 is 18 for 20 percent, the other part must be 72 for 80 percent. The diagram is as shown here.

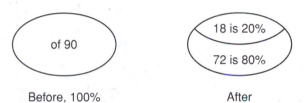

Before, 100%   After

Learning to draw the diagram is very important. The diagram lets us "see" the problem. We note that the "*of*" number is always the number in the first oval and always represents 100 percent. Then, in the second oval, the "*of*" number is separated into two parts, each with its own percent. We are using very simple problems to help us learn to draw the diagrams. The diagrams will be helpful when the problems get more involved.

example 7.2   Fifteen hundred is what percent of 250? Work the problem and then draw the completed diagram.

solution   Again we choose to use the fractional form of the equation.

$$\frac{WP}{100} \times 250 = 1500$$

To solve, we multiply both sides by $\frac{100}{250}$.

$$\frac{100}{250} \cdot \frac{WP}{100} \cdot 250 = 1500 \cdot \frac{100}{250} \quad \longrightarrow \quad WP = 600 \text{ percent}$$

The diagram shows 250 increased to 1500, which is 600 percent.

Before, 100%   After

Whenever the final number is greater than the "*of*" number, the final percent is greater than 100 percent. The final shape has only one part and is greater than the initial shape.

7.B
equations from geometry

We can devise problems that let us practice working with geometric concepts and that also let us practice solving equations. Please note that when we write the equations, we do not have to use the degree symbol.

If

$$A° + 10° = 14°$$

then *A* plus 10 must equal 14.

$$A + 10 = 14$$

example 7.3   Find *x*.

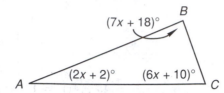

solution   We know that the sum of the measures of the angles in any triangle equals 180°. Thus

$$(2x + 2) + (7x + 18) + (6x + 10) = 180 \quad \text{equation}$$
$$15x + 30 = 180 \quad \text{simplified}$$
$$15x = 150 \quad \text{added } -30$$
$$x = \mathbf{10} \quad \text{solved}$$

Now we can use 10 for *x* and find the measures of the angles.

$$(2x + 2)° = (2 \cdot 10 + 2)° = 22°$$
$$(7x + 18)° = (7 \cdot 10 + 18)° = 88°$$
$$(6x + 10)° = (6 \cdot 10 + 10)° = \underline{70°}$$
$$180° \quad \text{check}$$

example 7.4   Find *x*. Then find the measure of a small angle and the measure of a large angle.

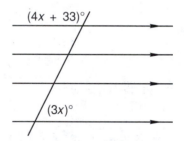

solution   When parallel lines are cut by a transversal, the sum of the measures of a small angle and a large angle must equal 180°. This lets us write

$$4x + 33 + 3x = 180 \quad \text{equality}$$
$$7x = 147 \quad \text{simplified}$$
$$x = \mathbf{21} \quad \text{solved}$$

We can use 21 for *x* and find the values of $4x + 33$ and $3x$.

LARGE ANGLE		SMALL ANGLE
$(4x + 33)°$		$(3x)°$
$[(4(21) + 33)]°$	let *x* equal 21	$[3(21)]°$
117°	simplified	**63°**

We note that $117° + 63° = 180°$.

example 7.5   The measures of angles *A*, *B*, *C*, and *D* are in the ratio of 1:2:4:2. Find the measure of each angle.

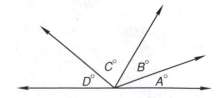

solution   If we relabel the angles as having measures of $x°$, $(2x)°$, $(4x)°$, and $(2x)°$, the solution is easy.

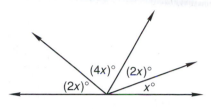

$$x + 2x + 4x + 2x = 180 \quad \text{equation}$$
$$9x = 180 \quad \text{simplified}$$
$$x = 20 \quad \text{solved}$$

So $x° = 20°$, $(2x)° = 40°$, $(4x)° = 80°$, and $(2x)° = 40°$.

practice   Solve:

a.   Ninety-three is 30 percent of what number? Work the problem and draw the completed diagram.

b.   Seventy-eight hundred is what percent of 390? Work the problem and then draw the diagram.

c.   Find x, A, and B.

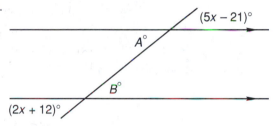

problem set 7

1.   Twenty-six is 20 percent of what number? Draw the diagram.

2.   Fourteen hundred is what percent of 350? Draw the diagram.

3.   Twenty percent of what number is 460? Draw the diagram.

4.   What percent of 20 is 680? Draw the diagram.

5.   Three hundred eighty is 1900 percent of what number? Draw the diagram.

6.   Find three consecutive odd integers such that 7 times the sum of the first and the third is 120 less than 10 times the opposite of the second.

7.   Find three consecutive even integers such that 6 times the sum of the first and the third is 8 less than 14 times the second.

8.   If twice the opposite of a number is increased by 5, the result is the opposite of the number. What is the number?

9.   Find x.

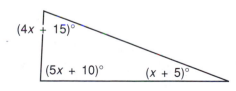

10.   Find x.

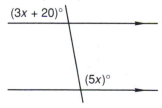

11.   The measures of angles a, b, c, and d are in the ratio of 1:3:6:2. Find the measure of each angle.

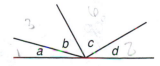

12. Whenever we see the word **equilateral**, we should think **equal sides** and **60° angles**. In this figure we show three equilateral triangles with a common vertex. Find the sum of the measures of angles x, y, and z. A suggested first step is to copy the figure and write 60° where the angles are 60° angles.

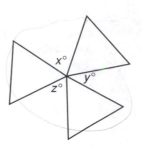

Solve:

13. $-3p(-2 - 3) + p - 2^2 = -(2p + 4) - p^0$

14. $0.005x - 0.07 = 0.02x + 0.0032$

15. $2\dfrac{1}{5} + 3\dfrac{1}{8} + 2\dfrac{1}{2}x = 4\dfrac{3}{20}$

16. $3x - 2 - 2^0(x - 3) - 2^0 + 2^2 = 5(-x - 2) + 3^0$

Expand:

17. $\dfrac{2xyp}{y^{-2}}\left(\dfrac{x^{-1}}{y^3 p} - \dfrac{3x}{yyp^2}\right)$

18. $\dfrac{4x^{-2}y}{k}\left(\dfrac{2kx^2}{y} - \dfrac{3xy}{k}\right)$

19. $\dfrac{(2x^2 y^3)^{-3} y}{(4xy)^{-2}(x^{-2}y)^3 y}$

20. $\dfrac{xx^{-2}y(x^{-3})^2 xy^0}{(2xy)^{-2}x^2(y^{-3})^2}$

Simplify by adding like terms:

21. $-\dfrac{3x^2 xy}{p} + \dfrac{7xyp^{-1}}{x^{-2}} - \dfrac{2xxxp^{-1}}{y^{-1}}$

22. $-4xp^2 + \dfrac{3xxp^4}{p^2 x^2} - \dfrac{2xp}{p^{-1}}$

Evaluate:

23. $-a(a - b)$ if $a = -\dfrac{1}{2}$ and $b = \dfrac{1}{3}$

24. $-xy(-x^2 - y)$ if $x = -\dfrac{1}{2}$ and $y = \dfrac{1}{4}$

25. $x^3 - x(xy - y)$ if $x = -2$ and $y = -4$

26. $x - a(a - xa)$ if $x = 2$ and $a = -\dfrac{1}{2}$

Simplify:

27. $-2\{[-2^0 - 3(-2)] - [-2(-3 - 2)(-2)]\}$

28. $-2^0 - 2 - 2^2 - (-2)^3 - 2(-2 - 2) - 2$

29. $3^0(-2 - 3)(-2 + 5)(-2) - (-3 + 7)(-4^0 - 3^0)$

30. $2\left[(-2^0 - 1)(-2^0 - 15^0) - (-2)^2 - 3^0\right] - 2$

LESSON 8   *Polynomials* • *Graphing linear equations* • *Intercept-slope method*

8.A
polynomials It is convenient to have a word to describe the simplest kind of algebraic expressions. These expressions have coefficients that are real numbers and variables that have whole numbers as exponents. No fractional exponents or negative exponents are allowed. The following are

examples of these very simple expressions.

$$-4 \qquad \frac{1}{2} \qquad 2x \qquad -3x^2 \qquad -5x^2 + 6x + 2$$

Unfortunately, we use an intimidating word to designate these simple expressions, and the word is **polynomial.** It would have been helpful had we called them **"simplenomials"** instead, but we can think "simplenomial" whenever we hear the word polynomial.

A polynomial in one variable has a real number for a coefficient and has one of the numbers 0, 1, 2, 3, . . ., etc., as the exponent of the variable. Thus, all of the following are polynomials. They are also called **monomials** because they have only one term.

$$\text{(a)} \ -4 \qquad \text{(b)} \ 2x^2 \qquad \text{(c)} \ 3x^{14} \qquad \text{(d)} \ 0.004x^5 \qquad \text{(e)} \ \sqrt{2}x$$

The first one, (a), can be thought of as $-4x^0$, and since x^0 equals 1, this expression fits the definition of a polynomial. The rest of the expressions have real number coefficients and whole number exponents, so they are all polynomials.

Polynomials of two terms are called **binomials,** and polynomials of three terms are called **trinomials.**

$$\text{(f)} \ \ x + 2 \qquad \text{(g)} \ \ x^4 + 2x \qquad \text{(h)} \ \ 2x^2 + 3x + 2$$

Thus, (f) and (g) are binomials, and (h) is a trinomial.

The degree of a polynomial is the same as the degree of the highest-degree term of the polynomial. Thus, (f) is a first-degree polynomial because the exponent of x is 1. The polynomial (g) is a fourth-degree polynomial because the exponent of x^4 is 4. Using the same reasoning, the polynomial (h) is a second-degree polynomial because the greatest exponent is 2.

An equation that contains only polynomial terms is called a **polynomial equation.** The degree of a polynomial equation is the same as the degree of the highest-degree term in the equation. Thus, the equations

$$2x + 3y = 6 \qquad 3x - 2y = 0 \qquad -3x = 2y + 4$$

are all first-degree polynomial equations. If we use two number lines to form a coordinate plane, we can graph the set of ordered pairs of x and y that satisfy one of these equations. **The graph of a first-degree polynomial equation in two unknowns is a straight line.**

8.B

graphing linear equations

To find two or more ordered pairs of x and y that satisfy the equation of a line, we often use five steps.

1. Solve the equation for y.
2. Make a table and select convenient values of x.
3. Use these values of x in the equation to find the matching values of y.
4. Complete the table.
5. Graph the ordered pairs and draw the line.

example 8.1   Graph the equation $2x + 3y = 6$.

solution   We will use the five steps listed above.

     1. First we solve the equation for y.

$$2x + 3y = 6 \ \longrightarrow \ 3y = -2x + 6 \ \longrightarrow \ y = -\frac{2}{3}x + 2$$

     2. Next we make the table and select 0, 6, and -6 as values for x.

x	0	6	-6
y			

3. Now we find the matching values of y.

WHEN $x = 0$:   WHEN $x = 6$:   WHEN $x = -6$:

$y = -\dfrac{2}{3}(0) + 2$ $y = -\dfrac{2}{3}(6) + 2$ $y = -\dfrac{2}{3}(-6) + 2$

$y = 2$ $y = -2$ $y = 6$

4. Next we complete the table.

x	0	6	−6
y	2	−2	6

5. Finally, we graph the points and draw the line.

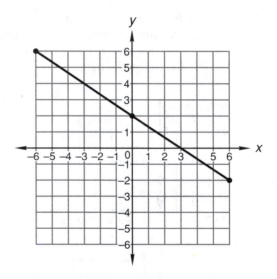

8.C

intercept-slope method

The method of graphing a line shown in example 8.1 is exact and will always work. However, the method is time-consuming and there is a quicker way to graph a line that is just as accurate.

Recall that two points are all that is needed to graph a line. We can use the y intercept as one of the points and use the slope to find another point. Since we use the intercept first, we call this method the **intercept-slope method.** To demonstrate, we will graph the same equation again.

example 8.2   Use the intercept-slope method to graph the equation $2x + 3y = 6$.

solution   The first step is the same. We solve the equation for y.

$$2x + 3y = 6 \quad \longrightarrow \quad 3y = -2x + 6 \quad \longrightarrow \quad y = -\dfrac{2}{3}x + 2$$

The equation has two numbers. The first number is $-\dfrac{2}{3}$ and is the slope. The second number is 2 and is the y intercept. This is the value of y when x equals zero. We graph the intercept, which is $(0, 2)$.

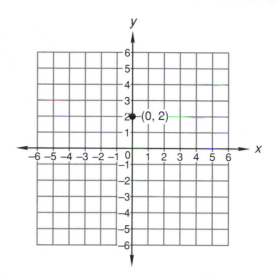

Next we write the slope, $-\frac{2}{3}$ as either (a) $\frac{-2}{+3}$ or (b) $\frac{+2}{-3}$. We remember that the slope is the rise over the run. Thus, from the point we have graphed, to find a second point we can (a) take a rise of -2 and a run of $+3$, or (b) take a rise of $+2$ and a run of -3. Both ways are shown below: (a) on the left and (b) on the right.

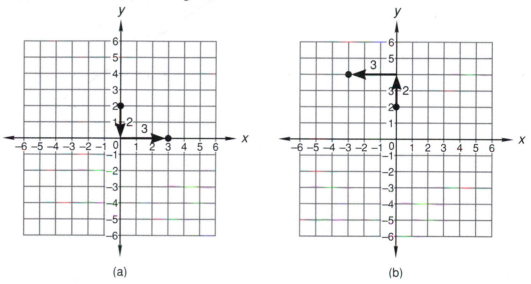

(a)   (b)

Both pairs let us draw the same line, as shown here.

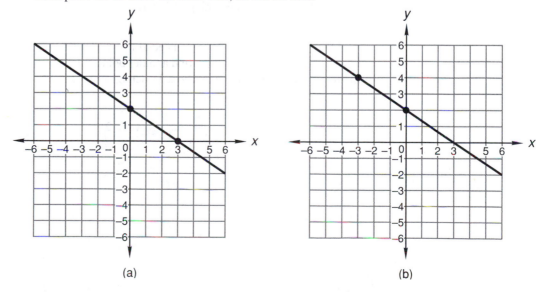

(a)   (b)

example 8.3   Use the intercept-slope method to graph the line $y = -3x - 3$.

   solution   The intercept is $y = -3$. First we graph the point $(0, -3)$.

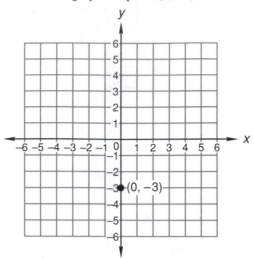

The slope can be written as (a) $\frac{-3}{+1}$ or as (b) $\frac{+3}{-1}$. Thus, from the intercept we can (a) take a rise of -3 and a run of $+1$, or (b) take a rise of $+3$ and a run of -1.

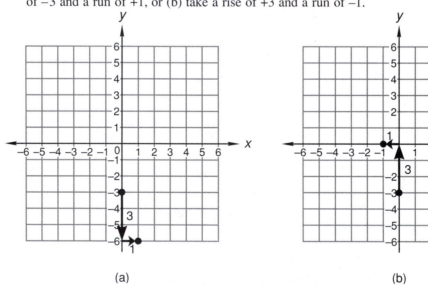

All three points lie on the line, as shown here.

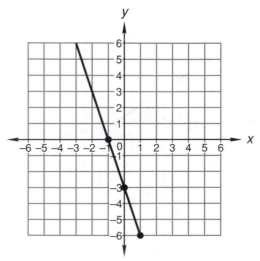

practice Use the intercept-slope method to graph the equation $3x + 4y = 12$.

problem set 8

1. When the tournament began, only 0.36 of the knights wore new armor. If 828 knights wore new armor, how many knights participated in the tournament?

2. Sir Lancelot found four consecutive even integers such that 10 times the sum of the first and the fourth was 24 greater than 9 times the sum of the second and fourth. What were his integers?

3. When the Danes multiplied their secret number by 3 and then added −7, the result was 72 less than twice the opposite of the number. What was the secret number of the Danes?

4. Only seven-sixteenths of the warriors necessary to defend the castle answered the call to arms. If 420 answered the call, how many were required to defend the castle?

5. The defenders thought of consecutive odd integers while waiting for the fusillade from the trebuchet. Their integers were three in number and were such that 5 times the sum of the first and third was 108 greater than twice the opposite of the second. What were the integers?

6. Eighty-six is 20 percent of what number? Draw a diagram of the problem.

7. What number is 340 percent of 56? Draw a diagram of the problem.

8. If the sum of the measures of two angles is 90°, the angles are complementary. Thus, if the measure of an angle is $A°$, the measure of the complement is $(90 - A)°$. Find an angle whose measure is 3 greater than twice the measure of its complement.

9. Graph $3x + 4y = 8$ on a rectangular coordinate system.

10. Find x.

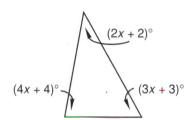

11. **The diagonals of a rectangle have equal lengths.** The area of the circle shown is $25\pi \, m^2$. Find the approximate radius of the circle. For the rectangle shown, what is the length of diagonal BD? What is the length of diagonal AC?

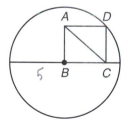

12. Find x.

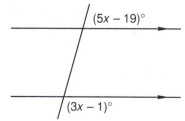

13. Find x and y.

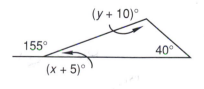

Solve:

14. $0.003x + 0.02x - 0.03 = 0.177$

15. $2\frac{1}{3}x + 1\frac{3}{5} = 7\frac{2}{5}$

16. $-4^0 - 2^2 - (-2)^3 - 3 - (2 - 2x) - 4 = -3(-2 + 2x)$

17. $-\frac{1}{2} + 2\frac{3}{8} - 7\frac{1}{4} + 3\frac{1}{2}x = 4\frac{1}{16}$

Expand:

18. $\dfrac{x^{-2}y}{p}\left(\dfrac{-3x^2p}{y} - \dfrac{4xy^2}{p^2}\right)$

19. $\dfrac{-3^0x^0}{p^0}\left(-3x + \dfrac{5xy}{p^{-2}}\right)$

Simplify:

20. $\dfrac{(4x)^{-2}y^0(y^{-2})^2y}{32^{-1}x^2(yx^0)^{-3}}$

21. $\dfrac{5x^{-2}(y^2x^3)^{-3}}{2^{-2}xyx^2(x^{-2}y)}$

Simplify by adding like terms:

22. $3x^{-2}y + 5x^2y^{-1} - \dfrac{3}{xxy^{-1}}$

23. $\dfrac{2xxy^{-3}y}{xy} + \dfrac{2y^{-1}y^{-2}}{x^{-1}} - \dfrac{7xy^2}{y}$

Evaluate:

24. $ax - a(x^2)$ if $a = -\dfrac{1}{2}$ and $x = \dfrac{1}{4}$

25. $ab^2(a - ab)$ if $a = \dfrac{1}{2}$ and $b = -\dfrac{1}{3}$

26. $mx - (m^2 - x)$ if $m = -\dfrac{1}{3}$ and $x = \dfrac{1}{2}$

Simplify:

27. $-3[(-2^0 - 4) - (-2)(-3)] - [(-6^0 - 2) - 2^2(-3)]$

28. $-3 - 3^0 - 3^{-2} + \dfrac{1}{9} - 3^0(-3 - 3)$

29. $-|-2^0| - 2^{-2} - (-2)^{-2}$

30. $(-1)^{-3} - 1^{-2} - 1^2 - (-1)^3$

LESSON 9   *Percent word problems*

To compute a given percent of a number, we can first divide the number into 100 parts. Then 30 percent of the number means 30 of these parts, 193 percent of the number means 193 of these parts, etc. To demonstrate, let us begin with the number

$$242$$

Now, if we divide 242 into 100 parts, we find that each part equals 2.42.

$$\frac{242}{100} = 2.42$$

Thus, 30 percent of 242 means 30 of these parts, or 30 times 2.42.

(a) $30 \times 2.42 = $ **72.6**

Likewise, 193 percent of 242 means 193 parts, or 193 times 2.42.

(b) $193 \times 2.42 = $ **467.06**

If we use the percent equations, we get the same answers.

(a) $\dfrac{30}{100} \times 242 = $ **72.6**   (b) $\dfrac{193}{100} \times 242 = $ **467.06**

Percent word problems fall into several different categories, and it is very helpful if a diagram of the problem is drawn as the first step. A diagram allows the visualization of the problem and will help prevent mistakes. Some students believe that drawing diagrams is childish, for they can work the problems without drawing diagrams. The author believes that making preventable mistakes is childish and that drawing pictures which will prevent these mistakes is an indication of maturity. Check this opinion with an engineer or a graduate physicist or graduate mathematician before making up your mind. There is no excuse for making errors that can be prevented by drawing a picture of the problem!

example 9.1   The wood nymphs and the maids gamboled and frolicked before the banquet began. If 70 percent of those present were wood nymphs and 120 maids were present, how many wood nymphs came to the banquet?

solution   We use the following diagrams to visualize the problem

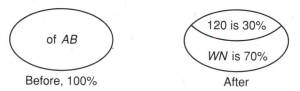

of *AB*   120 is 30%

                                            WN is 70%

Before, 100%   After

The diagram shows *AB* for all at the banquet and *WN* for *Wood Nymphs*. Since 70 percent were wood nymphs, 30 percent were maids. We see that 120 is 30 percent of all at the banquet.

$$\frac{30}{100} \times AB = 120$$

To solve, we multiply both sides by $\frac{100}{30}$.

$$\frac{100}{30} \cdot \frac{30}{100} \cdot AB = 120 \cdot \frac{100}{30} \longrightarrow AB = 400$$

Since the guests totaled 400 and 120 were maids, there must have been **280 wood nymphs.**

example 9.2   The harvest was cornucopian, as it was 120 percent greater than last year. If the yield was 140,800 bushels, how many bushels were harvested last year?

solution   Again we find that a diagram is helpful.

of *LY*   140,800 is 220%

Before, 100%   After

We see that we begin with 100 percent last year and that a 120 percent increase means 220 percent this year.

$$\frac{220}{100} \times LY = 140{,}800$$

We complete the solution by multiplying both sides by $\frac{100}{220}$.

$$\frac{100}{220} \cdot \frac{220}{100} \cdot LY = 140{,}800 \cdot \frac{100}{220}$$

$$LY = 64{,}000 \textbf{ bushels}$$

practice   **a.**   Sixty percent of those present at the performance were neophytes. If 200 present were not neophytes, how many people attended the performance?

b.   Production increased by 145 percent. If the number of widgets produced this year was 171,500, how many were produced last year?

problem set 9

1. The wood nymphs and the maids gamboled and frolicked before the banquet began. If 60 percent of those present were wood nymphs, and 160 maids were present, how many wood nymphs came to the banquet?

2. The harvest was cornucopian, as it was 45 percent greater than last year. If the yield was 140,795 bushels, how many bushels were harvested last year?

3. Twenty percent of the income was used to pay for raw materials. If $78,000 was spent for other purposes, what was the total income?

4. Find three consecutive integers such that the product of -5 and the sum of the first two is 43 less than twice the second.

5. When the war tocsin sounded, 84 percent of the soldiers staggered to their feet. If 40,000 did not get up, how many soldiers were present?

6. Diomedes peered into the darkness and saw 1400 Trojans. If he could not see seven-eighths of the Trojans, how many Trojans were there?

7. Geometry problems sometimes contain a geometric figure and additional information about the figure. A good first step in these problems is to sketch the figure and insert the additional information.

Given: $x = 60$

$y = 40$

$k = 50$

Find m and p.

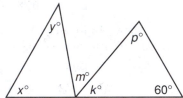

To solve this problem, sketch the figure and replace $x°$ with $60°$, replace $y°$ with $40°$, and replace $k°$ with $50°$. Then find m and p.

8. Find x and y.

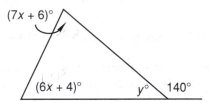

$(7x + 6)°$

$(6x + 4)°$ $y°$ $140°$

9. First find x. Then find $5x + 10$. Then find y. Then find z.

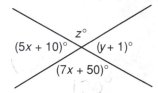

$z°$

$(5x + 10)°$ $(y + 1)°$

$(7x + 50)°$

10. The area of a circle is 9π square meters. What is the radius of the circle? What is the circumference of the circle?

Graph on a rectangular coordinate system:

11. $y - 2x + 3 = 0$

12. $3y + 6 = -x$

Solve:

13. $0.02 - 0.003x + x = 5.005$

14. $-3\frac{1}{5}x + 7\frac{1}{10} = 4\frac{2}{9}$

15. $-2[(2 - 3)x + 7(2^0 - 1)] = -3(x - 2)$

16. $-2^0(2x - 3) - 4 = 2x - 3^0$

Expand:

17. $\dfrac{x^0 y^2}{p^{-2}}\left(\dfrac{p^2}{y^2} - \dfrac{y^{-2}}{p^{-2}}\right)$

18. $\dfrac{ak^{-2}}{a^{-3}}\left(\dfrac{2k^4}{a^4} - 3k\right)$

Simplify:

19. $\dfrac{(2x^2\,ya)^{-3}\,ya^3}{x^2\,y(ay)^{-2}\,y}$

20. $\dfrac{(-2xyz)^{-3}}{(x^2z^{-3})^{-3}}$

Simplify by adding like terms:

21. $3x - \dfrac{2xy^2}{y} + \dfrac{4xx^{-2}}{(x^2)^{-1}}$

22. $\dfrac{2xy}{p} - \dfrac{5xxx}{(x^{-2})^{-1}y^{-1}} + \dfrac{3xp^{-1}}{y^{-1}}$

Evaluate:

23. $-a^2b - a$ if $a = -\dfrac{1}{2}$ and $b = \dfrac{1}{4}$

24. $a(a - ab)$ if $a = -\dfrac{1}{2}$ and $b = -\dfrac{1}{8}$

25. $a(a - b)(ab - b)$ if $a = -2$ and $b = 3$

26. $a^2(x - ax^2)$ if $a = -2$ and $x = -4$

Simplify:

27. $-2(-3 - 2^0) - 2^0(-2^2 - 2)$

28. $-3[(-5 + 2)(-2) - (3^0 - 2) - 2]$

29. $-2^0(-2 - 3^0) - (-2)^3 - |-3|$

30. $-\dfrac{1}{-2^{-3}} + \dfrac{1}{-(-2)^{-3}} - 3^2$

LESSON 10 *Pythagorean theorem*

Thus far, we have discussed properties and definitions. Properties are the way things are because they are. Definitions are things we have agreed on. For instance,

PROPERTIES	DEFINITIONS
(a) $3 + 2 = 2 + 3$	(c) $x^{-2} = \dfrac{1}{x^2}$
(b) If $a = b$, then $a + c = b + c$	(d) $4 + 3 \cdot 2 = 10$

Both properties and definitions can be called **rules.** Another kind of rule is a **theorem.** A theorem is just like a property because theorems tell us the way things are. The difference is that theorems can be proved by using properties and definitions. **The Pythagorean theorem states that the area of the square drawn on the hypotenuse of a right triangle equals the sum of the areas of the squares drawn on the other two sides.** On the left, we show a right triangle whose sides are 3, 4, and 5. The area of the square on the hypotenuse is 25 square units, which equals the sum of the areas of the squares drawn on the other two sides, because 9 plus 16 equals 25. In the center, we show a right triangle with sides a, b, and c, and on the right, we have drawn the squares on the sides of this triangle.

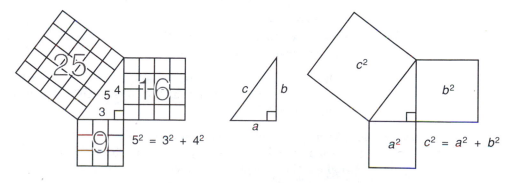

We see that the resulting algebraic equation is

$$c^2 = a^2 + b^2$$

where c is the length of the hypotenuse. This theorem is proved in geometry. For now, we will use it as if it were a property.

example 10.1   Use the Pythagorean theorem to find side a.

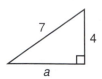

solution   Since $c^2 = a^2 + b^2$, we write

$$7^2 = a^2 + 4^2 \qquad \text{Pythagorean theorem}$$
$$49 = a^2 + 16 \qquad \text{squared 7 and 4}$$
$$33 = a^2 \qquad \text{added } -16$$
$$\sqrt{33} = a \qquad \text{solved}$$

example 10.2   Use the Pythagorean theorem to find the distance between the points $(4, 2)$ and $(-3, 4)$.

solution   We could use the distance formula, which is an algebraic statement of the Pythagorean theorem. However, the problem can be worked with fewer mistakes and more understanding by graphing the points, drawing the triangle, and using the theorem.

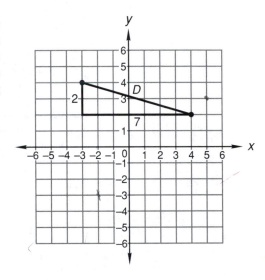

$$D^2 = 2^2 + 7^2 \qquad \text{Pythagorean theorem}$$
$$D^2 = 4 + 49 \qquad \text{squared 2 and 7}$$
$$D^2 = 53 \qquad \text{added}$$
$$D = \sqrt{53} \qquad \text{solved}$$

practice   a.   Use the Pythagorean theorem to find side a.

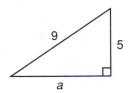

b.   Use the Pythagorean theorem to find the distance between $(5, 3)$ and $(-2, -3)$.

problem set   1.   Twenty percent of the people at the fair were in a festive mood. If 1400 were not in a
10   festive mood, how many attended the fair?

2.   When Julius crossed the Rubicon, he had with him $3\frac{1}{4}$ times as many soldiers as he needed to conquer Rome. If he had 26,000 soldiers with him, how many were needed to conquer Rome?

3.   When the Theban legion refused to obey the orders, it was decreed that the legion should be decimated—every tenth man killed. If 590 men were killed, how many men did the Theban legion have after it was decimated?

4. Find three consecutive odd integers such that 4 times the first is 8 less than 3 times the sum of the last two.

5. When Aegisthus found that his secret had been discovered, he upped the ante by 160 percent. If the ante was now 10,400 minas,† what was the original ante?

6. Atreus could see 4200 Argives marching toward the lion's gate. If he could see 14 percent of the Argives, how many were hidden from view?

7. The circumference of the circle is 6π cm. What is the area of the circle?

8. In the figure $AD = DB$. Find x, y, and z.

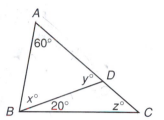

9. Find x, P, and Q.

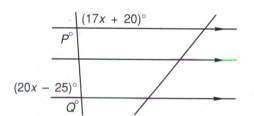

10. One-fourth of the square is shaded. The area of the shaded portion is 9 m². What is the area of the whole square? What is the length of one side of the square?

Graph on a rectangular coordinate system:

11. $2y = 3x + 2$

12. $y = -3$

13. Use the Pythagorean theorem to find p.

14. Find the distance between $(3, 2)$ and $(-4, 1)$.

Solve:

15. $\dfrac{3}{4}x - \dfrac{1}{5}x = 2\dfrac{3}{4}$

16. $-5.2 + 3y = 0.2(y + 2)$

17. $4x(2 - 3^0) + (-2)(x - 5) = -(3x + 2)$

Expand:

18. $\dfrac{4xy}{m^{-2}}\left(\dfrac{3y^{-1}}{m^2 x} - \dfrac{2x}{ym}\right)$

19. $\dfrac{2x^0 y}{p}\left(\dfrac{2p}{y} - \dfrac{3xy}{p}\right)$

Simplify:

20. $\dfrac{(3x^{-2})^{-2}xy}{3^{-3}x^{-2}(yx^0)^{-3}}$

21. $\dfrac{(x^2 p^2)^{-3}x^0 p^2}{x^{-2}px^0(xp)^{-3}}$

Simplify by adding like terms:

22. $3x + \dfrac{2x^2 x^{-3}}{x^{-2}y^0} - x^0$

23. $\dfrac{5x^2 y}{z} - \dfrac{3z^{-1}y}{x^{-2}} + \dfrac{7xxy^2 z}{yz^2}$

---

†In Greek currency before the fourth century B.C., it took 60 minas to equal 1 talent.

24. **If two tangents are drawn to a circle from a point outside the circle, the lengths of the tangent segments are equal**. In the figure on the left, x must equal 4 because the length of the other tangent segment is 4.

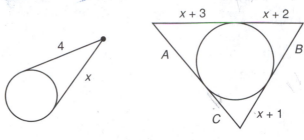

Perimeter = 30 cm

In the figure on the right, the length of segment A must be $x + 3$, the length of segment B must be $x + 2$, and the length of segment C must be $x + 1$. The perimeter of the triangle is 30 cm. Equate the sum of the lengths of the segments to 30 and solve for x.

Evaluate:

25. $b(ab - b)$ if $a = -\dfrac{1}{3}$ and $b = -\dfrac{1}{2}$

26. $ab - a^2b^2 - b$ if $a = -\dfrac{1}{2}$ and $b = -\dfrac{1}{2}$

27. $a^2(b - ab)b$ if $a = -\dfrac{1}{2}$ and $b = -\dfrac{1}{2}$

Simplify:

28. $-\dfrac{1}{2^{-3}} - \dfrac{1}{-2^{-3}} - (-3 - 2^0) - 2$

29. $-|-2| - |-2^0| - 3^2 - (-3)^3$

30. $-\dfrac{1}{2} - \left(\dfrac{1}{2}\right)^2 - \left(-\dfrac{1}{2}\right)^3 - \dfrac{1}{2}$

LESSON 11 *Addition of fractions • Inscribed angles*

11.A
addition of fractions

To add fractions whose denominators are equal, we add the numerators algebraically and record their sum over a common denominator.

$$\frac{1}{11} + \frac{5}{11} - \frac{4}{11} = \frac{1 + 5 - 4}{11} = \frac{2}{11}$$

In this example, each of the denominators is 11, and the sum of the numerators is 2. The same procedure is used when the fractions contain letters.

$$\frac{a}{b} + \frac{2}{b} + \frac{a + 3}{b} = \frac{2a + 5}{b}$$

In this example, each of the denominators is b, and the sum of the numerators is $2a + 5$.

When the denominators are not equal, the form of one or more of the fractions must be changed so that the denominators will be equal. One-half and three-fourths cannot be added because the denominators are not the same.

$$\frac{1}{2} + \frac{3}{4}$$

To make the denominators the same, we will change $\frac{1}{2}$ to $\frac{2}{4}$ by multiplying $\frac{1}{2}$ by $\frac{2}{2}$.

$$\frac{1}{2}\left(\frac{2}{2}\right) + \frac{3}{4} = \frac{2}{4} + \frac{3}{4} = \frac{5}{4}$$

If we wish to add the algebraic fractions

$$\frac{a}{b} + \frac{c}{x} + \frac{d+e}{4}$$

we need to change the forms of the fractions so that the denominators are equal.

$$\frac{a}{b}\left(\frac{4x}{4x}\right) + \frac{c}{x}\left(\frac{4b}{4b}\right) + \frac{d+e}{4}\left(\frac{bx}{bx}\right) = \frac{4ax + 4cb + bx(d+e)}{4bx}$$

We used a different multiplier for each fraction but this did not change the value of any of the fractions because each of the multipliers had a value of 1. The multipliers that we used were

$$\frac{2}{2} = 1 \qquad \frac{4x}{4x} = 1 \qquad \frac{4b}{4b} = 1 \qquad \frac{bx}{bx} = 1$$

The fact that the denominator and the numerator can be multiplied by the same nonzero quantity without altering the value of the fraction is often called the fundamental principle of fractions or the fundamental theorem of rational expressions. We will call it the **denominator-numerator same-quantity rule** because this name helps us remember what the rule is.

DENOMINATOR-NUMERATOR SAME-QUANTITY RULE

$$\frac{a}{b} = \frac{ac}{bc} \qquad \text{because} \qquad \frac{c}{c} = 1 \qquad (b, c \neq 0)$$

The denominator and the numerator of a fraction may be multiplied by the same nonzero quantity without changing the value of the fraction.

A three-step procedure can be used to add fractions whose denominators are different, as we will show in the next two examples.

example 11.1   Add: $\dfrac{k}{2a} + \dfrac{bc}{ax^2} - \dfrac{m}{ax^3}$

solution   The least common multiple of the denominators of these fractions is $2ax^3$. Thus, each new denominator will be $2ax^3$.

$$\frac{}{2ax^3} + \frac{}{2ax^3} - \frac{}{2ax^3}$$

We see that the original denominator of the first fraction has been multiplied by x^3. Thus, the numerator k must also be multiplied by x^3.

$$\frac{kx^3}{2ax^3} + \frac{}{2ax^3} - \frac{}{2ax^3}$$

The second denominator has been multiplied by $2x$, so the numerator bc must also be multiplied by $2x$.

$$\frac{kx^3}{2ax^3} + \frac{2xbc}{2ax^3} - \frac{}{2ax^3}$$

The multiplier in the last denominator is 2, so we must also multiply m by 2.

$$\frac{kx^3}{2ax^3} + \frac{2xbc}{2ax^3} - \frac{2m}{2ax^3}$$

Now the denominators are the same, so the numerators are added and their sum is recorded over a single denominator.

$$\frac{kx^3 + 2xbc - 2m}{2ax^3}$$

example 11.2   Add: $\dfrac{m}{k} - b + \dfrac{cx}{ak^2}$

solution   The new denominators will be ak^2.

$$\frac{}{ak^2} - \frac{}{ak^2} + \frac{}{ak^2}$$

Next we multiply each original numerator by the same quantity used as a multiplier for its denominator.

$$\frac{akm}{ak^2} - \frac{bak^2}{ak^2} + \frac{cx}{ak^2}$$

As the last step, we record the sum of the numerators over a single denominator.

$$\frac{akm - bak^2 + cx}{ak^2}$$

11.B

inscribed angles

We remember that the measure of an arc of a circle is the same as the measure of the central angle whose radii intercept the arc, as we have shown in the figure on the left.

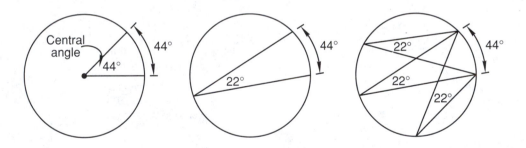

An angle whose sides are inside the circle and whose vertex is on the circle is called an **inscribed angle**. The measure of an inscribed angle equals half the measure of the intercepted arc. In the center figure, we use the same 44° arc, but we have moved the vertex to a point on the circle to form an inscribed angle. We note that the measure of the inscribed angle is half the measure of the intercepted arc. Any inscribed angle that intercepts an arc of 44° will have a measure of 22°. In the figure on the right we have drawn three angles so that each intercepts a 44° arc. Each of these angles has a measure of 22°.

The endpoints of a diameter of a circle intercept an arc whose measure is 180°. **Thus, any inscribed angle that intercepts a diameter is a 90° angle**, as we see in the following figures.

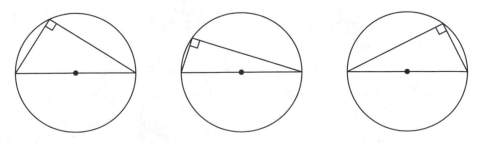

In a later lesson we will prove that the measure of an inscribed angle equals half the measure of the intercepted arc. For now we will content ourselves with an illustration that uses numbers. We begin with the circle on the left. The inscribed angle is angle B, and the arc is arc AC.

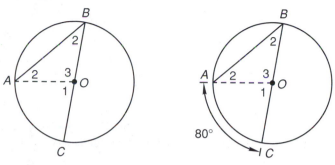

First we note that triangle ABO is an isosceles triangle because two of the sides are radii. Now, in the right-hand figure above, we are given that arc AC equals $80°$. This means that angle 1 must also be an $80°$ angle, as we show on the left below.

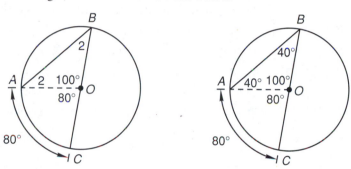

This means that angle 3 must be a $100°$ angle because $100° + 80° = 180°$. Since the angles marked 2 are equal, they must be $40°$ angles so that the three angles of the triangle will add to $180°$. Thus, $\angle ABC$ is a $40°$ angle, and $40°$ is half the measure of the $80°$ arc AC.

example 11.3   Find x.

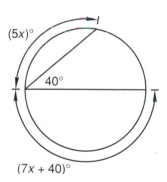

solution   The arc opposite the $40°$ angle is an $80°$ arc. The full measure of the circle is $360°$, so we can write

$$5x + 7x + 40 + 80 = 360 \qquad 360° \text{ in a circle}$$
$$12x + 120 = 360 \qquad \text{simplified}$$
$$12x = 240 \qquad \text{added } -120 \text{ to both sides}$$
$$x = 20 \qquad \text{divided}$$

practice   Add:

a. $\dfrac{m}{3b} + \dfrac{ak}{bz^3} - \dfrac{y}{bz^4}$

b. $\dfrac{z}{b} - k + \dfrac{2mn}{ab^3}$

c. Find x and y.

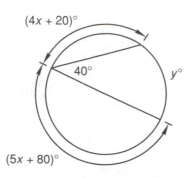

1. Thirteen percent of the people believed in lycanthropes. If 5220 did not believe in lycanthropes, how many believed?

2. Find four consecutive even integers such that the product of -2 and the sum of the first and the fourth is 20 less than the opposite of the third.

3. The recomputed price was $5599. If this was 120 percent greater than the original price, by how much had the original price been increased?

4. Thirty percent of the people refused to work and just sat around. If 1400 people worked, how many just sat around?

5. If Peter picked 240 percent more plums than Roger picked and if Peter picked 6800 plums, how many did both boys pick altogether?

6. Gilbreda listed four consecutive odd integers as she watched steam spew out. Her integers were such that the product of -4 and the sum of the first and the fourth was 10 greater than 10 times the opposite of the third. What were the first four integers on her list?

7. The inscribed angle is $32°$, as shown. Find z. Then find y. Then find x. Then find p.

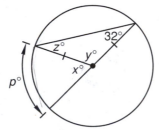

8. The sides of the square are 4 units long. The area of the shaded region equals the area of the square minus the areas of the three triangles. What is the area of the shaded region?

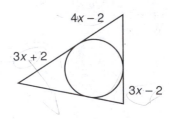

9. The radius of the circle is π cm. The base of the triangle is π cm. The area of the circle equals the area of the triangle. What is the height of the triangle?

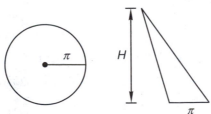

10. The perimeter of this figure is 36 meters. Find the length of the segment $3x + 2$.

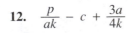

Add:

11. $\dfrac{k}{ax} + \dfrac{bc}{x^2} - \dfrac{m}{ax^3}$

12. $\dfrac{p}{ak} - c + \dfrac{3a}{4k}$

13. $\dfrac{m^2}{p} - \dfrac{3p}{cx} - \dfrac{5}{4c^2x}$

14. Find side x.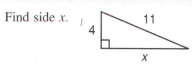

15. Find the distance between $(-3, -3)$ and $(5, 2)$.

16. Graph (a) $4x + 3y - 6 = 0$ and (b) $x = -3$ on the same rectangular coordinate system.

Solve:

17. $3\dfrac{1}{2} - 2\dfrac{1}{3}x = 3\dfrac{1}{4}$

18. $0.03x - x + 2 = -0.91$

19. $3x(2 - 3^0) - 7^0 = -2x(3 - 7^0) + 2$

Expand:

20. $\dfrac{a^{-2}}{y}\left(3a^2y - \dfrac{2y}{a^2}\right)$

21. $\dfrac{2x^0yp}{k}\left(\dfrac{3k}{yp} - \dfrac{2yk}{p}\right)$

Simplify:

22. $\dfrac{2p^2a^{-2}ap^0p^4}{(2pa^{-2})^{-3}ap^0}$

23. $\dfrac{xym^2m^{-4}xm}{(2x^2y)^{-3}xy^0x^{-3}y}$

Simplify by adding like terms:

24. $\dfrac{xp^{-3}}{y} - \dfrac{3y^{-1}}{x^{-1}p^3} + \dfrac{2x}{pppy}$

25. $-3ka + \dfrac{3k^2a^2}{ka} - \dfrac{5a^0k}{a^{-1}}$

Evaluate:

26. $-a - ax(a - x)$ if $a = -\dfrac{1}{2}$ and $x = \dfrac{3}{2}$

27. $-a^2(b - a)$ if $a = -\dfrac{1}{2}$ and $b = \dfrac{3}{2}$

Simplify:

28. $-2(-3 - 2^0 - 2)(-2 + 5)(-2)$

29. $-\dfrac{1}{-2^0} - \dfrac{1}{-2^2} - \dfrac{1}{-2^{-2}}$

30. $\left|-3^0\right| - \left|-2 - 3\right| + (-2^0)(-2 - 5)$

LESSON 12   *Equation of a line*

All first-degree equations in x and y, such as $2x + 3y - 6 = 0$, are called **linear equations,** and the graph of all the ordered pairs of x and y that satisfy such an equation is a straight line. There is an infinite number of ordered pairs of x and y that satisfy this equation. Two of them are $(-3, 4)$ and $(3, 0)$. We have graphed these points and drawn the line in the figure.

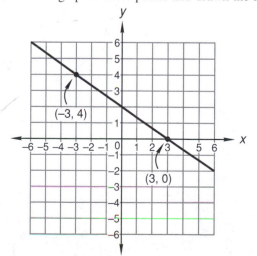

If we write the equation $2x + 3y - 6 = 0$ in the form

$$y = mx + b$$

called the **slope-intercept form,** we get

$$y = -\frac{2}{3}x + 2$$

The letter b in the general equation has been replaced with the number 2 in this equation. This number is the y value of the equation when x equals zero, and on the graph is the y coordinate of the point where the line crosses the y axis.

The letter m in the general equation has been replaced with the number $-\frac{2}{3}$ in this equation. This number is the ratio of the change in the y coordinate to the change in the x coordinate as we move along the line from one point to another. If we move from $(-3, 4)$ to $(3, 0)$, y changes from 4 to 0, a change of -4, and x changes from -3 to 3, a change of $+6$, so the slope is -4 over $+6$.

$$\text{Slope} = \frac{-4}{+6} = -\frac{2}{3}$$

The sign of the slope can be determined visually by using the little man and his car as a mnemonic device. **He always comes from the left side, as we show here.**

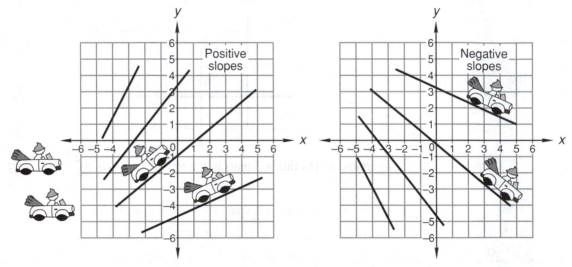

We can find the magnitude of the slope of a line by connecting two points on the line with a horizontal segment and a vertical segment to form a right triangle, as shown below. **The magnitude (absolute value) of the slope is the absolute value of the rise over the absolute value of the run,** or 4 over 6, which equals 2 over 3.

$$|\text{Slope}| = |m| = \left|\frac{\text{rise}}{\text{run}}\right| = \frac{4}{6} = \frac{2}{3}$$

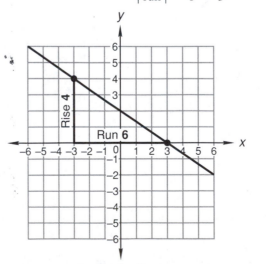

Thus, the line $y = -\frac{2}{3}x + 2$ graphed in this figure

    1. has a negative slope
    2. whose magnitude is $\frac{2}{3}$
    3. and has a *y* intercept of +2

all of which can be verified by looking at the graph. Test your understanding by covering the solution to the next problem until you have worked it. Remember that the equations of vertical and horizontal lines are special cases. The equation of a vertical line has the form $x = \pm k$, and the equation of a horizontal line has the form $y = \pm k$.

example 12.1   Write the equations of lines (a), (b), and (c).

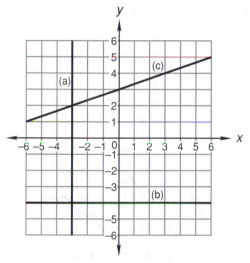

solution   (a)   Every point on this line is 3 units to the left of the *y* axis. The equation is

$$x = -3$$

    (b)   Every point on this line is 4 units below the *x* axis. The equation is

$$y = -4$$

    (c)   The *y* intercept is +3. The slope is positive, and the rise over the run for any triangle drawn is $\frac{1}{3}$. The equation is

$$y = \frac{1}{3}x + 3$$

practice   Write the equations of lines (a) and (b).

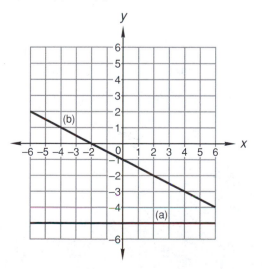

problem set 12

1. Forty percent of the vases were monochromatic and the rest were variegated. If 2400 vases were variegated, how many were monochromatic?

2. Find four consecutive integers such that twice the sum of the first, second, and fourth is 40 less than 3 times the opposite of the third.

3. For a given performance, $4\frac{1}{4}$ times as many tickets were sold as there were seats. If 5100 tickets were sold, how many could be seated in the auditorium?

4. Wilbur thought of a number. He calculated that $2\frac{1}{5}$ of his number was equal to 1. What was his number?

5. Five times the opposite of a number was increased by 25. This was exactly 90 greater than 8 times the number. What was the number?

6. The little train had completed 30 percent of the journey. If 6300 miles still remained, what was the total length of the journey?

7. $\overline{AB}$ is a segment drawn along one side of a triangle. If $x = 140$ and $y = 70$, find z.

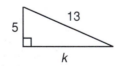

8. Find x and y.

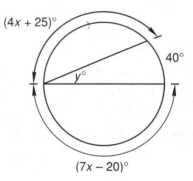

9. The area of a 60° sector of a circle is 36π cm². What is the diameter of the circle?

10. The measures of angles A, B, and C are in the ratio 3:2:1. What are the measures of the angles?

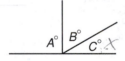

Add:

11. $m + \dfrac{x}{c} + \dfrac{c}{x^2 b}$

12. $\dfrac{a}{b} - \dfrac{3b}{a^2} - \dfrac{2}{abc}$

13. $1 + \dfrac{a}{b}$

14. Find side k.

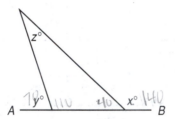

15. Find the distance between $(-2, 7)$ and $(-8, -2)$.

16. Graph (a) $3y + x - 9 = 0$ and (b) $x = 2$ on a rectangular coordinate system.

17. Find the equations of lines (a) and (b).

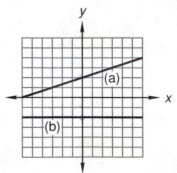

Solve:

18. $8\frac{1}{4} + 2\frac{1}{2}x = \frac{1}{8}$

19. $0.001 + 0.02x - 0.1 = 0.002x$

20. $-3(-2 - 2^0x) - (-2) - 2(-2 - 3x) = -2(x + 4)$

Expand:

21. $\dfrac{a^{-3}x}{y^{-3}}\left(\dfrac{xxx^{-2}}{y^{-2}yy} - 3\right)$

22. $\dfrac{x^0x^{-2}}{y}\left(x^2y - \dfrac{2x^2y}{x^4}\right)$

Simplify:

23. $\dfrac{x^{-2}(2x^{-3})y^2y^0}{x^{-3}yx^2y^{-7}}$

24. $\dfrac{a^0ba^2a^{-1}a}{(a^2b^{-2})^{-3}}$

Simplify by adding like terms:

25. $\dfrac{a}{x} - \dfrac{3a^2y^0x^{-1}}{a} + \dfrac{4x^{-1}}{aa^{-2}}$

26. $abc - \dfrac{5a^2c^3}{ab^{-1}c^2} - \dfrac{3}{a^{-1}b^{-1}c^{-1}}$

Evaluate:

27. $a^2(a^0 - ab)$ if $a = -\dfrac{1}{2}$ and $b = \dfrac{3}{4}$

28. $x(x^2 - x^2y)$ if $x = -\dfrac{1}{2}$ and $y = \dfrac{1}{2}$

29. $-ab(a - a^2b - b)$ if $a = 2$ and $b = -3$

30. Simplify: $-2^0(-2^0 - 3^0 - |-2|) - (-2)(-3) - \dfrac{1}{3^{-2}} - 3^2 - 3$

LESSON *13*   Substitution • *Area of an isosceles triangle*

13.A
substitution

In mathematics, equal quantities can always be substituted for one another. We identify this property by calling it the **substitution axiom.** The substitution axiom is stated in different ways by different authors. Three statements of this axiom that are frequently used are given here.

SUBSTITUTION AXIOM

1. Changing the numeral by which a number is named in an expression does not change the value of the expression.
2. For any numbers a and b, if $a = b$, then a and b may be substituted for each other.
3. If $a = b$, then a may replace b or b may replace a in any statement without changing the truth or falsity of the statement.

Definition 1 seems to apply only to individual expressions. Definition 2 is general enough but not sufficiently specific. Definition 3 seems to apply only to statements and not to individual expressions. We will use the following definition to state formally and exactly the thought that if two expressions have equal value, it is permissible to use either expression.

> SUBSTITUTION AXIOM
>
> If two expressions a and b are of equal value, $a = b$, then a may replace b, or b may replace a in another expression without changing the value of the expression. Also, a may replace b, or b may replace a in any statement without changing the truth or falsity of the statement. Also, a may replace b, or b may replace a in any equation or inequality without changing the solution set of the equation or inequality.

Thus, the substitution axiom applies to expressions, equations, and inequalities. We have been using this axiom in evaluation problems when we have replaced the variables with numbers. Now we will use the axiom to solve a system of first-degree linear equations in two unknowns.

example 13.1 Use substitution to solve: $\begin{cases} x = y + 5 \\ 3x + 2y = 5 \end{cases}$

solution We will replace x in the lower equation with $y + 5$ and then solve for y.

$$3(y + 5) + 2y = 5 \qquad \text{substituted}$$
$$3y + 15 + 2y = 5 \qquad \text{multiplied}$$
$$5y = -10 \qquad \text{simplified}$$
$$y = -2 \qquad \text{divided}$$

Now we replace y with -2 in the top equation and find that x equals 3.

$$x = y + 5 \qquad \text{top equation}$$
$$x = (-2) + 5 \qquad \text{substituted}$$
$$x = 3 \qquad \text{simplified}$$

Thus, the solution is the ordered pair **(3, −2)**.

example 13.2 Use substitution to solve: $\begin{cases} 3x - y = 11 \\ 2x + 3y = -11 \end{cases}$

solution We solve the top equation for y and get

$$y = 3x - 11$$

Then we substitute $3x - 11$ for y in the bottom equation and solve.

$$2x + 3y = -11 \qquad \text{bottom equation}$$
$$2x + 3(3x - 11) = -11 \qquad \text{substituted}$$
$$2x + 9x - 33 = -11 \qquad \text{multiplied}$$
$$11x = 22 \qquad \text{simplified}$$
$$x = 2 \qquad \text{divided}$$

Now we replace x with 2 in the bottom equation and solve for y.

$$2(2) + 3y = -11 \qquad \text{replaced } x \text{ with 2}$$
$$4 + 3y = -11 \qquad \text{multiplied}$$
$$3y = -15 \qquad \text{simplified}$$
$$y = -5 \qquad \text{divided}$$

Thus, our solution is **(2, −5).**

example 13.3   Use substitution to solve: $\begin{cases} x + y = 20 \\ 5x + 10y = 150 \end{cases}$

solution   We solve the top equation for x.

$$x + y = 20 \quad \longrightarrow \quad x = 20 - y$$

Now we substitute $20 - y$ for x in the bottom equation.

$$5(20 - y) + 10y = 150 \qquad \text{replaced } x \text{ with } 20 - y$$

$$100 - 5y + 10y = 150 \qquad \text{multiplied}$$

$$5y = 50 \qquad \text{simplified}$$

$$y = 10 \qquad \text{divided by 5}$$

Since $x + y = 20$,

$$x = 20 - 10 \quad \longrightarrow \quad x = 10$$

Thus, our solution is **(10, 10).**

13.B
area of an isosceles triangle

When we **bisect** an angle, we divide the angle into two angles whose measures are equal. When we **bisect** a line segment, we divide the line segment into two segments whose lengths are equal.

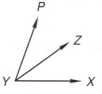

Ray *YZ* bisects angle *PYX* by dividing it into two equal angles. Point *R* bisects segment *TS* by dividing it into two equal segments.

We can find the altitude of an isosceles triangle by drawing a line segment that connects the midpoint of the base to the opposite vertex. Because one end of the segment is the midpoint of a side, the segment is called a **median**. The median to the base of an isosceles triangle is also a **perpendicular bisector** of the base. The median also bisects the angle at the opposite vertex.

example 13.4   Find the area of this triangle. Dimensions are in centimeters.

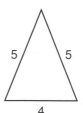

solution   Two sides have equal lengths, so the triangle is an isosceles triangle. First we connect the midpoint of the base to the opposite vertex. This segment is the altitude of the triangle.

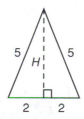

We can find the altitude H by using the Pythagorean theorem.

$$5^2 = H^2 + 2^2 \qquad \text{Pythagorean theorem}$$
$$25 = H^2 + 4 \qquad \text{multiplied}$$
$$21 = H^2 \qquad \text{added} -4 \text{ to both sides}$$
$$\sqrt{21} = H \qquad \text{square root of both sides}$$

Now we can find the area of the triangle.

$$\text{Area} = \frac{1}{2}BH = \frac{1}{2}(4)\sqrt{21} = \mathbf{2\sqrt{21} \ cm^2}$$

practice **a.** Use substitution to solve: $\begin{cases} x = y + 7 \\ 2x + 3y = 4 \end{cases}$

b. Find the area of this triangle. Dimensions are in meters.

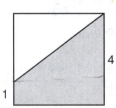

problem set 13

1. Sixty percent of the boats had blue sails. If 300 boats did not have blue sails, how many boats had blue sails?

2. Three-sixteenths of all the citizens in Rome thought Caligula was sane. If 93,750 citizens believed he was sane, how many citizens were there in Rome?

3. Find four consecutive even integers such that 3 times the sum of the first and the fourth is 14 greater than 5 times the third.

4. A number was increased by 14 and this sum was tripled. This result was 67 greater than twice the opposite of the number. What was the number?

5. Each side of the square is 4 units long. Find the area of the shaded region.

6. Find x, y, and P.

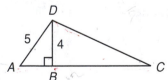

7. Use the Pythagorean theorem to find AB. If $AC = 12$, find DC.

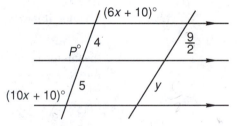

Solve by using substitution:

8. $\begin{cases} x = y + 1 \\ 3x + 2y = 8 \end{cases}$

9. $\begin{cases} 3x - y = 22 \\ 2x + 3y = -11 \end{cases}$

10. $\begin{cases} x + y = 20 \\ 5x + 10y = 200 \end{cases}$

11. $\begin{cases} x + y = 20 \\ 25x + 10y = 395 \end{cases}$

Add:

12. $4 + \dfrac{2}{a}$

13. $\dfrac{a^2}{k} + k + \dfrac{k}{4}$

14. $m^2 + \dfrac{m}{p} + \dfrac{m}{ap^2}$

15. Find the distance between $(-2, -2)$ and $(4, -6)$.

16. Graph (a) $y = -3$ and (b) $2x - 3y = 9$ on a rectangular coordinate system.

17. Find the equations of lines (a) and (b).

Solve:

18. $3\dfrac{1}{2}x + 4\dfrac{1}{3} = 7\dfrac{2}{9}$

19. $0.03 + 0.03x = 0.003$

20. $-3^0 - 3^2 - (-2x - |2|) = 7x$

Expand:

21. $\dfrac{x^2 a}{3}\left(\dfrac{9a^{-1}}{x^2} - \dfrac{2xa^2}{xa}\right)$

22. $\dfrac{-2a^0 p}{m}\left(\dfrac{mp}{a^{-2}} - \dfrac{2a^{-4}a}{p^2 m}\right)$

Simplify:

23. $\dfrac{xa^2(x^0 a^{-2})^4}{(2x^{-2})^{-2}}$

24. $\dfrac{m^2 pxx^{-4}(x^{-2})^2}{(3p^{-2})^{-2}xpx}$

Simplify by adding like terms:

25. $xa - \dfrac{3x^2 a^3}{xa^2} + \dfrac{2x}{a^{-1}}$

26. $\dfrac{amp^{-1}}{m^{-1}} - \dfrac{3a^2 m^2}{pa} + \dfrac{5pa}{m^2}$

Evaluate:

27. $-x^3 - x^2 - x(a - x)$ if $x = -\dfrac{1}{2}$ and $a = 2$

28. $a^2 - ax(x - ax)$ if $x = -3$ and $a = 4$

Simplify:

29. $\dfrac{1}{2^{-3}} - \dfrac{2}{-2^{-2}} - \dfrac{1}{(-2)^{-2}} - 2^0$

30. $3^0 - 2(-2) - |-2 - 4^0 - 3|$

LESSON 14   *Equation of a line through two points • Equation of a line with a given slope*

The easiest equations to solve are first-degree equations and second-degree equations, and it is fortunate that these are the equations most commonly encountered in everyday life and in science courses. Cubic equations (third-degree) and quartic equations (fourth-degree), such as

$$3x^3 + 2x^2 - 5x + 6 = 0 \quad \text{and} \quad 7x^4 + 2x + 6 = 0$$

are much more difficult to solve. Happily, they seldom occur in real-life problems.

Linear equations in two unknowns are also easy equations to understand, and they are useful equations that occur often. We will concentrate on mastering these equations to prepare for chemistry, physics, and advanced courses in mathematics. To make our task easier, we will use y to represent the dependent variable and x to represent the independent variable. Also, we will always try to use the slope-intercept form of the equation.

$$y = mx + b$$

Other forms of this equation are often used, but we will avoid them until we master the slope-intercept form.

We use the letter m to represent the slope and the letter b to represent the intercept. **When we are given the exact coordinates of two points on a line, we can find the exact value of the slope and the exact value of the intercept.**

example 14.1   Find the equation of the line that passes through $(-3, 2)$ and $(3, -3)$.

solution   We will graph the line to find the slope.

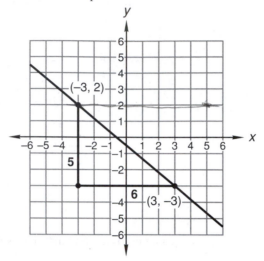

We see that the line has a negative slope whose magnitude is $\frac{5}{6}$. Since the slope is $-\frac{5}{6}$, we can write

$$y = -\frac{5}{6}x + b$$

This slope is exact because we were given the exact coordinates of both points. Now we can use this exact slope and the coordinates of one of the points to find the exact value of the intercept. We can use $(-3, 2)$ or $(3, -3)$ for x and y and solve algebraically for b. To demonstrate that both sets of coordinates will yield the same value for b, we will do the problem twice.

USING $(-3, 2)$:	USING $(3, -3)$:
$2 = -\frac{5}{6}(-3) + b$	$-3 = -\frac{5}{6}(3) + b$
$\frac{12}{6} = \frac{15}{6} + b \quad$ trick	$\frac{-18}{6} = \frac{-15}{6} + b \quad$ trick
$-\frac{3}{6} = b$	$-\frac{3}{6} = b$
$-\frac{1}{2} = b$	$-\frac{1}{2} = b$

Note that either way we get a value of $-\frac{1}{2}$ for b. Also, note the trick of writing 2 as $\frac{12}{6}$ and writing -3 as $-\frac{18}{6}$ to make finding the solution easier.

Now since $m = -\frac{5}{6}$ and $b = -\frac{1}{2}$, we can write the equation of the line as

$$y = -\frac{5}{6}x - \frac{1}{2}$$

The slope is exact and the intercept is exact; thus, this equation is an exact equation, not an estimate.

example 14.2   Find the equation of the line that passes through $(-4, 7)$ and has a slope of $-\frac{3}{5}$.

solution   This time we don't have to graph the points since we are told that the slope is $-\frac{3}{5}$. Thus, we can write

$$y = -\frac{3}{5}x + b$$

Now we can find the exact intercept algebraically by using – 4 for *x* and 7 for *y* in the equation. Note in the third line how we write 7 as $\frac{35}{5}$ to make the equation easier to solve for *b*.

$$y = -\frac{3}{5}x + b$$

$$7 = -\frac{3}{5}(-4) + b$$

$$\frac{35}{5} = \frac{12}{5} + b$$

$$\frac{23}{5} = b$$

Thus, our equation is

$$y = -\frac{3}{5}x + \frac{23}{5}$$

practice   a.   Find the equation of the line that passes through $(-2, 4)$ and $(3, -1)$.

   b.   Find the equation of the line that passes through $(-3, 6)$ and has a slope of $-\frac{2}{3}$.

problem set 14

1.   When the piper increased his volume, the number of the rats increased 160 percent. If he ended up with 6578 rats, how many rats did he have before the volume was increased?

2.   To pass the time at Loch Leven, Mary counted sheep. One day she counted 250 percent more than ever before. If the highest previous total had been 4900, how many sheep did she count this time?

3.   The number was doubled and then the product was increased by 7. This sum was multiplied by −3, and the result was 9 greater than 3 times the opposite of the number. What was the number?

4.   Find three consecutive odd integers such that 4 times the sum of the last two is 2 greater than 10 times the first.

Solve by using substitution:

5. $\begin{cases} 3x - 3y = 21 \\ 2x - y = 12 \end{cases}$

6. $\begin{cases} 4x - y = 22 \\ 2x + 3y = 4 \end{cases}$

7. $\begin{cases} x + y = 28 \\ 5x + 10y = 230 \end{cases}$

8. $\begin{cases} x + y = 22 \\ 100x + 25y = 2050 \end{cases}$

Add:

9. $x + \frac{x^2}{y} - \frac{3x}{cy^2}$

10. $\frac{m}{x} + 4$

11. $4 + \frac{c}{x} - cxy$

12.   Find the area of this isosceles triangle. Dimensions are in centimeters.

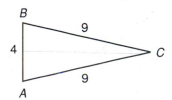

13.   Find the distance between $(-3, 2)$ and $(5, 2)$.

14.   Graph (a) $x = -4$ and (b) $3y + 2x = 6$ on a rectangular coordinate system.

15. Find the equations of lines (a) and (b).

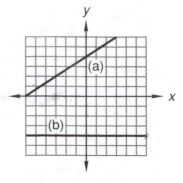

16. Find the equation of the line that passes through $(-3, 4)$ and $(3, -2)$.

17. Find the equation of the line that passes through $(-4, 6)$ and that has a slope of $-\frac{3}{4}$.

18. In this figure, x equals 134. Find A. Then find B and C.

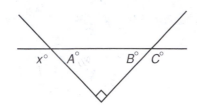

19. One central angle is 150°, as shown. Find A and B. Then find C. Then find D.

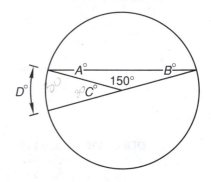

Solve:

20. $2\frac{1}{4}x - \frac{3}{5} = -1\frac{1}{20}$

21. $0.005x - 0.05 = 0.5$

22. $-2^0 - 3^2 = -2(x - 3^0) - 4(2x - 5)$

23. Expand: $\dfrac{-3^{-2}x}{y}\left(\dfrac{9y^0 x}{-x} - \dfrac{3x}{y}\right)$

24. Simplify: $\dfrac{(-2x)^{-2}xy^0 y^2(x)^{-2}}{(x^2 y)^{-2}xyxy^{-2}}$

25. $mp^2\left(m^{-1}p^{-2} - \dfrac{4m}{p^2}\right)$

26. Simplify by adding like terms: $-\dfrac{3xy}{m} + \dfrac{7x^2 x^{-1}m^{-1}}{y} - \dfrac{8m^{-1}x^{-1}}{x^{-2}y^{-1}}$

Evaluate:

27. $x^2 - y - xy^2$ if $x = -\frac{1}{3}$ and $y = \frac{1}{2}$

28. $a^2x - a(xa - a)$ if $a = -1$ and $x = -3$

Simplify:

29. $-5^0(-2 - 3^0) - 2^2 - \dfrac{1}{(-2)^{-2}}$

30. $5 - |-2 - 3| - 4^0 - 2|-2^0| - 3$

LESSON 15 *Elimination*

The addition rule for equations tells us that the same quantity can be added to both sides of an equation without changing the solution to the equation. If we look at the equation

$$x + 2 = 7$$

we know that the solution is 5. We can use the addition rule to get this solution if we add −2 to both sides.

$$\begin{array}{rcl} x + 2 &=& 7 \\ -2 &=& -2 \\ \hline x &=& 5 \end{array}$$

We can use a similar process to help us solve systems of equations. If we have the system of equations

$$\begin{cases} 2x + y = 11 \\ 3x - 2y = 6 \end{cases}$$

we can eliminate the variable y if we multiply the top equation by 2 and add the equations. Since we eliminate one variable by adding, this method is called the **elimination method**. It is also called the **addition method** or the **linear combination method**.

$$\begin{array}{ll} 2x + y = 11 \longrightarrow (2)^\dagger \longrightarrow & 4x + 2y = 22 \\ 3x - 2y = 6 \longrightarrow (1) \longrightarrow & \underline{3x - 2y = 6} \\ & 7x \qquad\quad = 28 \end{array}$$

$$x = 4$$

To use the addition method, we had to assume that $3x - 2y$ equaled 6. Since we made this assumption, we can say that we were adding equal quantities to both sides of the top equation.

Now we will use 4 for x in the bottom equation and solve for y.

$$3(4) - 2y = 6 \longrightarrow 12 - 2y = 6 \longrightarrow 6 = 2y \longrightarrow y = 3$$

Since we made an assumption, we must check the values $x = 4$ and $y = 3$ in both of the original equations.

Top Equation	Bottom Equation
$2x + y = 11$	$3x - 2y = 6$
$2(4) + (3) = 11$	$3(4) - 2(3) = 6$
$8 + 3 = 11$	$12 - 6 = 6$
$11 = 11$ Check	$6 = 6$ Check

example 15.1   Use elimination to solve: $\begin{cases} 3x + 2y = 23 \\ -2x + 3y = 2 \end{cases}$

solution   We decide to eliminate the x terms so we will multiply the top equation by 2 and the bottom equation by 3. Then we add the equations.

$$\begin{array}{ll} 3x + 2y = 23 \longrightarrow (2) \longrightarrow & 6x + 4y = 46 \\ -2x + 3y = 2 \longrightarrow (3) \longrightarrow & \underline{-6x + 9y = 6} \\ & 13y = 52 \end{array}$$

$$y = 4$$

Now we replace y with 4 in the top equation and solve for x.

$$3x + 2(4) = 23 \longrightarrow 3x + 8 = 23 \longrightarrow 3x = 15 \longrightarrow x = 5$$

Thus, our solution is the ordered pair **(5, 4)**. Now to check:

Top Equation	Bottom Equation
$3x + 2y = 23$	$-2x + 3y = 2$
$3(5) + 2(4) = 23$	$-2(5) + 3(4) = 2$

---

†The notation $\longrightarrow (2) \longrightarrow$ has no mathematical meaning. In this book we use this notation to help us remember the number by which we have multiplied.

$$15 + 8 = 23 \qquad\qquad\qquad -10 + 12 = 2$$
$$23 = 23 \quad \text{Check} \qquad\qquad 2 = 2 \quad \text{Check}$$

practice   Use elimination to solve: $\begin{cases} 4x + 3y = 17 \\ -3x + 4y = 6 \end{cases}$

problem set
15

1. Twenty percent of those interviewed were students. If in his lifetime the announcer interviewed 1400 students, how many people did he interview in all?

2. Sandra slammed everything she could reach. If she saw 1000 and slammed 200, what percent of what she saw did she slam?

3. The pharmacist calculated the correct number. She took 5 times the opposite of a number and added it to -7. If the result was 35 less than twice the correct number, what was the correct number?

4. Find three consecutive integers such that -4 times the sum of the first and the third is 12 greater than the product of 7 and the opposite of the second.

Use elimination to solve:

5. $\begin{cases} 3x + y = 11 \\ 3x - 2y = 2 \end{cases}$

6. $\begin{cases} 3x + 4y = 20 \\ -4x + 3y = 15 \end{cases}$

Use substitution to solve:

7. $\begin{cases} 3x + y = 16 \\ 2x - 3y = -4 \end{cases}$

8. $\begin{cases} x + 3y = -9 \\ 5x - 2y = 23 \end{cases}$

Add:

9. $y + \dfrac{x}{a^2} - \dfrac{mx}{3y^2}$

10. $4 - \dfrac{3a}{x}$

11. $c + \dfrac{c^2}{x} + ac^2$

12. Find side m.

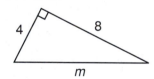

13. Find the distance between $(-4, 2)$ and $(3, -4)$.

14. Graph (a) $y = -3$ and (b) $3x - 4y = 8$ on a rectangular coordinate system.

15. Find the equations of lines (a) and (b).

16. Find the equation of the line through $(-2, -2)$ and $(4, 3)$.

17. Find the equation of the line through $(-2, 5)$ that has a slope of $-\frac{1}{7}$.

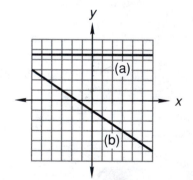

18. In this figure, x equals 134. Find A. Then find B and C. Then find k and y.

19. One central angle is 140°, as shown. Find A and B. Then find C. Then find D.

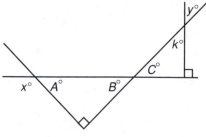

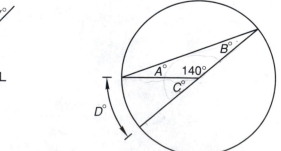

20. We know that P equals one-half of $9x - 17$ because P is the measure of an inscribed angle and $9x - 17$ is the measure of the intercepted arc. Also, P equals $3x + 2$ because vertical angles are equal. Find x. Then find P.

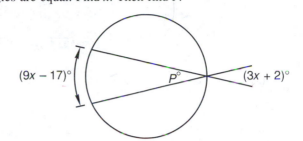

21. Expand: $\dfrac{-3x^{-1}y^2}{p}\left(\dfrac{-px}{y^2} - \dfrac{3x^2y}{p^{-3}}\right)$

Solve:

22. $3\dfrac{2}{5}x - \dfrac{4}{5} = 7\dfrac{3}{10}$ **23.** $0.07x - 0.02 = 0.4$

24. $-3x(-2 - 3^0) = -2^0(-x - 3x^0)$

25. Simplify: $\dfrac{x^2y^{-2}x^0(y^{-1})}{x^2y^2(x^0)^2}$

 26. Simplify by adding like terms: $\dfrac{8x^2yy}{m^{-2}x^2} + \dfrac{3y^2}{m^{-2}} - \dfrac{5m^2x^{-2}}{x^{-2}y^{-2}}$

Evaluate:

27. $axy - a^2x$ if $a = -\dfrac{1}{2}$, $x = 3$, and $y = -\dfrac{1}{3}$

28. $a^0(a^0x - ax)$ if $a = -2$ and $x = -3$

Simplify:

29. $-2^0(2^0 - 3^2) - [-2 - 3 - (-2)]$ **30.** $3^0 - \dfrac{3}{3^{-2}} + 2 - 5^0$

LESSON 16   *Multiplication of polynomials • Division of polynomials*

16.A

multiplication of polynomials

When algebraic expressions are multiplied, each term of one expression is multiplied by every term of the other expression. Then, like terms in the product are added. Some people prefer to use a vertical format, while others prefer a horizontal format. Neither format is more correct than the other, and many people use both formats, sometimes using one and sometimes using the other. The format is not important—the procedure is important. The simplest kind of algebraic expression is a polynomial, so we will demonstrate by multiplying polynomials.

example 16.1   Multiply: $(2x + 2)(3x^2 - 5x + 2)$

solution   We decide to use the vertical format. People tend to put the longer expression on the top; but this time, just to be different, we will put the longer expression on the bottom.

$$\begin{array}{rrrr} & & 2x & +\ 2 \\ & 3x^2 & -\ 5x & +\ 2 \\ \hline 6x^3 & +\ 6x^2 & & \\ & -\ 10x^2 & -\ 10x & \\ & & +\ 4x & +\ 4 \\ \hline \mathbf{6x^3} & \mathbf{-\ 4x^2} & \mathbf{-\ 6x} & \mathbf{+\ 4} \end{array}$$

example 16.2 Multiply: $(3x^2 - 2x - 2)(x - 2)$

solution We will use the horizontal format this time. Either expression can be first. We decide to use the same order as given in the problem.

$$\begin{aligned} (3x^2 - 2x - 2)(x - 2) &= 3x^3 - 6x^2 - 2x^2 + 4x - 2x + 4 \\ &= \mathbf{3x^3 - 8x^2 + 2x + 4} \end{aligned}$$

16.B division of polynomials

When we divide polynomials, we use a procedure that is very similar to the one used for long division with real numbers. To review this algorithm, we will divide 49 by 12.

$$\begin{array}{r} 4 \\ 12\overline{)49} \\ \underline{48} \\ 1 \end{array} \qquad \text{so} \qquad \frac{49}{12} = \mathbf{4\frac{1}{12}}$$

Note that the fraction is formed by writing the remainder over the divisor.

example 16.3 Divide $2 + 5x - 2x^2 + 5x^3$ by $-4 + x$.

solution It is customary to begin by writing both expressions in descending powers of the variable. This step is not absolutely necessary, but it helps to keep like terms below each other when doing the division.

$$\begin{array}{r} 5x^2 + 18x + 77 \\ x - 4\overline{)5x^3 - 2x^2 + 5x + 2} \\ \underline{5x^3 - 20x^2 \qquad\qquad} \end{array}$$

Now we **mentally** change the signs of $5x^3 - 20x^2$ to $-5x^3 + 20x^2$ and add.

$$\begin{array}{rl} \begin{array}{r} 5x^2 + 18x + 77 \\ x - 4\overline{)5x^3 - 2x^2 + 5x + 2} \\ \underline{5x^3 - 20x^2 \qquad\qquad} \\ 18x^2 + 5x \qquad \\ \underline{18x^2 - 72x \qquad} \\ 77x + 2 \\ \underline{77x - 308} \\ 310 \end{array} & \begin{array}{l} \\ \\ \\ \text{changed signs and added} \\ \\ \text{changed signs and added} \\ \\ \text{changed signs and added} \end{array} \end{array}$$

Thus, $$\frac{5x^3 - 2x^2 + 5x + 2}{x - 4} = \mathbf{5x^2 + 18x + 77 + \frac{310}{x - 4}}$$

We can check our division by writing the first three terms with denominators of $x - 4$ and adding.

$$\frac{5x^2(x-4)}{x-4} + \frac{18x(x-4)}{x-4} + \frac{77(x-4)}{x-4} + \frac{310}{x-4}$$

$$= \frac{5x^3 - 20x^2 + 18x^2 - 72x + 77x - 308 + 310}{x-4}$$

$$= \frac{5x^3 - 2x^2 + 5x + 2}{x-4} \qquad \text{Check}$$

example 16.4   Divide $-6 + x^3$ by $-2 + x$.

solution   Again we rewrite each expression in descending powers of the variable. Also, we insert $0x^2$ and $0x$ into the cubic expression to help us keep the proper spacing.

$$
\begin{array}{r}
x^2 + 2x + 4 \\
x - 2 \overline{)\, x^3 + 0x^2 + 0x - 6} \\
\underline{x^3 - 2x^2} \\
2x^2 + 0x \\
\underline{2x^2 - 4x} \\
4x - 6 \\
\underline{4x - 8} \\
2
\end{array}
$$

Thus,
$$\frac{x^3 - 6}{x - 2} = x^2 + 2x + 4 + \frac{2}{x - 2}$$

We will check our work by adding:

$$x^2\left(\frac{x-2}{x-2}\right) + 2x\left(\frac{x-2}{x-2}\right) + 4\left(\frac{x-2}{x-2}\right) + \frac{2}{x-2}$$

$$= \frac{x^3 - 2x^2 + 2x^2 - 4x + 4x - 8 + 2}{x-2} = \frac{x^3 - 6}{x-2} \qquad \text{Check}$$

practice   Divide $3 + 7x - 3x^2 + 4x^3$ by $-3 + x$.

problem set 16

1. Two and one-seventh times the acceptable number had crawled into the space provided. If 900 had crawled in, what was the acceptable number?

2. When the smoke cleared, only 0.016 of the microbes had disassociated. If 420,000 microbes were in the vial, how many had disassociated?

3. Only 14 percent of the reception guests were uninvited. If 903 of those present had been invited, how many guests attended the reception?

4. Of the angry crowd, 40 percent had been mollified, and now it was necessary to placate the rest. If 8600 had been mollified, how many had to be placated?

Use elimination to solve:

5. $\begin{cases} 4x - 3y = -1 \\ 2x + 5y = 19 \end{cases}$

6. $\begin{cases} 7x - 2y = 13 \\ 4x + 7y = 40 \end{cases}$

Use substitution to solve:

7. $\begin{cases} 3x + y = 2 \\ 2x - 5y = 7 \end{cases}$

8. $\begin{cases} x - 3y = 4 \\ 4x - 7y = 16 \end{cases}$

9. Multiply: $(2x + 3)(2x^2 - 4x + 3)$

Divide and check:

10. $2 + 4x - 2x^2 + 4x^3$ by $-3 + x$

11. $-8 + x^3$ by $-4 + x$

Add:

12. $2 + \dfrac{a}{2x^2}$

13. $\dfrac{4}{cx} + c - \dfrac{3}{4c^2 x}$

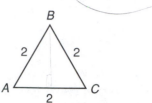

14. Find the area of this equilateral (also isosceles) triangle. Dimensions are in inches.

15. Find the distance between $(-3, -5)$ and $(-1, 2)$.

16. Graph (a) $x = -2\frac{1}{2}$ and (b) $x - 3y = 6$ on a rectangular coordinate system.

17. Find the equations of lines (a) and (b).

18. Find the equation of the line through $(-2, -4)$ and $(5, -6)$.

19. Find the equation of the line that passes through $(-4, 1)$ and has a slope of $\frac{3}{5}$.

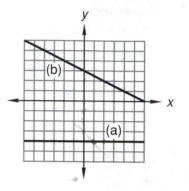

20. The area of the small circle is π cm². Find the radius of the small circle. Then find the radius of the large circle. Then find the area of the large circle.

21. One angle in the isosceles triangle is a 30° angle, as shown. Find A. Then find B. Then find C. Then find D.

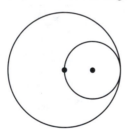

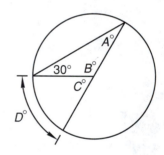

Solve:

22. $3\frac{1}{2}x - \frac{1}{5} = 3\frac{3}{20}$

23. $0.05m - 0.05 = 0.5$

24. $-4(-x - 2) - 3(-2 - 4) = -4^0(x - 2)$

25. Expand: $\dfrac{3^{-2}x^{-2}y}{z}\left(\dfrac{18x^2 z}{y} - \dfrac{3xyz^2}{x}\right)$

26. Simplify: $\dfrac{3^{-2}x^0(x^{-2}y^4)^{-3}}{xxy^2 y^0 (y^3)^{-4}}$

27. Simplify by adding like terms: $\dfrac{3x^2 xxy}{y^{-4}} - \dfrac{2x^2}{(x^{-1})^2 y^{-5}} - \dfrac{7xxyx^2}{(y^{-2})^2}$

28. Evaluate: $p^2 - xp - x(p - x)$ if $x = -2$ and $p = \dfrac{1}{2}$

Simplify:

29. $-2^0(-2 - 3) - (-2)^0 - 2[-1 - (-3 - 7) - 2(-5 + 3)]$

30. $2^0 - \dfrac{27}{3^2} - 3^1 - 3^0 - |-3^0 - 2| - (-3)$

LESSON 17   *Subscripted variables* • *Angle relationships*

17.A
subscripted variables

In first-degree equations in two unknowns, we often use the letter x to represent the independent variable and the letter y to represent the dependent variable. Thus, if we are given the linear equation

$$2x + 3y = 6$$

and are asked to graph the equation, we would first solve the equation for y and get

$$y = -\frac{2}{3}x + 2$$

This is the familiar slope-intercept form of the equation, and we can now graph the equation by choosing values for x (the independent variable) and seeing which values of y (the dependent variable) that this equation pairs with the chosen values of x. When we graph an equation, we always use the horizontal number line for x and the vertical number line for y. Thus, the graph of the equation is as follows:

x	0	−6	6
y	2	6	−2

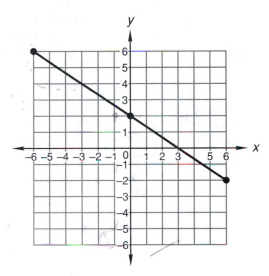

We now note that the intercept is +2 and that the sign of the slope is negative. If we visualize a triangle, we see that the slope is $-\frac{2}{3}$.

When we work word problems, the use of x and y as variables is often not helpful because it is difficult to remember what these letters represent. If we use subscripted variables, however, there is no difficulty in recognizing the variable. The large letter gives the general description, and the smaller letter supplies more specific information. To use subscripted variables to say that the sum of the number of nickels and the number of dimes was 40, we could write

$$N_N + N_D = 40$$

Here N_N means number of nickels and N_D means number of dimes.

example 17.1   Solve the system: $\begin{cases} N_N + N_D = 40 \\ 5N_N + 10N_D = 250 \end{cases}$

solution   We decide to use elimination, and we multiply the top equation by −5 and then add.

$$N_N + N_D = 40 \quad \longrightarrow \quad (-5) \quad \longrightarrow \quad \begin{array}{r} -5N_N - 5N_D = -200 \\ 5N_N + 10N_D = 250 \\ \hline 5N_D = 50 \end{array}$$

$$N_D = 10$$

Now since $N_N + N_D = 40$, N_N must equal **30.**

example 17.2   Solve the following system of equations.

$$R_M T_M + R_W T_W = 260 \qquad R_M = 40 \qquad R_W = 60 \qquad T_M + T_W = 5$$

solution   These equations come from a word problem in which a person walked part of a distance and rode a motorcycle the rest of the distance. Thus, R_M and T_M stand for the rate of the motorcycle and the time of the motorcycle, and R_W and T_W stand for the rate walking and the time walking. To solve, we substitute 40 for R_M and 60 for R_W and get

$$40 T_M + 60 T_W = 260 \qquad \text{substituted}$$

Now we rearrange the last equation.

$$T_M + T_W = 5 \quad \longrightarrow \quad T_M = 5 - T_W \qquad \text{rearranged last equation}$$

Next we substitute $5 - T_W$ for T_M and then solve.

$$40(5 - T_W) + 60 T_W = 260 \qquad \text{substituted}$$
$$200 - 40 T_W + 60 T_W = 260 \qquad \text{multiplied}$$
$$20 T_W = 60 \qquad \text{simplified}$$
$$T_W = 3 \qquad \text{divided}$$

Now since $T_M + T_W = 5$, we conclude that $T_M = 2.$

17.B

angle relationships   We can devise problems about angle relationships that let us practice the solutions of two equations in two unknowns.

example 17.3   Find x and y.

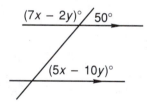

solution   We see that one small angle is a 50° angle. Thus, all small angles are 50° angles, and all large angles are 130° angles. We see that $7x - 2y$ is a large angle and that $5x - 10y$ is a small angle. This gives us two equations.

$$\text{(a)} \quad 7x - 2y = 130$$
$$\text{(b)} \quad 5x - 10y = 50$$

To solve, we will multiply equation (a) by -5 and add the equations.

$$
\begin{array}{llll}
\text{(a)} & 7x - 2y = 130 & \longrightarrow \quad (-5) \quad \longrightarrow & -35x + 10y = -650 \\
\text{(b)} & 5x - 10y = 50 & \longrightarrow \quad (1) \quad \longrightarrow & \underline{\quad 5x - 10y = \quad 50} \\
& & & -30x \qquad\quad = -600
\end{array}
$$

$$x = 20$$

Now we use equation (a) to solve for y.

$$7x - 2y = 130 \qquad \text{equation (a)}$$
$$7(20) - 2y = 130 \qquad \text{used 20 for } x$$
$$140 - 2y = 130 \qquad \text{multiplied}$$
$$-2y = -10 \qquad \text{simplified}$$
$$y = 5 \qquad \text{solved}$$

practice   **a.** Solve the system: $\begin{cases} N_N + N_O = 60 \\ 5N_N + 10N_O = 310 \end{cases}$

b. Solve the following system of equations:

$$R_M T_M + R_W T_W = 380 \qquad R_M = 50 \qquad R_W = 70 \qquad T_M + T_W = 5$$

c. Find x and y.

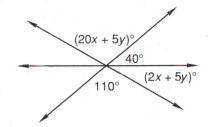

$(20x + 5y)°$

$40°$

$(2x + 5y)°$

$110°$

problem set   1. Only $\frac{2}{5}$ of the students doubted. If 600 students were nondoubters, how many
17   students were there in all?

2. Only $\frac{7}{13}$ of the teachers believed. If 210 teachers believed, how many teachers were
there?

3. The number of skeptics increased 260 percent overnight. If 400 were skeptical at sunset,
how many were skeptical at sunrise?

4. Find three consecutive integers such that 7 times the sum of the last two is 109 greater
than 10 times the first.

5. Use elimination to solve: $\begin{cases} N_N + N_D = 150 \\ 5N_N + 10N_D = 450 \end{cases}$

6. Use substitution to solve: $\begin{cases} N_P + N_D = 50 \\ N_P + 10N_D = 140 \end{cases}$

7. Multiply: $(2x + 3)(2x^2 + 2x + 2)$

8. Divide $3x^3 - 2$ by $x + 1$ and then check.

Solve for all variables:

9. $R_M T_M + R_W T_W = 250, R_M = 50, R_W = 80, T_M + T_W = 5$

10. $R_E T_E = R_W T_W, R_E = 200, R_W = 250, 9 - T_E = T_W$

11. $R_M T_M = R_R T_R, R_M = 8, R_R = 2, 5 - T_M = T_R$

Add:

12. $4x + \dfrac{3x}{a}$ **13.** $-\dfrac{2x}{y} - cx + \dfrac{7x^2 y}{np^2}$ **14.** Find the area of this triangle.
Dimensions are in meters.

15. Find the distance between $(2, 4)$ and $(6, -2)$.

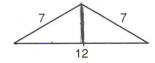

7   7

12

16. Graph: (a) $y = -3$
         (b) $3x - 5y = 10$

17. Find the equations of lines (a) and (b).

18. Find the equation of the line through
$(4, 2)$ and $(6, -3)$.

19. Find the equation of the line through
$(-3, 5)$ that has a slope of $-\frac{2}{7}$.

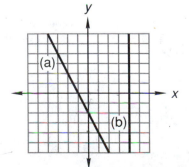

y

(a)

x

(b)

20. Solve for x, y, and z.

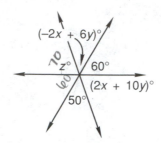

21. The radius of a right circular cylinder is 5 cm and the volume of the cylinder is 245π cm^3. What is the length of the cylinder?

Solve:

22. $2\frac{1}{5}x - 3\frac{1}{4} = \frac{7}{20}$

23. $0.003 - 0.03 + 0.3x = 3.3$

24. $3^0(2x - 5) + (-x - 5) = -3(x^0 - 2)$

25. Expand: $\dfrac{x^0 y^{-2} x}{x^3 y}\left(\dfrac{x^2 y}{m} - \dfrac{3x^4 y^2}{m^{-2}}\right)$

26. Simplify: $\dfrac{2^{-3} x^0 (x^2)}{x^{-3} xy^{-3} y}$

27. Simplify by adding like terms: $\dfrac{xy}{y^{-2}} - \dfrac{3x^4 y^4}{x^3 y} + \dfrac{7xy^{-2}}{xy^{-3}}$

28. Evaluate: $xy - x(x - y^0)$ if $x = 2$ and $y = -\dfrac{1}{2}$

Simplify:

29. $-(-3 - 2) + 4(-2) + \dfrac{1}{-2^{-3}} - (-2)^{-3}$

30. $-|-2 - 7| - |-2 - 4| - 3|-2 + 7|$

LESSON 18 *Ratio word problems* • *Similar triangles*

18.A
ratio word problems

A **ratio** is a comparison of two numbers that can be written as a fraction. If we write the fraction

$$\frac{3}{4}$$

we say that we have written the ratio of 3 to 4. All of the following ratios designate the same number and thus are equal ratios.

$$\frac{3}{4} \qquad \frac{6}{8} \qquad \frac{300}{400} \qquad \frac{15}{20} \qquad \frac{27}{36} \qquad \frac{111}{148}$$

An equation or other statement which indicates that two ratios are equal is called a **proportion**. Thus, we say that

$$\frac{3}{4} = \frac{15}{20}$$

is a proportion. We note that cross products of equal ratios are equal, as we show in the following example.

$$4 \times 15$$

$$4 \times 15 = 3 \times 20$$
$$60 = 60 \qquad \text{True}$$

$$3 \times 20$$

We can solve proportions that contain an unknown by setting the cross products equal and then dividing to complete the solution. To solve

$$\frac{4}{3} = \frac{5}{k}$$

we first set the cross products equal and get

$$4k = 15$$

We finish by dividing by 4.

$$\frac{4k}{4} = \frac{15}{4} \quad \longrightarrow \quad k = \frac{15}{4}$$

When we set the cross products equal, we say that we have *cross multiplied.*

In some ratio word problems the word "ratio" is used, but in other ratio word problems the word "ratio" is not used and it is necessary to realize that a constant ratio is implied.

example 18.1   The ratio of Arabians to mixed breeds in the herd was 2 to 17. If there were 380 horses in the herd, how many were Arabians?

solution   We are given 2 Arabians and 17 mixed breeds for a total of 19. Thus, we can write

$$2 \qquad \text{Arabians}$$
$$17 \qquad \text{mixed breeds}$$
$$19 \qquad \text{total}$$

Now by looking at what we have written, we see that three proportions are indicated.

$$\text{(a)} \ \ \frac{2}{17} = \frac{A}{M} \qquad \text{(b)} \ \ \frac{2}{19} = \frac{A}{T} \qquad \text{(c)} \ \ \frac{17}{19} = \frac{M}{T}$$

We are given a total of 380 and asked for the number of Arabians, so we will use proportion (b) and replace T with 380 and solve.

$$\frac{2}{19} = \frac{A}{380} \quad \longrightarrow \quad 2 \cdot 380 = 19A \quad \longrightarrow \quad 760 = 19A \quad \longrightarrow \quad A = \mathbf{40}$$

example 18.2   It took 600 kilograms (kg) of sulfur to make 3000 kg of the new compound. How many kilograms of other materials would be required to make 4000 kg of the new compound?

solution   This wording is typical of real-world ratio problems. The word "ratio" is not used in the statement of the problem, but the statement implies that the ratio of kilograms of sulfur to kilograms of the compound is constant. We will call the other component NS for "not sulfur." If 600 kg was sulfur, then 2400 kg must have been "not sulfur."

$$600 = S$$
$$2400 = NS$$
$$3000 = T$$

Thus, the three implied equations are:

$$\text{(a)} \ \ \frac{600}{2400} = \frac{S}{NS} \qquad \text{(b)} \ \ \frac{600}{3000} = \frac{S}{T} \qquad \text{(c)} \ \ \frac{2400}{3000} = \frac{NS}{T}$$

Since we are given 4000 kg of the compound and are asked for the amount of "not sulfur," we will use equation (c).

$$\frac{2400}{3000} = \frac{NS}{4000} \quad \longrightarrow \quad (2400)(4000) = 3000NS \quad \longrightarrow \quad \mathbf{3200 \text{ kg}} = NS$$

18.B

similar triangles

If two triangles have the same angles, the triangles have the same shape, and they look alike. We say that triangles that have the same angles are **similar triangles.**

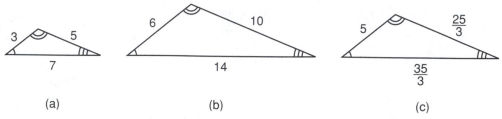

(a)   (b)   (c)

Triangles (a), (b), and (c) look alike but they are not the same size. They look alike because the angles in each triangle are equal angles, as indicated by the tick marks. The sides opposite equal angles in similar triangles are called **corresponding sides.** The ratios of the lengths of corresponding sides in similar triangles are equal. This implies that corresponding sides in similar triangles are related by a number called the **scale factor.** The scale factor going from triangle (a) to triangle (b) is 2 because each side in triangle (b) is twice as long as the corresponding side in triangle (a). Going from triangle (b) to triangle (a), we find that the scale factor is $\frac{1}{2}$ because each side in triangle (a) is half as long as the corresponding side in triangle (b). The scale factor between triangles (a) and (c) is $\frac{5}{3}$ because each side in triangle (c) is $\frac{5}{3}$ times the length of the corresponding side in triangle (a). Of course, if we go from triangle (c) to triangle (a), the scale factor is $\frac{3}{5}$.

example 18.3 The tick marks tell us that these two triangles are similar. Find the scale factor and then find x and y.

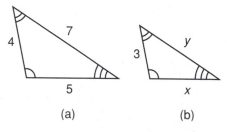

(a)   (b)

solution The side whose length is 4 in (a) and the side whose length is 3 in (b) are corresponding sides because they are opposite equal angles. We use these two sides to find the scale factor. Four times the scale factor equals 3. The arrow above SF shows this is the scale factor from left to right.

$$4\overrightarrow{SF} = 3 \qquad \text{equation}$$

$$\overrightarrow{SF} = \frac{3}{4} \qquad \text{divided}$$

Now we find x by using the left-to-right scale factor.

$$5\left(\overrightarrow{\frac{3}{4}}\right) = x \qquad \text{equation}$$

$$\frac{15}{4} = x \qquad \text{multiplied}$$

We can use the same procedure to solve for y.

$$7\left(\overrightarrow{\frac{3}{4}}\right) = y \qquad \text{equation}$$

$$\frac{21}{4} = y \qquad \text{multiplied}$$

practice   **a.**   The ratio of malefactors to good guys was 3 to 11. If there were 350 individuals investigated, how many were malefactors?

b.   It took 800 kilograms of sulfur to make 4000 kilograms of the new compound. How many kilograms of other materials would be required to make 5000 kilograms of the new compound?

c.   Find a and b.

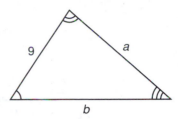

problem set   **1.**   The ratio of Arabians to mixed breeds in the herd was 2 to 19. If there were 420 horses
18   in the herd, how many were Arabians?

2.   It took 500 kg of sulfur to make 3000 kg of new compound. How many kilograms of other materials would be required to make 6000 kg of the new compound?

3.   The law of the land displeased 27 percent of the natives. If 54,000 natives were displeased, how many natives were there?

4.   The percentage for nonagenarians in the population was only 0.004 percent. If there were 40 nonagenarians in the village, what was the total population?

5.   Use elimination to solve: $\begin{cases} N_D + N_Q = 200 \\ 10N_D + 25N_Q = 2750 \end{cases}$

6.   Use substitution to solve: $\begin{cases} N_P + N_D = 30 \\ N_P + 10N_D = 291 \end{cases}$

7.   Multiply: $(2x + 4)(3x^2 - 2x - 10)$ **8.**   Divide $5x^3 - 1$ by $x - 2$ and check.

Solve for all unknown variables:

9. $R_F T_F = R_S T_S,\ T_S = 6,\ T_F = 5,\ R_F - 16 = R_S$

10. $R_M T_M = R_R T_R,\ R_M = 8,\ R_R = 2,\ T_R = 5 - T_M$

11. $R_G T_G + R_B T_B = 100,\ R_G = 4,\ R_B = 10,\ T_B = T_G + 3$

Add:

12. $7xyz + \dfrac{1}{xyz}$ **13.** $-\dfrac{3x}{y} - c + \dfrac{7c}{xy^3}$

14.   Use the Pythagorean theorem as required to find the area of this isosceles triangle. Dimensions are in centimeters.

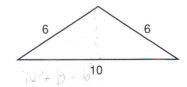

15.   Find the distance between $(-3, 5)$ and $(4, -2)$.

16.   Graph: (a) $y = 2$ (b) $y = 2x$

17.   Find the equations of lines (a) and (b).

18.   Find the equation of the line through $(-3, 5)$ and $(4, -2)$.

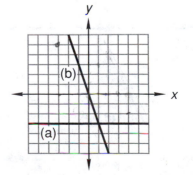

19. Find the equation of the line that has a slope of $\frac{5}{3}$ and passes through $(4, -2)$.

20. Find a and b.

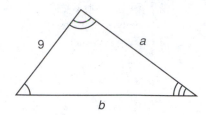

21. Find x and P. Remember that the measure of the arc is twice the measure of the inscribed angle.

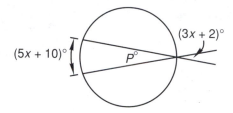

22. Two fences in a field meet at 120°. A cow is tethered at their intersection with a 15-foot rope, as shown in the figure. Over how many square feet may the cow graze?

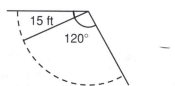

Solve:

23. $4\frac{1}{6}x - 2\frac{1}{12} = -\frac{5}{24}$

24. $-2[(x - 2) - 4x - 3] = -(-4 - 2x)$

25. Expand: $\dfrac{3x^0 y^{-2}}{z^2}\left(\dfrac{2xyz^{-1}}{p} - \dfrac{y^2}{(3x)^{-2}}\right)$

26. Simplify: $\dfrac{a^0 xy^2 (a^2)^{-2} (xy^{-2})^2}{ax^0 (y^{-2})^2}$

27. Simplify by adding like terms: $-\dfrac{2x^2}{y^2} - \dfrac{5y^{-2}p^0}{x^{-2}} + \dfrac{7xxy}{y^3}$

28. Evaluate: $ax - a(a - x)$ if $a = -\dfrac{1}{2}$ and $x = \dfrac{1}{4}$

Simplify:

29. $-2|-2 - 5| + (-3)|-2(-2) - 3| + 7$

30. $-\dfrac{1}{(-2)^{-3}} + \dfrac{3}{-3^{-2}}$

LESSON 19   *Value word problems • AA means AAA*

19.A

value word problems

Value word problems are a genre of problems in which one or more statements in the problem are about the value of items. The total value of one kind is the value of one item times the number of items of that kind. For example, every nickel has a value of 5 cents; thus, N_N nickels would have a value of $5N_N$ cents. In a like manner, the total value of N_D dimes would be $10N_D$ cents. We will begin value problems with problems that contain two statements of equality that lead to two equations in two unknowns. Either substitution or elimination can be used to solve these equations.

example 19.1 Karamagu had 50 nickels and dimes whose value was $4. How many of each kind of coin did he have?

solution If we use dollars as the basic value in the problem, we get the two equations

$$N_N + N_D = 50$$
$$0.05N_N + 0.1N_D = 4$$

We can avoid the decimal fractions at the outset if we use pennies as our basic value in the second equation. If we do this, our equations are

$$N_N + N_D = 50$$
$$5N_N + 10N_D = 400$$

We will solve the top equation for N_N and substitute $(50 - N_D)$ for N_N in the second equation.

$$5(50 - N_D) + 10N_D = 400 \qquad \text{substituted } (50 - N_D) \text{ for } N_N$$
$$250 - 5N_D + 10N_D = 400 \qquad \text{multiplied}$$
$$250 + 5N_D = 400 \qquad \text{simplified}$$
$$5N_D = 150 \qquad \text{added } -250 \text{ to both sides}$$
$$N_D = 30 \qquad \text{divided}$$

Since $N_N + N_D = 50$, it follows that $N_N = 20$.

example 19.2 The fishmonger sold codfish for 6 pence each and mussels for 1 pence each. If Harriet bought a total of 26 items and spent 86 pence, how many codfish did she buy?

solution The two equations are

$$N_C + N_M = 26$$
$$6N_C + 1N_M = 86$$

We will use elimination. We will multiply the top equation by -1 and then add the equations.

$$-N_C - N_M = -26$$
$$\underline{6N_C + N_M = 86}$$
$$5N_C = 60$$
$$N_C = 12$$

Thus, $N_M = 26 - 12 = 14$.

19.B
AA means AAA

When two angles in one triangle have the same measures as two angles in another triangle, the third angles are equal. This is easy to prove. To do the proof, we will use substitution and the fact that if two things equal the same thing, they are equal to each other. This is the sixth postulate of Euclid and will be discussed in Lesson 30. Consider these triangles.

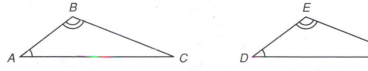

The tick marks show us that two angles in the left-hand triangle are equal to two angles in the right-hand triangle. Some authors say that this equality gives the third angles no choice but to be equal. They call this the **no-choice theorem.** We will do a simple three-step proof. The sum of the measures of the angles in both triangles is 180°.

$$\angle A + \angle B + \angle C = 180° \quad \text{and} \quad \angle D + \angle E + \angle F = 180°$$

Now we solve the equation on the left for angle C and the equation on the right for angle F.

$$\angle C = 180° - \angle A - \angle B \quad \text{and} \quad \angle F = 180° - \angle D - \angle E$$

In the left-hand equation, we will replace $\angle A$ with its equivalent, which is $\angle D$, and replace $\angle B$ with its equivalent, which is angle E.

$$\angle C = 180° - \angle D - \angle E \quad \text{and} \quad \angle F = 180° - \angle D - \angle E$$

Since angle C and angle F both equal the same sum, these two angles must be equal.

$$\angle C = \angle F$$

example 19.3 Find x and y.

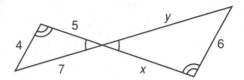

solution Two angles in the triangle on the left equal two angles in the triangle on the right, so the third angles are equal and the triangles are similar. To find the scale factor from left to right, we will use the sides opposite the angles with one tick mark.

$$4\overrightarrow{SF} = 6 \qquad \text{equation}$$

$$\overrightarrow{SF} = \frac{3}{2} \qquad \text{divided}$$

The sides marked 7 and y are corresponding sides because they are opposite angles with two tick marks. To find y, we multiply 7 by the left-to-right scale factor.

$$7\left(\frac{\overrightarrow{3}}{2}\right) = y$$

$$\frac{21}{2} = y$$

The sides marked x and 5 are corresponding sides because they are opposite the angles with no tick marks. To find x, we multiply 5 by the scale factor.

$$5\left(\frac{\overrightarrow{3}}{2}\right) = x$$

$$\frac{15}{2} = x$$

practice **a.** Florence had 80 nickels and dimes whose value was $6.50. How many of each kind of coin did she have?

b. Roses sold for 12 pence each and daffodils for 4 pence each. If Jim bought 35 flowers and spent 300 pence, how many daffodils did he buy?

c. Find x and y.

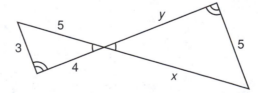

problem set 19 **1.** Karamuga had 60 nickels and dimes whose value was $5. How many of each kind of coin did he have?

2. The fishmonger sold codfish for 7 pence each and mussels for 1 pence each. If Harriet bought a total of 26 items and spent 86 pence, how many codfish did she buy?

3. Sulfur is mixed with other chemicals to make sulfuric acid. If it takes 16 tons of sulfur to make 49 tons of sulfuric acid, how many tons of other constituents are needed to make 294 tons of acid? (*Hint:* Sulfuric acid is the total.)

4. Nineteen percent of the nitric acid was used in the experiment. If 1134 liters remained, how much nitric acid had been available in the beginning? How much had been used?

5. A number was multiplied by -7 and this product was increased by -7. This sum was doubled, and the result was 4 greater than 5 times the opposite of the number. What was the number?

6. Use elimination to solve: $\begin{cases} 5x + 25y = -160 \\ -3x + 2y = -23 \end{cases}$

7. Multiply: $(x^2 - 2)(x^3 - 2x^2 - 2x + 4)$

8. Divide $-3x^3 - 2$ by $-2 + x$ and then check.

Solve for all unknown variables:

9. $R_H T_H + R_S T_S = 180, R_H = 70, R_S = 20, T_H = T_S$

10. $R_F T_F = R_S T_S, T_S = 6, T_F = 5, R_F - 10 = R_S$

11. $R_M T_M = R_R T_R, R_M = 8, R_R = 2, T_R = 5 - T_M$

Add:

12. $4 + \dfrac{3x^2}{7y^2 z}$

13. $\dfrac{a}{2x^2} - \dfrac{b}{x^2 y} - c$

14. Use the Pythagorean theorem to find z. Then use the scale factor to find x and y.

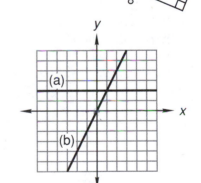

15. Find the distance between $(-3, -5)$ and $(2, 4)$.

16. Graph: (a) $x = -3$
            (b) $5x - 3y = 9$

17. Find the equations of lines (a) and (b).

18. Find the equation of the line through $(-3, -5)$ and $(2, 4)$.

19. Find the equation of the line that has a slope of $\frac{2}{7}$ and passes through $(-3, -5)$.

20. Find x, y, and k.

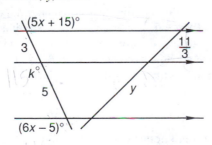

21. The arc measures $50°$, as shown. Find A, B, C, and D. If the radius of the circle is 3 cm, find the area of the $50°$ sector.

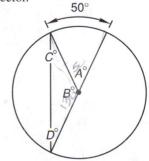

Solve:

22. $3\dfrac{1}{4}p - \dfrac{7}{3} = -\dfrac{5}{12}$

23. $0.07x - 0.7 = -7.7$

24. $-[(-2 - 6)(-2) - 2] = -2[(x - 2)2 - (2x - 3)3]$

25. Expand: $-\dfrac{3x^0 y^2}{p^{-2}}\left(\dfrac{x}{p^2 y^2} - \dfrac{3xy}{x^2 p}\right)$

26. Simplify: $\dfrac{x^2 x^0 (xx^{-3})}{x(y^0 y^2)^{-2}}$

27. Simplify by adding like terms: $\dfrac{2x^2 a}{y} - \dfrac{3xy^{-2} a}{y^{-1} x^{-1}} + \dfrac{4x^0 y^{-1}}{x^{-2} a^{-1}}$

28. Evaluate: $m - m^2 y(m - y)$ if $m = -\dfrac{1}{3}$ and $y = \dfrac{1}{6}$

Simplify:

29. $-4^0[(-5 + 2) - |-3 + 7| - 3(-1 - (-2)^0)]$

30. $-\dfrac{-1}{-(-2)^{-2}} + \dfrac{3}{-(-2)^{-2}}$

LESSON 20   *Simplification of radicals • Line parallel to a given line*

20.A
simplification of radicals

Hand-held calculators can often be used to demonstrate properties that would otherwise be difficult to demonstrate. The square root of 10 equals the square root of 2 times the square root of 5.

$$\sqrt{10} = \sqrt{2}\sqrt{5}$$

If we use our calculator to find decimal approximations of these numbers, we get

$$\sqrt{10} = 3.1622776 \quad\text{and}\quad \sqrt{2}\sqrt{5} = (1.4142136)(2.236068)$$

Then if we use the calculator to multiply the two numbers on the right, we get a product of

$$3.1622777$$

This number differs from our approximation of $\sqrt{10}$ on the left above by only 1 unit in the seventh decimal place. This exercise helps some to believe that $\sqrt{10}$ really does equal $\sqrt{2}\sqrt{5}$.

The rule that explains this can be proved and is really a theorem that we call the **product-of-square-roots theorem**. The proof of this theorem is so straightforward, however, that some people often use the word "property" for this rule and call it the product property of radicals. This rule is also applicable to higher-order roots such as cube roots, fourth roots, etc. For now, we restrict its use to square roots and state the rule formally here.

> **PRODUCT-OF-SQUARE-ROOTS THEOREM**
>
> If m and n are nonnegative real numbers, then
>
> $$\sqrt{m}\sqrt{n} = \sqrt{mn} \quad\text{and}\quad \sqrt{mn} = \sqrt{m}\sqrt{n}$$

This theorem can be generalized to the product of any number of factors. We say that the square root of any product of positive factors may be written as the product of the square roots of the factors. For example,

$$\sqrt{2 \cdot 5 \cdot 5} \qquad \text{can be written as} \qquad \sqrt{2}\sqrt{5}\sqrt{5}$$

and

$$\sqrt{3 \cdot 3 \cdot 3 \cdot 5} \qquad \text{can be written as} \qquad \sqrt{3}\sqrt{3}\sqrt{3}\sqrt{5}$$

We will use this theorem in the following problems to help us simplify radical expressions.

example 20.1   Simplify: $3\sqrt{50} - 5\sqrt{200}$

solution   As the first step, we write each radical as a product of prime factors.

$$3\sqrt{5 \cdot 5 \cdot 2} - 5\sqrt{2 \cdot 2 \cdot 2 \cdot 5 \cdot 5}$$

Now we can use the product-of-square-roots theorem to write

$$3\sqrt{5}\sqrt{5}\sqrt{2} - 5\sqrt{2}\sqrt{2}\sqrt{2}\sqrt{5}\sqrt{5}$$

We finish by remembering that $\sqrt{5}\sqrt{5} = 5$ and $\sqrt{2}\sqrt{2} = 2$. We get

$$15\sqrt{2} - 50\sqrt{2} = \mathbf{-35\sqrt{2}}$$

example 20.2   Simplify: $3\sqrt{2} \cdot 4\sqrt{12} \cdot 2\sqrt{3}$

solution   We multiply and get

$$24\sqrt{72}$$

Now we simplify $\sqrt{72}$ and multiply by 24.

$$24\sqrt{2}\sqrt{2}\sqrt{2}\sqrt{3}\sqrt{3} = \mathbf{144\sqrt{2}}$$

example 20.3   Simplify: $4\sqrt{3}(2\sqrt{3} - \sqrt{6})$

solution   Two multiplications are indicated. We do these and get

$$(4\sqrt{3})(2\sqrt{3}) + (4\sqrt{3})(-\sqrt{6}) = \mathbf{24 - 12\sqrt{2}}$$

20.B
line parallel to a given line

Parallel lines are lines that have the same slopes but different intercepts. If we look at the equations

$$y = \frac{1}{2}x + 2 \qquad y = \frac{1}{2}x - 1 \qquad y = \frac{1}{2}x - 4$$

we see that they all have a slope of $+\frac{1}{2}$, but that each line has a different intercept.

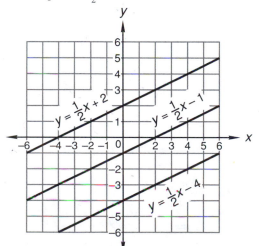

If we are asked to find the equation of a line that is parallel to one of these lines, its slope must also be $+\frac{1}{2}$. The only thing we need to find is the intercept.

example 20.4 Find the equation of the line that is parallel to the line $2y - x = 2$ and passes through the point $(3, -1)$.

solution If we write the equation of the given line in slope-intercept form, we get

$$y = \frac{1}{2}x + 1$$

The slope of the new line must also be $\frac{1}{2}$ if it is to be parallel to this line, so we have

$$y = \frac{1}{2}x + b$$

Now we use the coordinates 3 and -1 for x and y and solve algebraically for b.

$$-1 = \frac{1}{2}(3) + b \qquad \text{substituted}$$

$$-\frac{2}{2} = \frac{3}{2} + b \qquad \text{simplified}$$

$$-\frac{5}{2} = b \qquad \text{solved}$$

Thus, the intercept of the new line is $-\frac{5}{2}$, and the equation of the new line is

$$y = \frac{1}{2}x - \frac{5}{2}$$

practice Simplify:

    a. $4\sqrt{40} - 3\sqrt{140}$ **b.** $3\sqrt{2}(3\sqrt{2} - \sqrt{8})$

    c. Find the equation of the line that is parallel to the line $3y - x = 5$ and passes through $(3, 3)$.

problem set 20

    1. The formula required that 20 kilograms (kg) of carbon be used to get 160 kg of the compound. How many kilograms of other components were needed to make 640 kg of the compound?

    2. Fewer than half of the performers were virtuosos. In fact, only $\frac{3}{10}$ were in this category. If 28 were not virtuosos, how many performers were there? How many virtuosos were there?

    3. The expensive ones cost \$7 each, whereas the worthless ones sold for only \$2 each. Monongahela spent \$111 and bought three more expensive ones than worthless ones. How many of each kind did she buy?

    4. Only 40 percent of the combustibles burned. If 240 tons burned, how many tons did not burn?

    5. Find four consecutive even integers such that -4 times the sum of the first and fourth is 6 greater than the opposite of the sum of the second and third.

    6. Use substitution to solve: $\begin{cases} y - 2x = 8 \\ 2y + 2x = 40 \end{cases}$

    7. Divide $-2x^3 - x + 2$ by $-1 + x$ and check.

    8. Solve for the unknown variables: $R_G T_G + R_B T_B = 100, R_G = 4, R_B = 10, T_B = T_G + 3$

Simplify:

    9. $3\sqrt{3} \cdot 4\sqrt{12} - 5\sqrt{300}$ **10.** $4\sqrt{3}(2\sqrt{3} - \sqrt{6})$

    11. $5\sqrt{5}(2\sqrt{5} - 3\sqrt{10})$

Add:

12. $\dfrac{m^2}{x^2 a} + \dfrac{5}{ax} - \dfrac{m}{a}$

13. $\dfrac{a}{x^2} - a - \dfrac{3x}{2a^4}$

14. Find the area of this triangle. Dimensions are in feet.

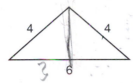

15. Find z and A.

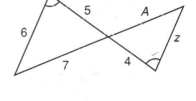
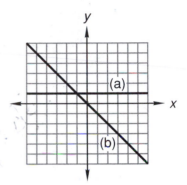

16. Graph: (a) $y = -2$
                  (b) $y = -2x - 2$

17. Find the equations of lines (a) and (b).

18. Find the equation of the line that is parallel to the line $3y - x = 3$ and passes through the point $(2, -1)$.

19. Find the equation of the line that has a slope of $-\frac{1}{12}$ and passes through $(2, 5)$.

20. In the figure, $x = 140$. Find A. Then find B and C. Then find P and y.

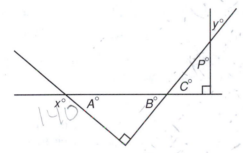

Solve:

21. $3\dfrac{1}{12} - \dfrac{1}{6}x = 2\dfrac{1}{24}$

22. $-0.04x - x - 0.2x = 6.2$

23. $-2[(-3)(-2 - x) + 2(x - 3)] = -2x$

24. Expand: $\dfrac{-x^0 y}{y^2 y^{-2}}\left(\dfrac{x}{y} - \dfrac{3xy^2}{y^3}\right)$

25. Simplify: $\dfrac{(-2x^0)^{-3} x^2 yy^0 y^3}{(xy^{-2})^2 (-2x^{-1})^{-2}}$

26. Simplify by adding like terms: $-\dfrac{3x}{a} + \dfrac{(a^0)a}{aax^{-1}} - \dfrac{5x^{-1}}{a}$

27. Evaluate: $k - kx(k^2 - x)$ if $k = -\dfrac{1}{4}$ and $x = \dfrac{1}{8}$

Simplify:

28. $-2^0\{(-7 + 3) - |-2 + 9| - 2[-2^0 - (-5)]\}$

29. $-2^0|-3 - 7| + |(-5^0)| - 2(-5) + 2$

30. $\dfrac{1}{-(-3)^{-3}} + \dfrac{2}{+(-3)^{-3}}$

LESSON 21 *Scientific notation • Two statements of equality*

21.A
scientific notation

We like to use the decimal system because base 10 numerals have a special advantage. When we multiply a number by a power of 10, the sole effect is to move the decimal point the number of places left or right that equals the power of 10. Multiplying by a positive integral power of 10 moves the decimal point to the right. To demonstrate, we will multiply 412.036 by 10^2 and by 10^4.

$$412.036 \times 10^2 = 41{,}203.6$$

$$412.036 \times 10^4 = 4{,}120{,}360$$

Multiplying by 10 to a negative integral power moves the decimal point to the left. If we multiply the same number by 10^{-2} and 10^{-4}, we move the decimal point two places to the left and four places to the left, as we show here.

$$412.036 \times 10^{-2} = 4.12036$$

$$412.036 \times 10^{-4} = 0.0412036$$

We find this property of decimal numerals very useful when we deal with very large or very small numbers. **We can put the decimal point anywhere we please as long as we follow the numeral with**

$$\times 10^{\pm b}$$

and use this notation to tell where the decimal point really should be. Thus,

$$304.162 \qquad 0.304162 \times 10^3 \qquad 304{,}162 \times 10^{-3}$$

all represent the same number. In the last two notations, the decimal point is not between the 4 and the 1, but that is all right because the notations

$$\times 10^3 \qquad \text{and} \qquad \times 10^{-3}$$

tell us where it should be placed. When we use this notation and place the decimal point just to the right of the first nonzero digit, we say that we have written the number in **scientific notation.**

example 21.1 Simplify: $\dfrac{(0.0003 \times 10^{-6})(4000)}{(0.006 \times 10^{15})(2000 \times 10^4)}$

solution We begin by writing all four numbers in scientific notation. Then we multiply and divide as indicated.

$$\frac{(3 \times 10^{-10})(4 \times 10^3)}{(6 \times 10^{12})(2 \times 10^7)} = \frac{3 \cdot 4}{6 \cdot 2} \times \frac{10^{-7}}{10^{19}} = \mathbf{1 \times 10^{-26}}$$

21.B
two statements of equality

Thus far, our study of problems that require two equations for their solution has been restricted to coin problems about nickels and dimes and to similar problems about the values of items that are not coins. Now we will begin our investigation of other types of problems that contain two statements of equality. The experience gained with value problems should make these new problems easy to understand.

example 21.2 The ratio of two numbers is 3 to 4 and their sum is 84. What are the numbers?

solution We decide to use N for the numerator and D for the denominator.

$$\text{(a)} \quad \frac{N}{D} = \frac{3}{4} \qquad \text{and} \qquad \text{(b)} \quad N + D = 84$$

We will cross multiply in (a), and then solve (b) for D and substitute into (a).

(a) $\dfrac{N}{D} = \dfrac{3}{4}$ → $4N = 3D$ (b) $N + D = 84$ → $D = 84 - N$

$4N = 3(84 - N)$ substituted

$4N = 252 - 3N$ multiplied

$7N = 252$ simplified

$N = 36$ divided

Since $N + D = 84$, then $D = 84 - 36 = 48$.

example 21.3   The sum of two numbers is 128 and their difference is 44. What are the numbers?

solution   A little thought in choosing the variables is often helpful. Here we have two numbers. One is greater than the other, so we will use S to represent the small number and L to represent the large number. The equations are

(a) $L + S = 128$ and   (b) $L - S = 44$

We will solve the equations by using elimination.

$$
\begin{array}{ll}
\text{(a)} & L + S = 128 \\
\text{(b)} & L - S = 44 \\
\hline
& 2L = 172 \\
& L = 86
\end{array}
$$

Thus,

$$S = 128 - 86 \quad \longrightarrow \quad S = 42$$

practice   **a.**   The ratio of two numbers is 4 to 5 and their sum is 108. What are the numbers?

b.   The sum of two numbers is 136 and their difference is 50. What are the numbers?

problem set 21

1.   The ratio of two numbers is 3 to 5 and their sum is 96. What are the numbers?

2.   The sum of two numbers is 200 and their difference is 66. What are the numbers?

3.   It took 900 kg of acetylene to make 2400 kg of the compound. How many kilograms of other components was required to make 3600 kg of the compound?

4.   Twenty percent of the nitrogen combined with the other elements. If 740 kg of nitrogen did not combine, how much did combine?

5.   The nickels and dimes had a value of $5.75. If there were 70 coins in all, how many were nickels?

6.   Find three consecutive integers such that -5 times the sum of the first and the third is 24 greater than 4 times the opposite of the second.

7.   Use elimination to solve: $\begin{cases} 8y - 3x = 22 \\ 2y + 4x = 34 \end{cases}$

8.   Multiply: $(4x^2 - 2x + 2)(-3 + 2x)$

9.   Solve for the unknown variables: $R_K T_K = R_N T_N,\ R_K = 6,\ R_N = 3,\ T_N - T_K = 8$

Simplify:

10. $3\sqrt{200} - 5\sqrt{18} + 7\sqrt{50}$ **11.** $2\sqrt{3} \cdot 2\sqrt{2}(6\sqrt{6} - 3\sqrt{2})$

Add:

12. $4x + \dfrac{1}{p}$ **13.** $\dfrac{m^2}{a^2 x^2} - \dfrac{3}{ax} - \dfrac{m}{x}$

14.   Simplify: $\dfrac{(0.0003 \times 10^8)(6000)}{(0.006 \times 10^{15})(2000 \times 10^5)}$

15. Find the distance between $(-3, -5)$ and $(4, -5)$.

16. Graph: (a) $x = 4$
    (b) $4x - 3y = 12$

17. Find the equations of lines (a) and (b).

18. Find the equation of the line through $(-3, -5)$ and $(4, -5)$.

19. Find the equation of the line that passes through the point $(2, 2)$ and is parallel to $y = -\frac{3}{7}x + 4$.

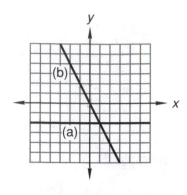

20. The triangle and the circle are tangent at three points, as shown. Find x and y.

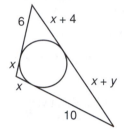

21. In this figure, $x = 130$. Find A. Then find B and C. Then find D and y.

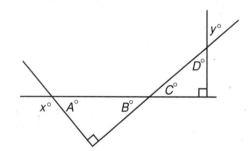

Solve:

22. $5\frac{1}{3}x - \frac{3}{4} = \frac{7}{8}$

23. $0.03(x - 4) = 0.02(x + 6)$

24. $-[(-4 - 1)(-3) - 6^0] = -2[(y - 4)3 - (2y - 5)]$

25. Expand: $\dfrac{5y^0 p^{-2}}{x^2}\left(-\dfrac{2p^2}{x^2} - \dfrac{p^2}{x}\right)$

26. Simplify: $\dfrac{x^2(y^{-2})^2(y^0)^2}{(2x^2 y^3)^{-2}}$

27. Simplify by adding like terms: $\dfrac{3x^2 a}{x} - 5xa + \dfrac{7x^{-2}}{x^{-3}a^{-1}}$

28. Evaluate: $x(x - ax)x$ \quad if $x = -\dfrac{1}{2}$ and $a = \dfrac{1}{3}$

Simplify:

29. $-7^0[(-2 + 3) - |-4 + 3|]$

30. $-\dfrac{1}{-2^{-2}} - \dfrac{1}{-(-2)^0}$

LESSON 22 *Uniform motion problems—equal distances •*
Similar triangles and proportions

22.A
uniform motion problems— equal distances

In this lesson we begin our study of the solution of uniform motion word problems. It is very probable that we will never encounter one of these problems in a science course or in everyday life. We study these problems because solving them helps us develop useful problem-solving skills.

Uniform motion problems involve statements about people or things that move at a constant velocity (speed)[†] or at an average velocity. The uniform motion problems on which we will concentrate will contain four statements about things that are equal or that differ by a specified amount. Each of these statements can be turned into an equation. We find that these equations are easy to write if we use four variables. In this book, we will use subscripted variables. It is often entertaining to use inventiveness and originality when choosing variables. For example, we could use either

$$R_W \quad \text{or} \quad R_{RW} \quad \text{or} \quad R_R$$

to stand for the rate that Ruby walked.

To solve uniform motion problems, we will write equations about rate or velocity, equations about time, and equations about distances traveled. Since the distance equations are the most difficult to write, we will consider these equations to be the key equations and we will write the distance equation first. **We will always draw a diagram to help us write the distance equation**. If both people or both things travel the same distance, the diagrams will have one of the two forms shown here.

$$D_1 = D_2 \quad \text{so} \quad R_1T_1 = R_2T_2$$

The distance equation is shown on the right. We always replace D_1 with R_1T_1 and D_2 with R_2T_2 since rate times time equals distance.

$$R_1T_1 = D_1 \quad \text{and} \quad R_2T_2 = D_2$$

example 22.1   Roger made the trip on Sunday, and Judy made the same trip on Monday. Roger traveled at 12 miles per hour. Judy traveled at 20 miles per hour, so her time was 2 hours less than Roger's time. How far did they travel?

solution   We begin by drawing the distance diagram and writing the distance equation.

$$D_R = D_J \quad \text{so} \quad R_RT_R = R_JT_J$$

Next we write the rate equations and the time equation.

$$R_R = 12 \qquad R_J = 20 \qquad T_J + 2 = T_R$$

To solve, we replace R_R in the distance equation with 12, replace R_J with 20, and replace T_R with $T_J + 2$.

$$
\begin{aligned}
12(T_J + 2) &= 20T_J & \text{substituted} \\
12T_J + 24 &= 20T_J & \text{multiplied} \\
24 &= 8T_J & \text{simplified} \\
3 &= T_J & \text{divided}
\end{aligned}
$$

Now since Judy's time was 3 hours, Roger's was 3 hours + 2 hours = 5 hours. To find the distances traveled, we multiply R_J by T_J and R_R by T_R.

$$D_J = R_JT_J \qquad\qquad D_R = R_RT_R$$
$$D_J = (20)(3) = \textbf{60 miles} \qquad D_R = (12)(5) = \textbf{60 miles}$$

[†]In science courses a distinction is made between the words "speed" and "velocity." In this book the words are synonymous.

example 22.2   Br'er Rabbit hopped off toward the briar patch at 10 kilometers per hour (kph) at 10 a.m. At noon Br'er Wolf began the chase from the same starting point. If he caught Br'er Rabbit at 2 p.m., how fast did he run?

solution   They traveled the same distances. The distance diagram and distance equation are as follows:

$$D_R = D_W \qquad \text{so} \qquad R_R T_R = R_W T_W$$

One rate was given and both times were given.

$$R_R = 10 \qquad T_R = 4 \qquad T_W = 2$$

To solve, we substitute the last three equations into the distance equation.

$$(10)(4) = R_W(2) \qquad \text{substituted}$$

$$40 = 2R_W \qquad \text{multiplied}$$

$$\textbf{20 kph} = \boldsymbol{R_W} \qquad \text{divided}$$

22.B

similar triangles and proportions

We know that the ratios of corresponding sides in similar triangles are proportional. We have been using the constant of proportionality (the scale factor) to find the lengths of the missing sides. We can also solve for the missing sides by using the ratios themselves. Consider this pair of similar triangles.

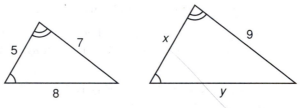

If we write the ratios of the sides opposite corresponding angles and put the sides from the right-hand triangle on top, we get

$$\frac{x}{5} \qquad \frac{9}{7} \qquad \text{and} \qquad \frac{y}{8}$$

Since the triangles are similar, all of the ratios are equal. When we connect two equal ratios with an equals sign, we say that we have written a proportion. We can solve for x by equating the first and second ratios.

$$\frac{x}{5} = \frac{9}{7} \qquad \text{proportion}$$

$$45 = 7x \qquad \text{cross multiplied}$$

$$\frac{45}{7} = x \qquad \text{divided by 7}$$

We can solve for y by equating the second and third ratios.

$$\frac{9}{7} = \frac{y}{8} \qquad \text{proportion}$$

$$7y = 72 \qquad \text{cross multiplied}$$

$$y = \frac{72}{7} \qquad \text{divided by 7}$$

We could have used the scale factor to find x and y. To find the left-to-right scale factor, we use the sides whose lengths are 7 and 9.

$$7\overrightarrow{SF} = 9$$

$$\overrightarrow{SF} = \frac{9}{7}$$

We note that the scale factor is one of the equal ratios. To use the scale factor to find x and y we multiply 5 by the left-to-right scale factor, and we multiply 8 by the left-to-right scale factor.

$$5\left(\frac{\overrightarrow{9}}{7}\right) = x \qquad\qquad 8\left(\frac{\overrightarrow{9}}{7}\right) = y$$

$$45 = 7x \qquad\qquad 72 = 7y$$

$$\frac{45}{7} = x \qquad\qquad \frac{72}{7} = y$$

practice   **a.** Elvira made a trip on Tuesday, and David made the same trip on Wednesday. Elvira traveled at 14 mph. David traveled at 21 mph, so his time was 3 hours less than Elvira's time. How far did they travel?

b. Use proportions to solve for x and y.

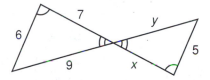

problem set 22

1. Eloise made the trip on Sunday, and Christian made the same trip on Monday. Eloise traveled at 15 miles per hour. Christian traveled at 30 miles per hour, so his time was 3 hours less than Eloise's time. How far did they travel?

2. Alonzo ran off toward Milano at 9 kilometers per hour at 8 a.m. At 10 a.m. Wilhelm began the chase from the same starting point. If he caught Alonzo at 12 p.m., how fast did he run?

3. The ratio of two numbers is 7 to 5. The sum of the numbers is 960, what are the numbers?

4. The federal tax was $500 more than the state tax. If the sum of the taxes was $6900, what was the amount of the federal tax?

5. Huckleberries cost $5 a peck, whereas whortleberries cost $13 a peck. Hortense spent $109 for a total of 9 pecks. How many pecks of whortleberries did she buy?

6. The chemist found that 30 grams of iodine was required to make 600 grams of the solution. How many grams of other things was required to make 5000 grams of the solution?

7. Use substitution to solve: $\begin{cases} 5x + y = 24 \\ 7x - 2y = 20 \end{cases}$

8. Divide $x^3 - 4x + 2$ by $-1 + x$ and check.

9. Use proportions to solve for x and y.

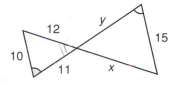

Simplify:

10. $2\sqrt{27} - 3\sqrt{75}$

11. $3\sqrt{2}(2\sqrt{2} - \sqrt{6}) \cdot 4\sqrt{3} + 2$

12. $2\sqrt{3}(5\sqrt{3} - 2\sqrt{6})$

Add:

13. $2 + \dfrac{1}{x}$

14. $\dfrac{5x^2}{y} + p^2 - \dfrac{3x}{py}$

15. Simplify: $\dfrac{(0.0035 \times 10^{-4})(200 \times 10^6)}{(700 \times 10^5)(0.00005)}$

16. Graph: (a) $x = -5$
            (b) $2x - y = 4$

17. Find the equations of lines (a) and (b).

18. Find the equation of the line that passes through $(6, 0)$ and $(-3, -3)$.

19. Find the distance between $(6, 0)$ and $(-3, -3)$.

20. Find the equation of the line that has a slope of $\frac{2}{5}$ and passes through $(3, -5)$.

21. In this figure, $x = 140$. Find A. Then find B and C. Then find k. Then find M.

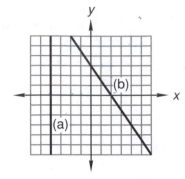

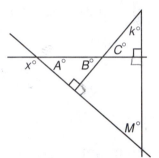

Solve:

22. $3\frac{2}{5}x - 4\frac{1}{10}x = 2\frac{1}{4}$

23. $0.02(p - 2) = 0.03(2p - 6)$

24. $-[(-3 - 6)(-1^0) - 6^0] = -4[(x - 3)2]$

25. Expand: $\dfrac{xy^{-2}}{z^0 p}\left(\dfrac{py^2}{x} - \dfrac{3xy^{-4}}{py}\right)$

26. Simplify: $\dfrac{(x^{-2}yp)^{-3}(x^0 yp)^2}{(2x^2)^{-2}}$

27. Simplify by adding like terms: $-\dfrac{3x^2 y}{xx} + \dfrac{2x^{-2}x^4}{y^{-1}x^2} - \dfrac{5xy^2}{xy}$

28. Evaluate: $ya(y - a)y$ if $y = -\dfrac{1}{2}$ and $a = \dfrac{1}{5}$

Simplify:

29. $-2^0[(5 - 7 - 2) - |-2 - 7| - 2^0] + 2$

30. $\dfrac{-3}{-3^{-2}} - \dfrac{2}{-2^{-3}}$

LESSON 23 *Graphical solutions*

We have been solving two equations in two unknowns by using either substitution or elimination. These equations can also be solved by graphing. To do this, we graph both of the equations and visually determine the coordinates of the point where the lines cross. The answer we get is an approximation because we must estimate the coordinates of the crossing point. The advantage of the graphical method is that we can see what we are doing. If a more exact answer is required, it can be obtained by using either substitution or elimination.

example 23.1   Solve this system by graphing. Check the solution by using either substitution or elimination.

$$\begin{cases} 3y - 2x = 6 & \text{(a)} \\ y + x = -1 & \text{(b)} \end{cases}$$

solution   To graph, we first solve each equation for *y* to get the slope-intercept form. Then we graph the equations.

(a) $y = \dfrac{2}{3}x + 2$

(b) $y = -x - 1$

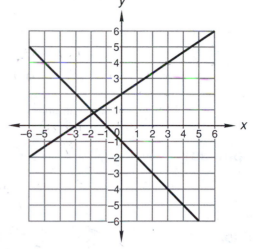

The lines appear to cross at (−1.8, 0.75). We will check this by using the original equations and using elimination.

$$
\begin{array}{rcl}
3y - 2x = 6 & \rightarrow \quad (1) \quad \rightarrow & 3y - 2x = 6 \\
y + x = -1 & \rightarrow \quad (2) \quad \rightarrow & \underline{2y + 2x = -2} \\
& & 5y = 4 \;\rightarrow\; y = \dfrac{4}{5}
\end{array}
$$

We use $y = \dfrac{4}{5}$ in equation (b) to solve for *x*.

$$\left(\dfrac{4}{5}\right) + x = -1 \qquad \text{substitution}$$

$$x = -1 - \dfrac{4}{5} \qquad \text{added } -\dfrac{4}{5} \text{ to both sides}$$

$$x = -\dfrac{9}{5} \qquad \text{simplified}$$

Thus the exact solution is $\left(-\dfrac{9}{5}, \dfrac{4}{5}\right)$, which is the same as (−1.8, 0.8).

example 23.2   Solve this system by graphing. Check the solution by using either substitution or elimination.

$$\begin{cases} y - 2x = 2 & \text{(a)} \\ y = -3 & \text{(b)} \end{cases}$$

solution   We rewrite equation (a) in slope-intercept form and graph both lines.

(a) $y = 2x + 2$

(b) $y = -3$

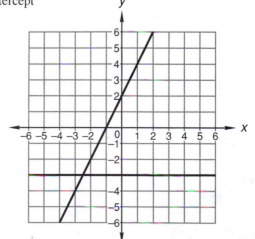

This time the graphical solution appears to be $(-2.5, -3)$. We will use substitution to check. We use -3 for y in equation (a).

$$(-3) = 2x + 2 \qquad \text{substitution}$$
$$-5 = 2x \qquad \text{added } -2 \text{ to both sides}$$
$$-\frac{5}{2} = x \qquad \text{divided}$$

Thus, the exact solution is $\left(-\frac{5}{2}, -3\right)$, which is the same as $(-2.5, -3)$.

practice   Solve this system by graphing. Check the solution by using either substitution or elimination.

$$\begin{cases} y = 2x + 1 \\ x = -1 \end{cases}$$

problem set 23

1. At the sound of the explosion, Mary began to run north at 600 feet per minute. Jim regained consciousness 4 minutes later and began to run after Mary at 800 feet per minute. How long did Jim run to catch Mary?

2. The fast freight made the trip in 10 hours while the slow freight took 12 hours for the same trip. How long was the trip if the fast freight was 10 kph faster than the slow freight?

3. The fraction had a value of $\frac{11}{12}$. The sum of the numerator and the denominator was 230. What was the fraction?

4. The vivandière sold viands and sandwiches. If she sold 300 total and 50 more viands than sandwiches, how many of each did she sell?

5. The value of the quarters and nickels was $5. If there were 40 more nickels than quarters, how many coins of each type were there?

6. Twenty percent of the compound was copper sulfate. If there were 400 tons of the compound in the warehouse, how much was not copper sulfate?

7. Use elimination to solve:
$$\begin{cases} 5x + 2y = 70 \\ 3x - 2y = 10 \end{cases}$$

8. Multiply: $(3x^3 - 2x)(2x^2 - x - 4)$

Simplify:

9. $4\sqrt{3} \cdot 3\sqrt{12} \cdot 2\sqrt{3}$

10. $3\sqrt{75} - 4\sqrt{48}$

11. $2\sqrt{5}(5\sqrt{5} - 3\sqrt{15})$

Add:

12. $3xy^2m + \dfrac{4}{x}$

13. $\dfrac{5x^2}{pm} - 4 + \dfrac{c}{p^2 m}$

14. Simplify: $\dfrac{(0.00003)(0.006 \times 10^{-6})}{(1800 \times 10^{15})(100,000)}$

Solve by graphing. Then get an exact solution by using either substitution or elimination.

15. $\begin{cases} 2y - 2x = 8 \\ y + x = -2 \end{cases}$

16. $\begin{cases} y - 2x = 1 \\ y = -2 \end{cases}$

17. Find the equation of the line that passes through $(0, 0)$ and $(4, 2)$.

18. Find the distance between $(0, 0)$ and $(4, 2)$.

19. Find the equation of the line that has a slope of $-\frac{3}{8}$ and passes through $(4, 4)$.

20. In this figure, M is halfway between B and C and divides $\overline{BC}$ into two equal segments. Thus M is the **midpoint** of $\overline{BC}$. We remember that we call $\overline{AM}$ a median because $\overline{AM}$ connects a vertex with the midpoint of the opposite side. Use the Pythagorean theorem to find BC. Then calculate CM. Then find the area of $\triangle ACM$. Dimensions are in meters.

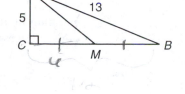

21. Use proportions to find x and y. Remember that vertical angles have equal measures.

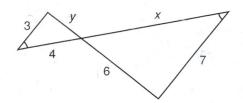

Solve:

22. $3\frac{1}{4}x + 5\frac{1}{3} = 2\frac{1}{6}$

23. $0.3(2p - 4) = 0.1(p + 3)$

24. $-[(-4 - 6)(-4) - 2] = -2^0(x - 4)$

25. Expand: $xy^{-2}\left(\dfrac{x^0 y^2}{x} - \dfrac{3x^0 y^2}{x^2}\right)$

26. Simplify: $\dfrac{(-3xy)^{-2} x^2 y^2}{(x^{-3})^2 yx^2}$

27. Simplify by adding like terms: $\dfrac{3m}{x} - \dfrac{2x^{-1}}{m^0 m^{-1}} + \dfrac{5x^2 m^2}{x^3 m}$

28. Evaluate: $xy - (x - y)$ if $x = -\dfrac{1}{2}$ and $y = 2$

Simplify:

29. $-3^0[(-2 - 3 + 8) - |-3 - 5| - 5^0]$

30. $\dfrac{-2^0}{-2^2} - \dfrac{(-2)^0}{(-2)^{-3}}$

LESSON 24   *Fractional equations • Overlapping triangles*

24.A

fractional equations

We have noted that the solution of an equation that contains decimal numbers can be simplified if we first multiply every term in the equation by a judiciously selected power of 10. For instance, the solution of this equation

$$0.03x + 0.02 = 0.2$$

can be made much easier if we first multiply every term by 10^2 (or 100 if you prefer). If we do this, we get

$$3x + 2 = 20 \qquad \text{multiplied by 100}$$
$$3x = 18 \qquad \text{added } -2$$
$$x = 6 \qquad \text{divided}$$

There is also a procedure that can be used to facilitate the solution of equations that contain fractions. **We can eliminate the fractions by multiplying every numerator by the least common multiple (LCM) of the denominators. Then the denominators will divide into the LCM, leaving the equation with integral coefficients.**

example 24.1   Solve: $4\frac{1}{2}x - \frac{3}{5} = -1\frac{1}{4}$

solution   As the first step, we will write the mixed numbers as improper fractions.

$$\frac{9}{2}x - \frac{3}{5} = -\frac{5}{4}$$

Next we multiply every numerator by 20, the LCM of the denominators.

$$20 \cdot \frac{9}{2}x - 20 \cdot \frac{3}{5} = 20\left(-\frac{5}{4}\right)$$

Next we simplify each term and get an equation with integral coefficients. Then we solve the equation.

$$90x - 12 = -25 \quad \text{simplified}$$
$$90x = -13 \quad \text{added } +12$$
$$x = -\frac{13}{90} \quad \text{divided}$$

example 24.2   Solve: $\dfrac{4x + 2}{5} - \dfrac{3}{2} = \dfrac{1}{3}$

solution   We begin by multiplying every numerator by 30, the LCM of the denominators.

$$30\frac{(4x + 2)}{5} - 30 \cdot \frac{3}{2} = 30 \cdot \frac{1}{3}$$

Now we simplify and solve.

$$24x + 12 - 45 = 10 \quad \text{simplified}$$
$$24x - 33 = 10 \quad \text{simplified}$$
$$24x = 43 \quad \text{added } +33$$
$$x = \frac{43}{24} \quad \text{divided}$$

example 24.3   Solve: $\dfrac{3x}{2} + \dfrac{8 - 4x}{7} = 3$

solution   We can eliminate the denominators if we multiply every numerator by 14.

$$14 \cdot \frac{3x}{2} + 14\frac{(8 - 4x)}{7} = 14(3)$$

Now we simplify and solve.

$$21x + 16 - 8x = 42 \quad \text{simplified}$$
$$13x + 16 = 42 \quad \text{simplified}$$
$$13x = 26 \quad \text{added } -16$$
$$x = 2 \quad \text{divided}$$

24.B
overlapping triangles

Sometimes the same angle can be an angle in two or more triangles.

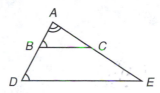

In this figure, angle A is an angle in $\triangle BAC$, and angle A is also an angle in $\triangle DAE$. If we draw the triangles separately, we can see this clearly.

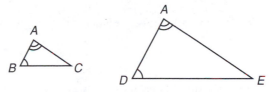

From the figures we see that two angles in the triangle on the left have the same measures as two angles in the triangle on the right, so the third angles must also be equal. Thus, the triangles are similar.

example 24.4   Find x and y.

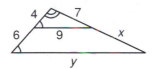

solution   It helps to draw the triangles separately.

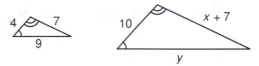

We have our choice of using the scale factor or using proportions. To use proportions, we write the ratios of the corresponding sides. We decide to put the sides from the right-hand triangle on top.

$$\frac{10}{4} \qquad \frac{y}{9} \qquad \frac{x+7}{7}$$

To find y, we equate the first two ratios.

$$\frac{10}{4} = \frac{y}{9} \qquad \text{proportion}$$

$$90 = 4y \qquad \text{cross multiplied}$$

$$\frac{45}{2} = y \qquad \text{solved}$$

To find x, we equate the first and third ratios.

$$\frac{10}{4} = \frac{x+7}{7} \qquad \text{proportion}$$

$$70 = 4x + 28 \qquad \text{cross multiplied}$$

$$42 = 4x \qquad \text{added } -28 \text{ to both sides}$$

$$\frac{21}{2} = x \qquad \text{divided and simplified}$$

Now we will solve the problem again by using the scale factor. To find the scale factor from left to right, we use the sides whose lengths are 4 and 10.

$$4\overrightarrow{SF} = 10$$

$$\overrightarrow{SF} = \frac{5}{2}$$

This is the same as the first of the three ratios listed above. To find y, we multiply 9 by the left-to-right scale factor.

$$9\left(\frac{\overrightarrow{5}}{2}\right) = y$$

$$\frac{45}{2} = y$$

To find x, we multiply 7 by the scale factor and solve the resulting equation.

$$7\left(\frac{\overrightarrow{5}}{2}\right) = x + 7$$

$$35 = 2x + 14 \qquad \text{multiplied both sides by 2}$$

$$21 = 2x \qquad \text{added } -14 \text{ to both sides}$$

$$\frac{21}{2} = x \qquad \text{divided}$$

practice   Solve:

    a. $\dfrac{2x + 3}{4} - \dfrac{4}{3} = \dfrac{1}{4}$ **b.** $\dfrac{4x}{3} + \dfrac{7 - 2x}{5} = 2$

    c. Find x and y.

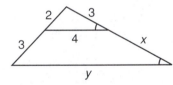

**problem set
24**

1. Scott noted that the fast freight left at noon and arrived there at 6 p.m. The next day he realized that the slow freight had made the same trip in 8 hours. What was the speed of the slow freight if the speed of the fast freight was 60 miles per hour?

2. Henry can ride his horse at 4 miles per hour and get to the battlefield on time. If he stops for 1 hour to make a speech to his troops, he must ride at 5 miles per hour for the whole trip to get to the battlefield on time. How far is it to the battlefield?

3. The fraction had a value of $\frac{5}{7}$. Amy and Zollie found that the sum of the numerator and the denominator was 120. What was the fraction?

4. Charles and Nelle picked 173 quarts of berries. How many did each pick if Charles picked 11 more quarts than Nelle picked?

5. Raisins were $700 for a measure, whereas plums cost $900 a measure. David and Bruce bought 50 measures and spent $41,000. How many measures of raisins did they buy?

6. It took 700 kilograms of potassium to make 49,000 kilograms of the new fertilizer. How many kilograms of other components did Gerd have to use to make 4200 kilograms of the new fertilizer?

7. Use substitution to solve: $\begin{cases} 7x + 9y = 119 \\ 2x + y = 23 \end{cases}$

8. Divide $3x^3 - 3$ by $-2 - x$ and check.

Simplify:

9. $4\sqrt{3} \cdot 5\sqrt{2} \cdot 6\sqrt{12}$

10. $4\sqrt{63} - 3\sqrt{28}$

11. $3\sqrt{2}(5\sqrt{2} - 6\sqrt{12})$

12. $2\sqrt{2}(5\sqrt{10} - 3\sqrt{2})$

Add:

13. $4m^2 yp + \dfrac{6}{m^2 y}$

14. $\dfrac{k^2}{2p} + c - \dfrac{4}{p^2 c}$

15. Simplify: $\dfrac{(0.0007 \times 10^{-23})(4000 \times 10^6)}{(0.00004)(7{,}000{,}000)}$

16. Solve by graphing. Then get an exact solution by using either substitution or elimination.

$$\begin{cases} 3x + 2y = 12 \\ 5x - 4y = 8 \end{cases}$$

17. Find the equations of lines (a) and (b).

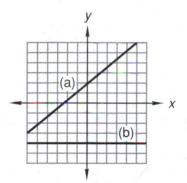

18. The arc measures 70°, as shown. Find A, B, C, and D. If the radius of the circle is 4 cm, find the area of the 70° sector.

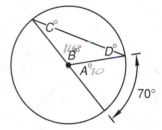

19. Find the area of this isosceles triangle. Dimensions are in inches.

20. Find the equation of the line that passes through $(-2, 5)$ and $(-6, -3)$.

21. Find the equation of the line through $(5, -3)$ that has a slope of $\frac{2}{9}$.

22. Find k.

Solve:

23. $\dfrac{4x + 2}{3} - \dfrac{3}{4} = \dfrac{1}{2}$

24. $\dfrac{3x}{2} + \dfrac{8 - 4x}{7} = 3$

25. $0.07 - 0.003x + 0.2 = 1.02$

26. $-2[x - (-2 - 4^0) - 3] + [(x - 2)(-3)] = -x$

27. Expand: $\dfrac{xy^{-2}}{p^2}\left(\dfrac{p^2 y^2}{x} + \dfrac{5x^2 y^3}{p^{-2}}\right)$

28. Simplify: $\dfrac{4x^2 y^{-2} x}{(-2x^0)^{-3}}$

29. Simplify by adding like terms: $\dfrac{3p^2 x^2}{xy} - \dfrac{5ppy^{-1}}{x^{-1}} - \dfrac{5p^2 x^2 x^2}{x^{-3} y}$

30. Evaluate: $xa(ax - a)$ if $a = -\dfrac{1}{3}$ and $x = \dfrac{1}{2}$

LESSON 25   *Monomial factoring* • *Cancellation* • *Parallel lines*

25.A
monomial factoring

The Latin word for maker or doer is the word *factor*. From this came the Old French word *facteur*, which led to the Middle English word *factour*. In modern English we use the same spelling as did the ancient Romans, and in mathematics the meaning is very close to the Latin meaning of maker or doer. If we multiply 3 and 2, we get 6.

$$3 \cdot 2 = 6$$

We say that 3 and 2 are **factors** of 6 because we can make 6 by multiplying these two numbers. If we can make an expression by multiplying two or more other expressions, we say that each of the expressions that is multiplied is a factor of the final expression.

> A factor is one of two or more expressions that are multiplied to form a product.

Thus, since we multiply $4x$ and $ax + y$ to get $4ax^2 + 4xy$,

$$4x(ax + y) = 4ax^2 + 4xy$$

we say that $4x$ and $ax + y$ are factors of $4ax^2 + 4xy$. **When we write a sum as the product of factors, we say that we are** *factoring*.

example 25.1   Factor: $4ax^2 + 4xy$

solution   We have been given a sum and asked to write this expression as a product. We begin by recording an empty set of parentheses.

$$(\qquad)$$

Now, in front of the parentheses, we write the greatest common factor of the original terms.

$$4x(\qquad)$$

Now we decide what must be recorded inside the parentheses so that when we multiply by $4x$ the result will be $4ax^2 + 4xy$, our original expression. The correct entry is $ax + y$.

$$\mathbf{4x(ax + y)}$$

We say that we have factored the expression by factoring out $4x$. We can check our work by multiplying.

$$4x(ax + y) = 4ax^2 + 4xy$$

example 25.2   Factor: $4x^2p^2k - 6k^2p^4x$

solution   We begin by writing a set of parentheses preceded by the greatest common factor of both terms.

$$2xp^2k(\qquad)$$

Next we find the proper entry for the parentheses so that the product will be our original expression.

$$\mathbf{2xp^2k(2x - 3kp^2)}$$

We can check our work by multiplying.

$$2xp^2k(2x - 3kp^2) = 4x^2p^2k - 6k^2p^4x$$

example 25.3   Factor: $4x^2y - 2xy + 10xy^2$

solution Again we begin with empty parentheses preceded by the greatest common factor of all the terms.

$$2xy(\qquad)$$

We finish by finding the proper entry for the parentheses.

$$2xy(2x - 1 + 5y)$$

25.B
cancellation

We say that multiplication and division are **inverse operations** because they undo each other. For example, if we begin with 7 and then multiply and divide by 2

$$\frac{7 \cdot 2}{2} = 7$$

the result is 7, the number with which we began. This procedure can sometimes be used to simplify expressions that appear rather formidable.

$$\frac{(4.062)(3.0176)}{4.062}$$

Here we can see that the result will be 3.0176 because multiplication and division by 4.062 undo each other. We can draw a line through each of these numbers and say that we have canceled.

$$\frac{(\cancel{4.062})(3.0176)}{\cancel{4.062}} = 3.0176$$

The only time cancellation is possible is when both the numerator and denominator are products of factors and each contains one or more common factors. Thus, we may cancel the 4's in the following expression

$$\frac{4(x + 2)}{4} = \frac{\cancel{4}(x + 2)}{\cancel{4}(1)} = x + 2$$

because the 4 in the denominator can be thought of as the product of 4 and 1. No cancellation is possible here because the numerator is a sum.

$$\frac{x + 4}{4}$$

We cannot cancel the 4's because addition and division are not inverse operations. We cannot cancel the 4's in the next expression

$$\frac{4x + \cancel{4}}{\cancel{4}} \qquad \text{incorrect}$$

in the present form, but if we factor out a 4 in the numerator, cancellation is possible because the numerator is a product.

$$\frac{4x + 4}{4} = \frac{\cancel{4}(x + 1)}{\cancel{4}} = x + 1$$

example 25.4 Cancel if possible: (a) $\dfrac{6x^3 + x^2}{x^2}$ (b) $\dfrac{6a^2 + x^2}{x^2}$

solution We can factor and cancel in (a), but no cancellation is possible in (b).

(a) $\dfrac{6x^3 + x^2}{x^2} = \dfrac{\cancel{x^2}(6x + 1)}{\cancel{x^2}} = 6x + 1$

(b) We cannot write the numerator as a product so we cannot cancel.

25.C
parallel lines

We are familiar with the similar triangles in the figures shown here.

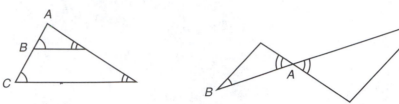

The large triangle and the small triangle in the figure on the left are similar because the angles are equal. Both triangles contain angle *A*, and angles *B* and *C* are also equal. The two triangles in the figure on the right are similar because angles *B* and *C* are equal, and the vertical angles are equal.

Parallel lines also tell us that angles are equal.

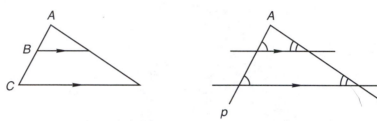

In the figure on the left, we see two parallel lines. On the right, we extend the segments and note that transversal *p* creates two equal small angles and that transversal *m* creates two more equal small angles.

In the figure on the left below, we see two parallel lines.

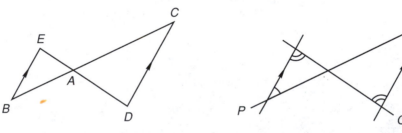

In the figure on the right, we extend the sides and find that the transversals *P* and *Q* give us two pairs of equal small angles. **Whenever a figure has parallel line segments, mentally extend the line segments and look for equal small angles and equal large angles.**

practice

a. Factor: $8m^2xy^5 + 6yx^2m^3 - 2xym^2$

b. Simplify: $\dfrac{4m^5 + m^2}{m^2}$

c. Find *x* and *y*.

d. Find *M* and *N*.

problem set 25

1. The *Mary Sue* stayed on the Grand Banks for 38 days and salted down 14,440 pounds of codfish. How long would she have had to stay to salt down 36,100 pounds?

2. Brown Bear made the trip in 40 hours. Flying Fish took only 30 hours to make the trip because his speed exceeded that of Brown Bear by 6 kilometers per hour. How long was the trip?

3. There were 5 times as many boys as girls at the party. Also, the number of boys was 100 less than 15 times the number of girls. How many boys and girls came to the party?

4. The day was a concatenation of disasters. If the ratio of minor disasters to major disasters was 5 to 2 and there were 980 disasters in all, how many were minor?

5. The machine broke open, and the quarters and half-dollars fell to the floor. The mysophobe grimaced but still retrieved them because there were 200 coins whose value was $75. How many of each kind were there?

6. Seventy percent of the compound was sodium chloride. If 660 grams of other chemicals were used, what was the total weight of the compound?

7. Divide $x^3 - 2$ by $x - 5$ and then check.

Factor:

8. $5x^2y^2 - 2xy + 10xy^2$

9. $x^2y^3m^5 + 12x^3ym^4 - 3x^2y^2m^2$

10. $16m^2p^3y - 8y^4mp^3 + 4m^2p^2y^2$

11. $x^3y^2z^3 + x^2yz^2 - 3x^3yz$

12. $p^5x^3 + p^4x^2 - p^3x$

Simplify:

13. $2\sqrt{3} \cdot 3\sqrt{6} \cdot 5\sqrt{12}$

14. $6\sqrt{18} + 5\sqrt{8} - 3\sqrt{50}$

15. $2\sqrt{5}(3\sqrt{15} - 2\sqrt{5})$

Add:

16. $a + \dfrac{a}{b}$

17. $\dfrac{ax^2}{m^2p} - c + \dfrac{2}{m}$

18. Simplify: $\dfrac{(38{,}000 \times 10^3)(300 \times 10^{-4})}{0.00019 \times 10^{-5}}$

19. Solve by graphing and then get an exact solution by using either substitution or elimination.

$$\begin{cases} 3x - 2y = 10 \\ y = -\dfrac{1}{2} \end{cases}$$

20. Find the equation of the line that passes through the point $(3, 5)$ and is parallel to the line $y = \dfrac{1}{6}x - 2$.

21. Find x, y, and P.

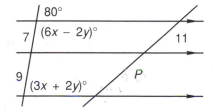

22. The radius of the circle is 12 cm. Find the length of arc ABC.

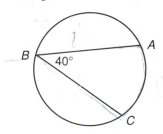

Solve:

23. $\dfrac{x + 1}{4} - \dfrac{3}{2} = \dfrac{2x - 9}{10}$

24. $\dfrac{4x - 8}{5} + \dfrac{2x - 4}{2} = 9$

25. $\dfrac{n + 3}{6} - \dfrac{1}{3} = \dfrac{2n - 2}{5}$

26. $\dfrac{5x + 3}{2} - \dfrac{3}{4} = \dfrac{5}{2}$

27. Cancel if possible: $\dfrac{8x^5 + 12x^3}{4x^3}$

28. Expand: $\dfrac{x^2 y^{-2}}{z^2}\left(\dfrac{z^2}{y^2(2x^{-2})^{-1}} - \dfrac{4x^2 y^0}{z^{-2}}\right)$

29. Simplify: $-3^0\big[(-2 - 4 - 2^2 - 2^0) - |-3 - 2|\big]$

30. Evaluate: $km(m^2k - k)$ if $k = \dfrac{1}{3}$ and $m = -\dfrac{1}{4}$

LESSON 26   *Trinomial factoring • Overlapping right triangles*

26.A

trinomial factoring   When we factor, we undo multiplication. Thus, to get rules for factoring trinomials, we look at the pattern that develops when we multiply binomials.

(a)
$$
\begin{array}{r}
x + 2 \\
x - 5 \\
\hline
x^2 + 2x \\
-5x - 10 \\
\hline
x^2 - 3x - 10
\end{array}
$$

(b)
$$
\begin{array}{r}
b + 4 \\
b - 1 \\
\hline
b^2 + 4b \\
-b - 4 \\
\hline
b^2 + 3b - 4
\end{array}
$$

(c)
$$
\begin{array}{r}
x - 3 \\
x - 4 \\
\hline
x^2 - 3x \\
-4x + 12 \\
\hline
x^2 - 7x + 12
\end{array}
$$

In each of the multiplication problems shown here, we note that in the trinomial products,

(a) the first term is the product of the first terms of the binomials.
(b) the last term is the product of the last terms of the binomials.
(c) the coefficient of the middle term is the sum of the last terms of the binomials.

Thus, to factor a trinomial whose lead coefficient is 1, we need to find two numbers whose sum and product meet the requirements of (b) and (c).

example 26.1   Factor: $x^2 + 6 - 5x$

solution   It is helpful if we begin by writing the trinomial in descending powers of the variable.

$$x^2 - 5x + 6$$

Now we write two sets of parentheses with x as the first entry in each set.

$$(x \quad)(x \quad)$$

The second entries are the two numbers whose sum is -5 and whose product is $+6$. The numbers are -3 and -2. Thus, our answer is

$$(x - 3)(x - 2)$$

example 26.2   Factor: $-x^2 + 5x + 14$

solution   When the x^2 term has a negative coefficient, it is helpful to first factor out a negative quantity. Here we will factor out -1.

$$(-1)(x^2 - 5x - 14)$$

Next we factor the trinomial and get

$$(-1)(x - 7)(x + 2)$$

We could leave the answer in this form, but many people like to multiply the (-1) by one of the other factors. If we do this, we can get

$$(-x + 7)(x + 2) \quad \text{or} \quad (x - 7)(-x - 2)$$

Either answer is acceptable.

example 26.3   Factor: $24 + 6x - 3x^2$

solution   First we rearrange the trinomial in descending powers of the variable.

$$-3x^2 + 6x + 24$$

Next we factor out the common negative factor, which is -3.

$$-3(x^2 - 2x - 8)$$

Then we factor the trinomial.

$$(-3)(x - 4)(x + 2)$$

example 26.4   Factor: $32x^2 + 12x^3 - 2x^4$

solution   We begin by writing the trinomial in descending powers of the variable.

$$-2x^4 + 12x^3 + 32x^2$$

Let's hope that this trinomial has a common factor, for we don't know how to factor quartic equations yet. The trinomial does have a common factor, which is $-2x^2$. Thus, we factor out $-2x^2$ and get

$$-2x^2(x^2 - 6x - 16)$$

Then we finish by factoring the trinomial.

$$-2x^2(x + 2)(x - 8)$$

26.B
overlapping right triangles

There are two right triangles in the figure shown here.

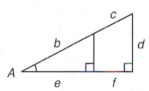

Angle A is an angle in both triangles. The triangles have six segments, as shown. By designating the lengths of some of the segments, we can create an interesting puzzle that can be solved by using the Pythagorean theorem and similar triangles. The puzzle can be varied by changing the lengths of the sides and by changing the segments whose lengths are designated. We will see this figure often in the problem sets.

example 26.5   Find A, B, and C.

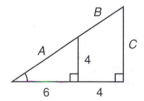

solution   It helps to draw the triangles separately.

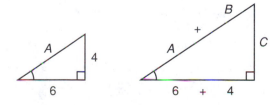

We will use the bottom lengths to find the scale factor from the small triangle to the large triangle.

$$6\overrightarrow{SF} = 10$$

$$\overrightarrow{SF} = \frac{10}{6} = \frac{5}{3}$$

Next we multiply 4 by the scale factor to find C.

$$4\left(\frac{5}{3}\right) = C$$

$$\frac{20}{3} = C$$

Now we use the Pythagorean theorem to find A.

$$A^2 = 6^2 + 4^2$$

$$A^2 = 52$$

$$A = \sqrt{52}$$

Now we have the following figure.

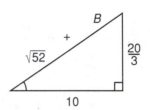

If we try to use the Pythagorean theorem to find B, we get a second-degree equation because we have to square the expression $\sqrt{52} + B$. So we will not use the Pythagorean theorem but will use the small-to-large scale factor of $\frac{5}{3}$ and the hypotenuses of both triangles.

$$\sqrt{52}\left(\frac{5}{3}\right) = \sqrt{52} + B \qquad \text{used small-to-large scale factor}$$

$$5\sqrt{52} = 3\sqrt{52} + 3B \qquad \text{multiplied both sides by 3}$$

$$2\sqrt{52} = 3B \qquad \text{added } -3\sqrt{52} \text{ to both sides}$$

$$\frac{4\sqrt{13}}{3} = B \qquad \text{simplified}$$

We remember that if we wish we can always use the proportion method instead of the scale factor when working with similar triangles.

practice Factor:

a. $x^2 - 7 - 6x$

b. $-x^2 + 6x + 16$

c. $24x^2 + 21x^3 - 3x^4$

d. Find A, B, and C.

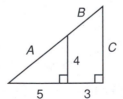

problem set 26

1. The *Silver Arrow* made the trip in 20 hours. The *Orange Blossom Special* made the same trip in only 8 hours because its speed was 60 mph greater than that of the *Silver Arrow*. How far was the trip?

2. The number of blues was 50 greater than 5 times the number of reds. Also, the number of reds was 210 less than the number of blues. How many were red and how many were blue?

3. The bag contained used $5 bills and used $10 bills. If there were 1200 bills and their total value was $7000, how many were $5 bills and how many were $10 bills?

4. Fourteen percent were belligerent while the rest were merely eristic. If 4300 were eristic, how many were belligerent?

5. When the compound contained 140 grams of calcium, its total weight was 1960 grams. If the total weight was measured to 2240 grams, how many grams of substances other than calcium would there be?

6. Ten percent of the chemicals being considered were nonabsorbent. If 7290 chemicals were absorbent, how many chemicals were being considered?

7. Divide $x^4 - 2$ by $x + 1$ and check.

Factor the greatest common factor:

8. $35x^7y^5m - 7x^5m^2y^2 + 14y^7x^4m^2$

9. $6x^2ym^5 - 2x^2ym + 4xym$

10. $4x^2y^4p^6 - 2xp^5y^7 + 8x^4p^5y^5$

Factor completely. Always factor the greatest common factor (GCF) as the first step.

11. $x^2 + x - 6$

12. $x^2 - 6x + 8$

13. $-2ab + abx + abx^2$

Simplify:

14. $\dfrac{6x^2y - xy}{xy}$

15. $3\sqrt{2} \cdot 2\sqrt{6} \cdot 3\sqrt{6}$

16. $-3\sqrt{12} + 5\sqrt{27} - 8\sqrt{25}$

17. $3\sqrt{2}(5\sqrt{3} - 2\sqrt{2})$

18. Add: $\dfrac{a^2m}{x^2} - x^2 - \dfrac{x^2}{c}$

19. Simplify: $\dfrac{(3000 \times 10^{-14})(0.00008)}{(0.0002 \times 10^5)(200{,}000)}$

20. Solve by graphing and then get an exact solution by using either substitution or elimination.
$$\begin{cases} 2y - x = 6 \\ y - 2x = -3 \end{cases}$$

21. Find the distance between $(4, -2)$ and $(-3, -5)$.

22. Find A, B, and C.

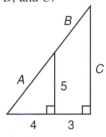

23. Angle ABC is a right angle. First find x. Then find y. Then find m. Then find z.

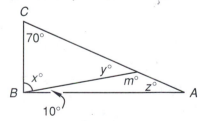

24. Find x and y. Note that angles B and D are equal small angles formed by parallel lines and a transversal. Also note that angle BCD is a straight angle.

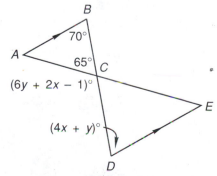

25. The area of the triangle is 54 m². Find *BC*. The radius of the circle equals *BC*. Find the area of the circle.

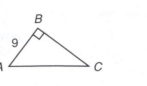

Solve:

26. $\dfrac{2x}{3} + \dfrac{6 - 3x}{4} = 7$

27. $\dfrac{4 - x}{3} + \dfrac{1}{6} = 5$

28. $-2x - \dfrac{3^0 - x}{2} + \dfrac{x - 5^0}{7} = 2$

29. Simplify: $\dfrac{x^2(yz^0)^{-2}}{(-2x^2y)^{-3}}$

30. Evaluate: $-\dfrac{1}{-(-3)^{-2}} + \dfrac{1}{-2^{-3}}(a - ba)$ if $a = -\dfrac{1}{2}$ and $b = 3$

LESSON 27 *Rational expressions*

A **ratio** can be written as a fraction. Thus,

$$\frac{3}{4}$$

can be described by saying three-fourths or by saying the ratio of 3 to 4. Whenever an algebraic ratio is written in the form of a fraction, such as

$$\frac{a}{b} \qquad \frac{a^2x}{4y} \qquad \frac{(m + p)^2}{k} \qquad \frac{3xy - 4}{7p}$$

we call it a **rational expression**. We have been adding rational expressions by using the least common multiple of the denominators as the new denominator. **When one of the denominators has a factor that is a sum, then this sum is one of the factors of the least common multiple.**

example 27.1 Add: $\dfrac{4}{x} + \dfrac{6}{x + a}$

solution The least common multiple is always a product of factors. The least common multiple of these denominators is $x(x + a)$.

$$\frac{}{x(x + a)} + \frac{}{x(x + a)} \qquad \text{new denominator}$$

Next we use the denominator-numerator same-quantity rule to help determine the new numerators. Then we add.

$$\frac{4(x + a)}{x(x + a)} + \frac{6x}{x(x + a)} = \frac{10x + 4a}{x(x + a)}$$

example 27.2   Add: $\dfrac{x + 2}{x + 4} + 6 + \dfrac{2}{x^2}$

solution   The least common multiple is $x^2(x + 4)$. We will use the same steps we used in the preceding example. First we write the new denominators.

$$\dfrac{}{x^2(x + 4)} + \dfrac{}{x^2(x + 4)} + \dfrac{}{x^2(x + 4)} \qquad \text{new denominators}$$

Now we determine the new numerators.

$$\dfrac{x^2(x + 2)}{x^2(x + 4)} + \dfrac{6x^2(x + 4)}{x^2(x + 4)} + \dfrac{2(x + 4)}{x^2(x + 4)} \qquad \text{new numerators}$$

Next we expand the numerators,

$$\dfrac{x^3 + 2x^2 + 6x^3 + 24x^2 + 2x + 8}{x^2(x + 4)} \qquad \text{expanded}$$

and then simplify as the last step.

$$\dfrac{7x^3 + 26x^2 + 2x + 8}{x^2(x + 4)} \qquad \text{simplified}$$

example 27.3   Add: $\dfrac{x + 2}{x^2 + 4x + 3} - \dfrac{1}{x(x + 1)}$

solution   Algebra books often have contrived problems which must be recognized or the solution is difficult. This is one of those problems. The denominator of the first term can be factored. If we do the factoring, the rest of the problem is straightforward.

$$\dfrac{x + 2}{(x + 3)(x + 1)} - \dfrac{1}{x(x + 1)} \qquad \text{factored}$$

$$= \dfrac{}{x(x + 3)(x + 1)} - \dfrac{}{x(x + 3)(x + 1)} \qquad \text{new denominators}$$

$$= \dfrac{x(x + 2)}{x(x + 3)(x + 1)} - \dfrac{(x + 3)}{x(x + 3)(x + 1)} \qquad \text{new numerators}$$

$$= \dfrac{x^2 + 2x - x - 3}{x(x + 3)(x + 1)} \qquad \text{added}$$

$$= \dfrac{x^2 + x - 3}{x(x + 3)(x + 1)} \qquad \text{simplified}$$

practice   Add:

a. $\dfrac{m + 3}{m + 4} + 5 + \dfrac{2}{m^2}$

b. $\dfrac{z + 3}{z^2 + 5z + 4} - \dfrac{1}{z(z + 4)}$

problem set 27

1.   The hiking club hiked to Robbers Cave State Park at 4 mph. They got a ride back to town in a truck that went 20 mph. If the round trip took 18 hours, how far was it from town to the park?

2.   Jojo set out on a hike. After walking for some time at 5 kph, he caught a ride back home in a truck that traveled at 20 kph. If the round trip took 10 hours, how far did he walk?

3.   There were 6 more girls than twice the number of boys. There were 36 boys and girls in all. How many boys and how many girls were there?

4.   The general expression for consecutive multiples of 7 is $7N$, $7(N + 1)$, $7(N + 2)$, etc., where N is some unspecified integer. Find three consecutive multiples of 7 such that the product of -3 and the sum of the first and third is 21 less than 5 times the opposite of the second.

5. Six hundred grams of barium was mixed with 2400 grams of other chemicals to form 3000 grams of the compound. If 9000 grams of compound was needed, how much barium was required?

6. Silver iodide made up 70 percent of the total. If the total weighed 2000 grams, how much was not silver iodide?

7. Divide $x^3 - 6$ by $x - 2$. Then check by adding.

Factor the greatest common factor:

8. $9m^2x^4p^2 + 3x^2p^6m^4 - 6x^4m^3p^2$ 　　　　9. $mx^4y - mx^2y^3 - 4mx^2y$

10. $a^2x^3p - 4a^3x^3p - a^2x^4p$

Factor completely. Always begin by looking for a common monomial factor.

11. $4ax + ax^2 - 5a$ 　　　　　　　　　12. $8x^2 - x^3 - 15x$

13. $24ax - 5ax^2 - ax^3$ 　　　　　　　14. $-ax^4 + 4ax^3 + 5ax^2$

15. $56p - 15px + px^2$

Simplify:

16. $\dfrac{4xa + 4x}{4x}$ 　　　　　　　　　17. $3\sqrt{2} - 2\sqrt{3} \cdot 3\sqrt{12}$

18. $-3\sqrt{20} + 2\sqrt{125} + 5\sqrt{45}$ 　　　19. $2\sqrt{3}(3\sqrt{2} - 3\sqrt{3})$

Add:

20. $\dfrac{2}{x} + \dfrac{3}{x + p}$ 　　　　　　　　21. $\dfrac{x + 3}{x + 6} + 5 + \dfrac{3}{x^2}$

22. Simplify: $\dfrac{(0.00056 \times 10^4)(7 \times 10^3)}{(0.00049 \times 10^{16})(0.00002 \times 10^{-5})}$

23. Solve by graphing and then get an exact solution by using either substitution or elimination.
$$\begin{cases} 3x + 2y = 12 \\ 8x - 2y = 10 \end{cases}$$

24. Find the equation of the line that passes through $(-3, -3)$ and is parallel to the line $y = -\frac{3}{8}x + 17$.

25. Find A, B, and C. 　　　　　　26. Find x and y.

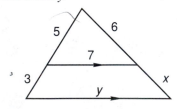

Solve:

27. $\dfrac{6 + x}{5} + \dfrac{1}{3} = 8$ 　　　　　　28. $\dfrac{3x - 2}{3} + 4 = \dfrac{1}{6}$

29. Simplify by adding like terms: $\dfrac{3a^{-2}y}{b} + 7b^{-1}ya^{-2} - \dfrac{5b^{-1}}{a^2y^{-1}}$

30. Evaluate: $-3^0(-3 - 2^2) - 4(-2)ax - a$ 　　　if $a = -2$ and $x = 4$

LESSON 28   Complex fractions • Rationalizing the denominator

28.A
complex fractions

Possibly the most useful rule in algebra is the denominator-numerator same-quantity rule. Because this rule is so important, we will write it again here.

DENOMINATOR-NUMERATOR SAME-QUANTITY RULE

The denominator and the numerator of a fraction may be multiplied by the same nonzero quantity without changing the value of the fraction.

This rule allows us to change the form of any fraction to a form that is more convenient. We have used this rule to help us add abstract fractions. To review the use of this procedure, we will add two fractions.

$$\frac{a}{b} + \frac{c}{d}$$

We will multiply the first fraction by d over d and multiply the second fraction by b over b. Then we can add the fractions.

$$\frac{a}{b}\left(\frac{d}{d}\right) + \frac{c}{d}\left(\frac{b}{b}\right) = \frac{ad + cb}{bd}$$

We also use the denominator-numerator same-quantity rule to help us simplify fractions of fractions.

$$\frac{\dfrac{a}{b}}{\dfrac{c}{d}}$$

If we multiply the denominator of this fraction by its reciprocal, which is $\frac{d}{c}$, the resulting denominator will be 1. If we do this, we must also multiply the numerator by the same quantity, $\frac{d}{c}$, so that the value of the expression will not be changed.

$$\frac{\dfrac{a}{b}\cdot\dfrac{d}{c}}{\dfrac{c}{d}\cdot\dfrac{d}{c}} = \frac{\dfrac{ad}{bc}}{1} = \frac{ad}{bc}$$

The denominator-numerator same-quantity rule can be used anywhere—and at any time—on a term in an equation or on a term that is not in an equation. In everyday language we can say,

Anytime, anywhere, the denominator and the numerator of an expression can be multiplied by the same quantity (not zero) without changing the value of expression. Only the form of the expression is changed.

example 28.1   Simplify: $\dfrac{\dfrac{a}{b}}{\dfrac{x + y}{b}}$

solution   We will multiply both the denominator and the numerator by b over $x + y$.

$$\frac{\dfrac{a}{b}}{\dfrac{x + y}{b}} \cdot \frac{\dfrac{b}{x + y}}{\dfrac{b}{x + y}} = \frac{a}{x + y}$$

example 28.2   Simplify: $\dfrac{\dfrac{a}{a + b}}{\dfrac{c}{a + b}}$

solution   We will multiply both the top and the bottom by $a + b$ over c.

$$\frac{\dfrac{a}{a + b}}{\dfrac{c}{a + b}} \cdot \frac{\dfrac{a + b}{c}}{\dfrac{a + b}{c}} = \frac{a}{c}$$

28.B

rationalizing the denominator

Often we encounter expressions such as

$$\frac{4}{\sqrt{7}}$$

that have a radical in the denominator. Many people like to change the form of these expressions so that the radical does not appear in the denominator. We remember that we can always change the form of an expression by multiplying both the denominator and the numerator by the same quantity. For this example, we choose to multiply by $\sqrt{7}$ over $\sqrt{7}$.

$$\frac{4}{\sqrt{7}} \cdot \frac{\sqrt{7}}{\sqrt{7}} = \frac{4\sqrt{7}}{7}$$

This new expression has the same value as the original expression, but the denominator is a rational number. This procedure is called **rationalizing the denominator.** The instructions for one of these problems will use the one word "simplify."

> **An expression that contains square root radicals is in simplified form when no radicand has a perfect-square factor and no radicals are in the denominator.**

example 28.3   Simplify: $\dfrac{3}{2\sqrt{5}}$

solution   We can eliminate the radical in the denominator by multiplying by $\sqrt{5}$ over $\sqrt{5}$. Of course, we still cannot get rid of $\sqrt{5}$ completely, for it will now appear in the numerator.

$$\frac{3}{2\sqrt{5}} \cdot \frac{\sqrt{5}}{\sqrt{5}} = \frac{3\sqrt{5}}{10}$$

example 28.4   Simplify: $\dfrac{2}{3\sqrt{12}}$

solution   We will multiply by $\sqrt{12}$ over $\sqrt{12}$ and then simplify the result.

$$\frac{2}{3\sqrt{12}} \cdot \frac{\sqrt{12}}{\sqrt{12}} = \frac{2\sqrt{12}}{36} = \frac{2(2\sqrt{3})}{36} = \frac{\sqrt{3}}{9}$$

practice   Simplify:

a. $\dfrac{\dfrac{m}{p}}{\dfrac{z + x}{p}}$

b. $\dfrac{\dfrac{m}{x + y}}{\dfrac{z}{x + y}}$

c. $\dfrac{4}{3\sqrt{2}}$

problem set 28

1. Bronson roared off on his motorcycle at 60 mph. Then, much to his chagrin, he ran out of petrol. He pushed the motorcycle all the way back at 3 mph. If the entire trip took 21 hours, how far did he push the motorcycle?

2. The general expressions for consecutive multiples of 3 are $3N$, $3(N + 1)$, $3(N + 2)$, etc., where N is some unspecified integer. Find four consecutive multiples of 3 such that 5 times the sum of the first and fourth is 6 less than 13 times the third.

3. The number of girls in the class was 1 less than 3 times the number of boys. There were 15 students in all. How many were boys, and how many were girls?

4. The class treasury contained $30 in nickels and dimes. If there were 500 coins, how many coins of each type were there?

5. Arthur found that for every 200 peasants only 10 had seen a Dane. If there were 150,000 peasants in the kingdom, how many had never seen a Dane?

6. Sixteen percent of the mixture was arsenic and the rest was silicon. Mendeleev knew 7350 kilograms of the mixture was silicon. How much arsenic was in this mixture? What did the entire mixture weigh?

7. Divide $x^3 - 7$ by $x - 5$ and then check.

Factor the greatest common factor:

8. $2x^2y - 8x^4y^4$

9. $4x^2y^3p^3 - 16x^2y^3p - x^4y^3p^4$

Factor completely. Always factor the GCF as the first step.

10. $-35xy + 2x^2y + x^3y$

11. $-8a - 7ax + ax^2$

12. $2m^2 + 3xm^2 + m^2x^2$

13. $-a^2 - a^2x^2 - 2xa^2$

Simplify:

14. $\dfrac{4x^2 + x}{x}$

15. $4\sqrt{27} - 3\sqrt{48} + 2\sqrt{75}$

16. $3\sqrt{5}(\sqrt{15} - 2\sqrt{5})$

17. $\dfrac{(0.00077 \times 10^{-3})(40 \times 10^6)}{(0.00011 \times 10^5)(140,000)}$

18. $\dfrac{\dfrac{x}{m + p}}{\dfrac{y}{m + p}}$

19. $\dfrac{3}{4\sqrt{15}}$

Add:

20. $\dfrac{4a}{a + x} + \dfrac{6}{a}$

21. $\dfrac{2x}{x^2 + 2x + 1} + \dfrac{3}{x + 1}$

22. Solve by graphing and then get an exact solution by using either substitution or elimination.
$$\begin{cases} 5x + 2y = 6 \\ y = \dfrac{1}{2}x \end{cases}$$

23. Find the equation of lines (a) and (b).

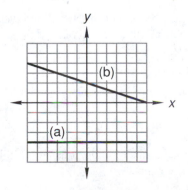

24. In this figure, X equals 150. Find A, B, C, D, E, and then find F.

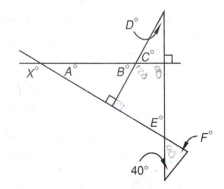

25. In this figure, AB is a diameter and is a straight angle. Find x and $2x$. If the radius of the circle is 2 meters, find the area of the shaded sector. Find the length of arc BC.

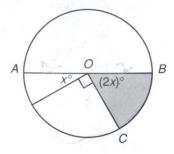

Solve:

26. $\dfrac{2x}{3} + \dfrac{x-3}{5} = 1$

27. $4\dfrac{1}{3} - 2\dfrac{1}{5}x = -\dfrac{3}{10}x$

28. Find the distance between $(-2, 5)$ and $(-6, -3)$.

29. Expand: $\dfrac{3x^{-2}y^2}{z}\left(\dfrac{x^2z}{3y^2} - \dfrac{4x^{-2}y}{z^{-3}}\right)$

30. Evaluate: $x^0 - xy^0(x - y)$ if $x = \dfrac{1}{3}$ and $y = -\dfrac{1}{4}$

LESSON 29 *Uniform motion problems:* $D_1 + D_2 = k$

We remember that the key equation in a uniform motion problem is the equation that describes the distances that have been traveled. In all the problems thus far, the persons or objects have traveled equal distances, so the distance diagrams for these problems have been similar to one of the two diagrams shown here.

$$D_1 = D_2 \qquad \text{so} \qquad R_1T_1 = R_2T_2$$

In the left-hand diagram both objects began at the same point and traveled in the same direction. In the right-hand diagram the objects began at different points and traveled in opposite directions, but again the distances traveled were equal. Thus, the same equation is applicable to both diagrams.

In some problems, the sum of one distance and another distance equals a certain number. If the number is 460, the distance diagram would look like one of the following.

In each diagram we know that the sum of distance 1 and distance 2 equals 460, so the same distance equation is applicable to all three diagrams.

$$D_1 + D_2 = 460 \qquad \text{so} \qquad R_1T_1 + R_2T_2 = 460$$

example 29.1 Napoleon walked part of the 60 miles to the site of the battle and rode the rest of the way on a caisson. He walked at 3 mph and rode at 9 mph. If the total time of the trip was 8 hours, for how long did he walk?

solution His distance walking plus his distance riding equaled 60 miles. This leads to the following distance diagram and distance equation.

$$\overset{D_W \quad\quad D_R}{\underset{60}{\longmapsto\!\!\longrightarrow}} \quad D_W + D_R = 60 \quad\text{so}\quad R_W T_W + R_R T_R = 60$$

We have four unknowns and only one equation. Thus, we need three more equations. We reread the problem and find that two are rate equations and one is a time equation.

$$R_W = 3 \qquad R_R = 9 \qquad T_W + T_R = 8$$

We finish by substituting into the distance equation.

$$3(8 - T_R) + 9T_R = 60 \qquad \text{substituted}$$
$$24 - 3T_R + 9T_R = 60 \qquad \text{multiplied}$$
$$24 + 6T_R = 60 \qquad \text{added}$$
$$6T_R = 36 \qquad \text{simplified}$$
$$T_R = 6 \qquad \text{divided}$$

Since he rode for 6 hours, and his total time was 8 hours, he must have walked for 2 hours.

$$T_W = \textbf{2 hours}$$

example 29.2   Edward Longshanks and Queen Eleanor were 54 miles apart at dawn. Edward began the journey to the meeting place at 8 a.m. at 3 mph; 2 hours later, the queen set out to meet him. If they met at 4 p.m., how fast did the queen travel?

solution   Between them they covered 54 miles, so the following distance diagram and distance equation apply.

$$\overset{D_L \quad\quad D_Q}{\underset{54}{\longmapsto\!\!\longleftarrow}} \quad D_L + D_Q = 54 \quad\text{so}\quad R_L T_L + R_Q T_Q = 54$$

We have one equation and four unknowns. Thus, we need three more equations. Two are time equations, and one is a rate equation.

$$T_L = 8 \qquad T_Q = 6 \qquad R_L = 3$$

We substitute these into the distance equation and solve.

$$(3)(8) + R_Q(6) = 54 \qquad \text{substituted}$$
$$24 + 6R_Q = 54 \qquad \text{multiplied}$$
$$6R_Q = 30 \qquad \text{added } -24$$
$$R_Q = 5 \qquad \text{divided}$$

Thus, Queen Eleanor traveled at 5 mph, which was a fast speed for the roads of thirteenth-century England.

example 29.3   At noon, Rocketman whizzed off toward Rocketland; 1 hour later, Moonfa whizzed off in the opposite direction at a speed 200 kph less than that of Rocketman. If they were 11,800 kilometers (km) apart at 5 p.m., how fast did each travel?

solution   Rocketman and Moonfa began at the same point and traveled in opposite directions. Together they covered 11,800 km. Thus, the distance diagram and distance equation are as follows.

$$\overset{D_R \quad\quad D_M}{\underset{11,800}{\longleftarrow\!\!\bullet\!\!\longrightarrow}} \quad D_R + D_M = 11{,}800 \quad\text{so}\quad R_R T_R + R_M T_M = 11{,}800$$

We need three more equations. Two are time equations and one is a rate equation.

$$T_R = 5 \qquad T_M = 4 \qquad R_M = R_R - 200$$

Now we substitute these equations into the distance equation and solve.

$$(R_M + 200)(5) + R_M(4) = 11{,}800 \qquad \text{substituted}$$

$$5R_M + 1000 + 4R_M = 11{,}800 \qquad \text{multiplied}$$

$$9R_M = 10{,}800 \qquad \text{simplified}$$

$$R_M = 1200 \text{ kph} \qquad \text{divided}$$

Since the rate of Rocketman was 200 kilometers per hour greater than that of Moonfa, Rocketman's rate was

$$R_R = 1400 \text{ kph}$$

practice   At 1 p.m. Chester left the roundup and began the 66-mile trip to Dodge City. At 2 p.m. Marshal Dillon left Dodge City to meet Chester. Marshal Dillon's speed was twice that of Chester, and they met at 9 p.m. How fast did each man ride?

problem set 29

1. Patton walked part of the 76 miles to the site of the battle and rode the rest of the way on a caisson. He walked at 4 mph and rode at 15 mph. If the total time of the trip was 8 hours, for how long did he walk?

2. Prince Charles and Princess Diana were 63 miles apart at dawn. Prince Charles began the journey to the meeting place at 7 a.m. at 3 mph. Three hours later, Princess Diana set out to meet him. If they met at 4 p.m., how fast did the princess travel?

3. At 5 p.m., Roger whizzed off from Asteroid. One hour later, Maryanne whizzed off from Asteroid in the opposite direction at a speed 400 kph less than that of Roger. If they were 7900 km apart at 11 p.m., how fast did each travel?

4. The number of roses was 15 greater than twice the number of prunes. If the roses and prunes totaled 255, how many roses were there?

5. For every 130 squirrels in the forest, there were 156,000 good places to hide. If there was a total of 3250 squirrels in the forest, how many good places to hide were there?

6. In a contiguous forest, 17 percent of the places to hide contained xenophobes. If 116,200 places to hide did not contain xenophobes, how many hiding places were there in the contiguous forest?

7. Divide $x^3 + 3x^2 + 7x + 5$ by $x + 1$ and then check.

Factor the greatest common factor:

8. $16x^3y^2z^3 - 8x^2y^2z^2$ 9. $2x^2yp^4 - 6x^3yp^3 - 2x^2yp^2$

Factor completely. Always factor the GCF as the first step.

10. $12a^2x + a^2x^2 + 35a^2$

11. $-2m^2x - m^2 - m^2x^2$ 12. $x^2k + 3kx - 40k$

Simplify:

13. $\dfrac{x^2 + ax^2}{x^2}$ 14. $2\sqrt{75} - 5\sqrt{48} + 2\sqrt{12}$

15. $2\sqrt{3}(3\sqrt{6} - 4\sqrt{3})$ 16. $\dfrac{(0.00052 \times 10^{-4})(5000 \times 10^7)}{(0.0026 \times 10^{21})(10{,}000 \times 10^{-42})}$

17. $\dfrac{\dfrac{m}{x}}{\dfrac{m + x}{x}}$ 18. $\dfrac{\dfrac{a}{m + x}}{\dfrac{b}{m + x}}$

19. $\dfrac{3}{5\sqrt{12}}$ 20. $\dfrac{14}{3\sqrt{75}}$

Add:

21. $\dfrac{4x}{x + 4} + \dfrac{6}{x + 2}$ 22. $\dfrac{3m}{m^2 + 3m + 2} - \dfrac{5m}{m + 1}$

23. Solve by graphing. Then get an exact solution by using either substitution or elimination.

$$\begin{cases} y - x = 3 \\ y + 2x = 6 \end{cases}$$

24. Find the equation of the line that passes through $(2, -3)$ and is parallel to $y = -\dfrac{3}{8}x + 2$.

25. The area of the circle was 2500π m². What was the radius of the circle? Jason walked along the circle through an arc of $36°$. How far did Jason walk?

26. Find A, B, and C.

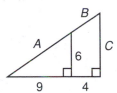

Solve:

27. $\dfrac{3x}{2} - \dfrac{5}{7} = \dfrac{x + 2}{3}$

28. $2\dfrac{1}{5} - \dfrac{1}{10}x = \dfrac{3}{15}$

29. Simplify: $\dfrac{(x^2 y^{-2} z)^{-3} x^0}{(x^0 y^{-3} z^2)^3}$

30. Evaluate: $x^2 - y^2(x - y)$ if $x = \dfrac{1}{2}$ and $y = \dfrac{1}{3}$

LESSON 30 *Deductive reasoning • Euclid • Vertical angles are equal • Corresponding interior and exterior angles • 180° in a triangle*

30.A
deductive reasoning

Deductive reasoning is a term that we apply to a process of reasoning from a nonreversible statement, called the **major premise,** to a result called the **conclusion.** The major premise is always an **if-then statement.** We identify nonreversible statements by using an arrow that points in just one direction.

$$A \;\longrightarrow\; B$$

This tells us that if A is true then B is true. It does not say that if B is true then A is true.

If a quadrilateral is a square, then the quadrilateral is also a rectangle.

If square $\longrightarrow$ then rectangle

We say that this premise is not reversible because we cannot say

If rectangle $\longrightarrow$ then square   **FALSE**

since all rectangles are not squares. A three-step deductive reasoning process consisting of a major premise, a minor premise, and a conclusion is called a **syllogism.** Note that the major premise can be stated without using the words *if* and *then*. The premise, "If a polygon is a square, then the polygon is also a rectangle," is stated without using *if* and *then* in the following syllogism.

(1) Major premise	All squares are rectangles.
(2) Minor premise	Quadrilateral *ABCD* is a square.
(3) Conclusion	Quadrilateral *ABCD* is a rectangle.

Syllogistic reasoning is tricky, and we must be careful. Consider the following syllogism.

(1) All poets are poor.
(2) Roger is poor.
‾‾‾‾‾‾‾‾‾‾‾‾‾‾‾‾‾‾‾‾‾‾‾
(3) Roger is a poet.   (Not valid)

This is not a valid conclusion because we have reversed the major premise. The premise we were given was

$$\text{Poet} \longrightarrow \text{Poor}$$

The major premise did not tell us

$$\text{Poor} \longrightarrow \text{Poet} \quad \textbf{FALSE}$$

An argument is a **valid argument** if we reason correctly. A valid argument does not lead to a true conclusion if one of the premises is false. Consider the following syllogism.

(1) Major premise	All chickens have three legs.
(2) Minor premise	Henny is a chicken.
(3) Conclusion	Henny has three legs. (Valid)

This is a **valid argument** because the reasoning process from (1) to (3) is correct. The conclusion is false because one of the premises is false. Now consider the following syllogism.

(1) Major premise	If it rains, I will go to town.
(2) Minor premise	It did not rain.
(3) Conclusion	I did not go to town. (Invalid)

This conclusion is false because the reasoning process is invalid. The major premise says that I go to town on the days that it rains. It makes no statements about what I will do on the days it does not rain. Therefore, the reasoning process is flawed and the conclusion is invalid. From this we see that we have to be careful when we try to use deductive reasoning.

30.B
Euclid

The first mathematics in the western world was that of the Egyptians and the Babylonians. Compared with the mathematics of their successors, the Greeks, the mathematics of the Egyptians and Babylonians was primitive at best.

The classical period of ancient Greece was from about 600 B.C. to 300 B.C., and the chief cultural center was Athens. The Greeks were the originators of philosophy and of the pure and the applied sciences. They were the first in political thought and institutions and were the first historians. Many new ideals, such as the freedom of the individual, are Greek contributions to western culture. Among the more important contributions of the Greeks were their emphasis on a human being's ability to reason and their belief in cause and effect as opposed to superstition and the supernatural. Their belief in reason allowed them to develop geometry as a deductive reasoning process.

We know the names of quite a few Greek mathematicians. The Pythagorean theorem is named after Pythagoras, a Greek believed to have been born on the isle of Samos and who later lived in Kroton in southern Italy circa 525 B.C. The first recorded work on geometric proofs is that of Hippocrates of Chios, circa 425 B.C. Other Greeks, including Eudoxos of Knidos, made major contributions to geometry.

Eukleides (whom we now call Euclid) was a Greek scholar who probably lived in Alexandria, Egypt, during the reign of Ptolemy I, the first Greek king of Egypt (323–285 B.C.). He compiled the work of his predecessors and expanded on it in his treatise on geometry called the *Elements*. In this treatise, Euclid stated that some facts about mathematics were true because they were true and that their truth could be accepted without proof. He called these self-evident truths **axioms** or **postulates**. Then he proved 467 other assertions by using deductive reasoning based on his self-evident truths. Because the reasoning was logical and was based on self-evident truths, the assertions that he proved were believed to be true even though their truth may not have been self-evident. These provable assertions are called **theorems**.

Definitions are the **names** that we give to ideas. Definitions are not proved. For example, Euclid defined parallel lines to be any two lines in the same plane that do not intersect. This definition does not imply the existence or the nonexistence of parallel lines. It just says that if two lines are in the same plane and if they do not intersect, we call the lines parallel lines. **It is important to remember that all definitions are reversible.** Thus, if we have a pair of parallel lines, they **must** be in the same plane and they **must not** intersect. Theorems, axioms, and postulates are not necessarily reversible. For instance, if two angles are right angles, their measures are equal. But, two angles whose measures are equal are not necessarily right angles.

Euclid was able to reduce his list of postulates or axioms to 10. The essence of Euclid's postulates is contained in the following statements. The wording of Postulate 5 shown here is attributed to John Playfair (1748–1819) and is the wording usually used in high school geometry texts in the early 1900s.

POSTULATE 1.   **Two points determine a unique straight line**.
POSTULATE 2.   **A straight line extends indefinitely far in either direction**.
POSTULATE 3.   **A circle may be drawn with any given center and any given radius**.
POSTULATE 4.   **All right angles are equal**.
POSTULATE 5.   **Given a line *n* and a point *P* not on that line, there exists in the plane of *P* and *n* and through *P* one and only one line *m*, which does not meet the given line *n*.**

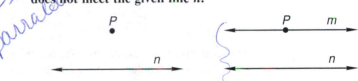

POSTULATE 6.   **Things equal to the same thing are equal to each other**.
POSTULATE 7.   **If equals be added to equals, the sums are equal**.
POSTULATE 8.   **If equals be subtracted from equals, the remainders are equal**.
POSTULATE 9.   **Figures which can be made to coincide are equal (congruent)**.
POSTULATE 10.   **The whole is greater than any part**.

The modern wording of some of these postulates is different, and mathematicians have found it necessary to add other postulates to the 10 postulates of Euclid. One of the postulates concerns betweenness and another concerns continuity. We will not discuss these additional postulates in this book, nor will we try to build a geometric structure based on postulates and proofs.

We will do a few simple proofs to familiarize the reader with the process of deductive reasoning. The major emphasis in this book, however, will be on learning and using the fundamental properties of geometric figures. Long-term practice with these fundamental properties will make the properties familiar, and then the proofs of the properties will be meaningful and easy, as you will see toward the end of the book.

30.C
vertical angles are equal

We can use the sixth and eighth postulates of Euclid to prove that vertical angles are equal.

Consider the following figure with angles x, y, z, and p whose measures are $x°$, $y°$, $z°$, and $p°$.

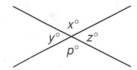

We see that $x°$ plus $y°$ equals $180°$. Also, we see that $x°$ plus $z°$ equals $180°$. So

$$x + y = 180 \qquad \text{and} \qquad x + z = 180$$

Both $x + y$ and $x + z$ equal 180, so they are equal to each other by Euclid's sixth postulate.

$$x + y = x + z$$

Postulate 8 tells us that if equals are subtracted from equals the results are equal. So we subtract x from both sides of this equality and find that y is equal to z.

$$
\begin{array}{rcl}
x + y = & & x + z \\
-x & & -x \\
\hline
y = & & z
\end{array}
$$

The same procedure can be used to prove that x equals p. Thus, we have used two postulates of Euclid and reasoned deductively from these postulates to prove a theorem.

30.D
corresponding interior and exterior angles

In Lesson 1, we postulated that when parallel lines are cut by a transversal that is perpendicular to one of the lines, all the angles formed are right angles, as we see in the left-hand figure.

If the angles are not right angles, we have postulated that half the angles are small angles whose measures are equal and half the angles are large angles whose measures are equal, as we see in the right-hand figure. To discuss this topic, it is customary to use only two parallel lines and to give the angles special names. The angles between the parallel lines are called **interior angles** and the angles outside the parallel lines are called **exterior angles**. Angles on opposite sides of the transversal are called **alternate angles**. In the figure on the left, we note that **alternate interior angles are equal.**

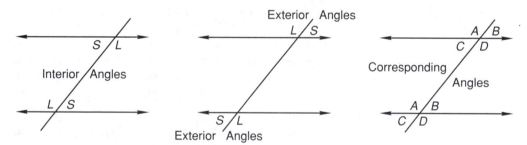

In the center figure, we note that alternate exterior angles are equal. In the figure on the right, we show four pairs of **corresponding angles**. Corresponding angles occupy corresponding positions in the figure, as indicated by the letters A, B, C, and D. Corresponding angles are equal.

Euclid used his postulates and deductive reasoning to develop a lengthy four-part argument that proves that if two parallel lines are cut by a transversal, the alternate interior angles are equal. The proof is above the level of this book. But, because the assertion can be proved, we call it a theorem.

THEOREM

When parallel lines are cut by a transversal, the pairs of alternate interior angles are equal.

This theorem permits us to label alternate interior large angles and small angles as having equal measures, as we do in the following figure.

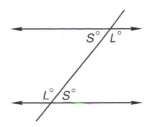

Now, because vertical angles are equal, we can label the other four angles.

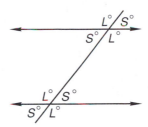

We can extend this procedure to any number of parallel lines cut by a transversal to prove that all the small angles are equal and all the large angles are equal.

30.E
180° in a triangle

The proof that the sum of the angles in a triangle is 180° is a simple proof that uses the knowledge we have about the angles formed when transversals intersect parallel lines. We can call this proof the *ABC* proof because we use these letters in the proof. On the left we show triangle *ABC*. On the right we draw a line through vertex *B* that is parallel to side *AC*, and we also extend the sides of the triangle as shown to form lines *m* and *n*.

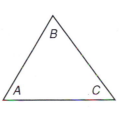

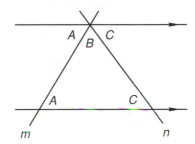

The angles marked *A* are equal because they are equal small angles (alternate interior) formed by the parallel lines and transversal *m*. The angles marked *C* are equal because they are equal small angles (alternate interior) formed by the parallel lines and transversal *n*. At the vertex we see that angles *A*, *B*, and *C* form a straight angle whose measure is 180°. Thus the three angles of the triangle *A*, *B*, and *C* also have a sum of 180°.

practice

Are the following arguments valid? Why?

a.   All dogs have three legs.
     Henry is a dog.
     _____
     Henry has three legs.

b.   All scholars are poor.
     Rita is poor.
     _____
     Rita is a scholar.

problem set 30

1.   When 240 grams of barium was mixed with 40 grams of sulfite, the desired reaction occurred. If a total of 3360 grams of barium and sulfite was to be used, how much should be sulfite?

2.   Fourteen percent of the mass was consumed in the reaction. If 430 grams remained, how much was the initial mass?

3.   At 10 a.m., Little Flower trotted off in one direction at 6 mph. At noon, Laughing Boy loped off in the opposite direction. If they were 68 miles apart at 4 p.m., how fast did Laughing Boy lope?

4. The daisies proliferated until 5 times the number of daisies equaled twice the number of prunes. If the daisies and prunes totaled 35, how many of each were there?

5. Yellow Basket found that she had 15 dimes and quarters and that their total value was $2.25. How many of each kind of coin did she have?

6. Find (a) the area and (b) the perimeter of a rectangular plot of land whose length is 40 feet and whose width is 120 inches. (*Hint*: Begin by converting inches to feet.)

7. Divide $x^3 + 2$ by $x + 1$ and then check.

Factor completely. Always factor the GCF as the first step.

8. $35a^2 + 2a^2x - a^2x^2$ 9. $30 + 3x^2 - 21x$

10. $-x^2ab - 25ab + 10axb$ 11. $14a^2b^2 + 9a^2xb^2 + x^2a^2b^2$

Simplify:

12. $\dfrac{p - 4px}{p}$ 13. $5\sqrt{18} - 10\sqrt{50} + 3\sqrt{72}$

14. $3\sqrt{12}(4\sqrt{2} - 2\sqrt{3})$ 15. $\dfrac{(0.00035)(5000 \times 10^{42})}{0.00025 \times 10^{-4}}$

16. $\dfrac{\frac{x}{y}}{\frac{x + y}{y}}$ 17. $\dfrac{\frac{a}{a + b}}{\frac{p}{a + b}}$

18. $\dfrac{2}{3\sqrt{6}}$ 19. $\dfrac{2}{5\sqrt{18}}$

Add:

20. $\dfrac{4a}{a + 4} + \dfrac{a + 2}{2a}$ 21. $\dfrac{4x}{x^2 + 5x + 6} + \dfrac{2}{x + 2}$

22. Solve by graphing and then get an exact equation by using either substitution or elimination.

$$\begin{cases} 3x + 2y = 8 \\ 2x + 3y = 6 \end{cases}$$

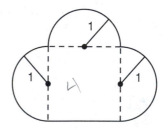

23. Find the equation of the line that passes through $(-2, 4)$ and $(1, 3)$.

24. The base of a right cylinder 10 m tall is shown. Find the lateral surface area of the cylinder. Dimensions are in meters.

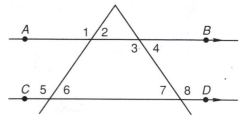

25. In the diagram shown, $\overline{AB}$ is parallel to $\overline{CD}$. Which one of the following statements is *not necessarily true*?

    (A) $\angle 1 + \angle 2 = 180°$
    (B) $\angle 4 = \angle 7$
    (C) $\angle 2 + \angle 3 = 180°$
    (D) $\angle 2 = \angle 6$

Solve:

26. $\dfrac{2x}{3} - \dfrac{2}{5} = \dfrac{x - 5}{2}$ 27. $3\dfrac{2}{3} + \dfrac{1}{5}x = \dfrac{1}{15}$ 28. $\dfrac{2x - 3}{2} = -\dfrac{2}{5}$

29. Simplify by adding like terms: $\dfrac{4a^2 x^2 y}{p} - \dfrac{3p^{-1} x^4 y}{a^{-2} x^2} + \dfrac{2xa}{a^{-1} y^{-1} p} - \dfrac{2a^2 yx^2}{p}$

30. Evaluate: $x^2 - y(x - y)$　　if $x = -\dfrac{1}{2}$ and $y = \dfrac{1}{4}$

LESSON 31　*Negative reciprocals　•　Perpendicular lines　•　Remote interior angles*

31.A
negative reciprocals

We remember that the reciprocal of a number is the number in inverted form. Thus,

$\dfrac{2}{5}$　　is the reciprocal of　　$\dfrac{5}{2}$

$\dfrac{5}{2}$　　is the reciprocal of　　$\dfrac{2}{5}$

$-\dfrac{1}{4}$　　is the reciprocal of　　-4

-4　　is the reciprocal of　　$-\dfrac{1}{4}$

3　　is the reciprocal of　　$\dfrac{1}{3}$

$\dfrac{1}{3}$　　is the reciprocal of　　3

We note that the sign of a number is the same as the sign of its reciprocal. This is not true for numbers that are negative reciprocals because each of these numbers is the inverted form of the other and the signs of the numbers are different. Thus,

$-\dfrac{2}{5}$　　is the negative reciprocal of　　$\dfrac{5}{2}$

$\dfrac{5}{2}$　　is the negative reciprocal of　　$-\dfrac{2}{5}$

$-\dfrac{1}{4}$　　is the negative reciprocal of　　4

4　　is the negative reciprocal of　　$-\dfrac{1}{4}$

-3　　is the negative reciprocal of　　$\dfrac{1}{3}$

$\dfrac{1}{3}$　　is the negative reciprocal of　　-3

31.B
perpendicular lines

We recall that lines that are parallel have equal slopes and different y intercepts. When a linear equation is written in slope-intercept form, the coefficient of the x term designates the slope of the line, and the constant term designates the y intercept. Note that the coefficient of the x term in both of the following equations is $-\frac{1}{2}$, and that the y intercepts are $+2$ and -3.

$$y = -\frac{1}{2}x + 2 \qquad\qquad y = -\frac{1}{2}x - 3$$

So the slope of both of these lines is $-\frac{1}{2}$, and the lines cross the y axis at $+2$ and -3, respectively. The graphs of these lines are shown in the figure on the left.

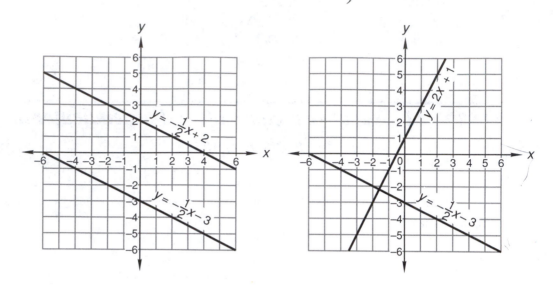

The slopes of lines that are perpendicular are negative reciprocals of each other. On the right we show the graphs of the equations

$$y = -\frac{1}{2}x - 3 \qquad \text{and} \qquad y = 2x + 1$$

and note that the lines appear to be perpendicular, and that the slopes are $-\frac{1}{2}$ and $+2$, numbers that are negative reciprocals. We can use this negative reciprocal relationship to help us write the equation of a line that is perpendicular to a given line and passes through a designated point.

example 31.1   Write the equation of the line that is perpendicular to $y = \frac{2}{5}x - 3$ and passes through the point $(3, 2)$.

solution   The new line is to be perpendicular to a line whose slope is $\frac{2}{5}$. Thus, the slope of the new line must be $-\frac{5}{2}$. This gives us

$$y = -\frac{5}{2}x + b$$

To finish the equation, we must find the value of the intercept, b. To do this, we will use 3 for x and 2 for y and solve algebraically for b.

$$2 = -\frac{5}{2}(3) + b \qquad \text{substituted}$$

$$\frac{4}{2} = -\frac{15}{2} + b \qquad \text{simplified}$$

$$\frac{19}{2} = b \qquad \text{solved}$$

Thus, the intercept is $\frac{19}{2}$, and the full equation of the perpendicular line is

$$y = -\frac{5}{2}x + \frac{19}{2}$$

31.C

remote interior angles

A polygon has one **interior angle** at each vertex. A polygon has two **exterior angles** at each vertex. An exterior angle is formed by extending one side of the polygon. The triangle shown on the left has three interior angles named *A*, *B*, and *C*.

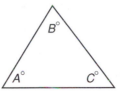
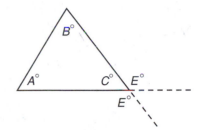

In the figure on the right, we extend the sides and form two equal exterior angles at *C*. The exterior angles are equal angles (have equal measures) because they are vertical angles. We can prove that the measure of each of these exterior angles equals *A* + *B*. Because *A* and *B* are on the other side of the triangle from the angle labeled *E*, these interior angles are called **remote interior angles.**

example 31.2   Prove that the measure of an external angle of a triangle equals the sum of the measures of the remote interior angles.

solution   We begin by drawing a triangle whose interior angles measure *A*°, *B*°, and *C*°. We draw an external angle at *C* and label it *E*.

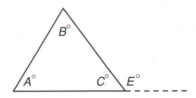

We remember that we must justify every step in a proof. We know that the sum of the measures of the angles in a triangle is 180. So

$$A + B + C = 180 \qquad 180° \text{ in a triangle}$$

We observe that angles *C* and *E* form a straight angle. So

$$C + E = 180 \qquad \text{straight angle}$$

Euclid's sixth postulate tells us that things that are equal to the same thing are equal to each other. This lets us write

$$A + B + C = C + E \qquad \text{equated equals}$$

Euclid's subtraction postulate tells us that if equals are subtracted from equals the results are equal. Thus, we subtract *C* from both sides.

$$
\begin{array}{rcl}
A + B + C & = & C + E \\
-\,C & = & -C \qquad \text{subtract equals} \\
\hline
A + B & = & E \qquad \text{QED}
\end{array}
$$

The letters **QED** are the first letters of the Latin words *quod erat demonstrandum,* which means **that which was to have been demonstrated.**

practice   Write the equation of a line that is perpendicular to $y = \frac{1}{4}x - 2$ and passes through (1, 2).

problem set 31

1.   The general expression for consecutive multiples of 5 is 5*N*, 5(*N* + 1), 5(*N* + 2), etc., where *N* is some unspecified integer. Find four consecutive multiples of 5 such that 6 times the first is 40 greater than 2 times the sum of the second and the fourth.

2. The bus headed north at 40 mph at 10 a.m. At noon the *Orange Blossom Special* headed south from the same station at 70 mph. What time was it when the train and the bus were 960 miles apart?

3. Susie jogged to the farm at 6 mph and rode back home in a truck traveling at 30 mph. How far was it to the farm if the entire trip took 12 hours?

4. Pansy plants were $4 a crate and tomato plants were $6 a crate. Sowega bought 70 crates for Patsy and spent $360. How many crates of each kind did he buy?

5. Ninety percent of the nitrogen combined. If 1200 kilograms did not combine, what was the total weight of the nitrogen?

6. Forty grams of potassium combined with 1400 grams of other elements to form the compound. If 4320 grams of the compound were needed, how many grams of potassium were required?

7. Write the equation of the line that is perpendicular to $y = \frac{1}{3}x - 1$ and passes through the point $(2, -3)$.

Simplify:

8. $\dfrac{(0.0006 \times 10^{-42})(2000 \times 10^{-4})}{0.004 \times 10^{-13}}$

9. $\dfrac{\frac{a}{b}}{\frac{a+b}{b}}$

10. $\dfrac{\frac{4}{x+y}}{\frac{m}{x+y}}$

11. $\dfrac{3}{2\sqrt{5}}$

12. $\dfrac{7}{3\sqrt{2}}$

13. $5\sqrt{75} - 3\sqrt{300} + 2\sqrt{27}$ 14. $3\sqrt{2}(5\sqrt{2} - 4\sqrt{6})$ 15. $\dfrac{x + 4x^2}{x}$

Add:

16. $\dfrac{x}{x+2} + \dfrac{3+x}{x^2 + 4x + 4}$

17. $\dfrac{x}{x-3} + \dfrac{2x}{x^2 - 3x}$

Factor completely. Always factor the GCF as the first step.

18. $-x^3 + 5x^2 - 6x$

19. $2ax^3 - 18ax^2 + 40ax$

20. $-3pax + pax^2 + 2pa$

21. $-10mc + 3mxc + mx^2c$

Solve:

22. $\dfrac{2x+3}{6} - \dfrac{x}{2} = 1$

23. $\dfrac{3x+2}{3} - \dfrac{2}{5} = \dfrac{x+2}{6}$

24. Solve by graphing and then get an exact solution by using either substitution or elimination.
$$\begin{cases} 2x + 3y = 18 \\ -12x + 6y = -18 \end{cases}$$

25. Divide: $x^4 - 2$ by $x + 1$

26. Use elimination to solve:
$$\begin{cases} 6x + 4y = 11 \\ 2x - 3y = -5 \end{cases}$$

27. Find c.

28. Solve: $-2(-x^0 - 3) + 4(-x - 5^0) = -3^0(2x - 5)$

29. Find A, B, C, D, E, F, and P.

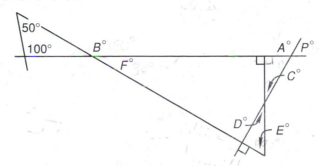

30. The sum of the areas of the four circles is 36π m². What is the area of one circle? What is the diameter of one circle? What is the area of the square?

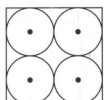

LESSON 32   *Quotient theorem for square roots • Congruency • Congruent triangles*

32.A

quotient theorem for square roots

We remember that the square root of a product can be written as the product of the square roots of its factors. Thus,

$$\sqrt{3 \cdot 2} = \sqrt{3}\sqrt{2}$$

A similar rule applies to the square root of a quotient (fraction), for the square root of a quotient can be written as a quotient of square roots. Thus,

$$\sqrt{\frac{3}{2}} = \frac{\sqrt{3}}{\sqrt{2}}$$

It is customary to rationalize the denominators of expressions that have radicals in the denominator. In this expression, we can rationalize the denominator by multiplying by $\sqrt{2}$ over $\sqrt{2}$. This fraction has a value of 1 and the multiplication changes the denominator from the irrational number $\sqrt{2}$ to the rational number 2.

$$\frac{\sqrt{3}}{\sqrt{2}} = \frac{\sqrt{3}}{\sqrt{2}}\frac{\sqrt{2}}{\sqrt{2}} = \frac{\sqrt{6}}{2}$$

The expression $\frac{\sqrt{6}}{2}$ has the same value as $\sqrt{\frac{3}{2}}$ but is in a different form.

example 32.1   Simplify: $\sqrt{\frac{3}{7}}$

solution   First we will write the root of the quotient as the quotient of the roots.

$$\frac{\sqrt{3}}{\sqrt{7}}$$

We finish by rationalizing the denominator.

$$\frac{\sqrt{3}}{\sqrt{7}}\frac{\sqrt{7}}{\sqrt{7}} = \frac{\sqrt{21}}{7}$$

example 32.2   Simplify: $\sqrt{\dfrac{2}{5}} + \sqrt{\dfrac{5}{2}}$

solution   This problem was contrived so that when the terms are rationalized, each term will contain $\sqrt{10}$. Then by using a common denominator, the terms can be added. Problems like this one are good problems for practice with quotients of radicals, so they will appear often in the problem sets. We begin our simplification by writing each term as a quotient of radicals.

$$\frac{\sqrt{2}}{\sqrt{5}} + \frac{\sqrt{5}}{\sqrt{2}}$$

Next we rationalize both denominators.

$$\frac{\sqrt{2}}{\sqrt{5}}\frac{\sqrt{5}}{\sqrt{5}} + \frac{\sqrt{5}}{\sqrt{2}}\frac{\sqrt{2}}{\sqrt{2}}$$

$$= \frac{\sqrt{10}}{5} + \frac{\sqrt{10}}{2}$$

We finish by using 10 as a common denominator and adding.

$$\frac{\sqrt{10}}{5}\left(\frac{2}{2}\right) + \frac{\sqrt{10}}{2}\left(\frac{5}{5}\right)$$

$$= \frac{2\sqrt{10}}{10} + \frac{5\sqrt{10}}{10}$$

$$= \frac{7\sqrt{10}}{10}$$

example 32.3   Simplify: $3\sqrt{\dfrac{3}{7}} - 5\sqrt{\dfrac{7}{3}}$

solution   We begin by writing each radical as a quotient of radicals.

$$\frac{3\sqrt{3}}{\sqrt{7}} - \frac{5\sqrt{7}}{\sqrt{3}}$$

Next we rationalize both denominators.

$$\frac{3\sqrt{3}}{\sqrt{7}}\frac{\sqrt{7}}{\sqrt{7}} - \frac{5\sqrt{7}}{\sqrt{3}}\frac{\sqrt{3}}{\sqrt{3}}$$

$$= \frac{3\sqrt{21}}{7} - \frac{5\sqrt{21}}{3}$$

We finish by changing each expression so that the denominators are 21. Then we add.

$$\frac{3\sqrt{21}}{7}\left(\frac{3}{3}\right) - \frac{5\sqrt{21}}{3}\left(\frac{7}{7}\right)$$

$$= \frac{9\sqrt{21}}{21} - \frac{35\sqrt{21}}{21}$$

$$= -\frac{26\sqrt{21}}{21}$$

example 32.4   Simplify: $2\sqrt{\dfrac{2}{7}} - 5\sqrt{\dfrac{7}{2}}$

solution   First we write each radical as a quotient of radicals.

$$\frac{2\sqrt{2}}{\sqrt{7}} - \frac{5\sqrt{7}}{\sqrt{2}}$$

Now we rationalize both denominators.

$$\frac{2\sqrt{2}}{\sqrt{7}}\frac{\sqrt{7}}{\sqrt{7}} - \frac{5\sqrt{7}}{\sqrt{2}}\frac{\sqrt{2}}{\sqrt{2}}$$

$$= \frac{2\sqrt{14}}{7} - \frac{5\sqrt{14}}{2}$$

Finally we find the common denominator and add.

$$\frac{2\sqrt{14}}{7}\left(\frac{2}{2}\right) - \frac{5\sqrt{14}}{2}\left(\frac{7}{7}\right)$$

$$= \frac{4\sqrt{14}}{14} - \frac{35\sqrt{14}}{14}$$

$$= -\frac{31\sqrt{14}}{14}$$

32.B
congruency

In mathematics only numbers are equal. The equation

$$4 = 2 + 2$$

is a true equation because 4 is a numeral that represents the number 4 and $2 + 2$ is another numeral that represents the same number. The two line segments shown here are 4 units long.

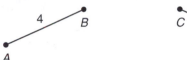

We say that these line segments are **congruent,** which means that the numbers used to describe their lengths are equal.

$$AB = CD \qquad \overline{AB} \cong \overline{CD}$$

The notation on the left is read, "*AB* is equal to *CD*." The notation on the right uses the symbol for congruent and is read, "Segment *AB* is congruent to segment *CD*."

The numbers used to describe the measures of the angles shown here are equal, so we say that the angles are congruent.

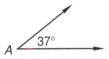

Thus, we can write any of the following.

$$\angle A = \angle B \qquad m\angle A = m\angle B \qquad \angle A \cong \angle B$$

If the measures of the angles in one polygon are equal to the measures of the angles in another polygon, and the sides opposite equal angles have equal lengths, then the scale factor is 1 and the polygons are congruent. Thus, these two triangles are congruent.

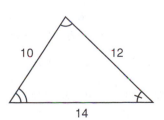 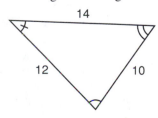

Thus, we see that *congruent* **is a word that means geometrically equal.** If we can mentally cut out one geometric figure, rotate it or flip it as necessary, and place it on another geometric figure so that it fits exactly, the two figures are congruent.

To place one geometric figure on top of a second figure, we can use translation, rotation, or "flipping." When we translate a figure, we slide it sideways as necessary, being careful not to rotate it.

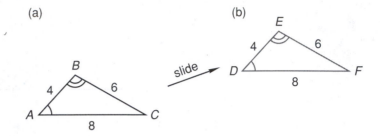

If we translate (slide) triangle (a) to the right and up, it will fit on top of triangle (b). Thus, the triangles are congruent. **When we write the statement of congruency, we are careful to list vertices whose angles are equal in the same order.**

$$\triangle ABC \cong \triangle DEF$$

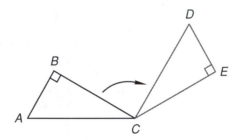

If we rotate triangle *ABC* about *C*, it will fit exactly on top of triangle *DEC*. So the triangles are congruent.

$$\triangle ABC \cong \triangle DEC$$

Sometimes it is necessary to flip a triangle to make it fit. We call a flip a **reflection** about a line.

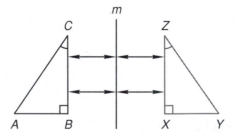

Point *X* is the same perpendicular distance from line *m* as point *B*. Point *Z* is the same perpendicular distance from line *m* as is point *C*. Point *Y* is the same perpendicular distance from line *m* as is point *A*. We say that $\triangle XYZ$ is the reflection of $\triangle BAC$ in line *m*. This means that $\triangle XYZ$ was "flipped" across line *m*. Thus,

$$\triangle ABC \cong \triangle YXZ$$

Note that corresponding vertices are listed in the same order. Sometimes a geometric figure must be translated, rotated, and flipped to place it on another geometric figure. If we do what is necessary and if the figure fits exactly, the two figures are congruent.

32.C

**congruent
triangles**

Congruent triangles are similar triangles whose scale factor is 1. Congruent means geometrically equal, so the angles in a pair of congruent triangles have equal measures and the sides opposite these equal angles have equal lengths. We can make this statement in one sentence by saying that corresponding parts of congruent triangles are congruent. We can abbreviate this statement by using the first letter of each word.

CPCTC means corresponding parts of congruent triangles are congruent.

example 32.5   Find x and p.

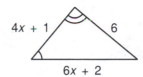 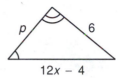

solution   We remember that if two angles are equal the third angles are equal, so the triangles are similar by AAA. The sides opposite the single-tick-marked angles are both 6 units long, so the scale factor is 1 and these triangles are congruent. Because a pair of corresponding sides have equal lengths, we add the letter S to AAA and say that the triangles are congruent by AAAS. Thus, the sides opposite the double-tick-marked angles also have equal lengths.

$$6x + 2 = 12x - 4 \qquad \text{equal lengths}$$

$$6 = 6x \qquad \text{simplified}$$

$$1 = x \qquad \text{divided}$$

If we replace x with 1, we can find that the side $4x + 1$ is 5 units long.

$$4(1) + 1 = 4 + 1 = 5$$

Since side p and the side labeled $4x + 1$ are corresponding sides, side p is also 5 units long. If we use 1 for x, we find that the sides labeled $12x - 4$ and $6x + 2$ are 8 units long.

$$12x - 4 \qquad\qquad 6x + 2$$

$$12(1) - 4 = 8 \qquad 6(1) + 2 = 8$$

$$\text{so} \qquad p = 5$$

practice   **a.** $\sqrt{\dfrac{3}{5}} + \sqrt{\dfrac{5}{3}}$ **b.** $2\sqrt{\dfrac{2}{7}} - 3\sqrt{\dfrac{7}{2}}$

c.   The two triangles shown here are similar by AAA. Two corresponding sides have a length of 4, so the scale factor is 1 and the triangles are congruent by AAAS. Find x. Then find p.

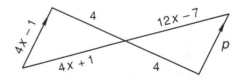

*problem set
32*

1.   Selby walked for a while at 4 mph and then jogged the rest of the way at 8 mph. If she covered 56 miles in 10 hours, how far did she walk and how far did she jog?

2.   Johnny found that the large ones totaled 30 more than 3 times the number of small ones. The ratio of the number of small ones to the number of large ones was 1 to 6. How many were large, and how many were small?

3.   Bruce discovered that the wishing well contained $9 in nickels and dimes, and that there were 30 more dimes than nickels. How many coins of each type were there?

4.   Sarah knew that forty percent of the mixture was calcium. If 300 kilograms of other elements was used, what was the total weight of the mixture?

5. Zollie noted that 4 grams of magnesium combined with 20 grams of the other elements to form the compound. If 1440 grams of the compound were required, how many grams of the other elements were required?

6. Hedonism was pandemic as 0.87 of the students were hedonists. If 1914 students were hedonists, how many students were there in all?

Simplify:

7. $\sqrt{\dfrac{2}{3}} + \sqrt{\dfrac{3}{2}}$

8. $2\sqrt{\dfrac{5}{7}} - 3\sqrt{\dfrac{7}{5}}$

9. $2\sqrt{\dfrac{3}{5}} - 5\sqrt{\dfrac{5}{3}}$

10. Each of the four circles is equal in area. The area of the square is 64 m². What is the length of one side? What is the radius of a circle? What is the area of one circle?

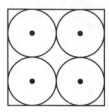

11. If A is the measure of an angle, $90 - A$ is the measure of the complement of the angle, and $180 - A$ is the measure of the supplement of the angle. Find an angle such that the supplement of the angle is 30 less than 3 times the complement of the angle.

12. Find the equation of the line that passes through the point (2, 2) and is perpendicular to the line $y = 3x - 5$.

Simplify:

13. $\dfrac{(0.0035 \times 10^{15})(0.002 \times 10^{17})}{7000 \times 10^{33}}$

14. $\dfrac{\dfrac{x + 4y}{y}}{\dfrac{x + y}{y}}$

15. $\dfrac{xy + 4x^2y^2}{xy}$

16. $3\sqrt{125} - 2\sqrt{45} + 3\sqrt{20}$

17. $4\sqrt{5}(2\sqrt{10} - 3\sqrt{5})$

Add:

18. $\dfrac{x}{x + 3} - \dfrac{2x - 2}{x^2 + 5x + 6}$

19. $\dfrac{m}{m - 5} - \dfrac{2}{m^2 - 5m}$

Factor completely. Always factor the GCF as the first step.

20. $-2x^3 + 8x^2 - 6x$

21. $-14x^3 + 5x^4 + x^5$

22. $7ax + ax^3 - 8ax^2$

23. $-12py + px^2y + 4xpy$

Solve:

24. $\dfrac{x + 2}{5} - \dfrac{3x - 3}{2} = 4$

25. $\dfrac{x}{2} - \dfrac{3x + 2}{4} = 7$

26. Solve by graphing and then get an exact solution by using either substitution or elimination.
$$\begin{cases} -x + 2y = 4 \\ x + y = -2 \end{cases}$$

27. Multiply: $(4x + 2)(x^3 - 2x + 4)$

28. Use substitution to solve: $\begin{cases} x + 2y = 5 \\ 3x - y = 7 \end{cases}$

29. Find the distance between $(-3, -2)$ and $(4, 5)$.

30. Simplify: $-3[(-2^0 - \overline{3^0} - 2)(-3^0) - 2^2] - |3 - 2^0|$

LESSON 33   *Major rules of algebra • Complex fractions*

33.A
major rules of algebra

There are three major rules for algebraic manipulations. Two of these rules require the presence of an equals sign for their use because these are the two rules we use for solving equations.

1. **The same quantity can be added to both sides of an equation without changing the solution to the equation.**
2. **Every term on both sides of an equation can be multiplied or divided by the same quantity (except zero) without changing the solution to the equation.**

$$x + 2 = 4 \qquad \frac{x}{3} = 5$$

We can use the first rule to solve the equation on the left and use the second rule to solve the equation on the right.

$$
\begin{array}{ll}
x + 2 = 4 & (3)\frac{x}{3} = (3)(5) \\
\underline{ - 2 = -2} & \\
x = 2 & x = 15
\end{array}
$$

These two rules cannot always be used, for if we have either of the expressions

$$\frac{4}{x + y} + \frac{xy}{x} \qquad \text{or} \qquad \frac{\dfrac{a}{x + y}}{\dfrac{b}{x}}$$

we cannot use the rules for solving equations since neither of these expressions contains an equals sign, and thus neither one is an equation.

When there is no equals sign in the expression, there is only one major rule that we can use, which is

> DENOMINATOR-NUMERATOR SAME-QUANTITY RULE

3. **The denominator and the numerator of an expression can be multiplied or divided by the same quantity (except zero) without changing the value of the expression. Only the form of the expression is changed.**

This is possibly the most important rule in all algebra and can be used even when there is no equals sign present. **When there is no equals sign present, as in the expression**

$$\frac{4}{x + y} + \frac{py}{x}$$

we cannot eliminate the denominators! All we can do is make the denominators the same so that the two terms can be added.

$$\frac{4}{x + y} + \frac{py}{x} = \frac{4}{x + y}\left(\frac{x}{x}\right) + \frac{py}{x}\left(\frac{x + y}{x + y}\right) = \frac{4x}{x(x + y)} + \frac{py(x + y)}{x(x + y)}$$

Now the terms can be added because the denominators are equal.

$$\frac{4x}{x(x + y)} + \frac{py(x + y)}{x(x + y)} = \frac{4x + py(x + y)}{x(x + y)}$$

Of course, either or both of the expressions in the numerator or denominator that contain parentheses can be multiplied out if we wish. If we do this, we get

$$\frac{4x + pxy + py^2}{x^2 + xy}$$

The denominator-numerator same-quantity rule can also be used to simplify the expression

$$\frac{\dfrac{a}{x + y}}{\dfrac{b}{x}}$$

We cannot eliminate the denominators because there is no equals sign in the expression. Therefore, the expression is not an equation, and the rules for equations cannot be used. But we can always (even in equations) use the denominator-numerator same-quantity rule.

Thus we can simplify this expression by multiplying both the numerator and the denominator by $\frac{x}{b}$, **which is the reciprocal of** $\frac{b}{x}$.

$$\frac{\dfrac{a}{x + y} \cdot \dfrac{x}{b}}{\dfrac{b}{x} \cdot \dfrac{x}{b}} = \frac{ax}{b(x + y)}$$

We did not eliminate the denominator, but now we have a simpler expression because we have only one fraction instead of one fraction divided by another fraction.

33.B

complex fractions

Fractions of fractions, such as the one just simplified, are called **complex fractions.** We also use these words to describe expressions such as

$$\frac{\dfrac{x}{y} + \dfrac{4}{yx}}{\dfrac{a}{y}}$$

This expression has a numerator composed of the sum of two fractions and a denominator that has only one fraction. **We will define complex fractions to be fractions that contain more than one fraction line.**

example 33.1   Simplify: $\dfrac{\dfrac{x}{y} + \dfrac{4}{yx}}{\dfrac{a}{y}}$

solution   **There is no equals sign, so we cannot use either of the rules for equations. The only rule that we can use is the denominator-numerator same-quantity rule.** We use it first to add the two terms in the numerator.

$$\frac{\dfrac{x}{y} + \dfrac{4}{yx}}{\dfrac{a}{y}} = \frac{\dfrac{x}{y}\left(\dfrac{x}{x}\right) + \dfrac{4}{yx}}{\dfrac{a}{y}}$$

$$= \frac{\dfrac{x^2}{xy} + \dfrac{4}{yx}}{\dfrac{a}{y}}$$

$$= \frac{\dfrac{x^2 + 4}{xy}}{\dfrac{a}{y}}$$

Now we will use the same rule again to multiply above and below by $\frac{y}{a}$, which is the reciprocal of the denominator, $\frac{a}{y}$.

$$\frac{\dfrac{(x^2 + 4)}{xy} \cdot \dfrac{y}{a}}{\dfrac{a}{y} \cdot \dfrac{y}{a}} = \frac{x^2 + 4}{xa}$$

example 33.2   Simplify: $\dfrac{\dfrac{a}{x + y} + \dfrac{m}{y}}{\dfrac{x}{a + m}}$

solution   **Again we note that an equals sign is not present. Thus, the only rule we can use is the denominator-numerator same-quantity rule.** First, we use it to help us add the two terms in the numerator.

$$\frac{\dfrac{a}{x + y} + \dfrac{m}{y}}{\dfrac{x}{a + m}} = \frac{\dfrac{a}{x + y}\left(\dfrac{y}{y}\right) + \dfrac{m}{y}\left(\dfrac{x + y}{x + y}\right)}{\dfrac{x}{a + m}}$$

$$= \frac{\dfrac{ay + mx + my}{y(x + y)}}{\dfrac{x}{a + m}}$$

Now we finish by using the same rule again. We multiply above and below by $\frac{a + m}{x}$, which is the reciprocal of $\frac{x}{a + m}$.

$$\frac{\dfrac{ay + mx + my}{y(x + y)} \cdot \left(\dfrac{a + m}{x}\right)}{\dfrac{x}{a + m} \cdot \left(\dfrac{a + m}{x}\right)} = \frac{(ay + mx + my)(a + m)}{xy(x + y)}$$

This answer is a little complicated but is closer to real-life answers than the answers to problems that are carefully contrived so that a lot of terms can be canceled.

practice   Simplify:

a. $\dfrac{\dfrac{m}{p} + \dfrac{3}{zp}}{\dfrac{s}{p}}$

b. $\dfrac{\dfrac{y}{a + m} + \dfrac{s}{m}}{\dfrac{a}{y + s}}$

problem set 33

1.   Four-seventeenths of the hedonists were also sybarites. If 104 were sybarites, how many hedonists were there in all?

2.   Mercury and phosphate were mixed in the ratio of 7 to 2. If 3600 grams of the mixture was required, how much mercury was needed?

3.   Twenty percent of the lithium did not combine. If 1620 grams did combine, how much lithium was there in all?

4.   There were 10 more reds than 8 times the number of blues. Also, the number of reds was 5 less than 11 times the number of blues. How many of each were there?

5.   The total distance was 540 miles. Part of the journey was on a motorcycle at 40 mph and part was in a car at 60 mph. What distance was covered by motorcycle if the total time of the journey was 11 hours?

Simplify:

6. $$\dfrac{\dfrac{m}{p} + \dfrac{3}{xp}}{\dfrac{y}{p}}$$

7. $$\dfrac{\dfrac{s}{a+b} + \dfrac{x}{b}}{\dfrac{a}{s+x}}$$

8. The central angle is 40° as shown. Find x, y, and z. If the radius of the circle is 6 cm, find the length of arc z.

9. Find x, y, and z. Remember that the measure of an arc is twice the measure of the inscribed angle.

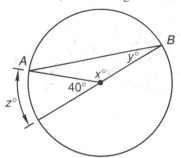

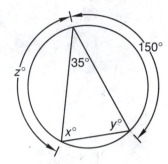

10. Find the equation of the line that is perpendicular to $y = -\frac{3}{5}x + 2$ and passes through the point $(-4, 2)$.

Simplify:

11. $3\sqrt{\dfrac{5}{2}} - 2\sqrt{\dfrac{2}{5}}$

12. $4\sqrt{\dfrac{5}{6}} - 2\sqrt{\dfrac{6}{5}}$

13. Find the perimeter of the isosceles triangle in Problem 8 if chord AB is 11.28 cm.

14. Find an angle such that 3 times the complement of the angle is 50° greater than the supplement of the angle.

Simplify:

15. $\dfrac{(0.0027 \times 10^{15})(500 \times 10^{-20})}{900 \times 10^{14}}$

16. $\dfrac{x + \dfrac{4xy}{x}}{\dfrac{1}{x} - y}$

17. $3\sqrt{18} + 2\sqrt{50} - \sqrt{98}$

18. $\dfrac{4x + 4xy}{4x}$

Add:

19. $\dfrac{a}{x(x+y)} + \dfrac{b}{x^2} + \dfrac{cx+4}{x+y}$

20. $\dfrac{4}{x+4} - \dfrac{6x-2}{x^2+2x-8}$

Factor completely. Always factor the GCF as the first step.

21. $5x^2 + 4x^3 - x^4$

22. $10k^2 - 7k^2x + k^2x^2$

23. $apx^2 - 20ap - apx$

Solve:

24. $\dfrac{x+2}{3} - \dfrac{2x-2}{4} = 5$

25. $\dfrac{3x-2}{2} - \dfrac{2x+3}{3} = 4$

26. Solve by graphing and then get an exact solution by using either substitution or elimination.
$$\begin{cases} 2x - y = -5 \\ x + y = 1 \end{cases}$$

27. Divide $2x^4 - x$ by $x - 2$ and check.

28. Evaluate: $-3^0 - x - y^0 - y^2(-2^0 - 3) - |-2 - x|$ if $x = -2$ and $y = -3$

29. Find the perimeter of a rectangle that measures 2 ft by 4 ft.

30. Find the area of the rectangle of Problem 29.

LESSON 34 *Uniform motion problems:* $D_1 + k = D_2$

The distance diagrams for the uniform motion problems encountered thus far have looked like one of the following:

(a)   (b)

In both diagrams in (a), the distance traveled by number 1 equals the distance traveled by number 2, so the distance equation for both diagrams is

$$R_1T_1 = R_2T_2$$

In all three diagrams in (b), the distance traveled by number 1 plus the distance traveled by number 2 equals 480, so the distance equation for all three diagrams is

$$R_1T_1 + R_2T_2 = 480$$

In some uniform motion problems, the distance traveled by one object exceeds by a specified amount the distance traveled by another object. The diagram for one of these problems usually looks like one of the following:

In the diagram on the left, both A and B began at the same point. For some reason, B traveled 40 more miles (or kilometers or whatever) than A. In the diagram on the right, A started 40 units in front of B, but they ended up at the same place. In both diagrams, the sum of 40 and the distance traveled by A equals the distance traveled by B.

$$\text{Distance } A + 40 = \text{distance } B$$

Now since rate times time equals distance, we can write the distance equation for both diagrams as

$$R_AT_A + 40 = R_BT_B$$

example 34.1   Millicent began the journey at 6 a.m. at 50 kilometers per hour. Beauregard began to chase her at 10 a.m. at 60 kilometers per hour. What time was it when Beauregard got within 40 kilometers of Millicent?

solution   Beauregard and Millicent began at the same point, but Millicent traveled 40 kilometers farther than Beauregard traveled. Thus, the distance diagram and distance equation are as follows:

$$R_BT_B + 40 = R_MT_M$$

We have one equation in four unknowns. Thus we need three more equations. They are

$$R_M = 50 \qquad R_B = 60 \qquad T_M = T_B + 4$$

Now we substitute these equivalences into the distance equation and solve for T_B.

$$60T_B + 40 = 50(T_B + 4) \quad \text{substituted}$$
$$60T_B + 40 = 50T_B + 200 \quad \text{multiplied}$$
$$10T_B = 160 \quad \text{simplified}$$
$$T_B = 16 \quad \text{divided}$$

Thus, in 16 hours, Beauregard got within 40 kilometers of Millicent. Sixteen hours after 10 a.m. would be **2 o'clock the next morning**.

example 34.2 When the sheriff began his pursuit, Robin Hood was already 7 miles out of Nottingham. If the sheriff traveled at 6 miles per hour while Robin Hood's rate was $2\frac{1}{2}$ miles per hour, how long did it take the sheriff to catch up?

solution The distance diagram, the distance equation, and the rate equations are

$$R_R T_R + 7 = R_S T_S \qquad R_S = 6 \qquad R_R = 2\frac{1}{2}$$

When the problem began, Robin was already 7 miles out of town (how he got there or when is not part of this problem). Thus Robin and the sheriff began traveling at the same time and stopped at the same time, so the time equation is

$$T_S = T_R$$

Next we substitute for R_R, R_S, and T_S in the distance equation.

$$\frac{5}{2}T_R + 7 = 6T_R$$

We substituted $\frac{5}{2}$ for R_R, 6 for R_S, and T_R for T_S. We finish by eliminating the denominator by multiplying every term on both sides by 2.

$$5T_R + 14 = 12T_R \quad \text{multiplied by 2}$$
$$14 = 7T_R \quad \text{simplified}$$
$$2 = T_R \quad \text{divided}$$

Thus, the sheriff caught Robin in **2 hours** because his traveling time equaled that of Robin.

practice Zane began the journey at 5 a.m. at 30 kilometers per hour. Tricia began to chase him at 7 a.m. at 40 kilometers per hour. What time was it when Tricia got within 20 kilometers of Zane?

problem set 34

1. Elliot began the journey at 8 a.m. at 40 kilometers per hour. Benita began to chase him at 11 a.m. at 50 kilometers per hour. What time was it when Benita got within 60 kilometers of Elliot?

2. Roland had already covered 14 kilometers when Charlemagne headed out to catch him Charlemagne's rate was 11 kph, whereas Roland's rate was $7\frac{1}{2}$ kph. How long did it take Charlemagne to catch Roland?

3. Kay rode the bicycle into the country at 10 mph, and Yancy pushed it back to town at 3 mph. If the round trip took 13 hours, how far did Kay ride the bicycle into the country?

4. Thirty percent of the sulfur desiccated. If 42 tons did not desiccate, how much sulfur was there in all?

5. Fifty grams of sodium bicarbonate was mixed with other compounds to get 150 grams of mixture. If 300 grams of the other compounds was available, how much sodium bicarbonate was needed to make the mixture?

6. Find three consecutive integers such that 5 times the sum of the first and third is 14 greater than 8 times the second.

Simplify:

7. $\dfrac{\dfrac{y}{ab} - ab}{\dfrac{1}{a} - \dfrac{a}{b}}$

8. $\dfrac{\dfrac{1}{x} - b}{x}$

9. The shape shown is the base of a cone that is 8 meters tall. Find the volume of the cone. All angles are right angles.

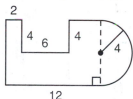

10. Find x.

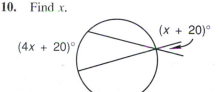

11. Find the equation of the line that is perpendicular to $2x + y = 4$ and passes through the point $(-2, -1)$.

Simplify:

12. $2\sqrt{\dfrac{2}{9}} - 3\sqrt{\dfrac{9}{2}}$

13. $-3\sqrt{\dfrac{2}{3}} + 2\sqrt{\dfrac{3}{2}}$

14. How many 1-cm-square floor tiles will it take to completely cover the shaded area in the figure shown? Dimensions are in centimeters.

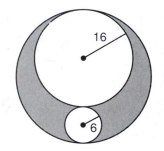

Simplify:

15. $\dfrac{(0.000032 \times 10^4)(700 \times 10^{-14})}{16,000}$

16. $\dfrac{4 + \dfrac{x}{y^2}}{3 - \dfrac{1}{y^2}}$

17. $8\sqrt{27} - 2\sqrt{75} + 2\sqrt{147}$

18. $\dfrac{x^2 y - 5x^2 y^2}{x^2 y}$

Add:

19. $\dfrac{a}{x(x + y)} + \dfrac{bx}{x^2(x + y)} + \dfrac{cx}{x^3}$

20. $\dfrac{x - 4}{x - 3} - \dfrac{2x - 1}{x^2 - 6x + 9}$

Factor completely. Always begin by factoring the GCF.

21. $-4x^2 + 2x^3 + 2x^4$ **22.** $ax^2 p - 8pa - 2axp$ **23.** $yx^2 - 4xy + 4y$

Solve:

24. $\dfrac{x - 3}{2} - \dfrac{3x + 4}{2} = 3$

25. $\dfrac{x}{3} - \dfrac{2x - 4}{2} = 5$

26. Solve by graphing and then get an exact solution by using either substitution or elimination.

$$\begin{cases} x - 3y = 6 \\ 2x + y = 2 \end{cases}$$

27. Multiply: $(x^2 + x)(x^2 + 2x + 3)$

28. Use substitution to solve: $\begin{cases} 3x - 2y = 2 \\ x - 3y = 4 \end{cases}$

29. Find the area of this isosceles triangle. Dimensions are in centimeters.

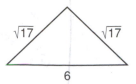

30. Evaluate: $-2^0[-2^0 - 2^2 - (-2)^3 - 2](-2^0 - 2) + x - xy$ if $x = -3$ and $y = -4$

LESSON 35   *Angles in polygons* • *Inscribed quadrilaterals* • *Fractional exponents*

35.A
angles in polygons

We remember that the sum of the interior angles in a triangle is 180°.

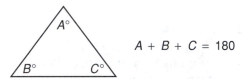

$$A + B + C = 180$$

A quadrilateral can be triangulated (divided into triangles) from any vertex into two triangles. Thus, the sum of the interior angles of any quadrilateral is 360°.

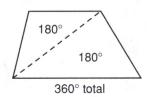

A pentagon is a 5-gon and can be triangulated from any vertex into three triangles. Thus, the sum of the interior angles of a pentagon is 540°.

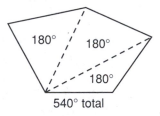

    To find the sum of the interior angles of any polygon, we triangulate the polygon from one vertex and multiply the number of triangles by 180°.

    It is poor practice to memorize a formula when the formula can be developed quickly. If we make a list, we can develop a formula for the sum of the interior angles of any polygon.

$$3 \text{ sides} \longrightarrow 1 \times 180 = 180°$$
$$4 \text{ sides} \longrightarrow 2 \times 180 = 360°$$
$$5 \text{ sides} \longrightarrow 3 \times 180 = 540°$$

Each time we add a side to the polygon, we can draw another triangle.

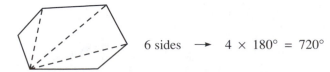

$$6 \text{ sides} \longrightarrow 4 \times 180° = 720°$$

From this we see that the sum of the measures of the interior angles of a polygon of *n* sides is

$$(n - 2) \times 180°$$

We can also find the sum of the measures of the exterior angles. **The sum of the measures of the exterior angles of any polygon is 360°.** We can see this clearly if we let our stick man walk around this polygon.

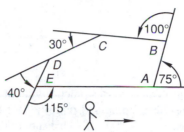

At *A,* he turns through an angle of 75°, and at *B* he turns through an angle of 100°. At *C, D,* and *E,* he turns through angles of 30°, 40°, and 115°. When he gets back to the starting point, he has completed a full circle, and so the sum of these five angles must be 360°. We can see that he will turn through 360° when he walks around any polygon, regardless of the number of sides the polygon has. This shows that the sum of the measures of the exterior angles of any polygon is 360°.

example 35.1   Find *x* and *y.*

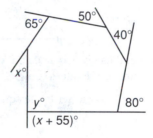

solution   The sum of the measures of the exterior angles of a polygon is 360°. So

$$x + (x + 55) + 80 + 40 + 50 + 65 = 360 \qquad \text{exterior angles} = 360°$$
$$2x + 290 = 360 \qquad \text{added}$$
$$2x = 70 \qquad \text{added} -290 \text{ to both sides}$$
$$x = 35$$

Since *x* + 55 plus *y* equals a straight angle, or 180°, we can find *y.*

$$x + 55 + y = 180 \qquad \text{equation}$$
$$35 + 55 + y = 180 \qquad \text{substituted}$$
$$y = 90 \qquad \text{solved}$$

35.B

inscribed quadrilaterals

If a quadrilateral is inscribed in a circle, the sum of the measures of any pair of opposite angles is 180°.

To see why this is true, we remember that the measure of an inscribed angle is half the measure of the intercepted arc. In the figure on the left, the measure of angle *A* is half the measure of arc *BCD.*

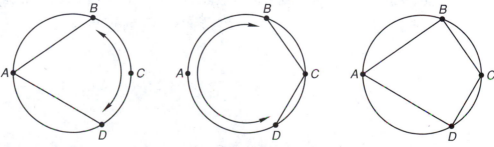

In the center figure, the measure of angle C is half the measure of arc BAD. Since the two arcs go all the way around the circle, the measure of the sum of the two arcs is 360°. Thus, the sum of the measures of angles A and C is half of 360°, or 180°.

example 35.2   Find x, y, and z.

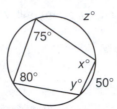

solution   This problem gives us practice with the fact that an intercepted arc has twice the measure of the inscribed angle and with the fact that the sum of two opposite angles in an inscribed quadrilateral is 180°. Also the sum of the measures of the interior angles of any quadrilateral is 360°. First we note that the arc of the 80° angle equals $z° + 50°$. Thus,

$$z + 50 = 2 \times 80 \qquad \text{arc}° = 2 \times \text{(inscribed angle)}°$$

$$z + 50 = 160 \qquad \text{multiplied}$$

$$z = 110 \qquad \text{solved}$$

Since opposite angles in an inscribed quadrilateral sum to 180°, we can solve for x and y.

$$x + 80 = 180 \qquad\qquad\qquad y + 75 = 180$$

$$\text{so} \quad x = 100 \qquad\qquad\qquad \text{so} \quad y = 105$$

35.C
fractional exponents

There are two ways to write the square root of 2.

$$\sqrt{2} \qquad \text{and} \qquad 2^{1/2}$$

There is nothing to understand, for this is a definition of what we mean when we use the square root radical sign or the fractional exponent $\frac{1}{2}$. We use the fractional exponent $\frac{1}{3}$ to designate the cube root, the fractional exponent $\frac{1}{4}$ to designate the fourth root, etc.

$$2^{1/2} = \sqrt{2} \qquad 2^{1/3} = \sqrt[3]{2} \qquad 2^{1/4} = \sqrt[4]{2}$$

The rules for exponents are the same for fractional exponents as they are for integral exponents. Thus, both the product theorem and power theorem also apply to fractional exponents.

$$x^{5/3} \cdot x^{8/3} = x^{13/3} \qquad \text{and} \qquad (x^{5/3})^{8/3} = x^{40/9}$$

example 35.3   Simplify: $4^{-1/2}$

solution   Negative exponents are not operational indicators, so our first step is to write the expression with a positive exponent. Then we simplify.

$$4^{-1/2} = \frac{1}{4^{1/2}} = \frac{1}{2}$$

example 35.4   Simplify: $-27^{-1/3}$

solution   If the minus sign is unprotected by a parenthesis, we may cover it up with a finger.

$$\text{\ulcorner}\text{D}\ 27^{-1/3}$$

Now we simplify $27^{-1/3}$.

$$\text{\ulcorner}\text{D}\ 27^{-1/3} = \text{\ulcorner}\text{D}\ \frac{1}{27^{1/3}} = \text{\ulcorner}\text{D}\ \frac{1}{3}$$

Now we remove our finger to uncover the minus sign and see that the answer is

$$-\frac{1}{3}$$

example 35.5   Simplify: $16^{3/2}$

solution   This expression can be simplified in two ways.

$$\text{(a)}\quad (16^3)^{1/2} \qquad \text{(b)}\quad (16^{1/2})^3$$

Both (a) and (b) have the same value as the original expression. Next we simplify both expressions within the parentheses and get

$$\text{(a)}\quad (4096)^{1/2} \qquad \text{(b)}\quad (4)^3$$

Both of these have a value of 64, but it is much easier to raise 4 to the third power than it is to take the square root of 4096. For this reason, it is recommended that the fraction always be left inside the parentheses when doing the first step of the simplification. Also we note that it is necessary to learn to recognize that these problems are contrived problems designed to give practice in working with fractional exponents. Not all expressions with fractional exponents can be simplified, for if we have

$$15^{3/2}$$

we can do nothing, for neither

$$(15^{1/2})^3 \qquad \text{nor} \qquad (15^3)^{1/2}$$

can be simplified without using logarithms or a calculator.

example 35.6   Simplify: $-8^{-2/3}$

solution   Since the minus sign is not "protected" by a parenthesis, we mentally cover up this sign with a finger and get

$$8^{-2/3}$$

This can be rewritten as

$$\frac{1}{8^{2/3}}$$

Next we write the exponent as a product that has the fraction inside the parentheses.

$$\frac{1}{(8^{1/3})^2}$$

Now we simplify

$$\frac{1}{2^2}$$

and bring back the minus sign to get

$$-\frac{1}{4}$$

practice   **a.**   Find the sum of the interior angles in a 14-gon.

b.   Find the sum of the exterior angles in a 14-gon.

Simplify:

c. $64^{-1/2}$ **d.** $64^{2/3}$ **e.** $-125^{-2/3}$

problem set 35

1. Matthew was 1200 yards ahead when Lowe began his pursuit. If Lowe ran 3 times as fast as Matthew and overtook him in 30 minutes, how fast did each boy run?

2. Cheryl and Judy trudged at 4 miles per hour until their packs got too heavy. Then they dropped their packs and continued at the brisk pace of 6 mph. If the total trip of 56 miles took 12 hours, how long did they trudge? How long did they walk briskly?

3. The dhow made the trip in 12 hours while the brigantine made the same trip in 4 hours. If the speed of the brigantine was 6 miles per hour greater than the speed of the dhow, what was the distance traveled by each?

4. Blues were $5 each and yellows cost $8 each. Penelope spent $82, and the number of blues she bought was 2 greater than twice the number of yellows. How many of each kind did she buy?

5. Sixty percent of the aluminum fused as it should have. If 40 tons did not fuse, how much aluminum was there in all?

6. Twenty grams of vanadium was melted with other metals to make 40 grams of the alloy. If 400 grams of the other metals was available, how much vanadium should be used?

7. The figure shown is the base of a right solid that is 10 meters high. Corners that look square are square. Find the lateral surface area of the solid. Dimensions are in meters.

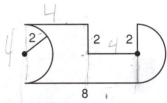

Simplify:

8. $16^{-1/2}$

9. $27^{-1/3}$

10. $9^{3/2}$

11. $-64^{-2/3}$

12. Find x, y, and A.

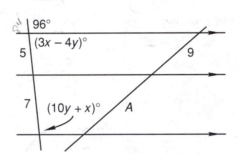

Simplify:

13. $3\sqrt{\dfrac{7}{5}} + 2\sqrt{\dfrac{5}{7}}$

14. $2\sqrt{\dfrac{2}{5}} - 9\sqrt{\dfrac{5}{2}}$

15. $\dfrac{\dfrac{a}{b} - 4}{\dfrac{xy}{b}}$

16. $\dfrac{\dfrac{x}{x+y} + 6}{\dfrac{4}{x+y}}$

17. Find x, y, and k.

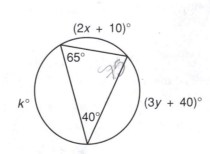

18. Find x. Then find y.

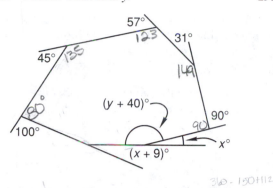

19. Find x, y, and p.

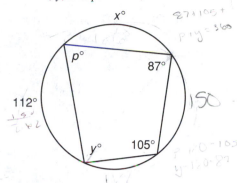

20. Find the equation of the line that passes through the points $(-2, 3)$ and $(4, -5)$.

Add:

21. $\dfrac{4}{x^2(x + y)} + \dfrac{2x - 2}{x(x + y)}$

22. $\dfrac{3x}{x - 2} - \dfrac{2x}{x^2 + x - 6}$

Factor completely:

23. $35a - ax^2 - 2xa$

24. $8x^2 - 2x^3 - x^4$

Simplify:

25. $2\sqrt{3} \cdot \sqrt{12} - 3\sqrt{2} \cdot \sqrt{6} + 4\sqrt{2}(3\sqrt{2} - \sqrt{6})$

26. $\dfrac{(7000 \times 10^{14})(0.0002 \times 10^{-11})}{1400 \times 10^{-10}}$

Solve:

27. $\dfrac{x - 3}{7} - \dfrac{2x}{4} = 5$

28. $0.002x = 0.02 + 0.04$

29. $2\dfrac{1}{3}x - 2x^0 = 3\dfrac{1}{4}$

30. Find the distance between $(-4, 2)$ and $(7, 3)$.

LESSON 36   *Contrived problems • Multiplication of rational expressions • Division of rational expressions*

36.A
contrived problems

In beginning algebra courses, we work at developing algebraic skills that can be used in advanced mathematics and in chemistry, physics, and other mathematically based disciplines. We certainly can find no way to apply immediately our skill of adding abstract expressions such as

$$\frac{a}{b} + \frac{c}{x} + \frac{b}{x + y}$$

but without this skill, many problems that will be encountered in advanced algebra and trigonometry would be difficult, if not impossible. In this lesson, we will investigate multiplication and division of factorable rational expressions that have common factors. These problems have no immediate application but are good problems for practicing factoring and for practicing canceling factors that are sums.

36.B

multiplication of rational expressions

We remember that fractions are multiplied by multiplying the numerators to form the new numerator and by multiplying the denominators to form the new denominator.

$$\frac{3}{5} \cdot \frac{7}{8} = \frac{21}{40} \qquad\qquad \frac{x}{y} \cdot \frac{m}{y + k} = \frac{xm}{y^2 + yk}$$

Another name for a fraction is a **ratio**, and we remember that this is the reason that we often call fractional expressions **rational expressions.**

example 36.1   Multiply: $\dfrac{x^2 + x - 12}{x^2 - x - 20} \cdot \dfrac{x^2 + 2x - 35}{x^2 + 9x + 14}$

solution   This is not a multiplication problem but is a contrived problem designed to provide practice in factoring and canceling. Thus, we will begin by factoring all four expressions, and then we will cancel the common factors.

$$\frac{\cancel{(x + 4)}(x - 3)}{\cancel{(x - 5)}\cancel{(x + 4)}} \cdot \frac{\cancel{(x + 7)}\cancel{(x - 5)}}{(x + 2)\cancel{(x + 7)}} = \frac{x - 3}{x + 2}$$

36.C

division of rational expressions

We remember that the denominator of a fraction of fractions can be changed to 1 by multiplying it by the reciprocal of the denominator. Thus,

$$\frac{\dfrac{x}{y}}{\dfrac{k}{p}}$$

can be simplified by multiplying above and below by $\frac{p}{k}$.

$$\frac{\dfrac{x}{y} \cdot \dfrac{p}{k}}{\dfrac{k}{p} \cdot \dfrac{p}{k}} = \frac{\dfrac{xp}{yk}}{1} = \frac{xp}{yk}$$

If the problem had been stated as

$$\frac{x}{y} \div \frac{k}{p}$$

the same result could be obtained by inverting the divisor and then multiplying.

$$\frac{x}{y} \div \frac{k}{p} = \frac{x}{y} \cdot \frac{p}{k} = \frac{xp}{yk}$$

This procedure will be used in the next example.

example 36.2   Simplify: $\dfrac{x^2 - 6x + 8}{x^2 + 3x - 28} \div \dfrac{x^2 - 2x - 15}{x^2 + 2x - 35}$

solution   This is also a contrived problem designed to give practice in factoring and canceling. We begin by inverting the divisor and changing the division symbol to a dot that indicates multiplication.

$$\frac{x^2 - 6x + 8}{x^2 + 3x - 28} \cdot \frac{x^2 + 2x - 35}{x^2 - 2x - 15}$$

We finish by factoring all four expressions and canceling, just as we did in the preceding example.

$$\frac{(x - 2)\cancel{(x - 4)}}{\cancel{(x + 7)}\cancel{(x - 4)}} \cdot \frac{\cancel{(x + 7)}\cancel{(x - 5)}}{\cancel{(x - 5)}(x + 3)} = \frac{x - 2}{x + 3}$$

example 36.3   Simplify: $\dfrac{x^2 + x - 6}{x^3 - 2x^2 - 35x} \div \dfrac{x + 3}{x^2 - 7x}$

solution   This time we will factor and invert the divisor in the same step. Then we finish by canceling common factors.

$$\frac{\cancel{(x+3)}(x - 2)}{x(x + 5)\cancel{(x-7)}} \cdot \frac{x\cancel{(x-7)}}{\cancel{(x+3)}} = \frac{x - 2}{x + 5}$$

practice   Simplify:

   a. $\dfrac{x^2 - 6x + 9}{x^2 + 5x - 24} \div \dfrac{x^2 - 5x + 6}{x^2 - x - 72}$ **b.** $\dfrac{x^2 + 2x - 8}{x^3 - 4x^2 - 21x} \div \dfrac{x - 2}{x^2 - 7x}$

problem set 36

1. The students whose phrasing was pleonastic used 240 percent more words than were necessary. If 400 words were necessary, how many words did these students use?

2. Fats cost 4 cents each and leans cost 21 cents each. Moxley and Rachel bought a total of 30 and spent \$2.90. How many of each kind did they buy?

3. The fraction had a value of $\frac{3}{5}$. The sum of the numerator and the denominator was 40. What was the fraction?

4. Don made the trip in 10 hours. Hazel drove 10 miles per hour faster than Don, so she made the trip in only 8 hours. How many miles long was the trip?

5. Brett and Julie headed north at 8 a.m. By noon, Brett was 80 miles ahead of Julie. What was Brett's speed if Julie's speed was 30 miles per hour?

6. Find four consecutive multiples of 7 such that 4 times the sum of the first multiple and 2 is 15 greater than 3 times the third.

7. Find x if the perimeter of the quadrilateral is 31 inches.

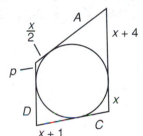

Simplify:

8. $\dfrac{x^2 + x - 20}{x^2 - x - 2} \cdot \dfrac{x^2 + 10x + 16}{x^2 + 3x - 40}$ 9. $\dfrac{x^2 - 3x - 18}{x^2 - 4x - 32} \div \dfrac{x^2 - 2x - 24}{x^2 - x - 20}$

10. $\dfrac{1}{-3^{-2}}$ 11. $-27^{-2/3}$ 12. $\dfrac{1}{81^{-3/4}}$ 13. $(-27)^{-2/3}$

14. $\dfrac{\dfrac{1}{x} + \dfrac{4}{y}}{3 + \dfrac{1}{xy}}$ 15. $\dfrac{\dfrac{4}{x} - 3}{\dfrac{7}{x} + 2}$ 16. $3\sqrt{\dfrac{5}{3}} - 2\sqrt{\dfrac{3}{5}}$

17. $\dfrac{(6000 \times 10^{14})(300 \times 10^{-22})}{0.00018 \times 10^{-5}}$ 18. $2\sqrt{\dfrac{7}{3}} - 3\sqrt{\dfrac{3}{7}}$

19. $4\sqrt{12}(3\sqrt{2} - 4\sqrt{3})$ 20. $2\sqrt{28} - 3\sqrt{63} + 2\sqrt{175}$

Add:

21. $\dfrac{6}{x^2(x + 2)} - \dfrac{3}{x^2 + 3x + 2}$ 22. $\dfrac{p}{ax^2} + \dfrac{cx + a}{ax^3} + \dfrac{mx + b}{a^2 x^4}$

Solve:

23. $\dfrac{3x - 2}{7} - \dfrac{x}{4} - \dfrac{x - 3}{2} = 1$

24. $3x^0 - 2(x - 3^0) - |-11 - 2| = 4x(2 - 5^0) - 7x$

25. Find the equations of lines (a) and (b).

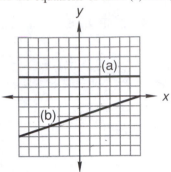

26. In the figure, X equals 140. Find M by first finding A, B, C, D, and E.

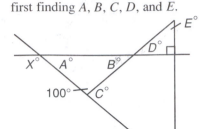

27. Divide $x^4 - 2$ by $x + 1$ and check.

28. Find the distance between $(-3, 7)$ and $(4, -2)$.

Simplify:

29. $\dfrac{x^{-2}y}{zx^4}\left(3x^6y^{-1}z - \dfrac{3x^{-3}y^2}{z^{-1}x^{-4}}\right)$

30. $-3^0(2 - 4^0) - |-3| - 2 - 4^2 - (-2)^3 - 2$

LESSON 37 *Chemical compounds • Parallelograms*

37.A
chemical compounds

In chemistry courses, some problems deal with the weight relationships of chemical compounds. Most of these problems are straightforward ratio and percent problems, and it is necessary to know only two things about chemistry to be able to work them.

The first thing is that atoms always combine in the same combinations when they unite to form a specific compound. The chemical formulas for the compounds tell us the number of each kind of atom in a molecule of the compound. For example,

H_2O is the formula for water. Each molecule of water has two hydrogen (H) atoms and one oxygen (O) atom.

Zn_3N_2 is the formula for zinc nitride. Each molecule of zinc nitride has three zinc (Zn) atoms and two nitrogen (N) atoms.

NaOH   is the formula for sodium hydroxide. Each molecule of sodium hydroxide has one sodium (Na) atom, one oxygen (O) atom, and one hydrogen (H) atom.

The second thing we need to know is that every kind of atom has a different weight. We express the weights of atoms in grams. The **gram atomic weight** of oxygen is 16 grams, and the gram atomic weight of hydrogen is 1 gram. We will give the gram atomic weights of the elements in parentheses. In a problem about oxygen and hydrogen, we would note the gram atomic weights as (O, 16; H, 1).

When elements combine to form a compound, we say that the molecules of the compound have a **gram molecular weight** that can be found from the gram atomic weights. Since the formula for water is

$$H_2O$$

we find that the gram molecular weight of water is

$$2(1 \text{ gram}) + 1(16 \text{ grams}) = 18 \text{ grams}$$

example 37.1   The chemical formula for water is H_2O. If we have 3600 grams of water, what is the weight of the oxygen? (O, 16; H, 1).

solution   We follow the same procedure we have been using for ratio problems. We note the weight of each element and the total weight. In a molecule of water the weights are

Hydrogen: $2 \times 1 = 2$

Oxygen: $1 \times 16 = 16$

Total: $= 18$

Thus, the three possible ratios are

$$(a) \quad \frac{H}{Ox} = \frac{2}{16} \qquad (b) \quad \frac{H}{Total} = \frac{2}{18} \qquad (c) \quad \frac{Ox}{Total} = \frac{16}{18}$$

The chemical symbol for oxygen is O, but we have used Ox so that O for oxygen won't be confused with 0 for zero. We have been told that the total weight was 3600 grams and were asked for the weight of the oxygen, so we will use ratio (c) and replace total with 3600.

$$\frac{Ox}{Total} = \frac{16}{18} \quad \longrightarrow \quad \frac{Ox}{3600} = \frac{16}{18}$$

To solve, we cross multiply and then divide both sides by 18.

$$18 \cdot Ox = 16 \cdot 3600 \quad \longrightarrow \quad \frac{18 \cdot Ox}{18} = \frac{16 \cdot 3600}{18} \quad \longrightarrow \quad \mathbf{Ox = 3200 \text{ grams}}$$

example 37.2   The chemical formula for ammonia is NH_3. This tells us that each ammonia molecule contains one nitrogen atom and three hydrogen atoms. If we have 510 grams of ammonia, how much does the nitrogen weigh? (H, 1; N, 14)

solution

Nitrogen: $14 \times 1 = 14$

Hydrogen: $1 \times 3 = 3$

Total: $= 17$

Thus, the three ratios are

$$(a) \quad \frac{N}{H} = \frac{14}{3} \qquad (b) \quad \frac{N}{T} = \frac{14}{17} \qquad (c) \quad \frac{H}{T} = \frac{3}{17}$$

We have been told that the total is 510 grams and have been asked for the weight of the nitrogen, so we will use (b). We will replace T with 510, cross multiply, and then divide by 17.

$$\frac{N}{T} = \frac{14}{17} \quad \longrightarrow \quad \frac{N}{510} = \frac{14}{17} \quad \longrightarrow \quad 17N = 14 \cdot 510 \quad \longrightarrow \quad \frac{\cancel{17}N}{\cancel{17}} = \frac{14 \cdot 510}{17}$$

$$\mathbf{N = 420 \text{ grams}}$$

example 37.3   The formula for ammonium chloride is NH_4Cl. This means that in one molecule of ammonium chloride there is one atom of nitrogen, four atoms of hydrogen, and one atom of chlorine. How many grams of chlorine are there in 1060 grams of ammonium chloride? (N, 14; H, 1; Cl, 35).

solution   We begin by finding the molecular weights of each element in a molecule of the compound.

Nitrogen: $1 \times 14 = 14$

Hydrogen: $4 \times 1 = 4$

Chlorine: $1 \times 35 = 35$

Total: $= 53$

We see that the ratio of the weight of the chlorine to the weight of the total solution is 35 to 53.

$$\frac{Cl}{T} = \frac{35}{53}$$

We replace T with 1060 and then solve:

$$\frac{Cl}{1060} = \frac{35}{53} \quad \longrightarrow \quad 53Cl = 35 \cdot 1060 \quad \longrightarrow \quad \frac{53Cl}{53} = \frac{35 \cdot 1060}{53}$$

$$Cl = 700 \text{ grams}$$

37.B

parallelograms A parallelogram is a quadrilateral that has two pairs of parallel sides.

Parallelograms have four special properties.

    1. The pairs of parallel sides have equal lengths (are congruent).
    2. Angles opposite each other have equal measures (are congruent).
    3. The sum of any two adjacent angles is 180°.
    4. The diagonals bisect each other.

We can prove properties 1 and 2 by drawing a diagonal and noting that the two triangles formed are congruent.

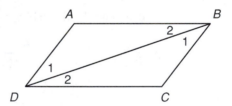

The angles marked 2 have equal measures because they are alternate interior angles formed by the long parallel sides and the diagonal. The angles marked 1 have equal measures because they are the angles formed by the short parallel sides and the diagonal. Two angles in each triangle are equal, so angles A and C must also be equal and the triangles are similar by AAA. The scale factor is 1 because the diagonal is the side opposite angles A and C. Thus the triangles are congruent by AAAS. When we write the statement of congruency, we are careful to list corresponding vertices in the same order.

$$\triangle DAB \cong \triangle BCD$$

Thus,

$$AB \cong DC \qquad \text{CPCTC}$$

$$DA \cong CB \qquad \text{CPCTC}$$

This proves that the pairs of opposite sides in a parallelogram have equal lengths.

    From the congruent triangles we also see that angle A has the same measure as angle C. We could draw the other diagonal and use the same procedure to prove that the other pair of angles in the parallelogram have equal measures. This proves that the angles opposite each other in a parallelogram have equal measures, as we show on the left on the next page. In the figure on the right, we label the equal measures as having measures of $x°$ and $y°$.

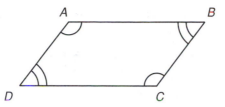 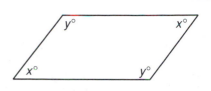

Any quadrilateral can be divided into two triangles, so the sum of the interior angles of a quadrilateral is 360°. Thus

$$2x + 2y = 360 \qquad 360° \text{ in a quadrilateral}$$

$$x + y = 180 \qquad \text{divided by 2}$$

This proves that the sum of the measures of any two adjacent angles in a parallelogram is 180°. In Lesson 39, we will prove that the diagonals of a parallelogram bisect each other.

example 37.4   *ABCD* is a parallelogram and $m\angle BAD$ is 65°. Find $m\angle ADC$ and $m\angle DCB$. Find x and y.

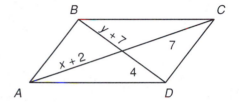

solution   The measures of opposite angles are equal so $m\angle DCB$ is 65°. The sum of the measures of two adjacent angles is 180°, so $m\angle ADC$ must be 115°. The diagonals of a parallelogram bisect each other. This gives us two equations.

$$y + 7 = 4 \qquad x + 2 = 7$$

$$y = -3 \qquad\quad x = 5$$

example 37.5   *ABCD* is a parallelogram. Find x and y.

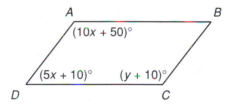

solution   The sum of a pair of adjacent angles in a parallelogram is 180°. Thus,

$$(10x + 50) + (5x + 10) = 180 \qquad \text{adjacent angles}$$

$$15x + 60 = 180 \qquad \text{simplified}$$

$$15x = 120 \qquad \text{added} -60 \text{ to both sides}$$

$$x = 8$$

To find y, we first evaluate $(10x + 50)°$.

$$10x + 50 \qquad \text{angle } A$$

$$10(8) + 50 \qquad \text{substituted}$$

$$130 \qquad \text{simplified}$$

Angles A and C are a pair of opposite angles in a parallelogram, so they have equal measures.

$$A = C \qquad \text{equal angles}$$

$$130 = y + 10 \qquad \text{substituted}$$

$$120 = y \qquad \text{solved}$$

practice   **a.** The chemical formula for Freon-12 (dichlorodifluoromethane) is CCl_2F_2. If we have 1200 grams of Freon-12, how much does the carbon weigh? (C, 12; Cl, 35; F, 19)

b. $ABCD$ is a parallelogram and $m\angle ABC$ is 125°. Find $m\angle BCD$ and $m\angle DAB$. Find x and y.

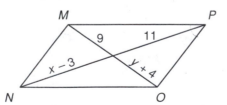

c. $ABCD$ is a parallelogram. Find x and y.

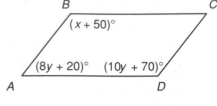

problem set 37

1. The chemical formula for water is H_2O. If we have 5400 grams of water, what is the weight of the oxygen? (O, 16; H, 1)

2. The chemical formula for ammonia is NH_3. This tells us that each ammonia molecule contains one nitrogen atom and three hydrogen atoms. If we have 850 grams of ammonia, how much does the nitrogen weigh? (N, 14; H, 1)

3. The formula for ammonium chloride is NH_4Cl. This means that in one molecule of ammonium chloride there is one atom of nitrogen, four atoms of hydrogen, and one atom of chlorine. How many grams of chlorine are there in 795 grams of ammonium chloride? (N, 14; H, 1; Cl, 35)

4. The bus and the train left the same town headed south at 8 a.m. At noon, the bus was 100 miles behind the train. How far did each one travel if the speed of the train was twice the speed of the bus?

5. His imitation turned into a travesty because he used $2\frac{3}{4}$ times as many gestures as were required for a reasonable imitation. If he used 550 gestures, how many were required for a reasonable imitation?

Simplify:

6. $\dfrac{x^2 + x - 6}{x^3 + 7x^2 + 12x} \cdot \dfrac{x^3 + 5x^2 + 4x}{x^2 + 2x - 8}$

7. $\dfrac{ax^3 - ax^2 - 12ax}{x^2 + 7x + 12} \div \dfrac{ax^2 - 4ax}{x^2 + 2x - 8}$

8. $-8^{-4/3}$ 9. $(-27)^{-4/3}$ 10. $-27^{-4/3}$ 11. $\dfrac{-3}{-9^{-3/2}}$

12. $MNOP$ is a parallelogram. The measure of $\angle MPO$ is 85°. Find $m\angle PON$. Find x and y.

13. Find x, y, and z.

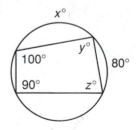

Simplify:

14. $\dfrac{x + \dfrac{1}{x^2}}{x^2 - \dfrac{2}{x^2}}$

15. $\dfrac{a + \dfrac{y}{x}}{a - \dfrac{my}{x}}$

16. $3\sqrt{\dfrac{2}{7}} - 5\sqrt{\dfrac{7}{2}}$

17. $2\sqrt{\dfrac{11}{3}} - 5\sqrt{\dfrac{3}{11}}$

18. Find the equation of the line that passes through $(-4, 2)$ and is perpendicular to the line that goes through $(-4, 6)$ and $(5, 2)$.

19. Find x and y.

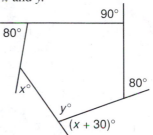

20. Solve by graphing and then get an exact solution by using either substitution or elimination.

$$\begin{cases} 3x - 3y = -6 \\ 3x + y = 6 \end{cases}$$

Add:

21. $\dfrac{x}{x + 5} - \dfrac{3x}{x^2 + 4x - 5}$

22. $\dfrac{4}{x(x + 2)} + \dfrac{6}{x}$

Solve:

23. $\dfrac{3x + 2}{5} - \dfrac{x - 3}{7} = 2$

24. $\dfrac{4x}{3} - \dfrac{2x}{4} + x = 5$

Simplify:

25. $3\sqrt{2}\left(5\sqrt{12} - 2\sqrt{2}\right)$

26. $4\sqrt{20}\left(3\sqrt{2} - 2\sqrt{5}\right)$

27. $\dfrac{(x^{-2})^{-1}(y^{-2}x)^{-3}x^0 y}{(x^{-2})^2 x^4 xx^0 y^0 x^2}$

28. $-2^2 - (-2)^3 - (-2) - 2^0 - 2$

29. $\dfrac{(0.00035 \times 10^{-14})(0.003 \times 10^5)}{21{,}000 \times 10^{-40}}$

30. Multiply: $\dfrac{4x^{-2}y^{-2}}{z^2}\left(\dfrac{x^2 y}{z^{-2}} - \dfrac{3x^2 y^2 z^2}{p}\right)$

LESSON 38 *Powers of sums • Solving by factoring • Only zero equals zero*

38.A
powers of sums

We know that x^2 means to multiply x by x.

$$x^2 = x \cdot x$$

In the same way, $(x^4 y^{-3} z^2)^2$ means to multiply $x^4 y^{-3} z^2$ by $x^4 y^{-3} z^2$.

$$(x^4 y^{-3} z^2)^2 = (x^4 y^{-3} z^2) \cdot (x^4 y^{-3} z^2) = x^8 y^{-6} z^4$$

The same result can be obtained by using the power theorem for exponents by multiplying the exponents, as shown here.

$$(x^4 y^{-3} z^2)^2 = x^8 y^{-6} z^4$$

We must be careful, however, when the notation indicates that a sum is to be raised to a power, because

$$(2x^2 + 3y)^2$$

means that $2x^2 + 3y$ must be multiplied by itself, and the power theorem cannot be used as in the last example. We will use the vertical format to multiply.

$$
\begin{array}{r}
2x^2 + 3y \\
2x^2 + 3y \\
\hline
4x^4 + 6x^2y \\
6x^2y + 9y^2 \\
\hline
4x^4 + 12x^2y + 9y^2
\end{array}
$$

example 38.1 Expand $(x + 3)^3$.

solution This notation indicates that $x + 3$ is to be used as a factor three times.

$$(x + 3)(x + 3)(x + 3)$$

We will use two steps to find the product of the three factors.

$$(x + 3)(x + 3) = x^2 + 3x + 3x + 9 = x^2 + 6x + 9$$

Now we multiply this product by $x + 3$.

$$
\begin{array}{r}
x^2 + 6x + 9 \\
x + 3 \\
\hline
x^3 + 6x^2 + 9x \\
3x^2 + 18x + 27 \\
\hline
x^3 + 9x^2 + 27x + 27
\end{array}
$$

We must try to remember that the power theorem can be used only when a product is raised to a power. The power theorem cannot be used when a sum is raised to a power.

38.B
solving by factoring

If a product of two factors equals zero, then one of the factors must be zero. For instance, if we have the notation

$$(4)() = 0$$

the only possible correct entry for the second set of parentheses is 0.

$$(4)(0) = 0$$

In the same way, if we have an indicated multiplication of two factors equal to zero,

$$()() = 0$$

then either the first factor must equal zero, or the second factor must equal zero. This fact is so important that we give it a name, the **zero factor theorem.** The formal statement of this theorem is as follows.

ZERO FACTOR THEOREM

If p and q are any real numbers and if $p \cdot q = 0$, then either $p = 0$ or $q = 0$, or both.

We use this theorem to help us solve equations by factoring.

example 38.2 Solve $-x + x^2 = 6$.

solution First we rewrite the equation so that the three given terms are on the left side and are in descending powers of the variable.

$$x^2 - x - 6 = 0$$

Next we factor the trinomial and get

$$(x - 3)(x + 2) = 0$$

Now the zero factor theorem tells us that in this case there are just two possibilities. Either $x - 3$ equals zero or $x + 2$ equals zero.

$$\begin{aligned} \text{If } x - 3 &= 0 \\ + 3 &= +3 \\ \hline x &= 3 \end{aligned} \qquad \begin{aligned} \text{If } x + 2 &= 0 \\ - 2 &= -2 \\ \hline x &= -2 \end{aligned}$$

We will check both solutions in the original equation.

If $x = 3$:

$$-(3) + (3)^2 = 6$$
$$-3 + 9 = 6$$
$$6 = 6 \qquad \text{Check}$$

If $x = -2$:

$$-(-2) + (-2)^2 = 6$$
$$2 + 4 = 6$$
$$6 = 6 \qquad \text{Check}$$

The zero factor theorem can be extended to the product of any number of factors by saying that if the product of two or more factors equals zero, then one or more of the factors must equal zero.

EXTENSION OF THE ZERO FACTOR THEOREM

If a, b, c, d, etc., represent real numbers and if

$$a \cdot b \cdot c \cdot d \cdot e \cdot f \ldots = 0$$

then one or more of the factors equals zero.

We will use this fact in the next example.

example 38.3   Solve $-35x = -2x^2 - x^3$.

solution   Again the first step is to write the equation in standard form.

$$x^3 + 2x^2 - 35x = 0$$

Next we factor out an x to get

$$x(x^2 + 2x - 35) = 0$$

and then we factor the trinomial.

$$x(x - 5)(x + 7) = 0$$

Now, by the extension of the zero factor theorem, one of these factors must equal 0.

$$\begin{aligned} \text{If } x &= 0 \\ x &= 0 \end{aligned} \qquad \begin{aligned} \text{If } x - 5 &= 0 \\ x &= 5 \end{aligned} \qquad \begin{aligned} \text{If } x + 7 &= 0 \\ x &= -7 \end{aligned}$$

We finish by checking all three solutions in the original equation.

If $x = 0$:

$$-35(0) = -2(0)^2 - (0)^3$$
$$0 = 0 - 0$$
$$0 = 0 \qquad \text{Check}$$

If $x = 5$:

$$-35(5) = -2(5)^2 - (5)^3$$
$$-175 = -50 - 125$$
$$-175 = -175 \qquad \text{Check}$$

If $x = -7$:

$$-35(-7) = -2(-7)^2 - (-7)^3$$
$$245 = -98 + 343$$
$$245 = 245 \qquad \text{Check}$$

38.C

only zero equals zero

If we solve the following two equations by factoring,

$$\text{(a)} \quad x^3 + 2x^2 - 35x = 0 \qquad \text{(b)} \quad 2x^2 + 4x - 70 = 0$$

the results of the factoring are similar.

$$\text{(a)} \quad x(x - 5)(x + 7) = 0 \qquad \text{(b)} \quad 2(x - 5)(x + 7) = 0$$

In both cases, we have the product of three factors equal to zero, and it would appear to some that each equation has three roots. In equation (a), we can set each of the factors equal to zero and find that the solutions to the equation are

$$\textbf{0, 5, and } -7$$

Equation (b) is different because although it has three factors, only two of the factors can ever be equal to zero. The factor $(x - 5)$ equals zero if x equals 5 and the factor $(x + 7)$ equals zero if x equals -7, but there is no way that 2 can equal zero. The number 2 equals only the number 2.

Some teachers try to help students avoid a mistake by having them record the solution for (b) as

$$x = 5 \qquad x = -7 \qquad 2 \neq 0$$

practice **a.** Expand: $(x + 2)^3$ **b.** Solve: $-36x = 16x^2 - x^3$

problem set 38

1. Verruca counted protrusions. She found that 3 times the number of protrusions was 15 less than -4 times the opposite of the number of protrusions. How many protrusions did she count?

2. Two kilograms of iron was melted with 7 kilograms of other metals to make the alloy. If 1440 kilograms of the alloy was required, how many kilograms of iron should be used?

3. Charles and Matthew knew that the formula for sulfuric acid was H_2SO_4. If they had 196 grams of sulfuric acid, what was the weight of the sulfur? (H, 1; S, 32; O, 16)

4. The ratio of the two numbers was 7 to 2. When Sir Richard and Marion multiplied the denominator by 10, they found that the result was 84 greater than twice the numerator. What were the numbers?

5. Jerry and Milton set out for a ride in the hill country at 30 mph. Their car broke down, and they caught a ride back home in a truck at 20 mph. If they were gone for 10 hours, how far from home did the breakdown occur?

Expand:

6. $(x + 5)^3$ 7. $(x + 4)^3$

Solve:

8. $-x + x^2 = 12$ 9. $-48x = -2x^2 - x^3$ 10. $2x^2 + 2x - 112 = 0$

Simplify:

11. $\dfrac{x^2 + 7x + 10}{14x + 9x^2 + x^3} \div \dfrac{x^2 + 2x - 15}{x^3 + 11x^2 + 28x}$

12. $-16^{-1/4}$ 13. $-16^{-3/4}$ 14. $8^{2/3}$ 15. $(-8)^{1/3}$

16. Find the area of this triangle. Dimensions are in inches.

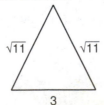

17. The radius of this circle is 20 m. Find the length of $\overset{\frown}{DEF}$.

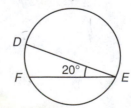

Simplify:

18. $\dfrac{x^2 - \dfrac{a}{x}}{a^2 - \dfrac{a}{x}}$

19. $\dfrac{\dfrac{mp^2}{4} - 5}{4p^2 - \dfrac{p^2}{4}}$

20. $\dfrac{(0.00007 \times 10^{-23})(3000 \times 10^{-10})}{700,000 \times 10^{-30}}$

21. $3\sqrt{\dfrac{2}{11}} + 5\sqrt{\dfrac{11}{2}}$

22. $-2\sqrt{\dfrac{11}{3}} + 7\sqrt{\dfrac{3}{11}}$

23. $3\sqrt{24}\left(2\sqrt{6} - 3\sqrt{12}\right)$

24. Find the equation of the line that goes through the points $(-2, 5)$ and $(3, 4)$.

25. Solve: $\dfrac{5x - 2}{3} - \dfrac{x}{4} = 7$

26. Divide $x^4 - 1$ by $x + 1$.

27. Find the distance between $(-2, 5)$ and $(3, 4)$.

28. Add: $\dfrac{1}{x + 3} + \dfrac{3x}{x + 2} + \dfrac{2x + 1}{x^2 + 5x + 6}$

29. Simplify: $\dfrac{(p^2 y^{-2})^{-3} p^{-2}(y^0)^{-2}}{(p^{-2} p^0 py)^{-3}(yp^{-2})^{-4} p}$

30. $\left|3^0 - 3^2\right| - \dfrac{1}{2^{-2}} - 3^0(-3^3 - 3^2)$

LESSON 39   *Difference of two squares • Parallelogram proof • Rhombus*

39.A
difference of two squares

If both terms of a two-term algebraic sum are identical except that the sign of the second term is different, we say that each expression is the **conjugate** of the other expression. Thus,

$-2x - 3y$	is the conjugate of	$-2x + 3y$
$-2x + 3y$	is the conjugate of	$-2x - 3y$
$3x + 4$	is the conjugate of	$3x - 4$
$3x - 4$	is the conjugate of	$3x + 4$

When conjugates are multiplied, the product does not have a middle term, as seen here.

$$
\begin{array}{r}
-2x - 3y \\
-2x + 3y \\
\hline
4x^2 + 6xy \\
-6xy - 9y^2 \\
\hline
4x^2 \quad\quad - 9y^2
\end{array}
\qquad
\begin{array}{r}
3x - 4 \\
3x + 4 \\
\hline
9x^2 - 12x \\
+ 12x - 16 \\
\hline
9x^2 \quad\quad - 16
\end{array}
\qquad
\begin{array}{r}
a + b \\
a - b \\
\hline
a^2 + ab \\
- ab - b^2 \\
\hline
a^2 \quad\quad - b^2
\end{array}
$$

We see that each of these products of conjugates can be written as the difference of two squared expressions.

$$(2x)^2 - (3y)^2 \qquad (3x)^2 - (4)^2 \qquad (a)^2 - (b)^2$$

These expressions can be factored only by recognizing that the expression is the *difference of two squares*. There is no procedure to follow.

example 39.1   Solve $4x^2 - 9 = 0$ by factoring.

solution   **We recognize that the left-hand side of the equation is the difference of two squares.** Thus, we factor as follows:

$$(2x - 3)(2x + 3) = 0$$

We finish the solution by using the zero factor theorem.

$$\text{If } 2x - 3 = 0 \qquad \text{If } 2x + 3 = 0$$
$$2x = 3 \qquad\qquad 2x = -3$$
$$x = \frac{3}{2} \qquad\qquad x = -\frac{3}{2}$$

example 39.2   Solve $81m^2 - 25 = 0$ by factoring.

solution   **We recognize that the left-hand side of the equation is a difference of two squares,** which we factor as

$$(9m + 5)(9m - 5) = 0$$

We complete the solution by setting each factor equal to zero.

$$\text{If } 9m + 5 = 0 \qquad \text{If } 9m - 5 = 0$$
$$9m = -5 \qquad\qquad 9m = 5$$
$$m = -\frac{5}{9} \qquad\qquad m = \frac{5}{9}$$

39.B
parallelogram proof

We remember that a parallelogram is defined to be a quadrilateral with two pairs of parallel sides. Also recall that a parallelogram has four other properties, which are listed here.

1. The sides opposite each other have equal lengths.
2. The angles opposite each other have equal measures.
3. The sum of the measures of any two consecutive angles is 180°.
4. The diagonals bisect each other.

We proved properties 1, 2, and 3 in Lesson 37. Now we will prove property 4. Since the pairs of opposite sides in a parallelogram have equal lengths, we can use identical tick marks to indicate that the long sides of parallelogram *ABCD* below are equal.

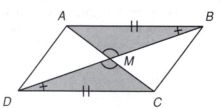

The small, shaded angles at *B* and *D* have equal measures because they are alternate interior angles formed by diagonal *BD* and the parallel long sides. The vertical angles marked at the intersection of the diagonals are equal. Thus, the third angles in each triangle are equal, and the triangles are similar by AAA. The scale factor is 1 because the corresponding long sides have equal lengths, so the triangles are congruent by AAAS.

$$\triangle CMD \cong \triangle AMB$$

Also,
$$CM \cong MA \qquad \text{CPCTC}$$
$$DM \cong MB \qquad \text{CPCTC}$$

This proves that the diagonals of a parallelogram bisect each other.

39.C
rhombus

A **rhombus** is a parallelogram that has three additional properties. The first is the definition of a rhombus. The other two properties are theorems and can be proved.

1. A rhombus is a parallelogram whose four sides have equal lengths.
2. The diagonals of a rhombus bisect the angles of the rhombus.
3. The diagonals of a rhombus are perpendicular bisectors of each other.

To prove that the diagonals of a rhombus bisect the angles of the rhombus, we draw diagonal AC in rhombus $ABCD$. We will prove that diagonal AC bisects $\angle A$ and $\angle C$.

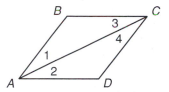

Since $ABCD$ is a rhombus then side AB is congruent to side BC. Also, we know that in a triangle, the angles opposite sides of equal lengths have equal measures. Therefore, $\angle 1 = \angle 3$. Because the rhombus is a parallelogram then side BC is parallel to side AD, and side AB is parallel to side DC. Therefore, $\angle 2 = \angle 3$, because they are alternate interior angles formed by diagonal AC and parallel sides BC and AD. Also, $\angle 1 = \angle 4$, because they are alternate interior angles formed by diagonal AC and parallel sides AB and DC. Therefore, $\angle 1 = \angle 3$, $\angle 2 = \angle 3$ and $\angle 1 = \angle 3$, $\angle 1 = \angle 4$. By the sixth postulate of Euclid, we know that things equal to the same thing are equal to each other. Thus, $\angle 1 = \angle 2$ and $\angle 3 = \angle 4$. Therefore, diagonal AC bisects $\angle A$ and $\angle C$. We can also show that diagonal BD bisects $\angle B$ and $\angle D$ in the same way. **This proves that the diagonals of a rhombus bisect the angles of the rhombus.**

Because the rhombus is a parallelogram, we already know that the diagonals bisect each other. To prove that the angles formed by the diagonals are right angles, we use the fact that the diagonals bisect the angles of the rhombus and the fact that the sum of the measures of two consecutive angles in a parallelogram is 180°, as we remind ourselves in the figure on the left.

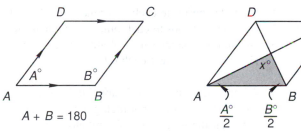

In the rhombus on the right, we see that the base angles in the shaded triangle measure half of A and half of B. Since $A + B$ equals 180, half of that sum equals 90, and $x + 90$ must equal 180. Thus, x must equal 90 so that the sum of the measures of the three angles is 180°.

example 39.3 $DEFG$ is a rhombus. The measure of the reflex angle is 280°. Find A, X, B, K, and C.

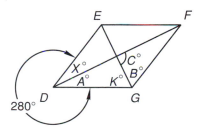

solution The sum of 280, X, and A equals 360. Thus, X plus A must equal 80.

$$X + A = 80$$

Angles X and A have equal measures because $\overline{DF}$ bisects the angle at D. Thus, both X and A equal 40.

$$X = 40 \qquad A = 40$$

The sum of two adjacent angles in a parallelogram is 180°, so K plus B equals 100.

$$K + B = 100$$

But EG bisects the angle at G, so K equals B.

$$K = 50 \qquad B = 50$$

Angle C is a 90° angle because the diagonals of a rhombus are perpendicular. Thus,

$$C = 90$$

practice

a. Solve $144s^2 - 36 = 0$ by factoring.

b. Solve $121m^2 - 64 = 0$ by factoring.

c. *MNOP* is a rhombus. The measure of the reflex angle is 306°. Find X, A, and Z.

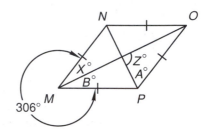

problem set 39

1. Some of the sophomores were ingenuous, but $\frac{13}{16}$ had ulterior motives for their actions. If 420 did not have ulterior motives, how many did have ulterior motives?

2. Some of the teacher's proclamations were democratic, but 72 percent were ukases. If 9576 were democratic, how many were ukases?

3. The chemical formula for carbon dioxide is CO_2. If the reaction produced 528 grams of carbon dioxide, what was the weight of the carbon produced? (C, 12; O, 16)

4. The symbol for strontium is Sr. If the mixture was 14 percent strontium and 688 grams of the mixture was not strontium, what was the weight of the strontium?

5. Jimmy and Gary ran from the disaster at 5 mph. Then they reconsidered and walked back at 3 mph. If they were gone for 8 hours, how far did they run from the disaster?

6. Find an angle such that 7 times the complement of the angle is 110° greater than twice the supplement of the angle.

7. The area of the triangle is 52 ft². Find MP. The radius of the circle equals MP. Find the area of the circle.

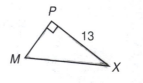

Solve:

8. $x^2 - 9 = 0$

9. $36x^2 - 36 = 0$

10. $24x = -11x^2 - x^3$

11. Expand: $(x - 1)^3$

Simplify:

12. $\dfrac{x^3 + 6x^2 + 5x}{x^2 + 2x - 15} \div \dfrac{7x + 8x^2 + x^3}{x^2 + 5x - 14}$

13. $\dfrac{2^0}{-4^{-3/2}}$

14. $\dfrac{-3^0}{-27^{-2/3}}$

15. $\dfrac{ax^2 - \dfrac{4}{a}}{\dfrac{x^2}{a} + 6}$

16. $\dfrac{\dfrac{m^2 p}{x} - 6}{m^2 p - \dfrac{4}{x}}$

17. $\dfrac{(3000 \times 10^{-41})(0.0008 \times 10^{10})}{2{,}400{,}000 \times 10^8}$

18. $2\sqrt{\dfrac{3}{13}} - 5\sqrt{\dfrac{13}{3}}$

19. $5\sqrt{\dfrac{3}{2}} - 2\sqrt{\dfrac{2}{3}}$

20. $5\sqrt{45} - 2\sqrt{75} + 2\sqrt{108}$

21. Find x and y.

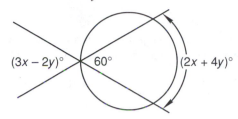

22. *WXYZ* is a parallelogram. The measure of angle *YZW* is 68°. Find $m\angle WXY$. Find x and y.

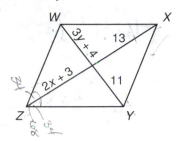

23. Simplify: $3\sqrt{12}(4\sqrt{3} - 3\sqrt{3})$

24. Find the equation of the line that passes through $(2, 4)$ and is perpendicular to the line that passes through $(2, 4)$ and $(-3, -2)$.

Solve:

25. $2\frac{1}{4}x - 3\frac{1}{2} = -\frac{1}{16}$

26. $0.002x - 0.02 = 6.6$

27. Add: $\dfrac{3}{x + 1} + \dfrac{2x}{y(x + 1)} + \dfrac{3x + 2}{x^2 + 2x + 1}$

28. Simplify: $\dfrac{4xy + 4x^2 y^2}{4xy}$

29. Expand: $\dfrac{4x^{-2} y^{-2}}{z^2}\left(\dfrac{3x^2 y^2 z^2}{4} + \dfrac{2x^0 y^{-2}}{z^2 y^2}\right)$

30. Evaluate: $-2^0 - 3^0(-2 - 5^0) - \dfrac{1}{-2^{-2}} + x^2 y - xy$ if $x = -2$ and $y = 3$

LESSON 40 *Abstract fractional equations*

We have noted that the easiest way to solve fractional equations is to eliminate the denominators as the first step. Thus, to solve this equation

$$\frac{4 + x}{5} + \frac{7}{3} = 4$$

we begin by multiplying every numerator by 15, which is the least common multiple of the denominators. This will permit the elimination of the denominators.

$$\frac{15(4 + x)}{5} + \frac{15(7)}{3} = 4(15) \qquad \text{multiplied by 15}$$

$$12 + 3x + 35 = 60 \qquad \text{eliminated denominators}$$

$$3x = 13 \qquad \text{simplified}$$

$$x = \frac{13}{3} \qquad \text{solved}$$

We also eliminate the denominators as the first step in the solution of abstract equations, as we will demonstrate in the following examples. (**In the solution of abstract equations, we will assume that no variable or combination of variables in any denominator equals zero.**)

example 40.1   Solve for x: $\dfrac{a}{x} + \dfrac{m}{a} = c$

solution   We will solve in three steps. **The first step will be to multiply every numerator by the least common multiple of the denominators, which is ax. Then we will cancel the denominators.**

Step 1: $\dfrac{a}{x} \cdot ax + \dfrac{m}{a} \cdot ax = c \cdot ax$ multiplied by ax

                $a^2 + mx = cax$ canceled denominators

Next we move all terms that contain x to the same side of the equation (either side) and factor out the x.

Step 2: $a^2 = cax - mx$ added $-mx$ to both sides

                $a^2 = x(ca - m)$ factored out x

We finish by dividing both sides by $(ca - m)$, which is the coefficient of x.

Step 3: $\dfrac{a^2}{ca - m} = \dfrac{x(ca - m)}{ca - m} \longrightarrow \dfrac{a^2}{ca - m} = x$ divided

example 40.2   Solve for m: $\dfrac{x}{m} + c = \dfrac{y}{a}$

solution   **Again as the first step, we will eliminate the denominators.** We begin by multiplying every numerator by ma.

Step 1: $(ma)\dfrac{x}{m} + (ma)c = (ma)\dfrac{y}{a}$ multiplied

                $ax + mac = my$ canceled

Next we move all terms that contain an m to the same side of the equation and factor out the m.

Step 2: $ax = my - mac$ added $-mac$ to both sides

                $ax = m(y - ac)$ factored out m

We finish by dividing both sides by $y - ac$, and we get

Step 3: $\dfrac{ax}{y - ac} = m$

In step 2, if we had placed all terms that contained m on the left side, our answer would have been

$$m = \dfrac{-ax}{-y + ac}$$

which is the same answer as the one above except that all signs above and below are different. We remember that we can always multiply the denominator and the numerator by the same nonzero quantity. If we use (-1) as the multiplier, we can change this last answer to the first form of the answer.

$$\dfrac{-ax}{-y + ac}\dfrac{(-1)}{(-1)} = \dfrac{ax}{y - ac}$$

Both answers are equally correct, and neither is preferred.

example 40.3   Solve for p: $\dfrac{6}{p} - ax = \dfrac{m}{y} + k$

solution   Again we begin by canceling the denominators.

$$(py)\dfrac{6}{p} - (py)ax = (py)\dfrac{m}{y} + (py)k$$ multiplied by py

$$6y - pyax = pm + pyk \qquad \text{canceled}$$

Next we place all terms that contain p on the same side and factor out the p.

$$6y = pm + pyk + pyax \qquad \text{added } +pyax$$

$$6y = p(m + yk + yax) \qquad \text{factored}$$

Again we finish by dividing both sides by the coefficient of the variable for which we are solving.

$$\frac{6y}{m + yk + yax} = p$$

practice   **a.**   Solve for m: $\dfrac{5}{m} - ay = \dfrac{p}{x} + s$ **b.**   Find x: $\dfrac{m}{x} + \dfrac{p}{m} = z$

problem set 40

1. Many students took a foreign language to increase their vocabularies and improve their grammar. These students earned 64 percent more than the others. How much did they make if the others earned $1,200,000?

2. The more successful professionals had vocabularies that were 280 percent larger than those who were less successful. If those who were less successful knew 4800 words, how many words did those who were more successful know?

3. The formula for iron sulfide is FeS. If the iron (Fe) in a batch of iron sulfide weighed 448 grams, how much did the iron sulfide weigh? (Fe, 56; S, 32)

4. Shoes cost $20 a pair and boots cost $60 a pair. Arlene and Jerry spent $8000 and bought 3 times as many boots as shoes. How many pairs of each did they buy?

5. Amy had a 120-mile head start on Kathy. Kathy drove at 15 miles per hour, and Amy ran at 3 miles per hour. How far did Kathy have to drive to catch Amy?

6. Find a: $\dfrac{c}{x} + \dfrac{m}{a} = c$ 7. Find a: $\dfrac{x}{m} + c = \dfrac{y}{a}$

8. Find y: $\dfrac{6}{p} - ax = \dfrac{m}{y} + k$ 9. Find k: $\dfrac{a}{c} - b = \dfrac{m}{k}$

10. Expand: $(x - 3)^3$

11. $ABCD$ is a rhombus. Find X, Y, and Z.

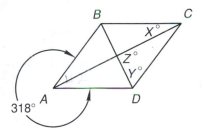

Solve:

12. $4x^2 - 49 = 0$ 13. $x^3 + 3x^2 = 18x$

Simplify:

14. $\dfrac{10x - 7x^2 + x^3}{x^2 + 4x - 12} \div \dfrac{x^3 - 8x^2 + 15x}{x^2 + 3x - 18}$ 15. $\dfrac{(-4^0)^2}{4^{-3/2}}$

16. $\dfrac{1}{16^{-1/4}}$ 17. $\dfrac{\dfrac{4x^2 a}{y^2} + \dfrac{1}{a^2}}{2 - \dfrac{2}{y^2 a^2}}$

18. $\dfrac{\dfrac{xy^2}{p} - 4}{a^2 y - \dfrac{1}{p}}$ 19. $\dfrac{(4000 \times 10^{14})(0.007 \times 10^{-23})}{14{,}000 \times 10^{-20}}$

20. $3\sqrt{50} - 2\sqrt{72} + 3\sqrt{162}$

21. $3\sqrt{\dfrac{3}{7}} + 2\sqrt{\dfrac{7}{3}}$

22. Find A, B, and C.

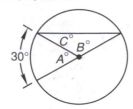

23. The altitude of a circular cylinder is 8 cm. The volume of the cylinder is 32π cm^3. What is the radius of the cylinder?

24. Find the equations of lines (a) and (b).

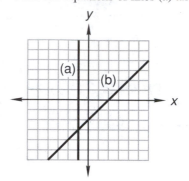

Solve:

25. $\dfrac{3 - 2x}{4} + \dfrac{x}{3} = 5$

26. $0.004x - 0.02 = 2.02$

27. Add: $\dfrac{3}{x} + \dfrac{2}{x + 2} + \dfrac{3x}{x^2 + 3x + 2}$

28. Simplify: $\dfrac{x + 4x}{x}$

29. Expand: $\dfrac{x^{-2}y}{p}\left(\dfrac{x^2 p}{y} - \dfrac{3x^2 y}{p}\right)$

30. Evaluate: $x^2 - xy - x^3$ if $x = \dfrac{1}{2}$ and $y = \dfrac{1}{3}$

LESSON 41 *Units • Unit multipliers*

41.A
units

When we attach words to numbers as shown here,

4 ft   6 centimeters   42.78 mph   93 liters

we often call the resulting combinations **denominate numbers.** The Latin prefix for "completely" is *de-*, and the Latin word for "to name" is *nominare*. Thus, *denominate* literally means "completely named." The words are called the **units** of the denominate numbers. Thus, the units in the denominate numbers above are feet, centimeters, miles per hour, and liters.

We find it convenient to use exponential notation to handle units. If we do this, we can handle units the same way we handle numbers or variables. This is especially useful when we multiply or divide units, as we see in these examples.

(a) ft^2 · ft = **ft^3**   (b) $\dfrac{\text{cm}^3}{\text{cm}}$ = **cm^2**

(c) $\dfrac{\text{yd}^3}{\text{yd}}$ = **yd^2**   (d) $\dfrac{\text{in.}^3}{\text{in.}^2}$ = **in.**

41.B
unit multipliers

We remember that any nonzero quantity divided by itself has a value of 1.

$\dfrac{x^2}{x^2} = 1$ $\dfrac{24.123}{24.123} = 1$ $\dfrac{6 \text{ ft}^2}{6 \text{ ft}^2} = 1$ $\dfrac{4 \text{ in.}}{4 \text{ in.}} = 1$

Furthermore, we remember that the product of any quantity and 1 is the quantity itself.

$x^2ym(1) = x^2ym$ $\left(4\dfrac{\text{ft}}{\text{sec}}\right)(1) = 4\dfrac{\text{ft}}{\text{sec}}$ $(3 \text{ in.})(1) = 3 \text{ in.}$

We know that 12 inches equals 1 foot, so if we write either

$$\frac{12 \text{ in.}}{1 \text{ ft}} \quad \text{or} \quad \frac{1 \text{ ft}}{12 \text{ in.}}$$

we have written an expression whose value is 1. We can multiply any expression by either of these terms without changing the value of the expression. We call these terms **unit multipliers** for two reasons: One reason is that the expressions contain units, and the second reason is that the expressions have a value of unity (1). Unit multipliers are very helpful when we want to change one set of units to another set of units.

example 41.1   Use a unit multiplier to change 600 inches to feet.

solution   We will use one of the unit multipliers above. We choose the one on the left.

$$600 \text{ in.} \times \frac{12 \text{ in.}}{1 \text{ ft}} = \frac{7200 \text{ in.}^2}{\text{ft}}$$

This answer is not incorrect, but it is not what we want. Let's try again and use the other unit multiplier.

$$600 \text{ in.} \times \frac{1 \text{ ft}}{12 \text{ in.}} = \textbf{50 ft}$$

This time the inches canceled and we found the desired answer.

example 41.2   Use unit multipliers to convert 44 square feet to square inches.

solution   We write what was given and use 1 for a denominator.

$$\frac{44 \text{ ft}^2}{1}$$

We note that square feet (ft^2) is in the numerator. Thus, we will use the unit multiplier that has the abbreviation ft in the denominator. We must use two unit multipliers because we are converting from square feet (ft^2) to square inches (in.^{2}).

$$\frac{44 \text{ ft}^2}{1} \times \frac{12 \text{ in.}}{1 \text{ ft}} \times \frac{12 \text{ in.}}{1 \text{ ft}} = \textbf{(44)(12)(12) in.}^2$$

The multiplication in this example is relatively easy, but some unit conversion problems will result in very complicated multiplications and divisions. We suggest that these answers be left in the form above or that a pocket calculator be used to get the final numerical answer. These problems are designed to teach unit conversions and are not designed for practice in arithmetic.

We have begun our study of unit conversions with problems that can be solved mentally without the use of unit multipliers. It is recommended that the use of unit multipliers not be avoided in these simple problems because the unit conversion problems that we encounter later will be rather involved. The use of unit multipliers will make these involved conversions straightforward, and the experience that we gain by doing simple problems will prove to be valuable.

example 41.3   Use unit multipliers to convert 42 square yards to square inches.

solution   We could use the fact that 1 yard equals 36 inches, but instead we will go from square yards (yd^2) to square feet (ft^2) to square inches (in.^{2}), a procedure that is recommended because shortcuts can lead to errors.

$$42 \text{ yd}^2 \times \frac{3 \text{ ft}}{1 \text{ yd}} \times \frac{3 \text{ ft}}{1 \text{ yd}} \times \frac{12 \text{ in.}}{1 \text{ ft}} \times \frac{12 \text{ in.}}{1 \text{ ft}} = \textbf{42(3)(3)(12)(12) in.}^2$$

example 41.4   Use unit multipliers to convert 16 cubic miles (mi^3) to cubic inches (in.^{3}).

solution   We will go from cubic miles (mi³) to cubic feet (ft³) to cubic inches (in.³).

$$16 \text{ mi}^3 \times \frac{5280 \text{ ft}}{1 \text{ mi}} \times \frac{5280 \text{ ft}}{1 \text{ mi}} \times \frac{5280 \text{ ft}}{1 \text{ mi}} \times \frac{12 \text{ in.}}{1 \text{ ft}} \times \frac{12 \text{ in.}}{1 \text{ ft}} \times \frac{12 \text{ in.}}{1 \text{ ft}}$$

$$= (16)(5280)(5280)(5280)(12)(12)(12) \text{ in.}^3$$

It seems that there are quite a few cubic inches in 16 cubic miles.

practice   a.   Use one unit multiplier to change 840 inches to feet.

b.   Use two unit multipliers to change 90 square feet to square inches.

c.   Use six unit multipliers to convert 30 cubic miles to cubic inches.

problem set   **1.**   While Dr. Andy operated, he thought of consecutive odd integers. His integers were
41   such that 4 times the sum of the first and fourth was 12 greater than 3 times the sum of
the second and third. What were the first four integers on Dr. Andy's list?

2.   The formula for chromium chloride is $CrCl_3$. What would be the weight of the chlorine
(Cl) in 1256 grams of chromium chloride? (Cr, 52; Cl, 35)

3.   The delivery truck unloaded 184 percent more silicon than was required for the experiment.
If 1136 tons were unloaded, how many tons were required for the experiment?

4.   Bob and Judy found that their horde of nickels and dimes was worth $7. If they had a
total of 100 coins, how many coins of each kind did they have?

5.   Larry and Shadid rode on their motor scooters at 16 mph until Larry ran out of petrol.
Then they walked the rest of the way at 4 mph. If the entire trip was 76 miles and it took
a total of 7 hours, how far did they walk and how far did they ride?

Use unit multipliers to convert:

6.   87 ft² to square inches   **7.**   61 yd² to square inches   **8.**   32 mi³ to cubic inches

9.   Find p: $\dfrac{x}{p} - \dfrac{k}{m} = c$ **10.**   Find p: $\dfrac{xy}{p} - \dfrac{k}{c} = m$

11.   Find c: $\dfrac{4p}{x} - \dfrac{xk}{c} = \dfrac{y}{m}$

12.   Find sides m and p.   **13.** $\triangle BDE$ is an isosceles triangle. $\triangle DEF$
is an equilateral triangle. Find X, Y,
and K.

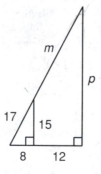

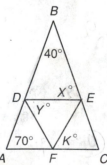

Solve:

14. $16x = -x^3 + 10x^2$ **15.** $4x^2 - 9x = 0$

16.   Divide $2x^3 - 1$ by $x - 2$.   **17.**   The area of this figure equals the sum
of the areas of the triangle and the
semicircle. Find the area of the figure.

Simplify:

18. $\dfrac{x^3 + 8x^2 + 15x}{x^2 - 4x + 4} \cdot \dfrac{x^2 + 2x - 8}{12x + x^3 + 7x^2}$

19. $32^{-2/5}$

20. $\dfrac{a^2 x - \dfrac{a}{x}}{ax - \dfrac{4}{x}}$

21. $\dfrac{(21{,}000 \times 10^{-42})(500{,}000)}{0.00015 \times 10^{-7}}$

22. $3\sqrt{\dfrac{5}{7}} - 6\sqrt{\dfrac{7}{5}}$

23. $4\sqrt{24}(2\sqrt{6} - 3\sqrt{2})$

24. Find the equation of the line that is perpendicular to $3y + x = -2$ and passes through the point $(-2, -5)$.

Solve:

25. $\dfrac{-3 - x}{2} - \dfrac{x}{2} = 7$

26. $2\dfrac{1}{3}x - \dfrac{1}{9} = -\dfrac{1}{18}$

27. Add: $\dfrac{2}{x^2(x - 2)} - \dfrac{2x + 2}{x^2 - 4}$

Simplify:

28. $\dfrac{4x + 8x^2}{4x}$

29. $\dfrac{x^2 x^0 x^{-1}(x^{-2})^2 yx^{-3}}{(x^2 y)^{-3} xyx^{-2} x^2}$

30. Evaluate: $x^3 - xy + x^2$ if $x = \dfrac{1}{2}$ and $y = \dfrac{1}{3}$

LESSON 42 *Estimating with scientific notation*

The scientific notation problems we have encountered thus far have been carefully designed so the numbers multiply and divide easily, and so the first part of the answer is an integer.

$$\dfrac{(0.0003 \times 10^{-6})(4000)}{(0.006 \times 10^{15})(2000 \times 10^4)} \quad \text{problem}$$

$$= \dfrac{(3 \times 10^{-10})(4 \times 10^3)}{(6 \times 10^{12})(2 \times 10^7)} \quad \text{scientific notation}$$

$$= \dfrac{12 \times 10^{-7}}{12 \times 10^{19}} = 1 \times 10^{-26} \quad \text{simplified}$$

These problems have been used to help us develop the skills required to handle both positive and negative integral exponents in scientific notation. Unfortunately, real-life problems contain numbers that are not so easy to handle. For instance, the answer to the last example in Lesson 41 was

$$16(5280)(5280)(5280)(12)(12)(12) \text{ in.}^3$$

Multiplying these numbers by hand would be tedious, and we might make a mistake. If we use a calculator that does not have scientific notation, we get an error notation early because the answer is a number too large for these calculators to handle. If we use a calculator that has scientific notation for this multiplication, we will get

$$4.0697 \times 10^{15}$$

However, we will find to our dismay that we often make mistakes when we use calculators for complicated operations such as this one. Thus we need to develop a way to see if this answer is reasonable, and we should be able to estimate the answer when a calculator is not available. In this problem, we should be able to estimate an answer between

$$4 \times 10^{14} \quad \text{and} \quad 4 \times 10^{16}$$

This would let us know that our calculator answer of

$$4.0697 \times 10^{15}$$

is a reasonable answer. **There would be no excuse for accepting an answer of**

$$4.0697 \times 10^{21}$$

and blaming the error on the calculator. A mistake that generates an answer

1,000,000 times

the correct answer is totally inexcusable.

example 42.1   Estimate the answer to

$$(16)(5280)(5280)(5280)(12)(12)(12)$$

solution   Let's begin by rounding each number to one digit and writing each number in scientific notation.

$$(2 \times 10^1)(5 \times 10^3)(5 \times 10^3)(5 \times 10^3)(1 \times 10^1)(1 \times 10^1)(1 \times 10^1)$$

We used 2×10^1 for 16, 5×10^3 for each 5280, and 1×10^1 for each 12. Multiplying, we get

$$(2)(5)(5)(5) \times 10^{13} = 250 \times 10^{13} \approx 3 \times 10^{15}$$

From this estimate we see that our calculator answer of 4.0697×10^{15} is probably correct.

example 42.2   Use scientific notation to help estimate the answer to this problem.

$$\frac{(3728)(470,165 \times 10^{-14})}{(278,146)(0.000713 \times 10^{-5})}$$

solution   We write each entry in scientific notation, rounding to one digit. Then we simplify.

$$\frac{(4 \times 10^3)(5 \times 10^{-9})}{(3 \times 10^5)(7 \times 10^{-9})} \approx 1 \times 10^{-2}$$

example 42.3   Use scientific notation to help estimate the answer to this expression.

$$\frac{(0.0418765 \times 10^{-14})(41,725 \times 10^{43})}{9764 \times 10^{-23}}$$

solution   We use an approximation in scientific notation for each entry. Then we simplify.

$$\frac{(4 \times 10^{-16})(4 \times 10^{47})}{1 \times 10^{-19}} = \frac{16}{1} \times \frac{10^{31}}{10^{-19}} \approx 2 \times 10^{51}$$

practice   a.   Estimate the answer to: $(13)(5280)(5280)(12)(12)$

Use scientific notation to estimate:

b. $\dfrac{(4353)(933,216 \times 10^{-11})}{(319,214)(0.01603 \times 10^{-31})}$ c. $\dfrac{(0.013926 \times 10^{-12})(27,153 \times 10^{21})}{6354 \times 10^{-31}}$

problem set 42

1.   Pelagic fish were not usually seen near the reef so the diver was surprised to find that $\frac{3}{17}$ of the fish sighted were pelagic. If the diver saw 2244 fish on the morning dive, how many were pelagic?

2.   Only 20 percent of the students were taciturn, as most of them had a penchant for prolixity. If 4800 had a penchant for prolixity, how many students were there in all?

3.   It took 600 grams of potassium chlorate to make a batch of 3600 grams of the aggregate. If 43,200 grams of aggregate was required, how much potassium chlorate was needed?

4. The chemical formula for potassium chlorate is $KClO_3$. If 488 grams of this compound were on the scales, how much did the potassium (K) weigh? (K, 39; Cl, 35; O, 16)

5. Bruce and Lynn found that the larger number was 2 greater than 4 times the smaller number. Also, the larger number was 6 smaller than 8 times the smaller number. What were the two numbers?

Estimate the answer to:

6. $(24)(5280)(5280)(5280)(12)(12)(12)$

7. $\dfrac{(2472)(570,185 \times 10^{-12})}{(243,195)(0.0003128 \times 10^{-6})}$

8. $\dfrac{(0.0319743 \times 10^{-15})(61,853 \times 10^{37})}{6934 \times 10^{-29}}$

9. Use unit multipliers to convert 40 cubic yards to cubic inches.

10. Find c: $\dfrac{m}{c} - x = \dfrac{p}{m}$

11. Find y: $\dfrac{ax}{y} + m = \dfrac{pc}{d}$

12. Divide $4x^3 - 1$ by $x + 2$.

13. The radius of the circle is 2 cm and $X = 120$. Find the area of the shaded region of the circle.

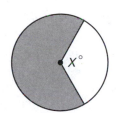

Solve:

14. $-20x = x^3 - 9x^2$

15. $4x^2 - 25 = 0$

Simplify:

16. $\dfrac{x^3 - 3x^2 + 2x}{x^2 + 3x - 4} \div \dfrac{x^3 - 6x + x^2}{15 + 8x + x^2}$

17. $-81^{-3/4}$

18. $\dfrac{\dfrac{m}{x} - 4}{6 - \dfrac{1}{x}}$

19. $2\sqrt{\dfrac{3}{11}} - 5\sqrt{\dfrac{11}{3}}$

20. $3\sqrt{6}(2\sqrt{6} - 4\sqrt{2})$

21. $3\sqrt{20} + 2\sqrt{45} - \sqrt{245}$

22. Find the length of side a and side b.

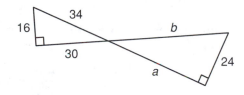

23. Find x and y.

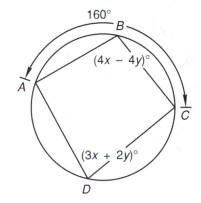

24. Find the equation of the line that passes through the point $(-3, -2)$ and is perpendicular to the line that passes through the points $(-3, -2)$ and $(4, 5)$.

Solve:

25. $\dfrac{-2x - 4}{2} - \dfrac{x}{3} = 5$

26. $3\dfrac{1}{4}x - \dfrac{1}{8} = \dfrac{3}{16}$

27. Add: $\dfrac{3}{x + 2} - \dfrac{4}{x^2 - 4} - \dfrac{3}{x - 2}$

Simplify:

28. $-\dfrac{x^{-2}}{y^2}\left(y^2 x^2 - \dfrac{3x^{-2}}{y^{-2}}\right)$

29. $\dfrac{4x^2 - 4x^4}{4x^2}$

30. Evaluate: $-3x - x^{-2} - x^{-3}$ if $x = \dfrac{1}{2}$

LESSON 43   *Sine, cosine, and tangent • Inverse functions*

43.A
sine, cosine, and tangent

Here we show three triangles. Each one is a right triangle, and each one contains a 30° angle.

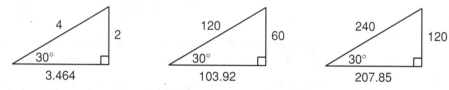

The longest side of a right triangle is always the side opposite the right angle, and this side is called the **hypotenuse.** The hypotenuses of these triangles from left to right are 4, 120, and 240 units long.

We remember that right triangles that also have one equal acute angle are similar triangles, and that the ratios of corresponding sides of similar triangles are equal. If, for each of the three triangles shown, we write the ratio of the side opposite the 30° angle to the hypotenuse, we get these expressions,

$$\frac{2}{4} \qquad \frac{60}{120} \qquad \frac{120}{240}$$

each of which has a value of 0.5. **We would find this to be true for this ratio in any right triangle with a 30° angle. In a right triangle we call this ratio, which is the ratio of the side opposite the angle to the hypotenuse, the** *sine of the angle.*

$$\sin 30° = \frac{\text{side opposite the 30° angle}}{\text{hypotenuse}}$$

This ratio always equals 0.5 for a 30° angle. Every angle has a unique value of the ratio of the side opposite to the hypotenuse.

In a right triangle, the ratio of the side adjacent to the angle to the hypotenuse is called the *cosine of the angle.* For these triangles, the ratios are

$$\frac{3.464}{4} \qquad \frac{103.92}{120} \qquad \frac{207.85}{240}$$

Each of these ratios has a value of approximately 0.8660, which we round off to 0.87. **This ratio has the same value in any right triangle with a 30° angle regardless of the size of the triangle.** So we say

$$\cos 30° = \frac{\text{side adjacent to 30° angle}}{\text{hypotenuse}} \approx 0.87$$

We call the ratio of the side opposite the angle to the side adjacent to the angle the *tangent of the angle.* Thus,

$$\tan 30° = \frac{\text{side opposite the 30° angle}}{\text{side adjacent to 30° angle}}$$

This ratio is approximately 0.5774, which we round off to 0.58.

$$\frac{2}{3.464} \approx 0.58 \qquad \frac{60}{103.92} \approx 0.58 \qquad \frac{120}{207.85} \approx 0.58$$

There is nothing sacrosanct about the words sine, cosine, and tangent, and there is no particular reason for them to be defined as they are. Learning which one has which definition is simply a matter of memorization, and mnemonics are always helpful for memorizing. On the left, we will write the abbreviations for sine, cosine, and tangent in that order. On the right, we will use the first letters of the words opposite, hypotenuse, and adjacent to form the first letters of a sentence that is easy to remember.

$$\sin A = \frac{\text{opposite}}{\text{hypotenuse}} \qquad \frac{\text{Oscar}}{\text{had}}$$

$$\cos A = \frac{\text{adjacent}}{\text{hypotenuse}} \qquad \frac{\text{a}}{\text{hold}}$$

$$\tan A = \frac{\text{opposite}}{\text{adjacent}} \qquad \frac{\text{on}}{\text{Arthur}}$$

Thus, if we can remember to write down sine, cosine, and tangent in that order and then write down "Oscar had a hold on Arthur," we have the definitions memorized.

The sines, cosines, and tangents of angles can be obtained by using the $\boxed{\sin}$, $\boxed{\cos}$, and $\boxed{\tan}$ keys on a scientific calculator. **To find the proper values when the angle is measured in degrees, the calculator must be in the *degree mode*.** Calculators will give an estimate of the value of a trigonometric function to more than five decimal places. To find the sine of 39.2°, first put the calculator in degree mode (press the key $\boxed{\text{DRG}}$ or $\boxed{\text{DEG}}$). Then

ENTER	DISPLAY
39.2	39.2
$\boxed{\sin}$	0.632029302

To find the cosine of 39.2°,

ENTER	DISPLAY
39.2	39.2
$\boxed{\cos}$	0.774944488

To find the tangent of 39.2°,

ENTER	DISPLAY
39.2	39.2
$\boxed{\tan}$	0.815580098

Often we must multiply the trigonometric function of an angle by a number. To find the value of 4 cos 57.2°, we can enter the 4 first or enter the cosine of 57.2° first. To enter the 4 first, we proceed as follows:

ENTER	DISPLAY
4	4
$\boxed{\times}$	4
57.2	57.2
$\boxed{\cos}$	0.54170821
$\boxed{=}$	2.166832841

To enter the cosine of 57.2° first, we proceed as follows:

ENTER	DISPLAY
57.2	57.2
cos	0.54170821
×	0.54170821
4	4
=	2.166832841

For convenience, when we transcribe these answers, we will round them to two decimal places.

$$4 \cos 57.2° \approx \mathbf{2.17}$$

example 43.1 Find (a) the sine of A, (b) the cosine of B, (c) the tangent of C. Round answers to two decimal places.

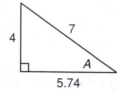 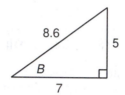 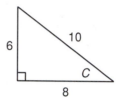

solution We will use our mnemonic to help remember the definitions of sine, cosine, and tangent. We will use a calculator to do the divisions, and we will round answers to two decimal places.

(a) $\sin = \dfrac{\text{Oscar}}{\text{had}}$ $\sin A = \dfrac{\text{opposite}}{\text{hypotenuse}} = \dfrac{4}{7} \approx \mathbf{0.57}$

(b) $\cos = \dfrac{\text{a}}{\text{hold}}$ $\cos B = \dfrac{\text{adjacent}}{\text{hypotenuse}} = \dfrac{7}{8.6} \approx \mathbf{0.81}$

(c) $\tan = \dfrac{\text{on}}{\text{Arthur}}$ $\tan C = \dfrac{\text{opposite}}{\text{adjacent}} = \dfrac{6}{8} = \mathbf{0.75}$

43.B

inverse functions

The inverse sine of a number is the angle whose sine is the number. Several examples will help.

(a) The sine of 30° is 0.5 so the inverse sine of 0.5 is 30°.
(b) The cosine of 30° is approximately 0.866 so the inverse cosine of 0.866 is 30°.
(c) The tangent of 30° is approximately 0.577 so the inverse tangent of 0.577 is 30°.

To find the inverse sine, cosine, and tangent of a number, we use the `inv` key followed by the trigonometric function key.

example 43.2 Find (a) the angle whose sine is 0.643, and (b) the angle whose cosine is 0.216.

solution (a)

ENTER	DISPLAY
0.643	0.643
`inv` `sin`	40.0158 ≈ 40°

(b)

ENTER	DISPLAY
0.216	0.216
`inv` `cos`	77.52579 ≈ 77.53°

practice   Use a calculator to evaluate:

    a.   4 cos 75.8°   **b.**   6 sin 37.42°

For these triangles, find sin *C*, cos *D*, and tan *E*:

  c. 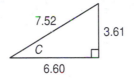   **d.** 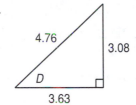   **e.**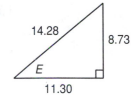

    f.   Find the angle whose tangent is 0.405.

    g.   Find the angle whose sine is 0.794.

problem set 43

1.   The oligarchs were in control, although they comprised only 0.0032 of the total population. If there were 1280 oligarchs, what was the total population?

2.   The boys believed that temerity was a desirable attribute, so the number of boys who were temerarious increased 132 percent in only one month. If the temerarious now numbered 9280, how many boys were temerarious last month?

3.   The chemical formula for sodium hydroxide is NaOH. What is the weight of the sodium (Na) in 320 grams of sodium hydroxide? (Na, 23; O, 16; H, 1)

4.   Twenty-three percent of the mixture was cadmium. If the total weight of the mixture was 3000 grams, what was the weight of the other constituents of the mixture?

5.   Connie was 20 miles ahead of Larry when he started after her. If he caught her in 5 hours and traveled twice as fast as she traveled, how far did he have to go to catch her?

6.   Use the mnemonic "Oscar had a hold on Arthur" as an aid in writing the definition of sine *A*, cosine *A*, and tangent *A*.

7.   Use a calculator to evaluate: (a) 417 cos 51.5°   (b) 32.6 tan 86.3°

8.   Use the definitions of sine, cosine, and tangent and the triangles shown to find to two decimal places: (a) sin *A*, (b) cos *B*, and (c) tan *C*.

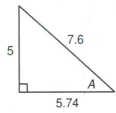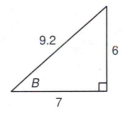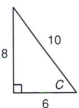

9.   Use unit multipliers to convert 4 ft³ to cubic inches.

10.   Find *b*: $\dfrac{a}{b} - x = \dfrac{p}{z}$ **11.**   Find *p*: $\dfrac{xz}{p} - k = \dfrac{m}{c}$

12.   Find *m*: $\dfrac{x}{m} - \dfrac{k}{c} = \dfrac{p}{z}$

13.   Solve for *a*, *b*, and *c*.

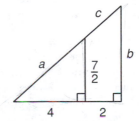

Solve by factoring:

14. $-x^3 - 9x^2 = 20x$

15. $4x^2 - 81 = 0$

16. Expand: $(x - 3)^3$

Simplify:

17. $\dfrac{5x - 6x^2 + x^3}{x^2 + 2x - 3} \cdot \dfrac{x^2 - 9}{x^3 - 8x^2 + 15x}$

18. $16^{-3/4}$

19. $\dfrac{4x + \dfrac{1}{x}}{\dfrac{ay^2}{x} - 4}$

20. $\sqrt{\dfrac{3}{7}} - 4\sqrt{\dfrac{7}{3}}$

21. $3\sqrt{2}(5\sqrt{12} - \sqrt{2})$

22. The altitude of this isosceles triangle is 4. Find A and B.

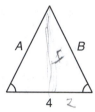

4

23. Find the equation of the line that passes through the point $(-2, -5)$ and has a slope of $-\dfrac{1}{7}$.

24. Solve: $\dfrac{5x - 7}{2} - \dfrac{3x - 2}{5} = 4$

25. Estimate: $\dfrac{(47{,}816 \times 10^5)(4923 \times 10^{-14})}{403{,}000}$

26. Solve the system by graphing and then get an exact solution by using either substitution or elimination.
$$\begin{cases} 2x - 3y = -9 \\ x + y = 2 \end{cases}$$

27. Add: $\dfrac{3}{x + 1} - \dfrac{2}{x^2(x + 1)} + \dfrac{3x + 2}{x^2 - 1}$

28. Simplify: $\dfrac{4x^{-2}y}{p^{-2}}\left(\dfrac{2p^{-2}x^2}{y} - \dfrac{4x^{-2}y}{p^2}\right)$

29. Find the distance between $(-2, 5)$ and $(4, 4)$.

30. Evaluate: $x^2 - yx - (x - y)$ if $x = \dfrac{1}{2}$ and $y = \dfrac{1}{4}$

LESSON 44 *Solving right triangles*

Sines, cosines, and tangents are ratios and are always the same for any given angle. For example, the sine, cosine, and tangent of 37° given by the author's calculator are as follows:

$$\sin 37° = 0.601815 \qquad \cos 37° = 0.7986355 \qquad \tan 37° = 0.753554$$

Since these ratios depend only on the size of the angle and do not depend on the size of the triangle, we can use them to find missing parts of right triangles. If we use these six- and seven-digit figures in our calculations, we tend to get lost in the arithmetic of the problems. This is true even when four-digit values are used. Thus, for the present we will round the numbers and concentrate on the process rather than on the exactness of the answers.

example 44.1 Find the missing parts of this triangle.

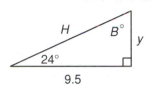

solution The sum of the interior angles of any triangle is 180°. This triangle has one 90° angle and one 24° angle. Thus, angle *B* must be 66° because $90° + 24° + 66° = 180°$.

    Now we can use either the 24° angle or the 66° angle to find *H* and *y*. We decide to use the 24° angle. Beginners often find it difficult to decide whether to use the sine, cosine, or tangent in a particular case. Some find it helpful to use all three and see which ones work out. We will do this. **Before we use a calculator, *we always estimate the answer*.** This will prevent us from making mistakes with the calculator and accepting an answer that is absurd.

We estimate first: *Y* is a number between 3 and 6.
              H is a number between 10 and 14.

Now we use the calculator to find the sine, cosine, and tangent of 24°. For convenience, we round these values to two decimal places.

$$\sin 24° \approx 0.41 \qquad \cos 24° \approx 0.91 \qquad \tan 24° \approx 0.45$$

Next we write all three ratios and substitute.

(a) $\sin 24° = \dfrac{\text{opposite}}{\text{hypotenuse}} \quad \longrightarrow \quad 0.41 \approx \dfrac{y}{H}$ (two unknowns)

(b) $\cos 24° = \dfrac{\text{adjacent}}{\text{hypotenuse}} \quad \longrightarrow \quad 0.91 \approx \dfrac{9.5}{H}$ (one unknown)

(c) $\tan 24° = \dfrac{\text{opposite}}{\text{adjacent}} \quad \longrightarrow \quad 0.45 \approx \dfrac{y}{9.5}$ (one unknown)

We see that we can go no further in (a) because we still have the two unknowns, *y* and *H*, but only one equation. However, equations (b) and (c) can be solved because each of these equations contains only one unknown.

(b) $0.91 \approx \dfrac{9.5}{H} \quad \longrightarrow \quad H \approx \dfrac{9.5}{0.91} \quad \longrightarrow \quad H \approx 10.44$

(c) $0.45 \approx \dfrac{y}{9.5} \quad \longrightarrow \quad (0.45)(9.5) \approx y \quad \longrightarrow \quad 4.28 \approx y$

From (b) we find that *H* is approximately equal to 10.44, which agrees with our estimate of between 10 and 14. Also, we see that *H* equals 9.5 divided by the cosine of 24°. We can use a calculator to compute this value without copying the value of the cosine and rounding. Thus,

$$H = \frac{9.5}{\cos 24°}$$

Enter	Display
9.5	9.5
÷	9.5
24	24
cos	0.913545457
=	**10.39904465** ≈ **10.40**

This answer of 10.40 differs from the 10.44 we got when we rounded the value of the cosine before we divided.

In (c) the y value of 4.28 agrees with our estimate of between 3 and 6. We can get a more exact answer by not rounding the value of the tangent and computing the value on the calculator.

$$y = 9.5 \tan 24°$$

If we do this, we get

$$y = 4.22967251 \approx \textbf{4.23}$$

example 44.2   Find the missing parts of the triangle.

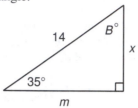

solution   We begin by looking at the figure and estimating the answers.

We estimate first. Segment m is a little shorter than 14, say between 8 and 12. Segment x is even shorter, say between 6 and 10.

Angle B must be 55° because $90° + 35° + 55° = 180°$. This time we decide to use the 35° angle. We use the calculator to find the values of the sine, cosine, and tangent of 35°.

$$\sin 35° \approx 0.57 \qquad \cos 35° \approx 0.82 \qquad \tan 35° \approx 0.70$$

Now we write all three ratios and substitute where possible.

(a) $\sin 35° = \dfrac{\text{opposite}}{\text{hypotenuse}} \longrightarrow 0.57 \approx \dfrac{x}{14}$ (one unknown)

(b) $\cos 35° = \dfrac{\text{adjacent}}{\text{hypotenuse}} \longrightarrow 0.82 \approx \dfrac{m}{14}$ (one unknown)

(c) $\tan 35° = \dfrac{\text{opposite}}{\text{adjacent}} \longrightarrow 0.70 \approx \dfrac{x}{m}$ (two unknowns)

This time we see that we can proceed no further with equation (c) because this equation has two unknowns, x and m. Equations (a) and (b), however, contain only one unknown and can be solved.

(a) $0.57 \approx \dfrac{x}{14} \longrightarrow (0.57)(14) \approx x \longrightarrow \textbf{7.98} \approx \textbf{\textit{x}}$

(b) $0.82 \approx \dfrac{m}{14} \longrightarrow (0.82)(14) \approx m \longrightarrow \textbf{11.48} \approx \textbf{\textit{m}}$

From (a) we see that $x = 14 \sin 35°$. If we use the calculator and multiply 14 by the sine of 35°, we get a slightly different value.

(a) $x = 14 \sin 35° = 8.030070109 \approx \textbf{8.03}$

If we calculate m directly, we get

(b) $m = 14 \cos 35° = 11.46812862 \approx \textbf{11.47}$

These answers agree with our estimates, so we accept them as probably being correct.

example 44.3   Find the missing parts of the triangle.

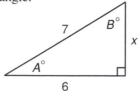

solution   When two sides of a right triangle are given, we can always find the third side by using the Pythagorean theorem.

$$7^2 = 6^2 + x^2$$
$$49 = 36 + x^2$$
$$13 = x^2$$
$$\sqrt{13} = x$$

Now, to find angle A, we will use trigonometric functions. We will try all three functions and see what happens.

(a) $\sin A = \dfrac{\text{opposite}}{\text{hypotenuse}} \longrightarrow \sin A = \dfrac{x}{7}$ (two unknowns)

(b) $\cos A = \dfrac{\text{adjacent}}{\text{hypotenuse}} \longrightarrow \cos A = \dfrac{6}{7}$ (one unknown)

(c) $\tan A = \dfrac{\text{opposite}}{\text{adjacent}} \longrightarrow \tan A = \dfrac{x}{6}$ (two unknowns)

We can't solve equations (a) and (c) because each of these equations has two unknowns. We can solve equation (b), however, and find the cosine of angle A.

(b) $\cos A = \dfrac{6}{7} \longrightarrow \cos A = 0.8571$

The angle whose cosine is 0.8571 is the inverse cosine of 0.8571. Thus, we use the [inv] [cos] keys to find the angle.

ENTER	DISPLAY
0.8571	0.8571
[inv] [cos]	0.8571
[=]	$31.0074861 \approx \mathbf{31.01°}$

Now since the sum of angle A and angle B is $90°$, we can find angle B by subtracting $31.01°$ from $90°$.

Angle $B \approx 90° - 31.01° \longrightarrow$ Angle $B \approx 58.99°$

practice   Use trigonometric functions as necessary to find the missing parts of these triangles:

a. 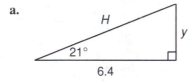   b.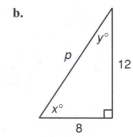

**problem set
44**
   1.   A secret number was unearthed in the Mayan ruins. When the number was increased by 5 and this sum multiplied by -2, the result was 18 greater than 6 times the opposite of the number. What was the secret number found in the ruins?

   2.   Bobby and Garnetta thought of four consecutive odd integers such that 5 times the sum of the first and the third was 22 greater than the product of 8 and the sum of the second and the fourth. Find the numbers.

   3.   The mixture was composed of antimony and tin, and 430 tons of antimony was required to make 700 tons of the mixture. How many tons of tin was required to make 2800 tons of the mixture?

4. Emelio and Nessie had 550 grams of freon in a container. What was the weight of the carbon (C) in the container given that the formula for freon is CCl_2F_2? (C, 12; Cl, 35; F, 19)

5. The trip was only 200 miles, so Bob and Rita rode at 25 mph for a while. Then they doubled their speed for the last part of the trip so that the total time for the journey would be 5 hours. How far did they drive before they increased their speed?

Use trigonometric functions as necessary to find the missing parts of these triangles:

6.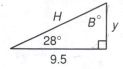

7.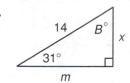

8.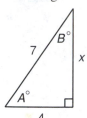

9. Use unit multipliers to convert 4 square miles to square feet.

10. Find m: $\dfrac{k}{m} - c = \dfrac{p}{d}$

11. Find m: $\dfrac{3k}{m} - \dfrac{d}{p} = \dfrac{\ell}{c}$

12. Find c: $\dfrac{a}{c} + d = \dfrac{x}{m}$

13. $ABCD$ is a parallelogram. The measure of $\angle BAD$ is $K°$. Find m$\angle ABC$. Find x and y.

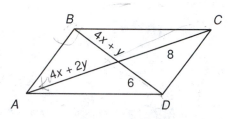

Solve by factoring:

14. $24x = -x^3 + 10x^2$

15. $25p^2 - 81 = 0$

16. Divide $x^4 - 1$ by $x + 4$.

Simplify:

17. $\dfrac{14x + 9x^2 + x^3}{x^2 - x - 6} \div \dfrac{x^3 + 12x^2 + 35x}{x^2 + 10x + 25}$

18. $\dfrac{1}{-32^{4/5}}$

19. This figure is the base of a cone that is 4 feet high. Find the volume of the cone. Dimensions are in inches.

Simplify:

20. $\dfrac{5x^2 + \dfrac{1}{x}}{\dfrac{pm^2}{x} + 5}$

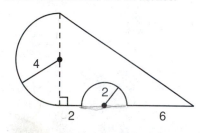

21. $3\sqrt{\dfrac{2}{17}} - 5\sqrt{\dfrac{17}{2}}$

22. $3\sqrt{6}(2\sqrt{6} - \sqrt{12})$

23. Find the equations of lines (a) and (b).

Solve:

24. $\dfrac{5x - 2}{3} - \dfrac{2x - 4}{2} = 6$

25. $3\dfrac{1}{5}k + \dfrac{2}{5} = \dfrac{1}{10}$

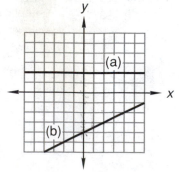

26. Solve the system by graphing and then get an exact solution by using either substitution or elimination.

$$\begin{cases} 2x - y = 5 \\ 4x + 3y = 9 \end{cases}$$

27. Add: $\dfrac{x}{x + 2} - \dfrac{2}{x^2 - 4}$

28. Estimate: $\dfrac{(51{,}463 \times 10^{-14})(748{,}600 \times 10^{-21})}{7{,}861{,}523}$

29. Evaluate: $a^2 - ab^2$ if $a = \dfrac{1}{2}$ and $b = -\dfrac{1}{2}$

30. Simplify: $-2^0 - 2^2 - (-2)^3 - |{-2} - 3^0| - (-2)^3$

LESSON 45   *Difference-of-two-squares theorem*

The two equations

$$\text{(a)} \quad x^2 = 4 \qquad \text{and} \qquad \text{(b)} \quad x^2 - 4 = 0$$

are two different forms of the same equation. In (b) the equation is written as the difference of two squares. It can be solved by factoring.

$$x^2 - 4 = 0 \qquad \text{equation}$$

$$(x + 2)(x - 2) = 0 \qquad \text{factored}$$

$$x = 2 \quad \text{or} \quad x = -2 \qquad \text{solved}$$

In the same way, any other equation that is in the form

$$p^2 = q^2$$

can be solved by rewriting the equation as the difference of two squares and factoring.

$$p^2 = q^2 \qquad \text{equation}$$

$$p^2 - q^2 = 0 \qquad \text{difference-of-two-squares form}$$

$$(p - q)(p + q) = 0 \qquad \text{factored}$$

$$p = +q \quad \text{or} \quad p = -q \qquad \text{solved}$$

It is important to note that we get two solutions to this equation. One is positive, and the other is negative.

This result is important and useful, and we call it the **difference-of-two-squares theorem.** We state it formally here.

> DIFFERENCE-OF-TWO-SQUARES THEOREM
>
> If p and q are real numbers and if $p^2 = q^2$, then
>
> $$p = q \quad \text{or} \quad p = -q$$

We find this theorem will permit quick solutions to problems in which a squared variable is equal to a constant, as shown here.

Equation	Solution
$x^2 = 3$	$x = \pm\sqrt{3}$
$x^2 = 5$	$x = \pm\sqrt{5}$

The same pattern holds if the squared term contains a constant. First, we take the square root of both sides, and then we solve the resulting equation.

Equation	Solution
$(x + 2)^2 = 3$	$x + 2 = \pm\sqrt{3} \longrightarrow x = -2 \pm \sqrt{3}$
$(p - 4)^2 = 5$	$p - 4 = \pm\sqrt{5} \longrightarrow p = 4 \pm \sqrt{5}$
$\left(m + \dfrac{3}{4}\right)^2 = 7$	$m + \dfrac{3}{4} = \pm\sqrt{7} \longrightarrow m = -\dfrac{3}{4} \pm \sqrt{7}$

example 45.1   Solve: $(x + 17)^2 = 2$

solution   We begin our solution by taking the square root of both sides. We remember that there are two answers.

$$(x + 17)^2 = 2 \qquad\qquad \text{equation}$$

$$x + 17 = \pm\sqrt{2} \qquad\qquad \text{square root of both sides}$$

$$x = -17 + \sqrt{2} \quad \text{and} \quad x = -17 - \sqrt{2} \quad \text{two solutions}$$

We will check both solutions in the original equation.

Check $-17 + \sqrt{2}$:   Check $-17 - \sqrt{2}$:

$$(-17 + \sqrt{2} + 17)^2 = 2 \qquad\qquad (-17 - \sqrt{2} + 17)^2 = 2$$

$$(\sqrt{2})^2 = 2 \qquad\qquad\qquad (-\sqrt{2})^2 = 2$$

$$2 = 2 \quad \text{Check} \qquad\qquad 2 = 2 \quad \text{Check}$$

example 45.2   Solve: $\left(x + \dfrac{2}{5}\right)^2 = 3$

solution   We begin by taking the square root of both sides of the equation.

$$\left(x + \frac{2}{5}\right)^2 = 3 \qquad\qquad \text{equation}$$

$$x + \frac{2}{5} = \pm\sqrt{3} \qquad\qquad \text{square root of both sides}$$

$$x = -\frac{2}{5} \pm \sqrt{3} \qquad \text{added } -\frac{2}{5} \text{ to both sides}$$

Now we check:

Check $-\dfrac{2}{5} + \sqrt{3}$:   Check $-\dfrac{2}{5} - \sqrt{3}$:

$$\left(-\frac{2}{5} + \sqrt{3} + \frac{2}{5}\right)^2 = 3 \qquad\qquad \left(-\frac{2}{5} - \sqrt{3} + \frac{2}{5}\right)^2 = 3$$

$$(\sqrt{3})^2 = 3 \qquad\qquad\qquad (-\sqrt{3})^2 = 3$$

$$3 = 3 \quad \text{Check} \qquad\qquad 3 = 3 \quad \text{Check}$$

practice   Use the difference-of-two-squares theorem to find the solution to each equation.

   a. $x^2 = 14$ **b.** $(x + 9)^2 = 11$ **c.** $\left(x + \dfrac{1}{7}\right)^2 = 8$

problem set 45

1. Since knowledge of chemistry is useful even in nonscientific fields of study, a majority of the students elected to take chemistry. If 38 percent did not take chemistry and 248 students did take chemistry, how many students were there in all?

2. The chemical formula for phosphine is PH_3. If the hydrogen (H) in a quantity of phosphine weighed 24 grams, what was the total weight of the phosphine? (P, 31; H, 1)

3. Sergio and Lolita found that bromides took up 36 percent of the storage space in the stockroom. If 5376 bottles did not contain bromides, how many did contain bromides?

4. Roses were $12 a bunch and carnations were $14 a bunch. Amy and Kathy bought a total of 11 bunches and spent $138. How many bunches of each did they buy?

5. The train made the trip in 8 hours, and the bus made all but 80 miles of the trip in 12 hours. What was the rate of each if the rate of the train was 30 mph greater than the rate of the bus?

Use the difference-of-two-squares theorem to find the solution to each equation.

6. $x^2 = 5$

7. $(x + 7)^2 = 11$

8. $(x + 12)^2 = 2$

9. $\left(x + \dfrac{3}{4}\right)^2 = 13$

10. Find side x.

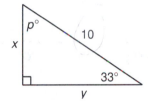

11. Find side p.

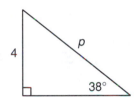

12. Find angles A and B.

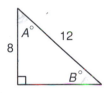

13. Use unit multipliers to convert 100,000 square miles to square inches.

14. Estimate: $\dfrac{(571,652)(40,316)}{214,000 \times 10^6}$

15. Find side d.

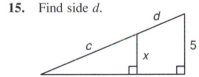

16. Find p: $\dfrac{mx}{4} + \dfrac{y}{p} = c$

17. Find p: $\dfrac{5x}{p} - k = \dfrac{m}{c}$

Solve by factoring:

18. $x^3 = -5x^2 + 50x$

19. $36x^2 - 25 = 0$

Simplify:

20. $\dfrac{x^3 + 6x^2 - 7x}{x^2 + 4x - 21} \div \dfrac{x^3 - 2x + x^2}{x^2 + 2x - 15}$

21. $-(-27)^{-5/3}$

22. The perimeter of this figure is $(24 + 4\pi)$ cm. Find the area of the shaded semicircle.

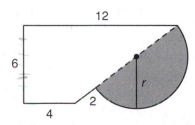

Simplify:

23. $\dfrac{4xp - \dfrac{1}{x}}{6p - \dfrac{p^2}{x}}$

24. $5\sqrt{\dfrac{2}{5}} + 3\sqrt{\dfrac{5}{2}}$

25. $2\sqrt{5}(3\sqrt{15} - 2\sqrt{10})$

26. Find the equation of the line that passes through $(-2, -5)$ and has a slope of $-\dfrac{2}{3}$.

27. Solve: $\dfrac{5x - 2}{7} - \dfrac{x - 3}{5} = 4$

28. Solve the system by graphing and then find an exact solution by using either substitution or elimination.

$$\begin{cases} 2x + 3y = 3 \\ x - 5y = 20 \end{cases}$$

29. Add: $\dfrac{2}{x - 2} - \dfrac{3}{x + 2} - \dfrac{2x}{x^2 - 4}$

30. Evaluate: $a^2 - ab^3$ if $a = -\dfrac{1}{2}$ and $b = \dfrac{1}{2}$

LESSON 46 *More on radical expressions • Radicals to fractional exponents*

46.A

more on radical expressions

We have been simplifying expressions such as

$$-\sqrt{\dfrac{2}{3}} + 3\sqrt{\dfrac{3}{2}}$$

by first using the quotient-of-square-roots theorem, then rationalizing the denominators by using the denominator-numerator same-quantity rule, and finishing by finding a common denominator and adding.

(a) $\quad -\dfrac{\sqrt{2}}{\sqrt{3}} + \dfrac{3\sqrt{3}}{\sqrt{2}}$ $\qquad\qquad\qquad$ quotient-of-square-roots theorem

(b) $\quad -\dfrac{\sqrt{2}}{\sqrt{3}}\dfrac{\sqrt{3}}{\sqrt{3}} + \dfrac{3\sqrt{3}}{\sqrt{2}}\dfrac{\sqrt{2}}{\sqrt{2}} = -\dfrac{\sqrt{6}}{3} + \dfrac{3\sqrt{6}}{2}$ $\qquad$ rationalized denominators

(c) $\quad \dfrac{-2\sqrt{6}}{6} + \dfrac{9\sqrt{6}}{6} = \dfrac{7\sqrt{6}}{6}$ $\qquad\qquad$ found common denominator and added

Problems such as this one are good practice problems but will provide even more practice if we add one more term whose simplification requires that we use the product-of-square-roots theorem.

example 46.1 Simplify: $3\sqrt{\dfrac{5}{2}} + 3\sqrt{\dfrac{2}{5}} - \sqrt{90}$

solution The last term was contrived so it will contain a $\sqrt{10}$, as will the first two terms. We begin by rationalizing the denominators and simplifying the last term.

$$\frac{3\sqrt{5}}{\sqrt{2}}\frac{\sqrt{2}}{\sqrt{2}} + \frac{3\sqrt{2}}{\sqrt{5}}\frac{\sqrt{5}}{\sqrt{5}} - \sqrt{9}\sqrt{10} \qquad \text{used rules for square roots}$$

$$\frac{3\sqrt{10}}{2} + \frac{3\sqrt{10}}{5} - 3\sqrt{10} \qquad \text{simplified}$$

Now we use 10 as the common denominator and add.

$$\frac{15\sqrt{10}}{10} + \frac{6\sqrt{10}}{10} - \frac{30\sqrt{10}}{10} = \frac{-9\sqrt{10}}{10} \qquad \text{simplified and added}$$

example 46.2   Simplify: $3\sqrt{\dfrac{2}{7}} - 5\sqrt{\dfrac{7}{2}} + 3\sqrt{56}$

solution   Again we begin by simplifying the radicals.

$$\frac{3\sqrt{2}}{\sqrt{7}}\frac{\sqrt{7}}{\sqrt{7}} - \frac{5\sqrt{7}}{\sqrt{2}}\frac{\sqrt{2}}{\sqrt{2}} + 3\sqrt{4}\sqrt{14}$$

$$\frac{3\sqrt{14}}{7} - \frac{5\sqrt{14}}{2} + 6\sqrt{14}$$

As the last step, we use 14 as the common denominator and add.

$$\frac{6\sqrt{14}}{14} - \frac{35\sqrt{14}}{14} + \frac{84\sqrt{14}}{14} = \frac{55\sqrt{14}}{14}$$

46.B
radicals to fractional exponents

Many radical expressions are easier to simplify if we first change the radical expressions to expressions that have fractional exponents.

example 46.3   Simplify: $\sqrt{3\sqrt{3}}$

solution   We begin by replacing both radicals with parentheses, brackets, and fractional exponents.

$$[3(3^{1/2})]^{1/2}$$

Next we use the power theorem and get

$$3^{1/2}(3^{1/4})$$

We multiply expressions with like bases by adding exponents, so we get as our answer

$$\mathbf{3^{3/4}}$$

example 46.4   Simplify: $\sqrt[3]{3\sqrt{3}}$

solution   First we replace the radicals with parentheses, brackets, and fractional exponents.

$$[3(3^{1/2})]^{1/3}$$

Next we use the power theorem to simplify.

$$3^{1/3}(3^{1/6})$$

Then we multiply the exponential expressions by adding the exponents.

$$3^{2/6}(3^{1/6}) = 3^{3/6} = \mathbf{3^{1/2}}$$

example 46.5   Simplify: $\sqrt{x^3 y^2}\,\sqrt[3]{xy}$ †

solution   We replace the radicals with parentheses and fractional exponents and get

$$(x^3y^2)^{1/2}(xy)^{1/3}$$

Next we use the power theorem.

$$x^{3/2}yx^{1/3}y^{1/3}$$

We finish by multiplying exponential expressions with like bases and we get

$$x^{3/2}x^{1/3}yy^{1/3} = x^{9/6}x^{2/6}y^{3/3}y^{1/3} = \mathbf{x^{11/6}y^{4/3}}$$

example 46.6   Simplify: $\sqrt[3]{x^5 y^3}\,\sqrt[4]{xy^5}$

solution   First we replace the radicals with parentheses and fractional exponents.

$$(x^5y^3)^{1/3}(xy^5)^{1/4}$$

Next we use the power theorem.

$$x^{5/3}yx^{1/4}y^{5/4}$$

Finally we simplify to get

$$x^{23/12}y^{9/4}$$

practice   Simplify:

a. $3\sqrt{\dfrac{7}{3}} + 5\sqrt{\dfrac{3}{7}} - \sqrt{84}$ b. $2\sqrt{\dfrac{3}{5}} - 5\sqrt{\dfrac{5}{3}} + \sqrt{135}$ c. $\sqrt{7\sqrt{7}}$

problem set 46

1.   Many did not consider a 140 percent increase to be excessive, as the total was still only 1440. What was the total before the increase?

2.   One thousand grams of beryllium and 3000 grams of other elements made up the mixture. If 24,000 kilograms of the mixture was needed, how much beryllium was required?

3.   The formula for ammonium chloride is NH_4Cl. If the chlorine (Cl) in a quantity of ammonium chloride weighed 140 grams, what was the weight of the ammonium chloride? (N, 14; H, 1; Cl, 35)

4.   The ratio of fairies to elves was 7 to 2, and the number of fairies was 11 greater than 3 times the number of elves. How many of each were there?

5.   Bruce and Maria headed north at 11 a.m. at full speed. After 4 hours Bruce was 16 miles ahead. What did Bruce consider to be full speed if Maria's speed was 16 mph?

Simplify:

6. $5\sqrt{\dfrac{3}{2}} + 2\sqrt{\dfrac{2}{3}} - \sqrt{600}$ 7. $3\sqrt{\dfrac{5}{7}} - 5\sqrt{\dfrac{7}{5}} + 3\sqrt{140}$

8. $\sqrt{5\sqrt{5}}$ 9. $\sqrt[3]{6\sqrt{6}}$ 10. $\sqrt{x^3 y^3}\,\sqrt[4]{xy}$ 11. $\sqrt[4]{x^5 y^3}\,\sqrt[3]{xy^5}$

Solve:

12. $\left(x - \dfrac{2}{5}\right)^2 = 7$ 13. $\left(x - \dfrac{1}{4}\right)^2 = 5$

---

†Even roots of even powers can be troublesome when the replacements for the variable are negative numbers. Thus, for the present, the domain for problems like this one will be assumed to be the set of positive real numbers.

14. Find angle B and side y.

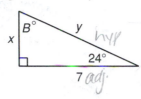

15. Find angle A and side C.

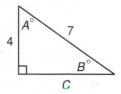

16. Estimate: $\dfrac{(476,800)(9,016,423 \times 10^4)}{408 \times 10^{10}}$

17. Find m: $\quad \dfrac{ax}{b} - c = \dfrac{k}{m}$

18. Find c: $\quad \dfrac{x}{m} - \dfrac{yb}{c} = p$

19. Find sides x and y.

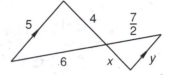

20. Solve $-40x = 13x^2 + x^3$ by factoring.

Simplify:

21. $\dfrac{x^2 + 45 + 14x}{x^3 + 10x^2 + 9x} \cdot \dfrac{x^3 - 2x - x^2}{x^2 - 2x - 35}$

22. $8^{-4/3}$

23. Find x, y, and z.

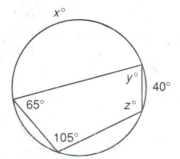

Simplify:

24. $\dfrac{\dfrac{m^2}{x} - x}{\dfrac{p^2}{x} + 2x}$

25. $3\sqrt{7}(2\sqrt{14} - \sqrt{7})$

26. Find the equation of the line that goes through $(-2, 4)$ and is perpendicular to the line that goes through $(-5, 7)$ and $(4, 2)$.

Solve:

27. $\dfrac{-x + 3}{2} - \dfrac{2x + 4}{3} = 5$

28. $\dfrac{-2x - 5}{4} - \dfrac{x - 2}{3} = 7$

29. Solve the system by graphing and then get an exact solution by using either substitution or elimination.
$$\begin{cases} x - y = -3 \\ x + 2y = -2 \end{cases}$$

30. Evaluate: $a^2 - a^3b - ab$ $\quad$ if $a = -\dfrac{1}{2}$ and $b = -\dfrac{1}{4}$

LESSON 47   **Rate unit conversions • More on fractional exponents**

47.A

rate unit conversions

Rate units such as miles per hour or feet per second contain both a distance unit and a time unit. To convert rate units, we use distance unit multipliers to convert the distance units and time unit multipliers to convert the time units.

example 47.1   Use unit multipliers to convert 42 inches per second to miles per hour.

solution   We begin by writing the unit "inches per second" in fractional form.

$$\frac{42 \text{ in.}}{\text{sec}}$$

We will convert the inches to miles first. We will go from inches to feet and from feet to miles.

$$\frac{42 \text{ in.}}{\text{sec}} \times \frac{1 \text{ ft}}{12 \text{ in.}} \times \frac{1 \text{ mi}}{5280 \text{ ft}}$$

Now our rate is miles per second. We continue by converting seconds to minutes and then minutes to hours.

$$\frac{42 \text{ in.}}{\text{sec}} \times \frac{1 \text{ ft}}{12 \text{ in.}} \times \frac{1 \text{ mi}}{5280 \text{ ft}} \times \frac{60 \text{ sec}}{\text{min}} \times \frac{60 \text{ min}}{\text{hr}} = \frac{(42)(60)(60)}{(12)(5280)} \frac{\text{mi}}{\text{hr}}$$

If we use a calculator to multiply and divide and then round the answer to two decimal places, we get

$$2.39 \frac{\text{mi}}{\text{hr}}$$

example 47.2   Use unit multipliers to convert 480 miles per hour to feet per second.

solution   Again we begin by writing the given units in fractional form.

$$\frac{480 \text{ mi}}{\text{hr}}$$

Now we will use one unit multiplier to convert miles to feet and two unit multipliers to convert hours to seconds.

$$\frac{480 \text{ mi}}{\text{hr}} \times \frac{5280 \text{ ft}}{\text{mi}} \times \frac{\text{hr}}{60 \text{ min}} \times \frac{\text{min}}{60 \text{ sec}} = \frac{(480)(5280)}{(60)(60)} \frac{\text{ft}}{\text{sec}}$$

Now a calculator can be used to get a decimal answer if desired.

example 47.3   Use unit multipliers to convert 15 inches per second to yards per hour.

solution   We will convert inches to feet to yards, and seconds to minutes to hours.

$$\frac{15 \text{ in.}}{\text{sec}} \times \frac{1 \text{ ft}}{12 \text{ in.}} \times \frac{1 \text{ yd}}{3 \text{ ft}} \times \frac{60 \text{ sec}}{\text{min}} \times \frac{60 \text{ min}}{\text{hr}} = \frac{(15)(60)(60)}{(12)(3)} \frac{\text{yd}}{\text{hr}}$$

47.B

more on fractional exponents

Some problems involving exponents are trick problems and cannot be simplified unless one realizes that they are trick problems.

example 47.4   Simplify: $\sqrt[3]{27\sqrt{3}}$

solution   This is a trick problem that cannot be simplified unless it is recognized. The trick is to write 27 as 3^3 so all the bases will be the same.

$$\sqrt[3]{3^3\sqrt{3}}$$

Next we replace the radicals with parentheses, brackets, and fractional exponents.

$$[3^3(3^{1/2})]^{1/3}$$

Now we use the power theorem and finish by adding the exponents.

$$3^1(3^{1/6}) = \mathbf{3^{7/6}}$$

To work these problems, we must always remember simplifications such as

$$4 = 2^2 \qquad 9 = 3^2 \qquad 16 = 2^4 \qquad 27 = 3^3$$

If we do not recognize the problem and make these replacements, the expressions cannot be simplified.

example 47.5   Simplify: $\sqrt[6]{4\sqrt{2}}$

solution   First we write 4 as 2^2.

$$\sqrt[6]{2^2\sqrt{2}}$$

Next we replace the radicals with parentheses, brackets, and fractional exponents.

$$[2^2(2^{1/2})]^{1/6}$$

Then we finish by using the power theorem and the product theorem.

$$2^{2/6}(2^{1/12}) = \mathbf{2^{5/12}}$$

practice   Simplify:

a. $\sqrt[3]{64\sqrt{4}}$ b. $\sqrt[6]{9\sqrt{3}}$

Use unit multipliers to convert:

c.   85 miles per second to miles per hour

d.   207 miles per hour to feet per second

problem set 47

1.   Three-sevenths of the assembled throng watched in horror as the tsunami approached. If 2800 did not watch in horror, how many were in the throng?

2.   Melissa giggled as she thought about four consecutive odd integers such that 5 times the sum of the first two was 10 less than 7 times the sum of the second and fourth. What were her four integers?

3.   Thirty percent of the girls often pondered their muliebrity, and the rest didn't even know what the word meant. If 1400 were ignorant of the meaning, how many girls were there in all?

4.   The chemical formula for sulfurous acid is H_2SO_3. If the oxygen in a quantity of sulfurous acid weighed 192 grams, what was the total weight of the sulfurous acid? (H, 1; S, 32; O, 16)

5.   Tacitus made the first leg of the trip to Londinium in his chariot at 8 kph and walked the last leg at 4 kph. If the total distance traveled was 48 km and the total time was 7 hours, how far did he walk and how far did he ride?

Use unit multipliers to convert:

6.   52 inches per second to miles per hour

7.   805 miles per hour to feet per second

8. 13 inches per second to yards per hour

Simplify:

9. $\sqrt[3]{25\sqrt{5}}$

10. $\sqrt[7]{9\sqrt{3}}$

11. $\sqrt{x^2 y}\sqrt[3]{y^2 x}$

12. $\sqrt{2\sqrt[3]{2}}$

13. $3\sqrt{\dfrac{2}{3}} - 5\sqrt{\dfrac{3}{2}} + 2\sqrt{24}$

14. $3\sqrt{\dfrac{5}{7}} - 2\sqrt{\dfrac{7}{5}} - \sqrt{315}$

15. Find x.

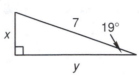

16. Find x.

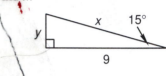

Solve:

17. $(x - 3)^2 = 5$

18. $\left(x + \dfrac{2}{7}\right)^2 = \dfrac{4}{49}$

19. Estimate: $\dfrac{(36{,}421 \times 10^5)(493{,}025)}{40{,}216 \times 10^7}$

20. Find x: $\dfrac{ay}{x} + p = \dfrac{m}{c}$

21. Find p: $\dfrac{a}{x} - \dfrac{c}{p} = b$

22. Find side C.

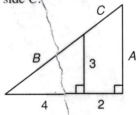

23. Given: $m\angle JAC = 150°$
Find: $x, y, z, s, p,$ and m.

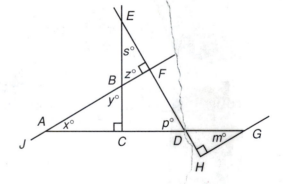

Simplify:

24. $\dfrac{\dfrac{x^2 p}{m} - m}{\dfrac{x}{m} - p}$

25. $\dfrac{-14x + x^3 + 5x^2}{x^2 + 12x + 35} \div \dfrac{x^3 - 2x^2 - 15x}{x^2 + 8x + 15}$

26. $-27^{-4/3}$

27. Solve the system by graphing and then get an exact solution by using either substitution or elimination.
$$\begin{cases} x - 4y = 8 \\ 2x - 3y = 9 \end{cases}$$

28. Find the equation of the line that goes through $(-2, 5)$ and has a slope of $\dfrac{2}{5}$.

29. Solve $20x = -12x^2 - x^3$ by factoring.

30. Evaluate: $-a - a^2 - a^3 - ab$ if $a = -\dfrac{1}{2}$ and $b = \dfrac{1}{5}$

LESSON 48 *Radical equations*

We have noted that the difference-of-two-squares theorem permits the solution of some equations by taking the square roots of both sides of the equation. We can use this theorem to find the roots of the equation

$$(x + 2)^2 - 5 = 0$$

First, we transform the equation by adding +5 to both sides, and we get

$$(x + 2)^2 = 5$$

Next we take the square root of both sides, remembering that we get two signs on the right-hand side.

$$x + 2 = \pm\sqrt{5} \qquad \text{square root of both sides}$$

$$x = -2 \pm \sqrt{5} \qquad \text{added } -2 \text{ to both sides}$$

If we wish, we can list the two roots separately by writing

$$x = -2 + \sqrt{5} \qquad \text{and} \qquad x = -2 - \sqrt{5}$$

We use another process to solve equations that contain a square root radical. First we isolate the radical, and then we square both sides of the equation. This process is permitted because it follows directly from the definition of the square root of a positive number.

DEFINITION OF SQUARE ROOT

If x is a positive real number, then $\sqrt{x}$ is the unique positive real number such that $(\sqrt{x})^2 = x$.

But when we do this, we must always check our answers in the original equation because squaring both sides of an equation sometimes generates an equation that has roots that are not roots of the original equation. To see how this is possible, let us begin with the equation

$$x = 2$$

whose only solution is the number 2. Now if we square both sides of this equation, we get the equation

$$x^2 = 4$$

Now we have an equation that can be satisfied by using either +2 or −2 as the replacement for x. **The number 2 is a solution to both the original equation and the new equation, but −2 is not a solution to the original equation. By squaring both sides, we have generated an equation that has more solutions than the original equation.**

example 48.1   Solve: $\sqrt{x - 4} - 2 = 6$

solution   We begin by **isolating the radical** by adding +2 to both sides of the equation, and we get

$$\sqrt{x - 4} = 8$$

Next we square both sides, remembering that $(\sqrt{x - 4})^2 = x - 4$.

$$x - 4 = 64$$

We finish by adding +4 to both sides.

$$x = 68$$

Now we must check this solution in the original equation to see if it satisfies the original equation.

$$\sqrt{x - 4} - 2 = 6 \quad \text{original equation}$$
$$\sqrt{68 - 4} - 2 = 6 \quad \text{substituted 68 for } x$$
$$\sqrt{64} - 2 = 6 \quad \text{simplified}$$
$$8 - 2 = 6 \quad \text{took square root}$$
$$6 = 6 \quad \text{check}$$

Thus, 68 is a solution to the original equation.

example 48.2 Solve: $\sqrt{x + 4} + 2 = -7$

solution We always begin by **isolating the radical.**

$$\sqrt{x + 4} = -9$$

Now we square both sides and solve.

$$x + 4 = 81 \quad \longrightarrow \quad \boldsymbol{x = 77}$$

Now we must check the original equation to see if 77 is a solution.

$$\sqrt{x + 4} + 2 = -7 \quad \text{original equation}$$
$$\sqrt{77 + 4} + 2 = -7 \quad \text{substituted}$$
$$\sqrt{81} + 2 = -7 \quad \text{simplified}$$
$$9 + 2 = -7 \quad \text{not true}$$

Since $9 + 2 \neq -7$, then 77 is not a solution to the original equation. Thus, we find that the original equation has **no real number solution**.

example 48.3 Solve: $\sqrt{x - 3} - 3 = 5$

solution We begin by **isolating the radical.**

$$\sqrt{x - 3} = 8$$

Now we square both sides and solve.

$$x - 3 = 64 \quad \longrightarrow \quad \boldsymbol{x = 67}$$

We finish by checking 67 in the original equation.

$$\sqrt{67 - 3} - 3 = 5 \quad \longrightarrow \quad \sqrt{64} - 3 = 5 \quad \longrightarrow \quad 8 - 3 = 5 \quad \longrightarrow \quad 5 = 5 \quad \text{Check}$$

example 48.4 Solve: $\sqrt{x^2 + 2x + 13} - 3 = x$

solution We begin by isolating the radical by adding +3 to both sides.

$$\sqrt{x^2 + 2x + 13} = x + 3$$

Now we square both sides and solve.

$$x^2 + 2x + 13 = x^2 + 6x + 9 \quad \text{squared both sides}$$
$$4 = 4x \quad \text{simplified}$$
$$\boldsymbol{1 = x} \quad \text{divided by 4}$$

We check by replacing x with 1 in the original equation.

$$\sqrt{(1)^2 + 2(1) + 13} - 3 = 1 \quad \text{replaced } x \text{ with 1}$$
$$\sqrt{16} - 3 = 1 \quad \text{simplified}$$
$$1 = 1 \quad \text{check}$$

practice   Solve:

a. $\sqrt{x - 5} - 3 = 7$ b. $\sqrt{x + 5} + 1 = -11$

c. $\sqrt{x^2 + 3x - 10} + 2 = x$

**problem set
48**

1. The licutenant noted that 0.68 of the frontiersmen at the conclave wore buckskin. If 512 did not wear buckskin, how many frontiersmen were at the conclave?

2. If 140 grams of germanium were required to make 1540 grams of the compound, how many grams of other elements were required to make 6160 grams of the compound?

3. If a crucible contained 972 grams of $FeBr_2$, what was the weight of the iron (Fe) in the crucible? (Fe, 56; Br, 80)

4. In a group of children, 5 times the number of boys was 17 greater than 3 times the number of girls. Also, 6 times the number of girls was 2 greater than the number of boys. How many were boys and how many were girls?

5. The freight train took 20 hours to make the same trip the express train made in 10 hours. Find the speed of each if the express train was 30 mph faster than the freight train.

Solve:

6. $\sqrt{x - 3} - 3 = 4$ 7. $\sqrt{x + 5} + 3 = -4$

8. $\sqrt{x^2 + 8x + 15} - 5 = x$

9. Use unit multipliers to convert 60 miles per hour to feet per second.

Simplify:

10. $\sqrt{2\sqrt[3]{2}}$ 11. $\sqrt{m^2 y \sqrt[3]{m^4 y}}$ 12. $\sqrt{8\sqrt[3]{2}}$

13. $3\sqrt{\dfrac{2}{5}} - \sqrt{\dfrac{5}{2}} + 2\sqrt{40}$ 14. $2\sqrt{\dfrac{7}{11}} - \sqrt{\dfrac{11}{7}} + 2\sqrt{308}$

Solve:

15. $\left(x - \dfrac{2}{7}\right)^2 = 4$ 16. $(x - 3)^2 = 16$

17. Find angle C and side b.   18. Find angle A.

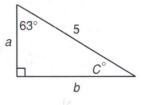

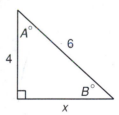

19. Estimate: $\dfrac{(4,071,623)(51,642 \times 10^5)}{200,000 \times 10^{-13}}$ 20. Find p: $\dfrac{b}{x} - \dfrac{c}{p} + k = \dfrac{m}{y}$

21. If the measure of an angle is $A°$, the measure of its supplement is $(180 - A)°$ and the measure of its complement is $(90 - A)°$. Find an angle such that twice its supplement is 40° greater than 4 times its complement.

22. Divide $3x^3 - 2$ by $x + 1$.

23. Find the distance between the points $(-4, -3)$ and $(-8, 2)$.

Simplify:

24. $\dfrac{x^2 - 3x + 2}{x^3 + 4x^2 - 5x} \div \dfrac{x^2 + x - 6}{35x + 12x^2 + x^3}$ 25. $-16^{-3/4}$

26. The figure is the base of a cone that is 4 inches high. The dimensions are in inches. Find the volume of the cone in cubic inches.

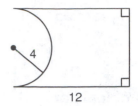

27. Find the distance between the points $(-2, 5)$ and $(7, -3)$.

28. Simplify: $\dfrac{x\left(\dfrac{x-1}{x}\right) - \dfrac{1}{x^2}}{\dfrac{x+2}{x^2} - 4}$

29. Find the equation of the line that passes through the points $(2, 5)$ and $(7, -3)$.

30. Solve: $\dfrac{2x-3}{2} - \dfrac{x}{2} = \dfrac{5}{3}$

LESSON 49 *Linear intercepts • Transversals*

49.A

linear intercepts

We remember that a line is completely identified if we know its slope and its intercept. The equation of the line in the figure on the left is

$$y = -\frac{2}{3}x + 2$$

The first number in the equation is $-\frac{2}{3}$. This number is the slope and tells us that the slope is negative and that the ratio of the rise to the run is 2 to 3. The last number in the equation is 2, and this number is the y intercept of the line, which is the y coordinate of the point where the line crosses the y axis.

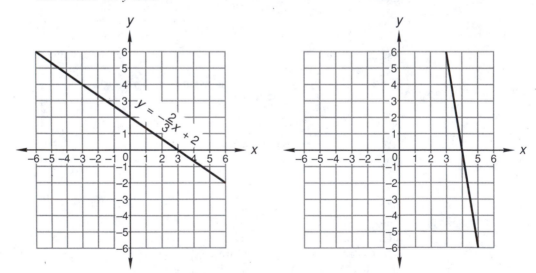

Sometimes it is necessary to find the equation of a line whose intercept is not shown on the graph. The line shown on the right is an example. By drawing a triangle, we can determine that the slope of this line is approximately -6. If we replace m with -6 in the equation $y = mx + b$, we get

$$y = -6x + b$$

Now, to determine the intercept, we must estimate the coordinates of one point on the line and use these numbers for *x* and *y* in the equation. It appears that the line goes through the point (4, 0). We will use these numbers for *x* and *y* and solve for *b*.

$$0 = -6(4) + b \quad \text{substituted}$$

$$0 = -24 + b \quad \text{multiplied}$$

$$24 = b \quad \text{solved}$$

Now that we have the intercept, we can write our estimate of the equation of the line as

$$y = -6x + 24$$

49.B
transversals

The next two problems are designed to give us practice both with right triangle trigonometry and with the relationships of the angles formed when transversals intercept parallel lines.

example 49.1   Lines (1) and (2) are parallel. Find sides *a* and *b*.

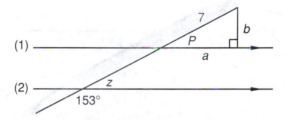

solution   Angle *z* equals 27° because it and the 153° angle together make a straight line, whose angle measure is 180°. Then angle *P* is 27° because $\angle P$ and $\angle z$ are corresponding angles.

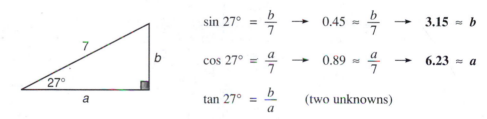

$$\sin 27° = \frac{b}{7} \longrightarrow 0.45 \approx \frac{b}{7} \longrightarrow \mathbf{3.15 \approx b}$$

$$\cos 27° = \frac{a}{7} \longrightarrow 0.89 \approx \frac{a}{7} \longrightarrow \mathbf{6.23 \approx a}$$

$$\tan 27° = \frac{b}{a} \quad \text{(two unknowns)}$$

example 49.2   Lines (1) and (2) are parallel. Find sides *M* and *N*.

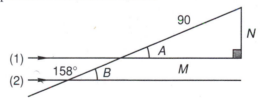

solution   Angle *B* is 22° because 22° + 158° = 180°. Thus, angle *A* equals 22° because $\angle A$ and $\angle B$ are corresponding angles.

$$\sin 22° = \frac{N}{90} \longrightarrow 0.37 \approx \frac{N}{90} \longrightarrow \mathbf{33.3 \approx N}$$

$$\cos 22° = \frac{M}{90} \longrightarrow 0.93 \approx \frac{M}{90} \longrightarrow \mathbf{83.7 \approx M}$$

$$\tan 22° = \frac{N}{M} \quad \text{(two unknowns)}$$

practice   **a.** Find the equation of the line shown.   **b.** Find N.

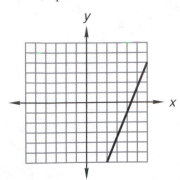

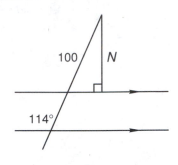

problem set   **1.** Five times the sum of a number and -13 is 92 less than 4 times the opposite of the
49   number. What is the number?

2. The number of celebrants increased 160 percent as the saturnalia drew to a close. If
there were 1092 celebrants at the end, what was the number before the increase?

3. The formula for beryllium fluoride is BeF_2. If the fluorine (F) in a quantity of this
compound weighs 95 grams, what is the total weight of the beryllium fluoride? (Be, 9;
F, 19)

4. In Saudi Arabian currency 1 riyal equals 20 qurush. Sheik Ahab had 15 coins whose
value was 205 qurush. How many coins of each kind were there?

5. Grant and Bruce went jogging. Grant was 200 yards ahead of Bruce after 10 minutes.
Find Bruce's speed if Grant's speed was 240 yards per minute.

6. Find the equation of the line shown.   **7.** Find side N.

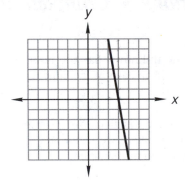

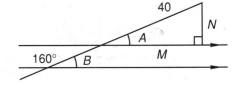

Solve:

8. $\sqrt{x - 4} - 3 = 5$ **9.** $\sqrt{x^2 - 2x + 5} = x + 1$

10. Use unit multipliers to convert 200 inches per minute to feet per second.

Simplify:

11. $\sqrt[3]{9\sqrt{3}}$ **12.** $\sqrt[4]{2}\sqrt[3]{2}$

13. $\sqrt{x^2 y^3}\sqrt{xy^4}$ **14.** $\sqrt[3]{xy}\sqrt{xy}$

15. $3\sqrt{\dfrac{7}{2}} + 2\sqrt{\dfrac{2}{7}} - 3\sqrt{56}$ **16.** $5\sqrt{\dfrac{2}{9}} + 3\sqrt{\dfrac{9}{2}} + \sqrt{162}$

Solve:

17. $\left(x - \dfrac{5}{3}\right)^2 = 5$ **18.** $(x + 2)^2 = 16$

19. Estimate: $\dfrac{(517{,}832 \times 10^{-14})(80{,}123)}{200{,}000 \times 10^{-42}}$

20. Find y: $\dfrac{x}{y} + c = \dfrac{m}{x} - d$

21. Find p: $\dfrac{2}{x} - \dfrac{3}{y} = \dfrac{m}{p}$

22. Expand: $(x + 1)^3$

23. Simplify: $\dfrac{x^2 - 6x + 5}{20x - 9x^2 + x^3} \div \dfrac{2 - 3x + x^2}{x^3 - 7x^2 + 12x}$

24. Find the distance between the points $(4, 2)$ and $(-5, 3)$.

25. Find the equation of the line that passes through $(4, 2)$ and $(-5, 3)$.

26. The measure of each of the exterior angles of a regular polygon is $12°$. How many sides does the polygon have?

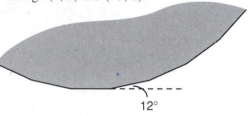

12°

Add:

27. $\dfrac{x}{ay^2} - \dfrac{3x}{a^2 y^3} - \dfrac{5}{y^4}$

28. $\dfrac{-3}{x + 3} - \dfrac{3x + 2}{x^2 - 9}$

Solve:

29. $\dfrac{x - 2}{2} - \dfrac{x - 3}{3} = 2$

30. $\dfrac{x}{2} - \dfrac{x + 2}{5} = 1$

LESSON 50   *Quadratic equations • Completing the square*

50.A
quadratic equations

A quadratic equation in one unknown has the number 2 as the highest power of the variable. The highest power of x in each of the three equations shown here is 2,

$$x^2 = -4x + 2 \qquad x - 3 = x^2 \qquad x^2 + 4x - 2 = 0$$

so these equations are all quadratic equations. We have solved quadratic equations thus far by two methods. To use the first method, we factor and then use the zero factor theorem. To review this method, we begin with the equation

$$x^2 + x - 2 = 0$$

First we factor and get

$$(x + 2)(x - 1) = 0$$

and then we set each of the factors equal to zero and solve.

$$\text{If } x + 2 = 0 \qquad \text{If } x - 1 = 0$$

$$x = -2 \qquad\qquad x = 1$$

The other method can be used when the equation is in the form

$$(x - 2)^2 = 3$$

We can solve equations that are in this form by taking the square root of both sides and remembering the $\pm$ sign that appears on the right.

$$(x - 2)^2 = 3 \qquad \text{equation}$$

$$x - 2 = \pm\sqrt{3} \qquad \text{square root of both sides}$$

$$x = 2 \pm \sqrt{3} \qquad \text{added +2 to both sides}$$

Unfortunately, the factoring method cannot always be used because some equations cannot be factored into binomials whose constants are all integers. For example, the equation

$$x^2 + 4x - 2 = 0$$

cannot be factored. **The inability to factor some quadratic equations is offset by the fact that any quadratic equation can be rearranged into the form**

$$(x + a)^2 = k$$

and then the equation can be solved by taking the square root of both sides, as we have just demonstrated. The method used to accomplish this rearrangement is called **completing the square.** If we rearrange the equation

$$x^2 + 4x - 2 = 0$$

by completing the square, we can change the equation into this form

$$(x + 2)^2 = 6$$

which we can solve by taking the square root of both sides.

$$x + 2 = \pm\sqrt{6} \qquad \text{square root of both sides}$$

$$x = -2 \pm \sqrt{6} \qquad \text{added } -2 \text{ to both sides}$$

50.B

completing the square

To complete the square, it is necessary to remember the form of a trinomial that is the square of some binomial. On the left, we show several binomials, and on the right we show the trinomials that result when the binomials are squared.

(a)	$x + 2$	(a′)	$(x + 2)^2 = x^2 + 4x + 4$
(b)	$x - 5$	(b′)	$(x - 5)^2 = x^2 - 10x + 25$
(c)	$x - 6$	(c′)	$(x - 6)^2 = x^2 - 12x + 36$
(d)	$x + 8$	(d′)	$(x + 8)^2 = x^2 + 16x + 64$
(e)	$x - 3$	(e′)	$(x - 3)^2 = x^2 - 6x + 9$

In each of these examples, we note that the last term of the trinomial is the square of one-half of the middle term of the trinomial. Thus,

	LAST TERM		$\left(\frac{1}{2} \text{ MIDDLE TERM}\right)^2$
In (a′)	4	is	$\left(\frac{4}{2}\right)^2$
In (b′)	25	is	$\left(-\frac{10}{2}\right)^2$
In (c′)	36	is	$\left(-\frac{12}{2}\right)^2$
In (d′)	64	is	$\left(\frac{16}{2}\right)^2$
In (e′)	9	is	$\left(-\frac{6}{2}\right)^2$

This pattern occurs every time we square a binomial. There is nothing to understand. It happens, so we will remember it and use it. This pattern is the key to completing the square.

Thus, if we have the expression

$$x^2 + 10x + \text{?}$$

and ask what number should replace the question mark if the trinomial is to be the square of some binomial, the answer is 25 because

$$\left(\frac{10}{2}\right)^2 = 25$$

So the completed expression is

$$x^2 + 10x + 25$$

Thus, the binomial that was squared to get this result had to be $x + 5$ because

$$(x + 5)^2 = x^2 + 10x + 25$$

example 50.1   Solve $x^2 + 6x - 4 = 0$ by completing the square.

solution   We want to rearrange the equation into the form

$$(x + a)^2 = k$$

We begin by enclosing the first two terms in parentheses and moving the -4 to the right-hand side as $+4$.

$$(x^2 + 6x \qquad) = 4$$

Note that we left a space inside the parentheses. Now we want to change the expression inside the parentheses so that it is a perfect square. To do this, we first divide the coefficient of x by 2 and square the result.

$$\left(\frac{6}{2}\right)^2 = 9$$

Then we add this number to both sides of the equation.

$$(x^2 + 6x + 9) = 4 + 9$$

Now the left-hand side can be written as $(x + 3)^2$ and the right-hand side as 13.

$$(x + 3)^2 = 13$$

We finish the solution by taking the square root of both sides and remembering the $\pm$ sign that will appear on the right.

$$x + 3 = \pm\sqrt{13} \qquad \text{square root of both sides}$$

$$x = -3 \pm \sqrt{13} \qquad \text{added } -3 \text{ to both sides}$$

example 50.2   Solve $2x + x^2 - 5 = 0$ by completing the square.

solution   We want to change the form of the equation to

$$(x + a)^2 = k$$

We begin by moving the constant term to the right-hand side and enclosing the other two terms in parentheses. We leave a space inside the parentheses.

$$(x^2 + 2x \qquad) = 5$$

Now we square $\frac{1}{2}$ of the coefficient of x

$$\left(\frac{2}{2}\right)^2 = 1$$

and add it to both sides.

$$(x^2 + 2x + 1) = 5 + 1$$

The left-hand side is a perfect square, and the right-hand side is 6.

$$(x + 1)^2 = 6 \qquad \text{simplified}$$

$$x + 1 = \pm\sqrt{6} \qquad \text{square root of both sides}$$

$$x = -1 \pm \sqrt{6} \qquad \text{added } -1 \text{ to both sides}$$

example 50.3   Solve $x^2 = 5x - 5$ by completing the square.

solution   We begin by placing the constant term on the right and enclosing the other two terms in parentheses, remembering to leave a space in the parentheses.

$$(x^2 - 5x \qquad) = -5$$

Next we divide -5 by 2 and square the result.

$$\left(\frac{-5}{2}\right)^2 = \frac{25}{4}$$

Now we add $\frac{25}{4}$ to both sides.

$$\left(x^2 - 5x + \frac{25}{4}\right) = -5 + \frac{25}{4}$$

Next we write the left-hand side as a perfect square and simplify the right-hand side.

$$\left(x - \frac{5}{2}\right)^2 = \frac{5}{4} \qquad \text{simplified}$$

$$x - \frac{5}{2} = \pm\sqrt{\frac{5}{4}} \qquad \text{square root of both sides}$$

$$x = \frac{5}{2} \pm \sqrt{\frac{5}{4}} \qquad \text{added } \frac{5}{2} \text{ to both sides}$$

$$x = \frac{5}{2} \pm \frac{\sqrt{5}}{2} \qquad \text{simplified}$$

practice   Solve by completing the square: $x^2 = 9x - 7$

problem set 50

1. The Two-Steppers were good recruiters as they numbered 10 more than 5 times the number of the Waltzers. Also, there were 10 times as many Two-Steppers as Waltzers. How many of each were there?

2. The first part of the trip was in a surrey at 8 mph and the last part was in a buckboard at 12 mph. If the total trip was 104 miles and took 10 hours, how much of the trip was made in each type of carriage?

3. What is the weight of the sodium (Na) in 348 grams of NaCl? (Na, 23; Cl, 35)

4. Richard and Lynn found three consecutive even integers such that 7 times the sum of the first and third was 48 less than 10 times the second. What were the integers?

5. Hadrian's soldiers increased their wall-building speed by 140 percent. If their new speed was 432 inches per day, what was their old wall-building speed?

Solve by completing the square:

6. $x^2 + 8x - 4 = 0$ 7. $12x + x^2 - 5 = 0$ 8. $x^2 = 7x - 3$

9. Find the equation of this line.   10.   Find angle C and side M.

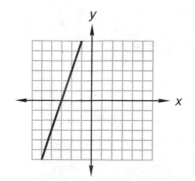

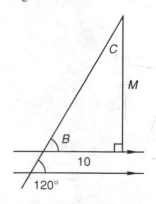

Solve:

11. $\sqrt{x^2 - 4x + 20} = x + 2$ **12.** $-5 = -\sqrt{x + 5} + 1$

13. Use unit multipliers to convert 400 yards per second to miles per hour.

Simplify:

14. $\sqrt[5]{2\sqrt[3]{2}}$ **15.** $\sqrt{9\sqrt{3}}$

16. $\sqrt{m^3 y^5}\,\sqrt[3]{m^2 y^2}$ **17.** $5\sqrt{\dfrac{3}{11}} + 2\sqrt{\dfrac{11}{3}} - \sqrt{297}$

18. Estimate: $\dfrac{(746{,}800 \times 10^{14})(703{,}916 \times 10^4)}{500{,}000}$

19. Find c: $\dfrac{mxc}{p} - k = \dfrac{2}{r}$ **20.** Find p: $\dfrac{4}{x} - \dfrac{3x}{p} = \dfrac{c}{m}$

21. Find c.   **22.** Divide $4x^3 + 3x + 5$ by $2x - 3$.

23. Solve $x^3 - 28x = 3x^2$ by factoring.

Simplify:

24. $\dfrac{x^2 + 49 - 14x}{x^3 - 13x^2 + 42x} \cdot \dfrac{x^3 - 4x^2 - 12x}{-35 - 2x + x^2}$ **25.** $-49^{3/2}$

26. The volume of a prism whose base is shown is $(600 - 50\pi)$ cm³. Find the height of the prism. Dimensions are in centimeters.

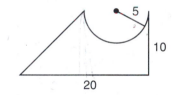

27. Find the equation of the line that passes through $(-2, 5)$ and is perpendicular to the line $3x - 5y = 2$.

28. Find the distance between the points $(-3, 5)$ and $(-3, 7)$.

29. Solve: $-2(x^0 - x - 3) - |-2^0 - 3| = x - 2(-2 - 4)$

30. Multiply: $\dfrac{x^2 y^{-2}}{z^2 p^0}\left(\dfrac{4p^0 x^{-5} z^2}{y^{-2}} - \dfrac{3x^{-4} y^2}{z^{-2} p}\right)$

LESSON 51   *Imaginary numbers • Product-of-square-roots theorem • Euler's notation • Complex numbers*

51.A
imaginary numbers

We know that $+2$ and -2 are both square roots of 4 because

$$(+2)^2 = 4 \quad \text{and} \quad (-2)^2 = 4$$

We remember that when we use the notation

$$\sqrt{4}$$

we are designating $+2$ because we reserve the notation $\sqrt{4}$ to designate the **principal** or **positive**

square root of 4. If we wish to designate –2, which is the negative square root of 4, we can write

$$-\sqrt{4}$$

and if we wish to designate both +2 and –2, we use both the plus sign and the minus sign and write

$$\pm\sqrt{4}$$

We remember that if the square root of a number is multiplied by itself, the result is the number. This is the definition of square root.

DEFINITION OF SQUARE ROOT

The square root of a number is that number which multiplied by itself equals the given number.

Thus, it follows that

$$\sqrt{3}\sqrt{3} = 3 \qquad \sqrt{0.046}\sqrt{0.046} = 0.046 \qquad \sqrt{3\frac{1}{2}}\sqrt{3\frac{1}{2}} = 3\frac{1}{2}$$

and it also follows that

$$\sqrt{-4}\sqrt{-4} = -4$$

But where on the number line can we locate $\sqrt{-4}$? Any positive or negative number times itself equals a positive number, not a negative number! There is no number on the number line that can be used for x in the following expression.

$$(x)(x) = -4$$

Thus, on the number line, we can locate both $+\sqrt{4}$ and $-\sqrt{4}$ as shown,

but we cannot locate $+\sqrt{-4}$ or $-\sqrt{-4}$ on the number line. For this reason, unfortunately, we call $\sqrt{-4}$ an **imaginary number.** We say unfortunately because all numbers are ideas and thus all numbers are really imaginary. The number $\sqrt{4}$ is an idea that we can graph on the number line, and the number $\sqrt{-4}$ is an idea that we cannot graph on the number line. **We call the square roots of negative numbers** *imaginary numbers* **because we cannot find them on the number line. The numbers that we can locate on the number line are called** *real numbers* **to distinguish them from numbers that cannot be found there.**

51.B
product-of-square-roots theorem

When one radicand (the number underneath the radical sign) is a negative number, we cannot always use the product-of-square-roots theorem as we have in the past. There are two versions of the product-of-square-roots theorem. One version applies when at least one radicand is positive.

PRODUCT-OF-SQUARE-ROOTS THEOREM
(AT LEAST ONE POSITIVE RADICAND)

If m and n are real numbers and one is a positive number or both are positive numbers, then

$$\sqrt{m}\sqrt{n} = \sqrt{mn}$$

Thus, $\sqrt{3}\sqrt{5} = \sqrt{15}$ and $\sqrt{-3}\sqrt{5} = \sqrt{-15}$.

The rule is different when both radicands are negative.

PRODUCT-OF-SQUARE-ROOTS THEOREM
(BOTH RADICANDS NEGATIVE)

If m and n are both negative real numbers, then

$$\sqrt{m}\sqrt{n} = -\sqrt{mn}$$

---

example 51.1   Simplify: (a) $\sqrt{3}\sqrt{2}$ (b) $\sqrt{3}\sqrt{-2}$ (c) $\sqrt{-3}\sqrt{-2}$

solution   In (a) and (b) at least one radicand is positive, so we simplify in both by multiplying the radicands.

$$\text{(a)} \quad \sqrt{3}\sqrt{2} = \sqrt{6} \qquad \text{(b)} \quad \sqrt{3}\sqrt{-2} = \sqrt{-6}$$

In (c) both radicands are negative, and we use a different rule.

$$\text{(c)} \quad \sqrt{-3}\sqrt{-2} = -\sqrt{6}$$

51.C
Euler's notation

The eighteenth-century Swiss mathematician Euler, pronounced "oiler," in addition to other major accomplishments, made four significant contributions to the notations used in mathematics. He introduced the Greek letter Σ (sigma) to stand for summation. He invented functional notation, and he was the first to use the letter e to represent the base of natural logarithms. In addition, he introduced the use of the letter i to represent $\sqrt{-1}$.

$$i = \sqrt{-1}$$

The first three concepts will be used in later courses. The concept of i will be used in this course.

It happens that the solutions of many quadratic equations are numbers that are wholly or in part the square roots of negative numbers. We can use the first part of the theorem on the previous page to write one of these as a real number times $\sqrt{-1}$. For example, if we have

$$\sqrt{-13}$$

we can write this as

$$\sqrt{13(-1)}$$

and use the product-of-square-roots theorem to write

$$\sqrt{13}\sqrt{-1}$$

Instead of writing $\sqrt{-1}$, Euler used the letter i. Thus, he would have written the above as

$$\sqrt{13}\,i$$

We surmise that Euler began to use i instead of $\sqrt{-1}$ because i can be made with only one stroke of the pen (if you leave off the dot) while $\sqrt{-1}$ requires three strokes of the pen.

example 51.2   Use Euler's notation to write: (a) $\sqrt{-13}$ (b) $\sqrt{-4}$ (c) $\sqrt{3}\sqrt{-2}$

solution

$$\text{(a)} \quad \sqrt{-13} = \sqrt{13(-1)} = \sqrt{13}\,i$$

$$\text{(b)} \quad \sqrt{-4} = \sqrt{4(-1)} = 2i$$

$$\text{(c)} \quad \sqrt{3}\sqrt{-2} = \sqrt{3}\sqrt{2}\,i = \sqrt{6}\,i$$

51.D

complex numbers   A complex number is a number that has a real part and an imaginary part. Thus,

$$4 + 5i \quad \text{and} \quad 5i + 4$$

are complex numbers. **We say that when we write the real part first, as we did on the left, we have written the complex number in** *standard form.* We add complex numbers by adding the real parts to find the real part of the sum and the imaginary parts to find the imaginary part of the sum.

example 51.3   Add: $4 + 6i + 3i - 2$

solution   We add real parts to real parts and imaginary parts to imaginary parts and get

$$2 + 9i$$

example 51.4   Add: $(4 - 3i) + (6 - 4i) - (2 - 5i)$

solution   First we remove the parentheses and get

$$4 - 3i + 6 - 4i - 2 + 5i$$

Now we add the like parts and get

$$8 - 2i$$

Since $\sqrt{-1}$ times $\sqrt{-1}$ equals -1, then ii equals -1 and i^2 equals -1. **We must keep this in mind when simplifying expressions that have** *i* **as a multiple factor.**

example 51.5   Simplify: $2ii - 3iii + 2i - 4 - \sqrt{-4}$

solution   We pair the i's as follows.

$$2(ii) - 3i(ii) + 2i - 4 - \sqrt{-4}$$

Now we remember that each $(ii) = i^2 = -1$, so we have

$$2(-1) - 3i(-1) + 2i - 4 - \sqrt{-4}$$

Now we simplify

$$-2 + 3i + 2i - 4 - 2i$$

and finish by adding like parts.

$$-6 + 3i$$

example 51.6   Simplify: $3i^3 + 2i^2 + 7i + 4 + 2i^5$

solution   We will begin by expanding and pairing the i's.

$$3i(ii) + 2(ii) + 7i + 4 + 2(ii)(ii)i$$

Next we remember that each (ii) equals -1.

$$3i(-1) + 2(-1) + 7i + 4 + 2(-1)(-1)i$$

Now we multiply and get

$$-3i - 2 + 7i + 4 + 2i$$

and lastly we simplify and get

$$2 + 6i$$

practice   Simplify:

a. $2ii - 7iii + 2i - 4 - \sqrt{-3}$ **b.** $3i^3 + 5i^2 + 5i + 7 + 2i^5$

problem set 51

1. Times were so hard that three-seventeenths of the knights did not even have a gauntlet to fling down. If 72 knights were in this category, how many knights were there in all?

2. Five hundred twenty-eight students had limited vocabularies and thus found that lucubration was almost impossible. If these students comprised 22 percent of the total, how many students were there in all?

3. The weight of the sodium (Na) in a quantity of Na_2SO_4 was 115 grams. What was the total weight of the Na_2SO_4? (Na, 23; S, 32; O, 16)

4. Horatio gleefully fingered his horde of quarters and dimes. If their total value was $6.50 and he had 35 coins, how many coins of each type did he have?

5. Buggs made the trip at 40 miles per hour during the first hiatus. Joan made the same trip at 50 miles per hour during the second hiatus. How long was the trip if it took Buggs 2 hours longer than it took Joan?

Simplify:

6. $5 + 6i - 3i - 2$

7. $5ii - 8iii + 2i - 4 - \sqrt{-5}$

8. $5i^3 + 3i^2 + 7ii + 4 + 2i^7$

9. $-3i^2 - 2i + i^3 - 3$

Solve by completing the square:

10. $x^2 = -x + 1$

11. $-4 = -x^2 - 3x$

12. Find the equation of this line.

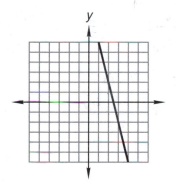

13. Find side m.

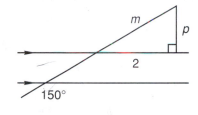

14. Solve: $\sqrt{x - 3} + 5 = 2$

15. Use unit multipliers to convert 20 inches per hour to miles per minute.

Simplify:

16. $\sqrt[5]{3\sqrt{3}}$

17. $\sqrt{9\sqrt[3]{3}}$

18. $\sqrt{x^2 y^2 m} \sqrt[4]{xym^2}$

19. $3\sqrt{\dfrac{2}{13}} + 3\sqrt{\dfrac{13}{2}} + 3\sqrt{104}$

20. $(-27)^{-5/3}$

21. Estimate: $\dfrac{(4,941,625)(7,041,683)}{0.00007142 \times 10^{-5}}$

22. Find p: $\dfrac{x}{p} - c = \dfrac{k}{m}$

23. Find R_2: $\dfrac{3}{p} - \dfrac{x}{R_1} = \dfrac{1}{R_2}$

24. **If two inscribed angles in a circle intercept equal arcs, the angles are equal.** Angles $\angle APB$ and $\angle ARB$ are equal because they both intercept arc AB. Find x, y, m, and z.

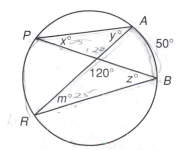

25. Find the equation of the line through $(-4, -2)$ that is perpendicular to the line $3x + 2y = 5$.

26. Find the distance between the points $(-2, -5)$ and $(-2, -8)$.

27. Multiply: $\dfrac{x^{-2}y}{p^0}\left(\dfrac{4p^2}{x^{-2}y} - \dfrac{2px^2y}{y^2}\right)$ 28. Divide $2x^3 - 2x + 4$ by $x + 2$.

Solve:

29. $\dfrac{-x-2}{4} - \dfrac{x+2}{3} = 3$

30. $-2(x^0 - x - 2) - |-2| + 30(-x - x^0) = -x$

LESSON 52 *Chemical mixture problems*

In chemical mixture problems two different constituents are mixed to get a desired result. Often one of the constituents is water, and the other is alcohol, or iodine, or antifreeze, or whatever else is mixed with the water. The key equation in one of these problems comes from either constituent. Thus, if the constituents are water and iodine, the statement can be made about water.

Water poured in + water dumped in = water total

Or the statement can be made about iodine.

Iodine poured in + iodine dumped in = iodine total

We will restrict our investigation of these problems to two basic types. The first type is discussed in this lesson and the second type will be discussed in Lesson 61.

example 52.1 A druggist has one solution that is 10% iodine and another that is 50% iodine. How much of each should the druggist use to get 100 milliliters (ml) of a mixture that is 20% iodine?

solution A picture is helpful.

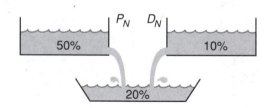

As shown in the figure, we will pour in some of the 50% solution and dump in some of the 10% solution. We can work the problem by considering either iodine or water. First we will work it by considering only the iodine. Our equation in words is

Iodine poured in + iodine dumped in = iodine total

Next we write three sets of parentheses that we will use as **"mixture containers."**

( ) + ( ) = ( )

Many people find that using "mixture containers" is helpful, and the use of these containers is recommended. Now we put the requisite **mixture** in each container. We use P_N in the first one, D_N in the second one, and 100 in the third one.

$(P_N) + (D_N) = (100)$

Next we multiply each set of parentheses by the proper decimal number so that each term represents only iodine.

$$\text{(a)} \quad 0.5(P_N) + 0.1(D_N) = 0.2(100)$$

This equation has two unknowns, so we need one more equation. The equation is

$$\text{(b)} \quad P_N + D_N = 100$$

We solve equation (b) and find that $P_N = 100 - D_N$. Then we substitute $100 - D_N$ for P_N in equation (a) and solve.

$0.5(P_N) + 0.1(D_N) = 0.2(100)$	equation (a)
$0.5(100 - D_N) + 0.1D_N = 0.2(100)$	substituted
$50 - 0.5D_N + 0.1D_N = 20$	multiplied
$0.4D_N = 30$	simplified
$D_N = 75 \text{ ml}$	divided by 0.4

And since $P_N + D_N = 100$,

$$P_N = 25 \text{ ml}$$

Thus, the druggist should use 25 ml of the 50% solution and 75 ml of the 10% solution. If we wish to work the problem by considering only water, the word statement would be

$$\text{Water poured in } + \text{ water dumped in } = \text{ water total}$$

The entries in the mixture containers are exactly the same,

$$(P_N) + (D_N) = (100)$$

but the decimal multipliers are different because this time each term represents water.

$$\text{(c)} \quad 0.5(P_N) + 0.9(D_N) = 0.8(100)$$

The second equation is the same as before.

$$\text{(b)} \quad P_N + D_N = 100$$

Either substitution or elimination can be used to solve equations (c) and (b), and the same answers will result.

$$P_N = 25 \text{ ml} \qquad D_N = 75 \text{ ml}$$

Thus, the druggist should use 25 ml of the 50% solution and 75 ml of the 10% solution.

example 52.2   A chemist has one solution that is 10% salt and 90% water and another solution that is only 2% salt. How many milliliters of each should the chemist use to make 1400 ml of a solution that is 6% salt?

solution   We can use the same diagram as in the preceding example. Only the percents are different.

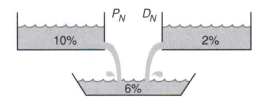

This time we decide to work the problem by considering water. Thus, our equation in words is

$$\text{Water poured in } + \text{ water dumped in } = \text{ water total}$$

Next we write the set of parentheses that we call mixture containers. Their use in these problems is always helpful.

$$(\quad) + (\quad) = (\quad)$$

Mixture containers always contain mixtures. We will use P_N and D_N to represent the unknown mixtures.

$$(P_N) + (D_N) = (1400)$$

This time we use decimal multipliers so that the decimal times the mixture equals water,

$$0.9(P_N) + 0.98(D_N) = 0.94(1400)$$

and we get

(a) $0.9P_N + 0.98D_N = 1316$

We have two unknowns, so we need another equation, which is

(b) $P_N + D_N = 1400$

We solve (b) for P_N and substitute into (a) and solve.

$$0.9(1400 - D_N) + 0.98D_N = 1316 \qquad \text{substitution}$$
$$1260 - 0.9D_N + 0.98D_N = 1316 \qquad \text{multiplied}$$
$$0.08D_N = 56 \qquad \text{added} -1260 \text{ to both sides}$$
$$D_N = \textbf{700 ml}$$

Thus, $P_N = 1400 - 700 = \textbf{700 ml}$. Thus, the chemist should use 700 ml of the 10% solution and 700 ml of the 2% solution.

practice   A chemist has one solution that is 25% salt and 75% water and another solution that is only 5% salt. How many milliliters of each should the chemist use to make 1600 ml of a solution that is 15% salt?

problem set 52

1.   A druggist has one solution that is 10% iodine and another that is 40% iodine. How much of each should the druggist use to get 100 ml of a mixture that is 25% iodine?

2.   A chemist has one solution that is 25% salt and 75% water and another solution that is only 5% salt. How many milliliters of each should the chemist use to make 1400 ml of a solution that is 10% salt?

3.   What is the weight of the sodium (Na) in 1580 grams of $Na_2S_2O_3$? (Na, 23; S, 32; O, 16)

4.   Four times the number of yellows equaled 76 reduced by 6 times the number of reds. If there were 4 more yellows than reds, how many of each were there?

5.   Queen Hatshepsut rode a litter at 2 kilometers per hour for the first part of the journey to the necropolis. She was going to be late, so she changed to a chariot traveling 8 kilometers per hour for the last part of the trip. If it was 28 kilometers to the necropolis and the trip took 8 hours, how far did she ride in the litter and how far did she ride in the chariot?

Simplify:

6. $4i^2 - 3i + 2$

7. $3i^5 - i + 5 - \sqrt{-9}$

8. $\sqrt{-16} - 2i^2 - 2i$

9. $2i^4 + \sqrt{-9} - 3i^3$

Solve by completing the square:

10. $2x + 5 = x^2$

11. $x^2 - 5x = 2$

12.   Find the equation of the line.

13.   Find P.

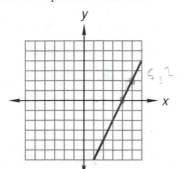

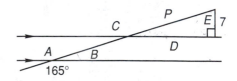

Solve:

14. $\sqrt{x^2 + 2x + 10} = x + 2$

15. $3 = -5 + \sqrt{x - 3}$

16. Use unit multipliers to convert 200 inches per hour to miles per minute.

Simplify:

17. $2\sqrt{2\sqrt[4]{2}}$

18. $3\sqrt{9\sqrt[4]{3}}$

19. $\sqrt{4x^3 y^5}\sqrt[3]{8xy^2}$

20. $3\sqrt{\dfrac{2}{13}} - 5\sqrt{\dfrac{13}{2}} + \sqrt{104}$

21. Estimate: $\dfrac{(987,612 \times 10^5)(413,280)}{(74,630)(400)}$

22. Find x: $\dfrac{x}{p} - c + \dfrac{a}{b} = m$

Simplify:

23. $\dfrac{x^2 + 3x - 28}{21x + 10x^2 + x^3}$

24. $-16^{5/4}$

25. $\dfrac{\dfrac{x^2 y}{p^2} - p}{\dfrac{m}{p} - \dfrac{1}{p^2}}$

26. Find x, y, z, and p.

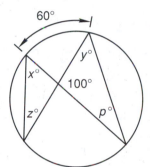

27. Add: $-\dfrac{3}{x - 3} + \dfrac{2x + 4}{x^2 - 9}$

28. Find the equation of the line that passes through the point $(-2, -7)$ and is parallel to the line $2x + 3y = 4$.

29. Divide: $3x^3 - x$ by $x - 4$

30. Solve: $\dfrac{3x + 4}{2} - \dfrac{2x - 5}{3} = 4$

LESSON 53 *Metric unit conversions • English units to metric units • Weight combination by percent*

53.A
metric unit conversions

The meter is the basic unit of length in the metric system. A meter is just a little longer than a yard. There are 100 centimeters in a meter, and there are 1000 meters in a kilometer, which is a little more than half a mile. From these relationships, we get two metric unit multipliers that can be used for distance and area conversion.

(a) $\dfrac{100 \text{ cm}}{1 \text{ m}}$ or $\dfrac{1 \text{ m}}{100 \text{ cm}}$ and   (b) $\dfrac{1000 \text{ m}}{1 \text{ km}}$ or $\dfrac{1 \text{ km}}{1000 \text{ m}}$

example 53.1 Use unit multipliers to convert 7560 centimeters to kilometers.

solution We will go from centimeters (cm) to meters (m) to kilometers (km).

$$7560 \cancel{\text{ cm}} \times \dfrac{1 \cancel{\text{ m}}}{100 \cancel{\text{ cm}}} \times \dfrac{1 \text{ km}}{1000 \cancel{\text{ m}}} = \dfrac{7560}{(100)(1000)} \text{ km} = \mathbf{0.0756 \text{ km}}$$

This problem illustrates the beauty of the metric system. All we have to do to change units in the metric system is to move the decimal point.

example 53.2   Use unit multipliers to convert 15,740,000 square centimeters to square kilometers.

solution   We will use two multipliers to go from square centimeters (cm²) to square meters (m²) and two more to go to square kilometers (km²).

$$15{,}740{,}000 \text{ cm}^2 \times \frac{1 \text{ m}}{100 \text{ cm}} \times \frac{1 \text{ m}}{100 \text{ cm}} \times \frac{1 \text{ km}}{1000 \text{ m}} \times \frac{1 \text{ km}}{1000 \text{ m}} = \frac{15{,}740{,}000}{1 \times 10^{10}} \text{ km}^2$$
$$= 0.001574 \text{ km}^2$$

Again we note that all we had to do was to move the decimal point.

53.B
English units to metric units

We need to know only one equivalence to change from English units of length to metric units of length.

$$2.54 \text{ cm} = 1 \text{ in.}$$

This equivalence is exact because 1 inch is exactly 2.54 centimeters long. This definition of the inch is official and has been official for almost 100 years. Thus, to go from English units to metric units, we first go to inches in the English system. Then we convert to centimeters.

example 53.3   Use unit multipliers to convert 32 yards to meters.

solution   **Some tables give a direct conversion from yards to meters. We disdain these conversions because the tables are not always available. We need only remember that 2.54 centimeters equals 1 inch to make any English to metric conversion involving length, area, or volume.** Thus, in this problem we will use unit multipliers as required to go from yards to feet to inches to centimeters to meters.

$$32 \text{ yd} \times \frac{3 \text{ ft}}{1 \text{ yd}} \times \frac{12 \text{ in.}}{1 \text{ ft}} \times \frac{2.54 \text{ cm}}{1 \text{ in.}} \times \frac{1 \text{ m}}{100 \text{ cm}} = \frac{(32)(3)(12)(2.54)}{100} \text{ m}$$

example 53.4   Use unit multipliers to convert 0.042 square kilometers to square miles.

solution   We need to use two unit multipliers in each step as we go from square kilometers to square meters to square centimeters to square inches to square feet to square miles.

$$0.042 \text{ km}^2 \times \frac{1000 \text{ m}}{1 \text{ km}} \times \frac{1000 \text{ m}}{1 \text{ km}} \times \frac{100 \text{ cm}}{1 \text{ m}} \times \frac{100 \text{ cm}}{1 \text{ m}} \times \frac{1 \text{ in.}}{2.54 \text{ cm}} \times \frac{1 \text{ in.}}{2.54 \text{ cm}}$$
$$\times \frac{1 \text{ ft}}{12 \text{ in.}} \times \frac{1 \text{ ft}}{12 \text{ in.}} \times \frac{1 \text{ mi}}{5280 \text{ ft}} \times \frac{1 \text{ mi}}{5280 \text{ ft}}$$
$$= \frac{(0.042)(1000)(1000)(100)(100)}{(2.54)(2.54)(12)(12)(5280)(5280)} \text{ mi}^2$$

53.C
weight combination by percent

Thus far, we have investigated the relative weights of the elements in chemical compounds by using ratios. For example, in a molecule of the compound

$$Na_2S_2O_3$$

there are two atoms of sodium (Na), two atoms of sulfur (S), and three atoms of oxygen (O). If we use the gram atomic weights of these elements, we can find the gram molecular weight of the molecule. (Na, 23; S, 32; O, 16)

Two atoms of sodium:	2 × 23 =	46 grams
Two atoms of sulfur:	2 × 32 =	64 grams
Three atoms of oxygen:	3 × 16 =	48 grams
Gram atomic weight of a molecule	=	158 grams

The decimal part of the total made up by each element can be found by dividing the atomic weight of the element by the molecular weight of the molecule.

$$\text{Sodium} = \frac{46}{158} = 0.29 \qquad \text{Sulfur} = \frac{64}{158} = 0.41 \qquad \text{Oxygen} = \frac{48}{158} = 0.30$$

The percent of each element by weight is found by moving each of these decimal points two places to the right. If we do this, we get

$$\text{Sodium} = 29\% \qquad \text{Sulfur} = 41\% \qquad \text{Oxygen} = 30\%$$

The sum of these percents is 100 percent. Sometimes, because we have rounded, the sum will not be exactly 100 percent.

example 53.5   What percent by weight of $Na_2S_2O_3$ is sodium (Na)?

solution   From the above, the compound is **29%** sodium.

practice   a.   What percent by weight of CCl_4 (carbon tetrachloride) is carbon (C)? (C, 12; Cl, 35)

b.   Use unit multipliers to convert 0.073 square kilometers to square miles.

problem set 53

1.   What percent by weight of $Na_2S_2O_3$ is sodium (Na)? (Na, 23; S, 32; O, 16)

2.   A chemist has one solution that is 20% alcohol and another that is 60% alcohol. How much of each should the chemist use to get 100 ml of a solution that is 52% alcohol?

3.   The hospital pharmacist wanted 250 ml of a solution that was 72% iodine. How many milliliters of a 40% solution should be mixed with how many milliliters of an 80% solution to get the desired result?

4.   The results totaled 80, and the number of good results totaled 8 more than 5 times the number of bad results. How many results were good and how many were bad?

5.   The automobile was traveling at 50 mph and had already gone 200 miles when the airplane set out in pursuit. If the airplane overtook the automobile in 4 hours, what was the speed of the airplane?

Use unit multipliers to convert:

6.   9350 centimeters to kilometers

7.   32 meters to yards

8.   16,480,000 square miles to square centimeters

9.   0.063 square kilometers to square miles

Simplify:

10. $-\sqrt{-4} + 2 + 2i^5$

11. $2i^2 + 5i + 4 + \sqrt{-9}$

12. $-4i^5 + 2\sqrt{-16}$

13. $2i^3 - i^4 + 3i^2$

Solve by completing the square:

14. $x^2 - 5 = 5x$

15. $-x^2 = -6x - 6$

16.   Find side m.

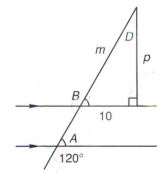

Solve:

17. $\sqrt{x - 11} - 1 = 6$

18. $\sqrt{x^2 + 2x + 5} - 3 = x$

Simplify:

19. $\sqrt[5]{2\sqrt[3]{2}}$

20. $\sqrt{81\sqrt[4]{3}}$

21. $\sqrt[5]{x^2 y}\sqrt[3]{xy^2}$

22. $-4^{-5/2}$

23. $3\sqrt{\dfrac{2}{9}} - 2\sqrt{\dfrac{9}{2}} - 2\sqrt{50}$

24. Estimate: $\dfrac{(2,135,820)(4,913,562)}{801,394,026}$

25. Find p: $\dfrac{x}{y} - \dfrac{m}{p} + \dfrac{k}{c} = 0$

26. Find x: $\dfrac{p}{x} + c = d$

27. Find B.

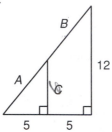

28. Find $x, y, P, Q,$ and R.

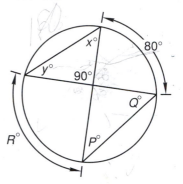

29. Divide $x^3 - 2x + 2$ by $x + 1$.

30. Find the solution to this system by graphing. Then find the exact solution by using either substitution or elimination.

$$\begin{cases} 2x - 3y = -9 \\ 5x + 3y = 3 \end{cases}$$

LESSON 54 *Polar coordinates • Similar triangles*

54.A
polar coordinates

We can use two different methods to describe the location of a point on the coordinate plane with respect to the origin. One method is to use one number to give the location of the point to the right or the left of the origin and another number to give the location of the point above or below the origin. When we do this, we say we are using **rectangular coordinates.** To associate the numbers with the proper directions, we can use parentheses and ordered pairs (x, y), or we can omit the parentheses and use letters (often $i, j,$ and k) to designate directions. In this book i always represents $\sqrt{-1}$, so we decide to use the letters R and U to indicate directions. We will use $+R$ for right, $-R$ for left, $+U$ for up, and $-U$ for down. We will demonstrate this notation by using rectangular coordinates to locate the four points on the following graph. For point (a) we write $4R + 3U$, for point (b) we write $-3R + 5U$, for point (c) we write $-3R - 3U$, and for point (d) we write $5R - 2U$.

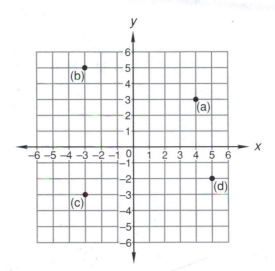

The other way to describe the location of a point is to use a distance at a given angle. **The angle is measured counterclockwise from the line that normally designates the +*x* axis.** We say that we are using **polar coordinates** when we designate the location of a point by using an angle and a distance.

It is interesting to note that seafarers and air navigators have always measured their angles clockwise from due north, while mathematicians, for some reason, measure their angles counterclockwise from due east!

We will use notations such as

$$4\underline{/57°} \qquad \text{and} \qquad 7\underline{/230°}$$

to designate the magnitudes and angles. The first number gives the magnitude and the second number gives the angle. Thus the above would be read from left to right as 4 at 57° and 7 at 230°. In many books the authors use ordered pairs to designate the magnitudes and the angles. In these books the magnitudes and angles shown above would be written as the ordered pairs

$$(4, 57°) \qquad\qquad\qquad\qquad (7, 230°)$$

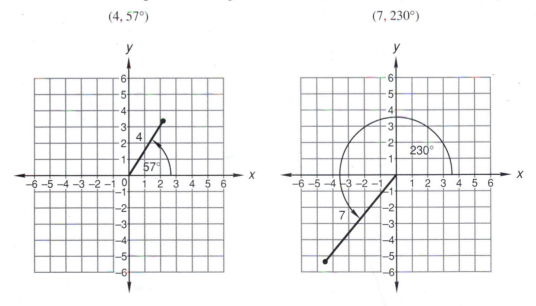

In mathematics and science, a **vector** is defined as a quantity that has both a magnitude and a direction. Since each of the line segments shown in the figures has a magnitude and a direction with respect to the origin, we can call these line segments *vectors*. In the figure on the left, we designate the point by using the vector $4\underline{/57°}$, and in the figure on the right, we designate the point by using the vector $7\underline{/230°}$.

We can use sines, cosines, and tangents to help us convert from rectangular coordinates to polar coordinates or from polar coordinates to rectangular coordinates. For the present we will concentrate on learning how to convert from polar coordinates to rectangular coordinates.

example 54.1   Change $4\underline{/57°}$ to rectangular coordinates.

solution   We draw the vector that designates the point and **then complete the triangle by drawing a vertical line from the point to the x axis.**

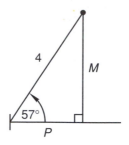

We can find M and P by using the sine and cosine.

(a) $\sin 57° = \dfrac{M}{4}$ (b) $\cos 57° = \dfrac{P}{4}$

We solve these equation for M and P and get

(a) $4 \sin 57° = M$ (b) $4 \cos 57° = P$

We use a calculator to get the sine and cosine and then multiply.

$$4(0.84) = M \qquad 4(0.54) = P$$
$$3.36 = M \qquad 2.16 = P$$

Thus the point is 2.16 to the right of the origin and 3.36 above the origin, so its location can be described in rectangular coordinates by writing

$$\mathbf{2.16R \ + \ 3.36U}$$

example 54.2   Change $7\underline{/230°}$ to rectangular coordinates.

solution   We draw the vector and then complete the triangle **by drawing a vertical line from the end of the vector to the x axis.** Since 230° is 50° more than 180°, the angle in our triangle is 50°.

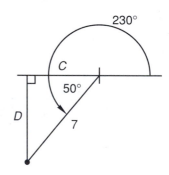

We can find D and C by using the sine and cosine.

(a) $\sin 50° = \dfrac{D}{7}$ (b) $\cos 50° = \dfrac{C}{7}$

We solve these equations for D and C and get

$$7 \sin 50° = D \qquad 7 \cos 50° = C$$

We get the sine and cosine from a calculator and then multiply.

$$7(0.77) = D \qquad 7(0.64) = C$$
$$5.39 = D \qquad 4.48 = C$$

Thus, our point is 4.48 to the left of the origin and 5.39 below the origin. We indicate this in rectangular coordinates by writing

$$\mathbf{-4.48R \ - \ 5.39U}$$

It is helpful if we note that only the sine and the cosine of the angle are used when we break a vector into its components. If *H* is the hypotenuse and θ is the angle, then one component is

$$H \cos \theta \quad \text{and the other component is} \quad H \sin \theta$$

example 54.3 Change $42\underline{/340°}$ to rectangular coordinates.

solution We draw the vector and then complete the triangle **by drawing a vertical line from the end of the vector to the *x* axis.** Since 340° is 20° less than 360°, the angle in the triangle is 20°.

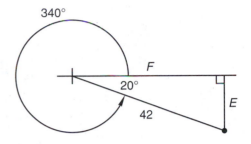

Again, we need to use only the sine and the cosine.

(a) $\quad \sin 20° = \dfrac{E}{42}$ (b) $\quad \cos 20° = \dfrac{F}{42}$

$$42 \sin 20° = E \qquad\qquad 42 \cos 20° = F$$

$$42(0.34) = E \qquad\qquad 42(0.94) = F$$

$$14.28 = E \qquad\qquad 39.48 = F$$

Thus, our point is 39.48 to the right and 14.28 down. We indicate this by writing

$$\textbf{39.48}\textit{R} \ - \ \textbf{14.28}\textit{U}$$

54.B

similar triangles

Sometimes the lengths of the sides in similar triangles are represented by letters. These triangles are similar.

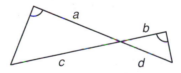

We can solve for one of the variables in terms of the other variables. The first step is to write the equal ratios. We decide to put the sides opposite the single-tick-marked angles on top.

$$\frac{c}{a} = \frac{d}{b}$$

The next step is to cross multiply. If we do, we get

$$bc = ad$$

Now we can solve for any one of the variables by dividing. If we want to solve for *d*, we would divide both sides by *a*.

$$\frac{bc}{a} = \frac{\not{a}d}{\not{a}} \qquad \text{divided both sides by } a$$

$$\frac{bc}{a} = d \qquad \text{canceled}$$

practice **a.** Change $28\underline{/310°}$ to rectangular coordinates.

b. The two triangles in this figure are similar. Use R, S, T, and V to write a proportion. Then solve for R in terms of S, T, and V.

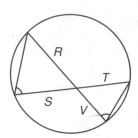

problem set 54

1. Sarah wants to mix 400 liters of a solution that is 57.5% iodine. She has two solutions available. One is 20% iodine and the other is 70% iodine. How many liters of each should Sarah use?

2. Selby was in the next stall and she needed 150 ml of a solution that was 30% glycerine. The two solutions available were 10% glycerine and 40% glycerine. How many milliliters of each should Selby use?

3. What percent by weight of potassium chlorate is potassium (K)? The chemical formula for potassium chlorate is $KClO_3$. (K, 39; Cl, 35; O, 16)

4. The ratio of ducks to geese was 5 to 4. Four times the number of ducks was 40 greater than 3 times the number of geese. How many of each kind of fowl were present?

5. The sports car was twice as fast as the truck and took 3 hours less to make the trip. If the truck traveled 50 miles per hour, how long was the trip?

6. Change $5\underline{/56°}$ to rectangular coordinates.

7. Change $8\underline{/212°}$ to rectangular coordinates.

8. Find T in terms of R, S, and V. Remember that inscribed angles that intersect the same arc are equal angles.

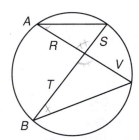

Use unit multipliers to convert:

9. 100 square kilometers to square centimeters

10. 100 square feet to square centimeters

11. 60 miles per hour to kilometers per second

Simplify:

12. $-\sqrt{-4} - 2i^3 + 4i^4$

13. $3i^3 + 2i - 4i^2 + \sqrt{-9}$

14. $-3i + 2i^2 - 2 + i$

15. $2i^2 + 2i - 2\sqrt{-25}$

Solve by completing the square:

16. $x^2 = 7 + 3x$

17. $-7x = -x^2 + 3$

18. Find the equation of the line shown.

Solve:

19. $\dfrac{4x - 2}{5} - \dfrac{x - 3}{2} = 7$

20. $\sqrt{x^2 - x - 5} + 1 = x$

21. $\sqrt{x - 2} - 11 = 1$

22. Find k: $\dfrac{x}{k} - cm = \dfrac{p}{c}$

23. Find m: $\dfrac{a}{p} - x + \dfrac{c}{m} = y$

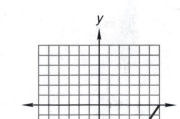

Simplify:

24. $\sqrt[3]{4\sqrt{2}}$ **25.** $\sqrt[5]{9\sqrt[3]{3}}$ **26.** $\sqrt[6]{xy}\sqrt[3]{xy^2}$ **27.** $-81^{1/4}$

28. $2\sqrt{\dfrac{3}{7}} - 5\sqrt{\dfrac{7}{3}} + 2\sqrt{84}$ **29.** $\dfrac{x^2 p - \dfrac{x}{p^2}}{\dfrac{x^2 y}{p^2} - x}$

30. Multiply: $\dfrac{4x^{-2}yp}{m^2 y^{-1}}\left(\dfrac{p^{-1}m^2 y}{16x^{-2}y} - \dfrac{2x^2 y^0 p}{m^{-2}}\right)$

LESSON 55   *Advanced abstract equations • Word problems and quadratic equations*

55.A
advanced abstract equations

We remember that equations that contain many variables and that have no numerical answer are often called **abstract equations.** We remember that when an equation has rational terms (fractions), the solution is facilitated if we begin by eliminating the denominators. Thus, to solve the equation

$$\frac{a}{x} + \frac{y}{m} = c$$

for x, we begin by multiplying every numerator by xm and then canceling the denominators.

$$\frac{a}{x} \cdot xm + \frac{y}{m} \cdot xm = c \cdot xm \qquad \text{multiplied by } xm$$

$$am + yx = cxm \qquad \text{canceled denominators}$$

$$am = cxm - yx \qquad \text{added } -yx$$

$$am = x(cm - y) \qquad \text{factored out } x$$

$$\frac{am}{cm - y} = x \qquad \text{divided}$$

The same procedure is used for more complicated equations. Parentheses sometimes help us to avoid mistakes.

example 55.1   Solve for m: $\dfrac{a + b}{x} + \dfrac{y}{m} = k$

solution   We can prevent some mistakes when we cancel if we enclose sums in the numerators in parentheses. **Then, after we cancel, it is important to eliminate the parentheses by multiplying.** In this problem, we begin by enclosing $a + b$ in parentheses. Then we multiply each term by mx, cancel, and eliminate the parentheses by multiplying.

$$\frac{(a + b)}{x}mx + \frac{y}{m}mx = kmx \qquad \text{multiplied by } mx$$

$$(a + b)m + yx = kmx \qquad \text{canceled}$$

$$am + bm + yx = kmx \qquad \text{multiplied}$$

$$yx = kmx - am - bm \qquad \text{added } -am - bm$$

$$yx = m(kx - a - b) \qquad \text{factored out } m$$

$$\frac{yx}{kx - a - b} = m \qquad \text{divided}$$

example 55.2 Solve for x: $\dfrac{mp}{c} + \dfrac{d+e}{x} = d$

solution We use parentheses around $d + e$ and then multiply by cx.

$$\dfrac{mp}{c}\,cx + \dfrac{(d+e)}{x}\,cx = dcx \qquad\qquad \text{multiplied}$$

$$mpx + dc + ec = dcx \qquad\qquad \text{canceled and multiplied}$$

$$dc + ec = dcx - mpx \qquad\qquad \text{added } -mpx$$

$$dc + ec = x(dc - mp) \qquad\qquad \text{factored out } x$$

$$\dfrac{dc + ec}{dc - mp} = x \qquad\qquad \text{divided}$$

example 55.3 Solve for a: $\dfrac{mx}{y} + \dfrac{d}{a+b} = p$

solution We begin by multiplying every numerator by $y(a + b)$ and canceling.

$$\dfrac{mxy(a+b)}{y} + \dfrac{dy(a+b)}{a+b} = py(a+b) \qquad\qquad \text{multiplied}$$

$$mxa + mxb + dy = pya + pyb \qquad\qquad \text{canceled and multiplied}$$

$$mxa - pya = pyb - mxb - dy \qquad\qquad \text{moved three terms}$$

$$a(mx - py) = pyb - mxb - dy \qquad\qquad \text{factored out } a$$

$$a = \dfrac{pyb - mxb - dy}{mx - py} \qquad\qquad \text{divided}$$

55.B

word problems and quadratic equations

Algebra books often contain problems whose solutions require the solutions of quadratic equations. Consecutive integer problems in which two of the integers are multiplied often lead to quadratic equations.

example 55.4 Find three consecutive integers such that the product of the first and the third is 4 greater than 4 times the second.

solution As always we begin by writing the integers as

$$N \qquad\quad N + 1 \qquad\quad N + 2$$

Now we write the equation.

$$N(N + 2) - 4 = 4(N + 1) \qquad\qquad \text{equation}$$

$$N^2 + 2N - 4 = 4N + 4 \qquad\qquad \text{multiplied}$$

$$N^2 - 2N - 8 = 0 \qquad\qquad \text{simplified}$$

$$(N - 4)(N + 2) = 0 \qquad\qquad \text{factored}$$

$$N = 4,\ -2 \qquad\qquad \text{solved}$$

Since we found two values of N, we get two sets of three consecutive integers. They are **4, 5, 6**, and **−2, −1, 0.**

practice **a.** Find p: $\dfrac{m+s}{p} + \dfrac{a}{x} = t$ **b.** Find m: $\dfrac{ay}{x} + \dfrac{c}{m+z} = s$

c. Find three consecutive integers such that the product of the first and third is 5 greater than 5 times the second.

problem set 55

1. Find three consecutive integers such that the product of the first and the second is equal to the product of -6 and the third.

2. Find three consecutive even integers such that the product of the first and the third is 24 less than 9 times the second.

3. One solution was 5% bromide and the other was 40% bromide. How much of each should be used to get 60 ml of a solution that is 12% bromide?

4. The formula for carbon tetrachloride is CCl_4. What is the weight of the carbon in 1368 grams of carbon tetrachloride? (C, 12; Cl, 35)

5. Find three consecutive multiples of 3 such that 6 times the first is 48 greater than 4 times the third.

6. Find x: $\dfrac{a+b}{x} + \dfrac{y}{m} = k$

7. Find e: $\dfrac{mp}{c} + \dfrac{d+e}{x} = d$

8. Find b: $\dfrac{mx}{y} + \dfrac{d}{a+b} = p$

Change to rectangular form:

9. $40\underline{/325°}$

10. $10\underline{/200°}$

Use unit multipliers to convert:

11. 4 cubic yards to cubic meters

12. 1000 feet per second to kilometers per minute

Simplify:

13. $4i^5 - 2\sqrt{-9} - 2i^4$

14. $4 + 2i^2 + 3i - \sqrt{-4}$

15. $2i^3 + 2i^4 + 2 - 2i$

16. $-3i^6 - 2i - 2 - 2i^2$

Solve by completing the square:

17. $x^2 = 5x + 5$

18. $x^2 - 6 = 6x$

Solve:

19. $\sqrt{x-2} + 4 = 2$

20. $\sqrt{x^2 - 2x + 14} - 12 = x$

Simplify:

21. $\sqrt{16\sqrt{2}}$

22. $\sqrt[4]{27\sqrt[3]{3}}$

23. $\sqrt[4]{x^2y}\sqrt{x^5y^2}$

24. $-81^{5/4}$

25. $4\sqrt{\dfrac{2}{11}} + 2\sqrt{\dfrac{11}{2}} - 4\sqrt{198}$

26. Estimate: $\dfrac{(4,183,256)(704,185 \times 10^{-42})}{802,164 \times 10^{30}}$

27. Find B.

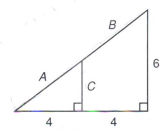

28. Find x in terms of y, p, and m.

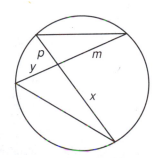

29. Find the volume of the prism shown. Dimensions are in feet.

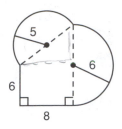 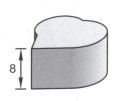

30. Simplify: $\dfrac{apx^2 - 10pa + 3xpa}{x^2 + 5x - 14} \cdot \dfrac{x^2 - 21 + 4x}{-20ap + xpa + ax^2 p}$

LESSON 56 *Angles in circles • Proofs*

56.A

angles in circles

We remember that the measure of an arc of a circle is the same as the measure of the central angle formed by the radii connecting the endpoints of the arc to the center of the circle. The endpoints of one arc are also the endpoints of another arc.

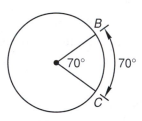

Minor arc is $\overparen{BC}$

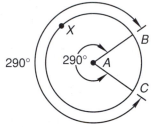

Major arc is $\overparen{BXC}$

If the two arcs are not equal, we call the longer arc the **major arc** and the smaller arc the **minor arc.** When we name the minor arc, we name the endpoints and use an arc sign. When we name the major arc, we name the endpoints and one other point on the major arc between the endpoints. We remember that if two circles have the same radii, we say that the circles are equal (congruent) circles.

If a line intersects a circle in two places, the line is called a **secant**. The part of the secant that is inside the circle is called a **chord**. If a line intersects (touches) a circle at only one point, the line is called a **tangent**, and the point where the line and the circle touch is called the **point of tangency**. A tangent is always perpendicular to the radius of the circle at the point of tangency.

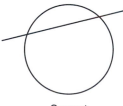   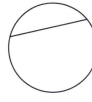   

Secant   Chord   Tangent

If two lines that are not parallel intersect a circle, the angles formed at the intersection of the lines are related to the arcs the lines intercept on the circle. The relationship is determined by the location of the vertex of the angles formed by the intersecting lines.

We remember that if the vertex is on the circle, the angle is called an **inscribed angle. We also remember that the measure of an inscribed angle equals half the measure of the intercepted arc.**

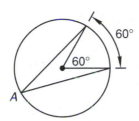
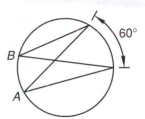

In the figure on the left, angle *A* is a 30° angle because both the central angle and the arc have a measure of 60°. In the figure on the right, both angle *A* and angle *B* have a measure of 30° because they both intersect a 60° arc.

If two lines intersect inside a circle, the measures of the angles formed equal one-half the sum of the measures of the intercepted arcs. If the lines intersect outside the circle, the measure of the angle formed equals one-half the difference of the measures of the intercepted arcs.

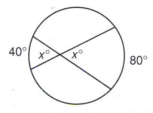
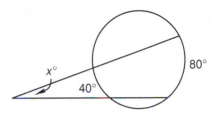

$$x = \frac{80 + 40}{2} = 60$$

$$x = \frac{80 - 40}{2} = 20$$

In the diagram on the right above, both lines are secants. The same rule applies if one or both lines are tangents.

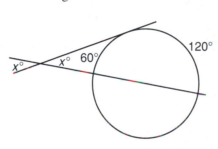
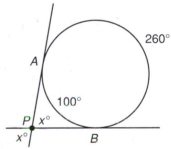

$$x = \frac{120 - 60}{2} = 30$$

$$x = \frac{260 - 100}{2} = 80$$

56.B

proofs We remember that in a triangle the measure of an exterior angle equals the sum of the measures of the remote interior angles.

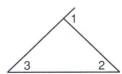
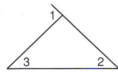

In both of the figures above, the angles labeled 1 are exterior angles and are equal to the sum of the angles labeled 2 and 3. In the left figure below, the angles labeled 1 are exterior angles and thus are equal to the sum of the angles labeled 2 and 3.

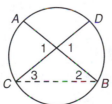
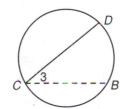
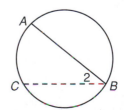

In the center on the preceding page, we see that angle 3 equals $\frac{1}{2}\widehat{DB}$, and, on the right, we see that angle 2 equals $\frac{1}{2}\widehat{AC}$. Angle 1 equals the sum of these angles, so

$$\angle 1 = \angle 3 + \angle 2 \qquad \text{exterior angle}$$

$$\angle 1 = \frac{1}{2}\widehat{DB} + \frac{1}{2}\widehat{AC} \qquad \text{substituted}$$

$$\angle 1 = \frac{1}{2}\left(\widehat{DB} + \widehat{AC}\right) \qquad \text{factored } \frac{1}{2}$$

This proves that the vertical angles formed when chords intersect equal half the sum of the intercepted arcs.

To prove that the angle formed by two secants that intersect outside a circle equals one-half the difference of the intercepted arcs, we begin with the figure on the left. Then we draw chord AD in the figure on the right.

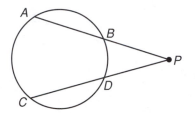 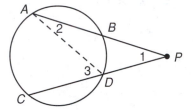

We note that angle 3 is an exterior angle and, therefore, equals the sum of angles 1 and 2. Thus, we can write

$$\angle 1 + \angle 2 = \angle 3$$

If we solve this equation for $\angle 1$, we get

$$\angle 1 = \angle 3 - \angle 2$$

The measure of angle 3 is $\frac{1}{2}\widehat{AC}$ and the measure of $\angle 2$ is $\frac{1}{2}\widehat{DB}$. We substitute and get

$$\angle 1 = \frac{1}{2}\widehat{AC} - \frac{1}{2}\widehat{DB}$$

$$= \frac{1}{2}\left(\widehat{AC} - \widehat{DB}\right) \qquad \text{QED}$$

The same reasoning process can be used if one or both of the intersecting lines is a tangent rather than a secant.

practice   Find the measures of the angles labeled x.

a.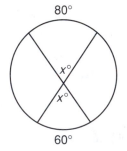

b.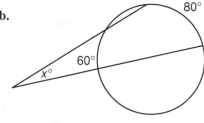

c.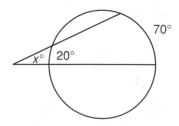

d.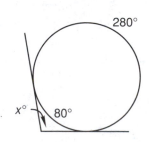

problem set 56

1. Find three consecutive even integers such that the product of the first and the second is 8 greater than the product of −10 and the third.

2. The chemical formula for methylene bromide is CH_2Br_2. If the bromine in a container of methylene bromide weighs 320 grams, what is the total weight of the methylene bromide? What percent by weight of the compound is bromine? (C, 12; H, 1; Br, 80)

3. The walk into the country at 4 mph was leisurely, but the ride back at 20 mph on a scooter was a little scary. If the total time of the trip was 12 hours, how far was the walk?

4. It was necessary to mix 1000 gallons that was 56% fluorine. If one solution was 20% fluorine and another was 80% fluorine, how much of each one should be used?

5. They just kept coming. Their approach was inexorable. Finally, there were $3\frac{3}{5}$ as many as were desired. If 1440 came, how many were desired?

6. Angles P and R are equal because both intercept $\overarc{AB}$. Find x and y.

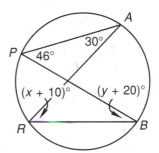

7. Find the measures of the angles labeled x.

(a)

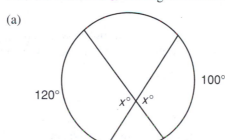

(b)

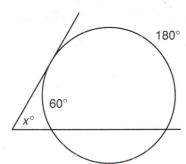

8. Write the ratios of the sides and find x in terms of m, y, and p.

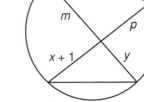

9. Find x: $\dfrac{a+x}{b} - \dfrac{c}{m} = \dfrac{p}{k}$

10. Find c: $\dfrac{m}{a+c} - \dfrac{x}{m} = p$

Convert to rectangular form:

11. $10\underline{/210°}$

12. $20\underline{/60°}$

Use unit multipliers to convert:

13. 60 kilometers per hour to inches per second

14. 400 cubic yards to cubic centimeters

Simplify:

15. $3i^5 + 2\sqrt{-25} - 3i^2$

16. $2i^4 - 3i^3 + 2i + 4$

17. $\sqrt[6]{4\sqrt[5]{2}}$

18. $\sqrt{y^4}\sqrt{xy^2}$

19. $\sqrt{25\sqrt[3]{5}}$

20. $4\sqrt{\dfrac{3}{4}} - 2\sqrt{\dfrac{4}{3}} - 2\sqrt{27}$

Solve by completing the square:

21. $x^2 = 7x + 7$

22. $-8x - 8 = -x^2$

Solve:

23. $\sqrt{x + 1} + 1 = 1$

24. $\sqrt{x^2 - 2x + 21} - 1 = x$

25. Divide $4x^4 - 1$ by $x - 3$.

26. Simplify: $\dfrac{\dfrac{ax}{y^2} - \dfrac{yp}{x}}{\dfrac{yp}{xy} - \dfrac{1}{y^2}}$

27. Find the equation of the line through $(-5, -5)$ that is perpendicular to the line that passes through $(5, -2)$ and $(-3, -3)$.

Solve:

28. $\dfrac{x - 3}{4} - \dfrac{2 - x}{8} = 6$

29. $\dfrac{a - 5}{2} - \dfrac{3 - a}{4} = 1$

30. Add: $\dfrac{x}{a^2 y} - \dfrac{3x + 2}{a^2 y(x - 1)} - \dfrac{4}{x^2 - 1}$

LESSON 57 *Ideal gas laws*

Ideal gas law problems are simple algebra problems that require the use of ratio equations for their solution. The equation that can solve any of these problems is called the **general ideal gas law equation.** In this equation,

$$\frac{P_1 V_1}{T_1} = \frac{P_2 V_2}{T_2}$$

P stands for pressure, V stands for volume, and T stands for temperature. If we omit one of these variables, the resulting equation is called Charles' law, Boyle's law, or Gay-Lussac's law. These names are not important, and we won't worry about which is which. The equations for these laws are

(a) $\dfrac{P_1}{T_1} = \dfrac{P_2}{T_2}$ (b) $\dfrac{V_1}{T_1} = \dfrac{V_2}{T_2}$ (c) $P_1 V_1 = P_2 V_2$

In (a) we have omitted the symbol V. In (b) we have omitted the symbol P, and in (c) we have omitted the symbol T. The units that we will use in these problems are:

For pressure: newtons per square meter, atmospheres, or millimeters of mercury

For volume: liters, milliliters, or cubic centimeters

For temperature: kelvins[†]

A **pascal** (abbreviated Pa) is the SI (metric) unit for pressure and is defined to be 1 newton per square meter. We will not dwell on the meanings of these units, for this is a topic for chemistry and physics.

However, we must always be careful to use the same units throughout a particular problem. The problems will be worded so that the units need not be considered. We will just use the numbers.

[†]A *kelvin* (abbreviated K) is a unit of temperature. To designate 400 of these units, we can write either 400 kelvins or 400 K. The word *degree* is not used with kelvin as it is with degrees Fahrenheit (°F) or degrees Celsius (°C). Absolute zero is 0 kelvin, which is approximately equal to –273°C.

example 57.1   Four liters of an ideal gas at a temperature of 800 kelvins had a pressure of 100 newtons per square meter. If the volume were increased to 10 liters and the temperature reduced to 600 kelvins, what would the pressure be?

solution   We begin by writing the general gas law equation.

$$\frac{P_1 V_1}{T_1} = \frac{P_2 V_2}{T_2}$$

The symbols P_1, V_1, and T_1 represent the original pressure, volume, and temperature; and the symbols P_2, V_2, and T_2 represent the final pressure, volume, and temperature. If we replace these letters with the proper numbers, we get

$$\frac{(100)(4)}{800} = \frac{P_2(10)}{600}$$

We can solve this equation easily by multiplying both sides by $\frac{600}{10}$.

$$\frac{600}{10} \cdot \frac{(100)(4)}{800} = \frac{P_2(10)}{600} \cdot \frac{600}{10} \quad \longrightarrow \quad 30 = P_2$$

Since P_1 was given in newtons per square meter, then P_2 will be in newtons per square meter because the same units must be used for the same variable everywhere in a problem. Thus, our final pressure is

$$P_2 = \textbf{30 newtons per square meter}$$

example 57.2   The initial pressure of a quantity of an ideal gas was 400 newtons per square meter and the initial temperature was 1200 kelvins. The volume was held constant. What was the pressure if the temperature was decreased to 900 kelvins?

solution   First we write the gas law as

$$\frac{P_1 V_1}{T_1} = \frac{P_2 V_2}{T_2}$$

If the volume is held constant, V_1 equals V_2. We can mentally divide both sides of the equation by V and eliminate volume. Now we have

$$\frac{P_1}{T_1} = \frac{P_2}{T_2}$$

Next we insert the given values for the variables.

$$\frac{400}{1200} = \frac{P_2}{900}$$

We can solve for P_2 by multiplying both sides of the equation by 900.

$$900 \cdot \frac{400}{1200} = \frac{P_2}{900} \cdot 900 \quad \longrightarrow \quad 300 = P_2$$

Since P_1 was given in newtons per square meter, then P_2 will be in newtons per square meter because the same units are always used for a variable everywhere in a problem. Thus, our answer is

$$P_2 = \textbf{300 newtons per square meter}$$

example 57.3   The temperature of a quantity of ideal gas was held constant in an experiment. The original pressure was 7 atmospheres and the original volume was 42 liters. If the volume was reduced to 10 liters, what was the final pressure?

solution   We write the gas law as

$$\frac{P_1 V_1}{T_1} = \frac{P_2 V_2}{T_2}$$

Since the temperature is constant, we can omit the symbol T from both sides of the equation and write

$$P_1V_1 = P_2V_2$$

Now we replace the symbols with the given numbers.

$$(7)(42) = (P_2)(10)$$

We solve by dividing by 10.

$$\frac{(7)(42)}{10} = \frac{P_2(10)}{10} \rightarrow P_2 = 29.4 \text{ atmospheres}$$

The final pressure is in atmospheres because atmospheres was the unit of pressure for P_1.

practice   a.   Eight liters of an ideal gas at a temperature of 1000 kelvins had a pressure of 200 newtons per square meter. If the volume were increased to 10 liters and the temperature reduced to 800 kelvins, what would the pressure be?

b.   The temperature of a quantity of ideal gas was held constant at 1400 kelvins in an experiment. The original pressure was 11 atmospheres and the original volume was 44 liters. If the volume were reduced to 4.4 liters, what would the final pressure be?

problem set   1.   Four liters of an ideal gas at a temperature of 800 kelvins had a pressure of 100 newtons
57   per square meter. If the volume were increased to 12 liters and the temperature reduced to 600 kelvins, what would the pressure be?

2.   The temperature of a quantity of ideal gas was held constant in an experiment. The original pressure was 7 atmospheres and the original volume was 42 liters. If the volume was increased to 49 liters, what was the final pressure?

3.   The initial pressure of a quantity of an ideal gas was 400 newtons per square meter and the temperature was 1200 kelvins. The volume was held constant. What was the pressure if the temperature was decreased to 300 kelvins?

4.   To get 1000 gallons of mixture that was 35.2% alcohol, it was necessary to mix some 20% alcohol solution with some 40% alcohol solution. How much of each solution was required?

5.   Find four consecutive odd integers such that the product of the third and fourth is 49 greater than the product of the first and the number 10.

6.   Find the measures of the angles labeled x.

(a)
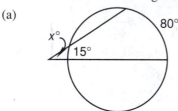
80°
$x°$
15°

(b)
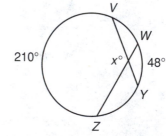
V
W
210°
$x°$
48°
Y
Z

7.   Find y: $\dfrac{a}{x - y} - \dfrac{c}{p} = m$ 8.   Find a: $\dfrac{x - a}{p} - c = \dfrac{k}{d}$

Convert to rectangular form:

9. $4\underline{/40°}$ 10. $40\underline{/330°}$

Use unit multipliers to convert:

11.   40 centimeters per second to miles per hour

12.   1000 cubic centimeters to cubic feet

Simplify:

13. $3i^3 - 2i^2 + i^4 - 5$ 14. $-2\sqrt{-9} - 3i^2 + 2i - 2$

Solve by completing the square:

15. $-5x - 6 = -x^2$

16. $-6x + x^2 = -5$

17. Find the equation of the line shown.

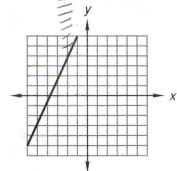

Solve:

18. $\sqrt{x - 2} + 2 = 3$

19. $\sqrt{x^2 - x + 13} - 1 = x$

20. $\dfrac{x - 2}{4} - \dfrac{x}{3} = 5$

21. $\dfrac{x}{4} - \dfrac{3x + 1}{2} = 3$

Simplify:

22. $\sqrt{9\sqrt[3]{3}}$

23. $\sqrt{x^4 \sqrt[3]{x^2 y}}$

24. $\sqrt[3]{4}\sqrt[5]{2}$

25. $3\sqrt{\dfrac{2}{5}} + 7\sqrt{\dfrac{5}{2}} - 2\sqrt{40}$

26. $16^{-5/4}$

27. $\dfrac{\dfrac{x^2 y}{p^5 z} - 1}{\dfrac{x}{p^5} - \dfrac{4}{z}}$

28. Solve: $-2(-x^0 - 4^0) - 3x(2 - 6^0) = (x)(-2 - 3^2 - 2) - x(-2 - 2^0)$

29. Solve $-28x + x^3 = 3x^2$ by factoring.

30. Add: $\dfrac{x}{a^2} - \dfrac{x + 2}{a(a + 2)}$

LESSON 58   *Lead coefficients* • *More on completing the square*

58.A
lead coefficients

The number in front of the x^2 term in a quadratic equation is called the **coefficient** of the x^2 term. When a quadratic equation is written in descending powers of the variable, the x^2 term comes first and is the lead term. Thus, the coefficient of this term is often called the **lead coefficient of the equation.**

$$\text{(a)}\quad 4x^2 + 3x + 5 = 0 \qquad \text{(b)}\quad x^2 - 3x + 5 = 0$$

In equation (a), we say that the lead coefficient is 4. The coefficient of the x^2 term in equation (b) is 1 and is not written because x^2 has an understood coefficient of 1. We say that equation (b) has a **unity lead coefficient** because mathematicians often say unity instead of saying 1.

58.B
more on completing the square

The coefficient of the x^2 term in the following equation is understood to be 1, and we say that this polynomial equation has a unity lead coefficient.

$$x^2 - 3x + 5 = 0$$

We have discussed the method of completing the square to solve quadratic equations whose lead coefficient is 1. **If the lead coefficient is not 1, then the first step is to divide every term on both sides of the equation by the lead coefficient. The result will be an equation with a unity lead coefficient which we can solve by completing the square.**

example 58.1   Solve $4x^2 + 3x - 3 = 0$ by completing the square.

solution   The coefficient of the lead term is not 1, but is 4. Thus, we begin by dividing every term by 4. Then we place parentheses around the first two terms and move the constant to the right-hand side of the equation.

$$x^2 + \frac{3}{4}x - \frac{3}{4} = 0 \qquad \text{divided by 4}$$

$$\left(x^2 + \frac{3}{4}x \qquad\right) = \frac{3}{4} \qquad \text{rearranged}$$

Next we multiply the coefficient of x by $\frac{1}{2}$ and square the result.

$$\left(\frac{3}{4} \cdot \frac{1}{2}\right)^2 = \left(\frac{3}{8}\right)^2 = \frac{9}{64}$$

Then we add $\frac{9}{64}$ to both sides of the equation and get

$$\left(x^2 + \frac{3}{4}x + \frac{9}{64}\right) = \frac{3}{4} + \frac{9}{64}$$

Now we write the left side as $\left(x + \frac{3}{8}\right)^2$, simplify the sum, and solve.

$$\left(x + \frac{3}{8}\right)^2 = \frac{57}{64} \qquad \text{simplified}$$

$$x + \frac{3}{8} = \pm\sqrt{\frac{57}{64}} \qquad \text{square root of both sides}$$

$$x = -\frac{3}{8} \pm \frac{\sqrt{57}}{8} \qquad \text{added } -\frac{3}{8} \text{ to both sides}$$

example 58.2   Solve $5x^2 - x - 2 = 0$ by completing the square.

solution   The first step is to divide every term by 5.

$$x^2 - \frac{1}{5}x - \frac{2}{5} = 0$$

Next we move the constant to the right-hand side and use parentheses.

$$\left(x^2 - \frac{1}{5}x \qquad\right) = \frac{2}{5}$$

Then we multiply $-\frac{1}{5}$ by $\frac{1}{2}$ and square the result.

$$\left(-\frac{1}{5} \cdot \frac{1}{2}\right)^2 = \left(-\frac{1}{10}\right)^2 = \frac{1}{100}$$

Next we add $\frac{1}{100}$ to both sides of the equation.

$$\left(x^2 - \frac{1}{5}x + \frac{1}{100}\right) = \frac{2}{5} + \frac{1}{100}$$

Now we simplify and solve.

$$\left(x - \frac{1}{10}\right)^2 = \frac{41}{100} \qquad \text{simplified}$$

$$x - \frac{1}{10} = \pm\sqrt{\frac{41}{100}} \qquad \text{square root of both sides}$$

$$x = \frac{1}{10} \pm \frac{\sqrt{41}}{10} \qquad \text{solved}$$

practice   Solve by completing the square:

    a. $3x^2 + 5x - 6 = 0$ **b.** $3x^2 - x - 1 = 0$

problem set 58

1. Six liters of an ideal gas at a temperature of 600 kelvins had a pressure of 4 atmospheres. If the volume was increased to 8 liters and the pressure decreased to 3 atmospheres, what was the final temperature?

2. The carbon in a container of C_6H_8NCl weighed 360 grams. What did the entire container of the compound weigh? (C, 12; H, 1; N, 14; Cl, 35)

3. In the compound of Problem 2, what percent by weight of the compound was chlorine (Cl)?

4. Two solutions are to be mixed to make 50 ml of a solution that is 16% bromine. One solution is 10% bromine and the other solution is 40% bromine. How much of each solution should be used?

5. After 3 hours the racer was 15 miles ahead of the trotter. How far did the trotter trot if the speed of the racer was 20 mph?

Solve by completing the square:

   6. $3x^2 + 4x - 3 = 0$ **7.** $4x^2 - x - 5 = 0$ **8.** $3x^2 - 4 = -2x$

9.   Find x, y, and z.

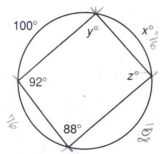

10.   Find p: $\dfrac{x-a}{p} - c = \dfrac{y}{k}$ **11.**   Find p: $\dfrac{a}{x-p} - c = \dfrac{y}{k}$

Convert to rectangular form:

12. $4\underline{/220°}$ **13.** $10\underline{/315°}$

Use unit multipliers to convert:

14.   70 meters per second to miles per hour   **15.**   40 cubic feet to cubic centimeters

Simplify:

16. $\sqrt[4]{9}\sqrt[5]{3}$ **17.** $\sqrt[7]{4\sqrt{2}}$

18. $\sqrt{x^2y^3}\sqrt[3]{xy^5}$ **19.** $-16^{-3/4}$

20. $5\sqrt{\dfrac{5}{11}} - 2\sqrt{\dfrac{11}{5}} + 3\sqrt{220}$ **21.** $5i^3 - 6i^8 + 2\sqrt{-25} - 4i^2$

22.   Estimate: $\dfrac{(40,621,857)(6,031,824)}{19,610 \times 10^{-24}}$

23.   Simplify: $\dfrac{x^2 + 16 + 10x}{x^2 + 11x + 24} \cdot \dfrac{x^2 + 8x + 15}{x^2 - 25}$

24.   The figure shown is the base of a cylinder 3 cm high whose volume is $(72 - 6\pi)$ cm³. Find the radius of the semicircle.

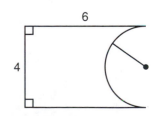

25.   Simplify: $\dfrac{\dfrac{x^2a^2}{m} - \dfrac{4}{p^3}}{\dfrac{xa}{mp^3} - 5}$

26. Solve by graphing and then find an exact solution by using either substitution or elimination.

$$\begin{cases} 3x + 4y = -4 \\ x - 5y = 10 \end{cases}$$

27. Divide $4x^3 - 5$ by $x + 2$.

Solve:

28. $\dfrac{3 - x}{2} - \dfrac{4x}{3} = 7$　　　　　　　　　**29.** $x^0 - 2x - 5(x - 3^0) = -2x^0 - 7$

30. Add: $\dfrac{b}{a(y + 1)} + \dfrac{cx}{a^2(y + 1)}$

LESSON 59　*Experimental data • Simultaneous equations with fractions and decimals • Rectangular form to polar form*

59.A
experimental data

Linear equations in two unknowns are important because they have applications in real-life problems. In chemistry and other advanced courses that use mathematics, we will find that the relationship between two variables can often be expressed by using equations. Fortunately, many of these equations are linear equations, and thus their graphs are straight lines. We can look at the graphs and get a pictorial representation of the relationship between the two variables being considered.

Laboratory experiments are used to confirm theories. The data points obtained in these experiments are graphed and circled to indicate that they are experimental data points. If the theory indicates that the relationship is linear, we estimate the line indicated by the data points. Then we write the equation of the line, using the variables of the data rather than x and y.

example 59.1　The data graphed below came from an experiment about bronze and copper. The line represented by the data points has been estimated. Write the equation of this line that gives bronze as a function of copper. Note that the two scales are different.

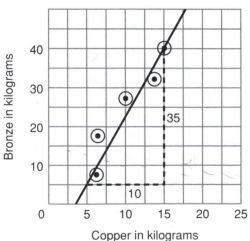

solution   Bronze is graphed vertically and copper is graphed horizontally, so in the equation we will replace y with B and x with C. Thus, our equation is

$$B = mC + b$$

From the triangle the slope is 3.5 so we have

$$B = 3.5C + b$$

Now we will use the coordinates of one point on the line so we can solve for b. We will use the point (15, 40). Then we will replace C with 15 and B with 40.

$$40 = 3.5(15) + b \quad \longrightarrow \quad 40 = 52.5 + b \quad \longrightarrow \quad b = -12.5$$

Thus, our final equation is

$$B = 3.5C - 12.5 \quad \text{kilograms}$$

This equation approximates the equation of the line indicated by the data points. The values of m and b that we have found are not exact.

example 59.2   The data graphed came from an experiment that involved nitrogen and sulfur. Write the equation that expresses nitrogen as a function of sulfur. Note the extreme difference in the horizontal and vertical scales.

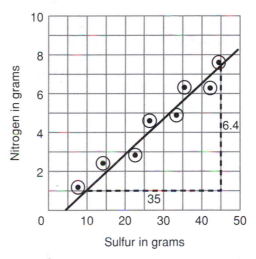

solution   Nitrogen is plotted vertically, so our equation is

$$N = mS + b$$

The slope is about $\frac{6.4}{35}$, which approximately equals 0.18, so we can write

$$N = 0.18S + b$$

We choose the point (10, 1) and replace S with 10 and N with 1 to get

$$(1) = (0.18)(10) + b \quad \longrightarrow \quad b = -0.8$$

so the complete equation is

$$N = 0.18S - 0.8 \quad \text{grams}$$

This equation approximates the equation of the line indicated by the data points. It is not exact. It is possible to get only approximate equations from experimental data. Another try would probably yield slightly different values for m and b.

59.B

simultaneous equations with fractions and decimals

Some systems of simultaneous equations that contain fractions and decimal numbers look very complicated. If we remember, as the first step, to change these equations to equations in which all numbers are integers, the solution will be much easier.

example 59.3   Solve:
$$\begin{cases} \text{(a)} \quad \dfrac{x}{2} + \dfrac{3y}{5} = -\dfrac{2}{5} \\[2mm] \text{(b)} \quad 0.06x - 0.2y = 1.04 \end{cases}$$

solution   We can eliminate the denominators in the fractional equation by multiplying every term by 10.

$$\text{(a)} \quad \frac{x}{2}(10) + \frac{3y}{5}(10) = -\frac{2}{5}(10) \quad \longrightarrow \quad \text{(a$'$)} \quad 5x + 6y = -4$$

And in equation (b) we can change the decimal coefficients to integers if we multiply every term by 100.

$$\text{(b)} \quad 0.06x(100) - 0.2y(100) = 1.04(100) \quad \longrightarrow \quad \text{(b$'$)} \quad 6x - 20y = 104$$

Now we can solve these equations by using the elimination method.

$$\begin{array}{llllll} \text{(a$'$)} & 5x + 6y = -4 & \longrightarrow & (-6) & \longrightarrow & -30x - 36y = 24 \\ \text{(b$'$)} & 6x - 20y = 104 & \longrightarrow & (5) & \longrightarrow & \underline{30x - 100y = 520} \\ & & & & & -136y = 544 \end{array}$$

$$\boldsymbol{y = -4}$$

We find x by replacing y with -4 in equation (a$'$).

$$\begin{array}{ll} 5x + 6y = -4 & \text{equation (a$'$)} \\ 5x + 6(-4) = -4 & \text{replaced } y \text{ with } -4 \\ 5x - 24 = -4 & \text{multiplied} \\ 5x = 20 & \text{added 24} \\ \boldsymbol{x = 4} & \text{divided} \end{array}$$

Thus, our solution is the ordered pair **(4, – 4).**

59.C

rectangular form to polar form

We can convert a number from rectangular form to polar form by using the rectangular coordinates to find the tangent of the angle. Then we use the inv tan keys on a calculator to find the angle. The hypotenuse of the triangle can be found by using trigonometric functions or by using the theorem of Pythagoras. The use of the sine and cosine to find the hypotenuse is not ideal because this method propagates errors made when the angle was determined.

example 59.4   Convert $-5R - 3U$ to polar form.

solution   We begin by locating the point on the coordinate plane and then drawing the triangle. We remember that $-5R$ means 5 to the left and $-3U$ means 3 down.

First we find the tangent of angle θ.

$$\tan \theta = \frac{\text{opposite}}{\text{adjacent}} \quad \longrightarrow \quad \tan \theta = \frac{3}{5} \quad \longrightarrow \quad \tan \theta = 0.6$$

Then we use a calculator to find that angle θ is approximately $31.0°$.

$$\text{Angle whose tangent is } 0.6 = 31.0°$$

Now we can use either the sine of $31°$ or the cosine of $31°$ to find H. We decide to use the sine.

$$\sin 31° = \frac{3}{H} \quad \rightarrow \quad H = \frac{3}{\sin 31°} \quad \rightarrow \quad H = \frac{3}{0.5150} \quad \rightarrow \quad H = 5.83$$

Finally we write the coordinates of the point in polar form and remember to add 180° so that the polar angle will be a third-quadrant angle.

$$-5R - 3U = \textbf{5.83} \underline{/211°}$$

If we use the Pythagorean theorem, we can find the exact length of H.

$$H = \sqrt{3^2 + 5^2} = \sqrt{34}$$

Thus, the location of the point can also be designated by writing

$$\sqrt{34} \underline{/211°}$$

practice

a. The data points shown come from an experiment that involved nitrogen (N) and fluorine (F). Write the equation that gives nitrogen as a function of fluorine: $N = mF + b$

b. Solve: $\begin{cases} \dfrac{x}{3} + \dfrac{2y}{7} = -\dfrac{3}{7} \\ 0.01x - 0.6y = 3.03 \end{cases}$

c. Convert $-3R - 5U$ to polar form.

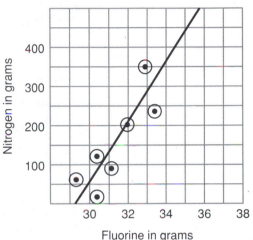

problem set 59

1. Find four consecutive integers such that the product of the first and the fourth is 22 less than the product of 10 and the opposite of the third.

2. What is the weight of the oxygen (O) in 460 grams of the compound whose formula is C_2H_6O? (C, 12; H, 1; O, 16)

3. Rasputin ran part of the way at 8 mph and walked the rest of the way at 3 mph. If the total trip was 41 miles and the total time was 7 hours, how far did he run and how far did he walk?

4. One solution was 90% iodine and the other was 70% iodine. How much of each should be used to get 100 liters of a solution that is 78% iodine?

5. In an experiment with a quantity of an ideal gas, the temperature was held constant. The initial pressure and volume were 14 newtons per square meter and 10 liters respectively. If the pressure was increased to 20 newtons per square meter, what was the final volume?

6. The data graphed are data points from an experiment. Write the equation that gives zirconium as a function of calcium: $Zr = mCa + b$

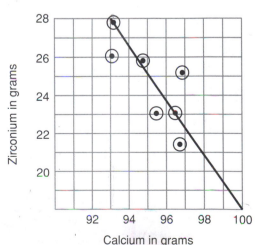

7. Find the measures of the angles labeled x.

    (a)

    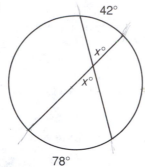

    (b)

    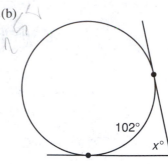

8. Solve: $\begin{cases} \dfrac{x}{3} + \dfrac{3y}{4} = -\dfrac{1}{4} \\ 0.04x - 0.2y = 1.13 \end{cases}$

9. Convert $-4R - 2U$ to polar form.

10. Convert $30\underline{/330°}$ to rectangular form.

Solve by completing the square:

11. $2x^2 = x + 5$ 12. $3x^2 - 5 = 2x$ 13. $3x^2 - 4 = x$

14. Find the area of this isosceles triangle. Dimensions are in centimeters.

15. Find m: $\dfrac{p - s}{m} - \dfrac{c}{4} + x = 0$

16. Find p: $\dfrac{m}{p + s} - \dfrac{c}{x} + 4 = 0$

Use unit multipliers to convert:

17. 40 inches per second to meters per hour

18. 18,000 cubic centimeters to cubic feet

Simplify:

19. $i^4 + 5 + 3\sqrt{-9} - 2\sqrt{-4}$

20. $3i^5 - 2i^2 - 4 - i$

Solve:

21. $\sqrt{x^2 - 4x + 39} + 1 = x + 4$

22. $\sqrt{x - 3} + 2 = -4$

Simplify:

23. $\sqrt[5]{3\sqrt{3}}$ 24. $\sqrt[4]{4\sqrt{2}}$ 25. $-16^{-3/4}$

26. $\sqrt{x^2 y}\sqrt[3]{y^5 x^4}$

27. $\sqrt{\dfrac{2}{13}} - 4\sqrt{\dfrac{13}{2}} + 3\sqrt{234}$

28. Solve: $\dfrac{x - 4}{2} - \dfrac{x}{3} - 2 = \dfrac{5}{2}$

29. Add: $\dfrac{x}{y} + \dfrac{x^2 + 2}{y^2 a} + \dfrac{x^3}{y(a + y)}$

30. Simplify: $\dfrac{(x^0 yp)^{-3} x^3 yp^0 p^{-2}}{x^2 p^2 y^0 (y^{-3} p)^2 x^{-3} y^0 p^2}$

LESSON 60   *Direct and inverse variation*

When the statement of a problem says that A **varies directly as** B or that A **is directly proportional to** B, the equation

$$A = kB$$

is implied. When the statement says that A **varies inversely as** B or that A **is inversely proportional to** B, the equation

$$A = \frac{k}{B}$$

is implied. The constant k is called the **constant of proportionality. Note that k is always in the numerator. In direct variation, both variables are in the numerator; and in inverse variation, one variable is in the numerator and the other is in the denominator.** The key to working variation problems is recognizing the equation implied by the statement. The following examples should be helpful.

STATEMENT	IMPLIED EQUATION
The number of boys varied directly as the number of girls.	$B = kG$
The price varied inversely as the number.	$P = \dfrac{k}{N}$
The resistance is directly proportional to the length.	$R = kL$
The number of revolutions per minute (RPM) is inversely proportional to the number of teeth.	$RPM = \dfrac{k}{N_t}$
The water produced varied directly as the amount of hydrogen burned.	$W = kH_B$

Direct and inverse variation problems are four-step problems. **The first step is recognizing that the words *varies directly* (is directly proportional to) and *varies inversely* (is inversely proportional to) imply equations of the forms**

$$A = kB \qquad \text{and} \qquad A = \frac{k}{B}$$

The next step is to find k. In order to find k, the problem must give sample values of A and B. The third step is to replace k in the equation with the proper number. The last steps are to reread the problem, make the final substitution, and then solve the equation.

example 60.1   The number of boys in every classroom of a school varies directly as the number of girls. In one room, there are 8 boys and 2 girls. If there are 5 girls in another room, how many boys are in this room?

solution   (1)   We write the equation implied by the words **varies directly.**

$$B = kG$$

(2)   We use 8 for boys and 2 for girls and solve for k.

$$(8) = k(2) \quad \longrightarrow \quad k = 4$$

(3)   We replace k in the equation with 4.

$$B = 4G$$

(4)   Now we use 5 for G and solve for B.

$$B = 4(5) \quad \longrightarrow \quad B = 20 \text{ boys}$$

example 60.2   The number of revolutions per minute (RPM) varies inversely as the number of teeth in the gear. If 40 teeth result in 100 RPM, what would be the RPM if the gear had 30 teeth?

solution   We will use the same four steps as in the last problem.

(1)   First we write the equation implied by the words **varies inversely.**

$$\text{RPM} = \frac{k}{N_t}$$

(2)   Next we find k.

$$100 = \frac{k}{40} \quad \longrightarrow \quad k = 4000$$

(3)   Then we replace k with 4000.

$$\text{RPM} = \frac{4000}{N_t}$$

(4)   Now we substitute 30 for the number of teeth and solve for RPM.

$$\text{RPM} = \frac{4000}{30} \quad \longrightarrow \quad \textbf{RPM} = 133\frac{1}{3}$$

example 60.3   The number of clowns was directly proportional to the number of performers. If there were 40 clowns when there were 20,000 performers, how many clowns would there be if there were 12,000 performers?

solution   We will again use four steps.

(1) $C = kP$ implied equation

(2) $40 = k(20{,}000) \quad \longrightarrow \quad k = \frac{1}{500}$ solved for k

(3) $C = \frac{1}{500}P$ replaced k with $\frac{1}{500}$

(4) $C = \frac{1}{500}(12{,}000) \quad \longrightarrow \quad C = \textbf{24 clowns}$ found C when P equals 12,000

practice   a.   The number of bluebirds in every tree in the grove varied directly as the number of redbirds. In one tree there were 12 bluebirds and 3 redbirds. If in another tree there were 6 redbirds, how many bluebirds were in this tree?

b.   The number of revolutions per minute (RPM) varies inversely as the number of teeth in the gear. If 60 teeth result in 150 RPM, what would be the RPM if the gear had 100 teeth?

problem set 60   1.   The number of boys in every classroom of a school varies directly as the number of girls. In one room there are 8 boys and 2 girls. If in another room there are 7 girls, how many boys are in this room?

2.   The number of revolutions per minute (RPM) varies inversely as the number of teeth in the gear. If 40 teeth result in 100 RPM, what would be the RPM if the gear had 25 teeth?

3.   The number of clowns was directly proportional to the number of performers. If there were 40 clowns when there were 20,000 performers, how many clowns would there be if there were 8000 performers?

4.   The volume of a quantity of an ideal gas was held constant. The initial pressure and temperature were 400 newtons per square meter and 500 kelvins respectively. What would the pressure be if the temperature were increased to 1000 kelvins?

5. There were 50 more pugnacious students than trepid students. In fact, twice the number of pugnacious exceeded 3 times the number of trepid by 60. How many students were in each category?

6. Convert $4\underline{/135°}$ to rectangular form.

7. Convert $-2R + 4U$ to polar form.

8. Solve: $\begin{cases} \dfrac{2}{5}x + \dfrac{3}{2}y = 34 \\ 0.02x + 0.3y = 6.2 \end{cases}$

Solve by completing the square:

9. $4x^2 - 3 = x$

10. $3x^2 = 2x + 1$

11. The data points shown came from an experiment involving sodium (Na) and carbon (C). Write the equation that expresses sodium as a function of carbon: $Na = mC + b$

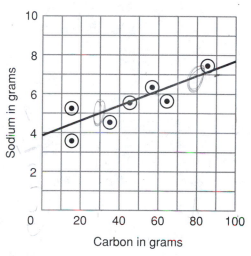

12. Find x and y.

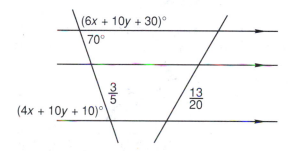

13. Find x: $\dfrac{a}{x + y} = c + \dfrac{m}{d}$

14. Find c: $\dfrac{xy + c}{m} + d = \dfrac{k}{z}$

Simplify:

15. $3i^3 - \sqrt{-4} + 3\sqrt{-9} - 2 + i^2$

16. $-3i^4 - 2i^2 + 2 - 3\sqrt{-25}$

Use unit multipliers to convert:

17. 60 centimeters per second to miles per hour

18. 1,400,000 cubic centimeters to cubic yards

Simplify:

19. $\sqrt[3]{27\sqrt{3}}$

20. $\sqrt[4]{81}\sqrt[3]{3}$

21. $81^{-3/4}$

22. $\sqrt{m^2 p}\,\sqrt[3]{m^5 p^4}$

23. $5\sqrt{\dfrac{2}{9}} + 3\sqrt{\dfrac{9}{2}} - 5\sqrt{8}$

24. Estimate: $\dfrac{(5,162,348)(0.0000165)}{0.003217642}$

25. Find K.

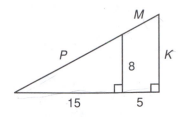

26. Given: $m\angle ABC = 140°$
Find x, k, and s.

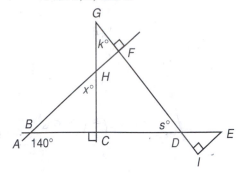

27. Find an exact solution by using either substitution or elimination:

$$\begin{cases} 2x - 3y = -3 \\ 2x + y = 8 \end{cases}$$

28. Solve $x^3 + 50x = 15x^2$ by factoring.

29. Find the equation of the line that has a slope of $\frac{2}{5}$ and passes through the point $(-40, 2)$.

30. Multiply: $\dfrac{3x^{-4}y^0y^2}{x^2x}\left(\dfrac{2x^2y^2}{y^4} - \dfrac{y^0x^0p^{-2}}{x^4y^4}\right)$

LESSON 61 *Chemical mixture problems, type B*

In the chemical mixture problems worked thus far, we have mixed two solutions of different percentage concentrations to get a mixture that has a different percentage concentration from either of the original solutions. It is interesting to note that the percentage concentration of the final mixture must fall between the percentage concentrations of the two solutions used. **For example, if we pour in some 15% iodine solution**

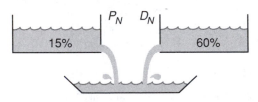

and dump in some 60% iodine solution, we can get a final mixture whose concentration is somewhere between 15% iodine and 60% iodine. It is impossible to mix these solutions and get a final concentration greater than 60% iodine or less than 15% iodine.

For want of a better name, we will call problems like the one above type A chemical mixture problems. Now we will look at another kind of chemical mixture problem. We will call these type B problems. In a type B problem, we will begin with a given mixture and add to this mixture or remove something from this mixture. **As in type A problems, the equation will make a statement about one of the components of the mixture.**

example 61.1   How much water must be evaporated from 100 gallons of a 10% brine solution to get a 40% brine solution?

solution   Usually, we would begin by deciding whether to work the problem in water or in salt. For this example, we will work the problem both ways to demonstrate that both ways give the same answer. First, we write the true statements that consider either water or salt.

CONSIDERING WATER	CONSIDERING SALT
Water one − water out = water final	Salt one − salt out = salt final
$W_1 - W_O = W_F$	$S_1 - S_O = S_F$

Now we write the mixture containers. Their use is very important in mixture problems.

$$(\ ) - (\ ) = (\ ) \hspace{3cm} (\ ) - (\ ) = (\ )$$

Next we make the entries in the containers, being careful to use neither W nor S as a variable. Our original mixture was 100 gallons, so this goes in the first container. We evaporated some, so E goes into the second container. The final mixture is the original mixture minus what was evaporated, so $100 - E$ goes into the last container.

CONSIDERING WATER	CONSIDERING SALT
$(100) - (E) = (100 - E)$	$(100) - (E) = (100 - E)$

Now we multiply each container by the proper decimal so that the containers times the decimals in the left-hand example represent water and those in the right-hand example represent salt. *No salt was evaporated,* so the multipliers for the center containers are 1 and 0.

CONSIDERING WATER	CONSIDERING SALT
$0.9(100) - (E) = 0.6(100 - E)$	$0.1(100) - 0(E) = 0.4(100 - E)$

Note that if a mixture is 0.9 water, then it is 0.1 salt; and if it is 0.6 water, then it is 0.4 salt. Now we solve.

$$90 - E = 60 - 0.6E \hspace{2cm} 10 = 40 - 0.4E$$
$$30 = 0.4E \hspace{2.5cm} -30 = -0.4E$$
$$\mathbf{75 = E} \hspace{2.8cm} \mathbf{75 = E}$$

Thus, both approaches give the same result. We must evaporate **75 gallons** of water to get a mixture that is 40% salt.

example 61.2   When Frank and Mark finished milking, they found that they had 900 pounds of milk that was 2 percent butterfat. How much butterfat did they have to add to raise the butterfat content to 8 percent? (Whole milk is a mixture of skim milk and butterfat.)

solution   This time we are adding something, so the final mixture will weigh more than the original mixture. We remember that the decimals for 2 percent and 8 percent are 0.02 and 0.08. We decide to work this problem in butterfat.

$$\text{Butterfat one } + \text{ butterfat added } = \text{ butterfat final}$$
$$B_1 + B_A = B_F$$

Now we write down the mixture containers and use P_N for the amount poured in.

$$(900) + (P_N) = (900 + P_N)$$

Now since we decided to work the problem in butterfat, our decimals are 0.02, 1, and 0.08.

$0.02(900) + 1(P_N) = 0.08(900 + P_N)$	equation
$18 + P_N = 72 + 0.08P_N$	multiplied
$0.92P_N = 54$	added $-0.08P_N$ and -18
$P_N = \mathbf{58.7 \text{ pounds}}$	divided and rounded

Thus, **58.7 pounds** of butterfat should be added to get a mixture that is 8 percent butterfat.

example 61.3   Virginia and Campbell had 100 kilograms of a 20% glycol solution. How much of a 40% glycol solution should they add to get a solution that is 35% glycol?

solution   We decide to make the statement about glycol.

$$\text{Glycol}_1 + \text{glycol added} = \text{glycol final}$$

Next we write the mixture containers. In the first we place 100. In the second we write P_N for "poured in," and in the last we write $100 + P_N$.

$$(100) + (P_N) = (100 + P_N)$$

Now we multiply each container by the proper percentage.

$$0.2(100) + 0.4(P_N) = 0.35(100 + P_N)$$

Then we multiply and solve.

$$20 + 0.4P_N = 35 + 0.35P_N \qquad \text{multiplied}$$
$$0.05P_N = 15 \qquad \text{simplified}$$
$$\mathbf{P_N = 300 \ kg} \qquad \text{divided}$$

Thus, if they pour in **300 kg** of a 40% glycol mixture, the result will be a mixture that is 35% glycol.

practice   How much water must be evaporated from 400 gallons of a 15% saline solution to get a 40% saline solution?

problem set
61

1. The rate of decomposition varied directly as the amount of substance present. When the amount was 5 kilograms, the rate was 0.005 kilogram per second. What was the rate of decomposition when the amount was 0.3 kilogram?

2. The volume of a quantity of ideal gas was kept constant in an experiment. The final temperature was 600 kelvins (K) and the final pressure was 300 newtons per square meter. What was the original pressure if the original temperature was 1000 K?

3. How much water must be evaporated from 100 gallons of a 10% brine solution to get a 20% brine solution?

4. When Frank and Mark finished milking, they found that they had 900 pounds of milk that was 2 percent butterfat. How much butterfat did they have to add to raise the butterfat content to 10 percent? (Whole milk is a mixture of skim milk and butterfat.)

5. Zollie and Beau had 100 kilograms of a 20% glycol solution. How much of a 30% glycol solution should they add to get a solution that is 25% glycol?

6. Convert $20\underline{/165°}$ to rectangular form.

7. Convert $6R - 2U$ to polar form.

8. Solve: $\begin{cases} \dfrac{2}{3}x - \dfrac{2}{5}y = -4 \\ 0.2x + 0.9y = 19.2 \end{cases}$

Solve by completing the square:

9. $-4x - 4 = -5x^2$

10. $-x = 7 - 2x^2$

11. The data points shown come from an experiment involving lead (Pb) and antimony (Sb). Write the equation that expresses lead as a function of antimony: $Pb = mSb + b$

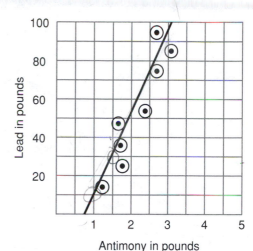

12. Find x and y.

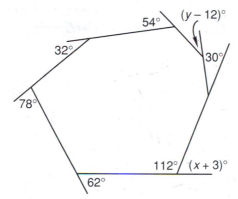

13. Find c: $\dfrac{(m + c + x)b}{k} + \dfrac{a}{d} = p$ 14. Find a: $\dfrac{4y}{2a + x} + \dfrac{m}{c} = d$

Simplify:

15. $-2i^2 - 3i - 4 - 2\sqrt{-4}$ 16. $8i^4 - 2i^3 - 2i - 6 - 4\sqrt{-16}$

17. Use unit multipliers to convert 40 cubic centimeters per second to cubic inches per hour.

18. Use unit multipliers to convert 4 cubic feet to cubic centimeters.

Simplify:

19. $\sqrt{32\sqrt{2}}$ 20. $\sqrt[3]{8\sqrt[3]{2}}$ 21. $-8^{-5/3}$

22. $\sqrt{x^5 y \sqrt[4]{y^2 x}}$ 23. $3\sqrt{\dfrac{2}{3}} - 4\sqrt{\dfrac{3}{2}} + 8\sqrt{24}$

24. Estimate: $\dfrac{(41{,}685{,}231)(0.0012846 \times 10^{-14})}{0.001998 \times 10^{-10}}$

25. Find C.   26. Solve for r in terms of s, t, and v.

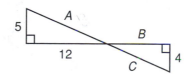

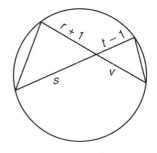

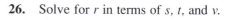

27. Solve by graphing and then find an exact solution by using either substitution or elimination:

$$\begin{cases} x - 2y = -6 \\ x + y = -1 \end{cases}$$

28. Solve $-15x = -x^3 + 2x^2$ by factoring.

29. Find the equation of the line that passes through $(-2, 4)$ and is perpendicular to $2x + 3y = 5$.

30. Solve: $-2^2 - 3^2 - (2^0)^2 - (-2)^0 = -x(-2x^0 - 5^0)2^2$

LESSON 62 *Complex roots of quadratic equations*

When we solve quadratic equations by completing the square, we will find that we get some solutions of the form

$$x = -3 \pm \sqrt{-5}$$

Now since the square root of a negative number can be written by using Euler's notation, we can write this answer as

$$x = -3 \pm \sqrt{5}i$$

In this lesson we will complete the square to solve quadratic equations whose solutions are complex numbers.

example 62.1 Solve: $-x + 3x^2 + 5 = 0$

solution As the first step we write the equation in standard form.

$$3x^2 - x + 5 = 0$$

Then we divide every term by 3 so that the coefficient of x^2 will be 1.

$$x^2 - \frac{1}{3}x + \frac{5}{3} = 0$$

Next we write the parentheses and move the constant term to the right-hand side.

$$\left(x^2 - \frac{1}{3}x \quad\right) = -\frac{5}{3}$$

Now we multiply the coefficient of x by $\frac{1}{2}$ and square the product.

$$\left(-\frac{1}{3} \cdot \frac{1}{2}\right)^2 = \frac{1}{36}$$

Then we add $\frac{1}{36}$ to both sides of the equation.

$$\left(x^2 - \frac{1}{3}x + \frac{1}{36}\right) = -\frac{5}{3} + \frac{1}{36}$$

Now we simplify and solve for x.

$$\left(x - \frac{1}{6}\right)^2 = -\frac{59}{36} \qquad \text{simplified}$$

$$x - \frac{1}{6} = \pm\sqrt{-\frac{59}{36}} \qquad \text{square root of both sides}$$

$$x = \frac{1}{6} \pm \frac{\sqrt{59}}{6}i \qquad \text{solved}$$

example 62.2 Solve: $-2x + 5x^2 = -3$

solution First we write the equation in standard form.

$$5x^2 - 2x + 3 = 0$$

Next we get a unity lead coefficient by dividing by 5.

$$\left(x^2 - \frac{2}{5}x \quad\right) = -\frac{3}{5}$$

Now we multiply $-\frac{2}{5}$ by $\frac{1}{2}$ and square this product.

$$\left(-\frac{2}{5} \cdot \frac{1}{2}\right)^2 = \frac{1}{25}$$

Now we add $\frac{1}{25}$ to both sides of the equation.

$$\left(x^2 - \frac{2}{5}x + \frac{1}{25}\right) = -\frac{3}{5} + \frac{1}{25}$$

Then we simplify and finish the solution.

$$\left(x - \frac{1}{5}\right)^2 = -\frac{14}{25} \qquad \text{simplified}$$

$$x - \frac{1}{5} = \pm\sqrt{-\frac{14}{25}} \qquad \text{square root}$$

$$x = \frac{1}{5} \pm \frac{\sqrt{14}}{5}i \qquad \text{solved}$$

practice Solve by completing the square: $-x + 5x^2 + 3 = 0$

problem set 62

1. The number of victories varied inversely as the skill of the opponents. The team won 8 games when the opponents had a skill factor of 2. How many victories could be expected when the opponents' average skill factor was 8?

2. The initial state for a quantity of an ideal gas was a pressure of 600 newtons per square meter, a temperature of 300 kelvins, and a volume of 2 liters. If the volume was increased to 4 liters and the pressure was decreased to 400 newtons per square meter, what was the final temperature?

3. The vat contained 40 liters of a 5% salt solution. How much of a 20% salt solution should be added to get a 10% salt solution?

4. Part of the journey was by sleigh at 8 mph and the rest was by truck at 20 mph. If the total distance of the journey was 152 miles and the total time was 10 hours, what part of the trip was by sleigh?

5. In the chemical compound CH_4ON_2, what percent of the total weight is nitrogen (N)? (C, 12; H, 1; O, 16; N, 14)

6. Convert $20\underline{/340°}$ to rectangular form.

7. Convert $-2R + 5U$ to polar form.

8. Solve: $\begin{cases} \dfrac{1}{3}x - \dfrac{2}{3}y = -1 \\ 0.02x + 0.4y = 2.58 \end{cases}$

Solve by completing the square:

9. $-x + 2x^2 + 3 = 0$

10. $-5x + 6x^2 = -3$

11. The data points shown came from an experiment that involved bismuth (Bi) and mercury (Hg). Write the equation that expresses bismuth as a function of mercury: $Bi = mHg + b$

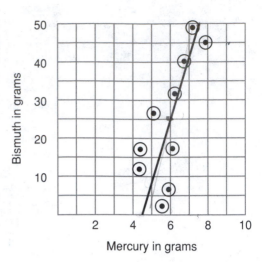

12. The volume of the right circular cylinder shown is $11,520\pi$ cubic inches. Find the radius. Dimensions are in inches.

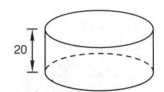

13. Find b: $\dfrac{(a + b)m}{c} - k = \dfrac{p}{r}$

14. Find y: $\dfrac{6x}{2y + 4a} - c = \dfrac{p}{r}$

Simplify:

15. $3i^3 - \sqrt{-4} - 2 - \sqrt{9} + 3$

16. $-3i^3 + 2i - 4 - 3i^2 - 2\sqrt{9}$

Use unit multipliers to convert:

17. 600 cubic centimeters per minute to cubic feet per second.

18. 20 cubic yards to cubic centimeters.

Simplify:

19. $\sqrt{16\sqrt{2}}$

20. $\sqrt[4]{4}\sqrt[5]{2}$

21. $-4^{5/2}$

22. $\sqrt{xy^7}\sqrt[3]{x^5 y}$

23. $3\sqrt{\dfrac{2}{5}} - 5\sqrt{\dfrac{5}{2}} - 3\sqrt{40}$

24. Estimate: $\dfrac{(0.000618427 \times 10^{14})(7,891,642)}{3,728,196,842}$

25. Find B.

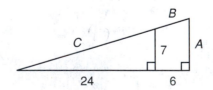

26. Find the measures of the angles labeled x.

(a)

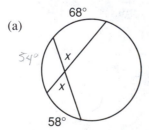

(b)

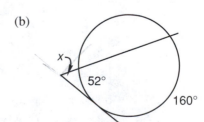

27. Solve by graphing and then find an exact solution by using either substitution or elimination:

$$\begin{cases} 2x - 3y = -9 \\ 2x + 3y = -3 \end{cases}$$

28. Solve $50x + x^3 = 15x^2$ by factoring.

29. Find the equation of the line whose slope is $-\frac{2}{7}$ and which passes through the point $(-5, -7)$.

30. Solve: $\dfrac{3 + x}{2} - \dfrac{2}{7} = 4$

LESSON 63   *Addition of vectors*

To add two vectors, we first write the vectors in rectangular form. Then we add the horizontal components to find the horizontal component of the sum and add the vertical components to find the vertical component of the sum. The process is easier to understand if we think of each vector as describing a journey, as in the following example.

example 63.1   Flying Arrow left the village and traveled 20 miles on a heading of 20°. From this point he travelled 40 miles on a heading of 210°. How far did he end up from the village?

solution   We need to add the vectors $20\underline{/20°}$ and $40\underline{/210°}$. **We remember that when we break up a vector k into components, one component is $k \sin \theta$ and the other component is $k \cos \theta$.**

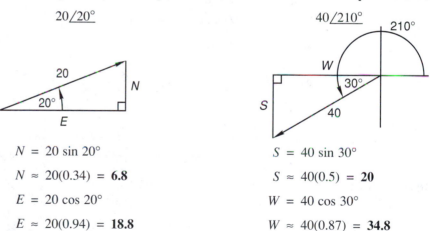

$$N = 20 \sin 20°$$
$$N \approx 20(0.34) = \mathbf{6.8}$$
$$E = 20 \cos 20°$$
$$E \approx 20(0.94) = \mathbf{18.8}$$

$$S = 40 \sin 30°$$
$$S \approx 40(0.5) = \mathbf{20}$$
$$W = 40 \cos 30°$$
$$W \approx 40(0.87) = \mathbf{34.8}$$

Thus, $20\underline{/20°} = 18.8R + 6.8U$ and $40\underline{/210°} = -34.8R - 20U$. Now we add the vectors by adding like components algebraically.

$$\begin{array}{r} 18.8R + 6.8U \\ -\ 34.8R - 20.0U \\ \hline \mathbf{-16R - 13.2U} \end{array}$$

Thus, Flying Arrow ended up 16 miles west and 13.2 miles south of the village.

example 63.2   Add $30\underline{/55°}$ and $10\underline{/170°}$.

solution   We can think of these as trips of 30 miles and 10 miles in the directions given. First we find the horizontal and vertical components of each vector.

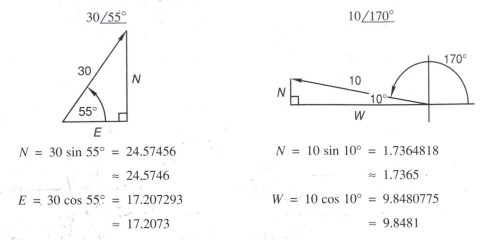

$30\underline{/55°}$

$10\underline{/170°}$

$N = 30 \sin 55° = 24.57456$ $N = 10 \sin 10° = 1.7364818$
≈ 24.5746 ≈ 1.7365

$E = 30 \cos 55° = 17.207293$ $W = 10 \cos 10° = 9.8480775$
≈ 17.2073 ≈ 9.8481

Next we write the vectors in rectangular form and add algebraically.

$$17.2073R + 24.5746U$$
$$-9.8481R + 1.7365U$$
$$\overline{7.3592R + 26.3111U}$$

If we round the sum of the vectors to two decimal places, we get **7.36R + 26.31U**.

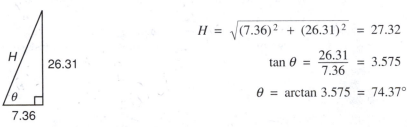

$H = \sqrt{(7.36)^2 + (26.31)^2} = 27.32$

$\tan \theta = \dfrac{26.31}{7.36} = 3.575$

$\theta = \arctan 3.575 = 74.37°$

Thus, the polar form of the sum of the vectors is

$27.32\underline{/74.37°}$

practice   Tracker left the cabin and traveled 25 miles on a heading of 20°. From this point she went 50 miles on a heading of 215°. How far did she end up from the village?

problem set 63

1. Odysseus found that his troubles varied directly as his distance from his home island of Ithaca. If he had 20 troubles when he was 400 miles from home, how many troubles did he have when he was only 60 miles from home?

2. The pressure of a container of ideal gas was held constant for an experiment. The initial temperature was 800 kelvins and the initial volume was 20 liters. If the final volume was reduced to 12 liters, what was the final temperature?

3. Somehow salt found its way into the rainbarrel, for the 50 gallons of water it contained was found to be 4% salt. How much pure water must be added to reduce the salt content to 1%?

4. Oedipus beat Rex to the goal by 4000 feet. If Rex ran at 20 feet per second and Oedipus ran at 40 feet per second, what was the length of the race course?

5. The weight of the carbon in a container of C_3H_7Cl was 48 grams. What was the total weight of the compound? (C, 12; H, 1; Cl, 35)

6. Running Bear left the village and traveled 30 miles on a heading of 30°. From this point he went 50 miles on a heading of 220°. How far did he end up from the village?

$$V = \frac{1}{3} B H$$

$$12\pi = \frac{1}{3} \left[\pi r^2 \right] H$$

$$12\pi = \frac{1}{3} \pi r^2 \, 6$$

$$\frac{12\pi}{2\pi} = \frac{2\pi r^2}{2\pi}$$

$$\sqrt{r^2} = \sqrt{6}$$

Algebra II	Bonus	May 7	8	9	10	11	14	15	16	17	18
Name:	Used	M	T	W	T	F	M	T	W	T	F
Albright, Kami		46		✓	✓	✓	✓	✓		✓	✓
Bollenbacher, Hayle		4		✓	✓	✓	✓	✓		✓	✓
Bransteter, Turner		14		✓	✓	✓	✓	✓		✓	✓
Caffee, Tia Kathlee		57		✓	✓	✓	✓	✓		✓	✓RR
Eddington, Shelbe Renee											
Eischen, Mia		37		✓	✓	✓	✓	Abs		✓	✓
Heindel, Caleb		54		✓	✓	✓	✓	✓		✓	✓
Jacobs, Caleb Arthu r		21		✓	✓	✓	✓	✓		✓	✓
Keinanen, Emmi So fi Susanna		41		✓	✓	✓	✓	✓	✓	✓	✓
Meuthen, Hannah *		58		✓	✓	✓	✓	✓	✓	✓	✓
Murphy, Elizabeth Ji llian		22				✓	✓	✓lock	✓	✓	Abs
Rollins, Whitney		11		✓			✓	✓		✓	✓
Statler, Autmn D											

7. Add $30\underline{/45°}$ and $10\underline{/160°}$.

8. Write $7.3R + 26.34U$ in polar form.

9. Solve: $\begin{cases} \dfrac{1}{5}x - \dfrac{5}{2}y = -48 \\ 0.4x + 0.05y = 5 \end{cases}$

Solve by completing the square:

10. $-x = -2x^2 - 5$

11. $3x^2 = -4 + 2x$

12. The data points shown came from an experiment that involved molybdenum (Mo) and zirconium (Zr). Write the equation that expresses molybdenum as a function of zirconium: $Mo = mZr + b$

13. Find the surface area in square centimeters of the prism shown. Dimensions are in meters.

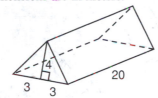

14. Find z: $\dfrac{p + zy}{m} - c = \dfrac{a}{b}$

15. Find m: $\dfrac{p + zy}{m} - c = \dfrac{a}{b}$

Simplify:

16. $4i^3 - i^5 + 2i^2 - \sqrt{-16}$

17. $3 - 2i^5 - 3i^4 + \sqrt{-4} - i$

Use unit multipliers to convert:

18. 400 centimeters per minute to yards per second.

19. 4 cubic miles to cubic kilometers.

Simplify:

20. $\sqrt[5]{3\sqrt[4]{3}}$

21. $\sqrt[5]{4\sqrt[4]{2}}$

22. $\dfrac{4}{(-27)^{-2/3}}$

23. $\sqrt[4]{xy^2}\,\sqrt{x^3y}$

24. $3\sqrt{\dfrac{2}{7}} + 5\sqrt{\dfrac{7}{2}} - 3\sqrt{126}$

25. Estimate: $\dfrac{(476{,}158 \times 10^{22})(79{,}318{,}642)}{(983{,}704)(514.0 \times 10^{-14})}$

26. A circular cone has an altitude of 6 meters. The volume of the cone is 12π cubic meters. What is the radius of the cone?

27. If the measure of an angle is $A°$, the measure of the supplement is $(180 - A)°$ and the measure of the complement is $(90 - A)°$. Find an angle such that 12 times its complement is 20 greater than its supplement.

28. Find the equation of the line that passes through $(2, -7)$ and is parallel to $5x + 4y = 7$.

Solve:

29. $-2^0 - 2^2 - 2^2(-2 - 1^0)x - 3x - 7x^0y^0 - 4 = 2$

30. $\dfrac{4x + 5}{3} - \dfrac{x}{7} = 2$

LESSON 64 *Complex fractions • Complex numbers*

64.A
complex fractions

Any fraction that contains more than one fraction line is called a **complex fraction**. Thus, this expression

$$\frac{\dfrac{1}{b} + x}{\dfrac{a}{b}}$$

is a complex fraction. We can simplify this fraction by adding in the numerator and then by multiplying the denominator and the numerator by $\frac{b}{a}$.

$$\frac{\dfrac{1}{b} + x}{\dfrac{a}{b}} \;\longrightarrow\; \frac{\dfrac{1 + bx}{b}}{\dfrac{a}{b}} \;\longrightarrow\; \frac{\dfrac{1 + bx}{b} \cdot \left(\dfrac{b}{a}\right)}{\dfrac{a}{b} \cdot \left(\dfrac{b}{a}\right)} \;\longrightarrow\; \frac{1 + bx}{a}$$

The next example is just a little more involved.[†]

example 64.1   Write the following as a simple fraction: $\quad a + \dfrac{1}{\dfrac{1}{b} + x}$

solution   We begin by simplifying the second term.

$$\frac{1}{\dfrac{1}{b} + x} \;\longrightarrow\; \frac{1}{\dfrac{1 + bx}{b}} \;\longrightarrow\; \frac{1 \cdot \dfrac{b}{1 + bx}}{\dfrac{1 + bx}{b} \cdot \dfrac{b}{1 + bx}} \;\longrightarrow\; \frac{b}{1 + bx}$$

Now we add the first term and the second term.

$$a + \frac{b}{1 + bx} \qquad\qquad\qquad \text{two terms}$$

$$\frac{}{1 + bx} + \frac{}{1 + bx} \qquad\qquad \text{new denominators}$$

$$\frac{a(1 + bx)}{1 + bx} + \frac{b}{1 + bx} = \frac{a + abx + b}{1 + bx} \qquad \text{added}$$

example 64.2   Write $\dfrac{a}{x} + \dfrac{4}{1 + \dfrac{b}{x}}$ as a simple fraction.

solution   First we simplify the second term.

$$\frac{a}{x} + \frac{4}{\dfrac{x + b}{x}} \;\longrightarrow\; \frac{a}{x} + \frac{4x}{x + b}$$

Now we use $x(x + b)$ as a common denominator and add.

$$\frac{}{x(x + b)} + \frac{}{x(x + b)} \qquad \text{new denominators}$$

$$\frac{a(x + b)}{x(x + b)} + \frac{4x^2}{x(x + b)} \qquad \text{new numerators}$$

$$\frac{ax + ab + 4x^2}{x(x + b)} \qquad \text{added}$$

[†]The arrows indicate successive steps in the simplification process. Equal signs could have been used instead of the arrows.

64.B
complex numbers

We remember that we say that any number that can be graphed on the number line is a real number.

On this line we have graphed $2\sqrt{2}$ and $-2\sqrt{2}$, so both of these numbers are real numbers. Square roots of negative numbers or numbers that have i as a factor, such as

$$2i \qquad \sqrt{-3} \qquad \sqrt{-1} \qquad \sqrt{-142} \qquad \sqrt{7}\,i$$

cannot be graphed on the number line. We remember that we call these numbers imaginary numbers to distinguish them from the real numbers. If a number has a real part and an imaginary part, such as

$$-4 + 2i \qquad 5 - 3i \qquad 7\sqrt{2} + 6i$$

we call the number a **complex number. If the real part is written first and the imaginary part is written second, we say that the complex number is written in** *standard form*. We define standard form to be the form

$$a + bi$$

where both a and b are real numbers. Since zero is a real number, then

$$0 + 2i \quad \text{which is} \quad 2i$$

and

$$7 - 0i \quad \text{which is} \quad 7$$

are complex numbers written in standard form. Thus, every real number is a complex number whose imaginary part is zero, and every imaginary number is a complex number whose real part is zero. For this reason, we can say that all of the following real numbers and imaginary numbers are also complex numbers.

$$-7 \qquad 2i \qquad -3\sqrt{2} \qquad -\sqrt{2}\,i \qquad 5 \qquad -\frac{3}{4}i$$

example 64.3   Simplify: $\sqrt{-2}\,\sqrt{-3}$

solution   **We remember that when both radicands are negative, the radicands cannot be multiplied to change the form of the expression. First we must use Euler's notation.**

$$\sqrt{2}\,i \cdot \sqrt{3}\,i \qquad \text{which is} \qquad \sqrt{6}\,i^2$$

But i^2 equals -1, so our answer is

$$-\sqrt{6}$$

example 64.4   Simplify: $4i^3 - 2i^4 + 2\sqrt{-9} + \sqrt{-3}\,\sqrt{-3}$

solution   Let's use two steps.

$$4(ii)i - 2(ii)(ii) + 2\sqrt{9}\,i + \sqrt{3}\,i\,\sqrt{3}\,i$$

Now we remember that $i^2 = -1$ and we write

$$-4i - 2 + 6i - 3$$

Now we add like parts and write the result in standard form.

$$-5 + 2i$$

example 64.5   Multiply: $(4 + 2i)(3 - 2i)$

solution   We will use the vertical format.

$$
\begin{array}{r}
4 + 2i \\
3 - 2i \\
\hline
12 + 6i \\
-8i - 4i^2 \\
\hline
12 - 2i - 4i^2
\end{array}
$$

Now since i^2 equals -1, then $-4i^2$ equals $+4$, so our answer is

$$16 - 2i$$

example 64.6   Multiply: $(5i - 2)(4 + 3i)$

solution   This time we use the horizontal format and get

$$(5i - 2)(4 + 3i) = 20i + 15i^2 - 8 - 6i$$

Now we simplify and remember that $15i^2$ equals -15, so we get

$$-23 + 14i$$

practice   Simplify:

a. $b + \dfrac{1}{\dfrac{1}{z} + 2m}$

b. $\dfrac{s}{z} + \dfrac{3}{5 + \dfrac{2s}{z}}$

c. $i^4 - 5i^3 - 3\sqrt{-3}\sqrt{-5}$

d. $(3 + 5i)(4 - 5i)$

problem set 64

1.   The total weight of the carbon varied inversely as the total weight of the fluorine. The carbon weighed 300 grams when the fluorine weighed only 2 grams. What was the weight of the carbon when the fluorine weighed 0.5 gram?

2.   An amount of an ideal gas was placed in a container whose volume was constant. The pressure was found to be 700 newtons per square meter and the temperature was 400 kelvins. If the pressure was increased to 2800 newtons per square meter, what was the final temperature?

3.   Two containers are on the shelf. The first one contains a 30% iodine solution and the other contains an 80% iodine solution. How much of each should be used to get 50 liters of a solution that is 40% iodine?

4.   The ride in was quick, as the speed was 400 kilometers per hour. The ride out at 100 kilometers per hour was much more relaxing. If the total time in and out was 40 hours, what was the distance in?

5.   The chemical formula for methyl iodide is CH_3I. What percent of the total weight of this compound is iodine (I)? (C, 12; H, 1; I, 127)

Simplify:

6. $m + \dfrac{2}{\dfrac{2}{c} + s}$

7. $\dfrac{m}{a} + \dfrac{3}{2 + \dfrac{s}{a}}$

8. $\dfrac{m}{c} + \dfrac{8}{2 + \dfrac{m}{c}}$

9. $9i^3 - 3i^4 + 2\sqrt{-4} + \sqrt{-2}\sqrt{-2}$

10. $\sqrt{-4} + \sqrt{-2}\sqrt{-2} - 4i^3$

11. $i^4 - 3i^2 - 2\sqrt{-2}\sqrt{-3}$

12. $2\sqrt{-9} - 3i^4 + 2\sqrt{3}\sqrt{-3} + i$

13. $(2 + 3i)(5 - 3i)$

14. $(3i - 5)(2 + 4i)$

15. $(2i - 4)(i + 2)$

16.   Add: $10\underline{/10°} + 30\underline{/150°}$

17.   Write $-4R - 6U$ in polar form.

18. Solve: $\begin{cases} \dfrac{3}{2}x - \dfrac{1}{5}y = 28 \\ 0.02x + 0.4y = 4.4 \end{cases}$

Solve by completing the square:

19. $3x^2 = -2x - 5$ **20.** $-3x + 2x^2 = -7$

21. The data points shown came from an experiment that involved tungsten (W) and iridium (Ir). Write the equation that expresses tungsten as a function of iridium: $W = mIr + b$

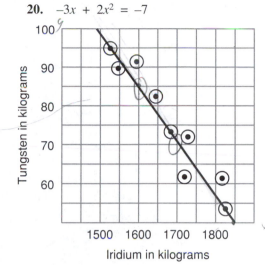

22. Find the area of this figure.

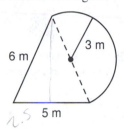

23. Use unit multipliers to convert 400 cubic centimeters per second to cubic inches per hour.

Simplify:

24. $\sqrt[3]{9\sqrt[3]{3}}$ **25.** $2\sqrt{\dfrac{7}{5}} - 3\sqrt{\dfrac{5}{7}} + 2\sqrt{140}$

26. Find x: $\dfrac{a(b + c)}{x} - m = \dfrac{d}{f}$ **27.** Find c: $\dfrac{a(b + c)}{x} - m = \dfrac{d}{f}$

28. Solve by graphing and then find an exact solution by using either substitution or elimination:
$$\begin{cases} x - 3y = -6 \\ 2x + 5y = 15 \end{cases}$$

29. Solve $56x = -15x^2 - x^3$ by factoring.

30. Estimate: $\dfrac{(146{,}842 \times 10^2)(0.0007892)}{(96{,}478 \times 10^{14})(0.000712 \times 10^{42})}$

LESSON 65   *Advanced substitution*

We have used substitution to solve systems of equations that are derived from uniform motion word problems. In the systems studied thus far, such as

$$R_W T_W + 210 = R_R T_R \qquad R_W = 4 \qquad R_R = 6 \qquad T_W + T_R = 5$$

the values of two of the variables have always been given. In this system of equations we have been told that R_W equals 4 and that R_R equals 6. Now we will investigate the solution of a system of four equations in which none of the values of the variables is given. In a later lesson, we will find that these equations will permit the solution of a new type of uniform motion word problem.

In the problem sets, the problems of this type will be exactly like the two problems that we will work here. The numbers and the subscripts will change, but otherwise the problems will be the same.

example 65.1 Find the values of all four variables in this system of equations.

(a) $R_W T_W = 6$ (b) $R_B T_B = 6$ (c) $R_B = 3R_W$ (d) $T_B = 2 - T_W$

solution We will use either equation (a) or equation (b) as our base equation and will substitute the other three equations into the base equation. Equation (c) determines which of the first two equations will be the base equation. Equation (c) has two forms:

(c) $R_B = 3R_W$ (c') $\dfrac{R_B}{3} = R_W$

If equation (b) is used as the base equation, we would use form (c) and replace R_B with $3R_W$. If equation (a) is used as the base equation, we would use form (c') and replace R_W with $\dfrac{R_B}{3}$. Either way will work, but the second approach introduces a fraction, which complicates the solution. Thus, we will use equation (b) as the base equation and replace R_B with $3R_W$ and T_B with $2 - T_W$.

(b) $R_B T_B = 6$ base equation

(d) $(3R_W)(2 - T_W) = 6$ substituted

(e) $6R_W - 3R_W T_W = 6$ multiplied

Now we have used three of the four equations.

$$R_W T_W = 6 \quad \cancel{R_B T_B = 6} \quad \cancel{R_B = 3R_W} \quad \cancel{T_B = 2 - T_W}$$

Next we use the remaining equation by replacing $R_W T_W$ in equation (e) with 6.

$6R_W - 3(6) = 6$ substituted

$6R_W - 18 = 6$ multiplied

$6R_W = 24$ added $+18$

$R_W = 4$ divided

We did not decide to solve for R_W when we began the solution. We just began substituting and simplifying, and it turned out that this approach produced the value of R_W. Now we will use this value of R_W to help us find the other three variables.

$R_B = 3R_W$ equation (c)

$R_B = 3(4) = \mathbf{12}$ substituted

Now we will use $R_W = 4$ and $R_B = 12$ in equations (a) and (b) to find T_W and T_B.

$R_W T_W = 6$ equation (a) $R_B T_B = 6$ equation (b)

$4T_W = 6$ substituted $12T_B = 6$ substituted

$T_W = \dfrac{3}{2}$ divided $T_B = \dfrac{1}{2}$ divided

example 65.2 Find the values of all four variables in this system of equations.

(a) $R_P T_P = 693$ (b) $R_C T_C = 165$ (c) $R_P = 3R_C$ (d) $T_P + T_C = 12$

solution This problem is just like the preceding one except that equation (d) must be rearranged before it can be used. We look at equation (c) and decide to use equation (a) as the base equation because we can replace R_P with $3R_C$.

$$R_P T_P = 693 \qquad \text{base equation}$$

$$(3R_C)(12 - T_C) = 693 \qquad \text{substituted}$$

$$36R_C - 3R_C T_C = 693 \qquad \text{multiplied}$$

Each time we get to this point in one of these problems, we will have a double variable. This time it is $R_C T_C$, and equation (b) tells us that $R_C T_C = 165$. We substitute 165 for $R_C T_C$ and solve.

$$36R_C - 3(165) = 693 \qquad \text{substituted 165}$$

$$36R_C - 495 = 693 \qquad \text{multiplied}$$

$$36R_C = 1188 \qquad \text{added 495}$$

$$R_C = 33 \qquad \text{divided}$$

Again, we note that we did not select the variable R_C in the beginning. We just substituted until we got a solution for one of the four variables. It just happened to be R_C. Now we find R_P.

$$R_P = 3R_C \qquad \text{equation (c)}$$

$$R_P = 3(33) = 99 \qquad \text{substituted}$$

Now we will use $R_P = 99$ and $R_C = 33$ to solve for T_P and T_C.

$R_P T_P = 693$	equation (a)		$R_C T_C = 165$	equation (b)
$99T_P = 693$	substituted		$33T_C = 165$	substituted
$T_P = 7$	divided		$T_C = 5$	divided

practice   Find the values of all four variables in this system of equations:

$$R_P T_P = 288 \qquad R_C T_C = 108 \qquad R_P = 3R_C \qquad T_P + T_C = 12$$

problem set 65

1. The weight of the silicon varied directly as the weight of the phosphorus. When the silicon weighed 400 kilograms, the phosphorus weighed only 100 kilograms. What was the weight of the silicon when the phosphorus weighed only 12 kilograms?

2. The temperature of a fixed amount of an ideal gas was held constant at 600 K.[†] The initial pressure and volume were 800 newtons per square meter and 2 liters, respectively. What was the final pressure if the volume was reduced to 0.1 liter?

3. The solution came up to the 500-ml mark on the beaker. If the solution was 84% alcohol, how much alcohol should be evaporated so that what is left would be only 80% alcohol?

4. The ignorant exceeded the erudite by 400. In fact, the ignorant numbered 100 more than 4 times the number of erudite. How many fell into each category?

5. Four percent of the phosgene combined with other chemicals. If 1920 kilograms did not combine, how much did combine?

Solve:

6. $R_W T_W = 8,\ R_B T_B = 8,\ R_B = 4R_W,\ T_B = 2 - T_W$

7. $R_P T_P = 600,\ R_C T_C = 105,\ R_P = 5R_C,\ T_P + T_C = 15$

8. $R_1 T_1 = 120,\ R_2 T_2 = 120,\ R_1 = 2R_2,\ T_1 + T_2 = 6$

[handwritten: $T_P = 15 - T_C$]

[handwritten: $5P_C(15 - T_C) = 600$]

[†]If the value of the temperature is constant, it can be eliminated from the equation. The fact that the temperature is 600 K is not relevant.

Simplify:

9. $x + \dfrac{1}{a + \dfrac{b}{c}}$

10. $\dfrac{4}{c} + \dfrac{1}{a + \dfrac{1}{b}}$

11. $x + \dfrac{a}{1 + \dfrac{1}{a}}$

12. $\sqrt{-4} - \sqrt{-3}\,\sqrt{-3} + 2i^5 - 4$

13. $(5i - 2)(2i - 3)$

14. $(-i - 3)(-2i + 4)$

15. Add: $20\underline{/45°} + 10\underline{/210°}$

16. Write $3R - 5U$ in polar form.

17. Solve: $\begin{cases} \dfrac{2}{7}x - \dfrac{1}{6}y = 1 \\ 0.3x + 0.07y = 0.84 \end{cases}$

Solve by completing the square:

18. $-2 = -3x^2 - 7x$

19. $2x^2 - 4 = -5x$

20. The data points shown came from an experiment that involved potassium (K) and radium (Ra). Write the equation that expresses potassium as a function of radium: $K = mRa + b$

21. Find the measure of arc AB in terms of y. Then find the measure of angle x in terms of y.

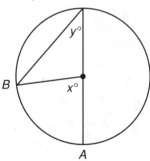

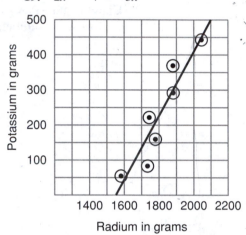

22. Use unit multipliers to convert 600 cubic feet per hour to cubic inches per minute.

Simplify:

23. $\sqrt{x^2\sqrt{y^3}}$

24. $\sqrt[6]{4\sqrt[5]{2}}$

25. $\sqrt{\dfrac{2}{9}} - 3\sqrt{\dfrac{9}{2}} - 2\sqrt{50}$

26. Find x: $\dfrac{x}{a(b + c)} - m = \dfrac{d}{f}$

27. Find b: $\dfrac{x}{a(b + c)} - m = \dfrac{d}{f}$

28. Simplify: $\dfrac{32 - 12x + x^2}{x^3 + x^2 - 20x} \div \dfrac{x^2 - x - 56}{45x + 14x^2 + x^3}$

29. Estimate: $\dfrac{(47{,}123 \times 10^5)(980)(476)}{(0.00134)(576 \times 10^5)}$

30. Add: $\dfrac{4}{x + 2} - \dfrac{3}{x^2 - 4}$

LESSON 66   *Signs of fractions • 30-60-90 triangles*

66.A

**signs
of fractions**

Every fraction has three signs. If one of the signs is not written, it is understood to be a plus sign. One of the signs is in front of the fraction, and the other two are above and below, as shown here.

$$-\frac{+3}{+4}$$

In this case, it is unnecessary to record the plus signs, for the fraction can be written with just one sign as

$$-\frac{3}{4}$$

Any two of the three signs of a fraction may be changed without changing the value of the fraction.

$$\text{(a)} \quad -\frac{+3}{+4} \qquad \text{(b)} \quad +\frac{-3}{+4} \qquad \text{(c)} \quad +\frac{+3}{-4} \qquad \text{(d)} \quad -\frac{-3}{-4}$$

Each of these four notations designates the same number, which is

$$-\frac{3}{4}$$

We find that the ability to change signs is often helpful when we add fractions.

example 66.1   Add: $\dfrac{1}{x-3} - \dfrac{7a}{-x+3}$

solution   The fractions can be added if we make the denominator of the second fraction $x - 3$. To do this, we must change both signs below and then we must change either the sign on top or the sign in front. We choose to change the sign in front.

$$\frac{1}{x-3} + \frac{7a}{x-3}$$

Now the denominators are the same, so we add and get

$$\frac{1+7a}{x-3}$$

example 66.2   Add: $\dfrac{4x+5}{x-3} + \dfrac{2x-3}{3-x}$

solution   We will change the second fraction by changing all signs below and all signs above. Thus, the + in front of the second fraction remains unchanged.

$$\frac{4x+5}{x-3} + \frac{-2x+3}{x-3}$$

Now the denominators are the same, so we add the numerators and get

$$\frac{2x+8}{x-3}$$

66.B

30-60-90 triangles

Right triangles whose acute angles are 30° and 60° are encountered often in physics and engineering. These triangles are often called "30-60-90 triangles." These triangles are all similar to this triangle.

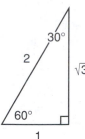

Sometimes we forget the lengths of the sides of this triangle and forget which length goes where. If we can remember to begin with an equilateral triangle whose sides are 2 units long, we can develop this triangle quickly.

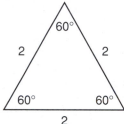

Next, on the left we draw the perpendicular bisector of the base and use the Pythagorean theorem to find the altitude.

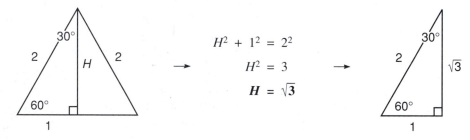

$$H^2 + 1^2 = 2^2$$
$$H^2 = 3$$
$$\mathbf{H} = \mathbf{\sqrt{3}}$$

example 66.3 Use similar triangles to find x and y.

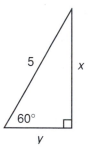

solution One acute angle is 60°, so this is a 30-60-90 triangle and is similar to the triangle on the right below.

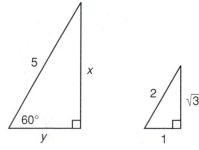

First we find the scale factor from right to left.

$$2 \overleftrightarrow{SF} = 5$$

$$\overleftrightarrow{SF} = \frac{5}{2}$$

Now we use this scale factor to find x and y.

$$\sqrt{3} \overleftrightarrow{SF} = x \qquad\qquad 1 \overleftrightarrow{SF} = y$$

$$\sqrt{3}\left(\frac{5}{2}\right) = x \qquad\qquad 1\left(\frac{5}{2}\right) = y$$

$$\frac{5\sqrt{3}}{2} = x \qquad\qquad \frac{5}{2} = y$$

practice   Add:

a. $\dfrac{1}{z - 2} - \dfrac{3m}{-z + 2}$

b. $\dfrac{3x + 7}{-x - 4} + \dfrac{x - 7}{x + 4}$

c.   Find a and b.

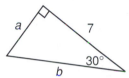

problem set 66

1. The amount of cobalt varied inversely as the amount of uranium. When there were 5 grams of cobalt, the mass of the uranium was 20 grams. How much cobalt was there when only 2 grams of uranium was present?

2. The initial pressure, volume, and temperature of a quantity of ideal gas were 400 newtons per square meter, 6 liters, and 200 kelvins, respectively. What was the temperature if the pressure was increased to 800 newtons per square meter and the volume increased to 60 liters?

3. The solution had to be exactly 36% arsenic. Two solutions were available. One was 60% arsenic, and the other was only 20% arsenic. How much of each should be used to get 200 liters of solution that is 36% arsenic?

4. Cleon had a 10-mile head start when Deborah set out in pursuit in a chariot. If Cleon was walking at 4 miles per hour and the speed of the chariot was 6 miles per hour, how long did it take Deborah to catch Cleon?

5. The laboratory assistant stumbled upon a container of methyl bromide, CH_3Br. If the methyl bromide weighed 950 grams, what did the bromine (Br) weigh? (C, 12; H, 1; Br, 80)

Add:

6. $\dfrac{1}{x + 3} - \dfrac{7a}{-x - 3}$

7. $\dfrac{x + 5}{x + 3} + \dfrac{2x - 3}{-3 - x}$

8. $\dfrac{4}{x^2 - 4} - \dfrac{2x}{x - 2}$

9. Three equal semicircles are drawn on the diameter of a circle with a center Q as shown in the diagram. If the area of the big circle circle Q is 9π square inches, find the area of the shaded region.

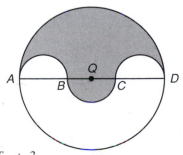

Solve:

10. $R_T T_T = 300, R_P T_P = 1200, R_P = 8R_T, T_T = T_P + 3$

11. $R_P T_P = 624, R_T T_T = 364, T_P = T_T - 4, R_P = 4R_T$

12. In the figure shown, if the small circle has a radius r meters and the larger circle has a diameter $6r$ meters, what is the area of the region inside the large circle and outside the small circle?

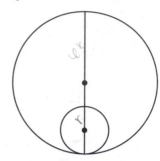

Simplify:

13. $ax + \dfrac{1}{a + \dfrac{1}{x}}$

14. $\dfrac{m}{x} + \dfrac{1}{x + \dfrac{1}{x}}$

15. $-\sqrt{-3}\sqrt{-2} + \sqrt{-4} - \sqrt{-3}\sqrt{-3} - 2i^3$

16. $(4i - 2)(3 + 5i)$

17. Add: $4\underline{/28°} + 10\underline{/35°}$

18. Write $8R + 4U$ in polar form.

19. Solve: $\begin{cases} \dfrac{3}{8}x - \dfrac{1}{4}y = -2 \\ 0.012x + 0.02y = 0.496 \end{cases}$

Solve by completing the square:

20. $-3 = -2x^2 - 6x$

21. $5x^2 - 4 = -5x$

22. Use similar triangles to find x and y.

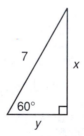

23. The data points shown came from an experiment that involved calcium (Ca) and magnesium (Mg). Write the equation that expresses magnesium as a function of calcium:
$Mg = mCa + b$

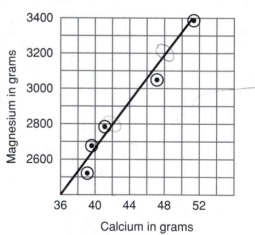

24. Use unit multipliers to convert 10 cubic inches per hour to cubic centimeters per minute.

Simplify:

25. $\sqrt[3]{x^{1/2}\sqrt{x^2}}$

26. $\sqrt[3]{x^{1/5}\sqrt[4]{x}}$

27. $3\sqrt{\dfrac{2}{5}} + 7\sqrt{\dfrac{5}{2}} - 6\sqrt{40}$

28. Find y: $\dfrac{px - y}{m} - c = \dfrac{k}{d}$

29. Find x: $\dfrac{m}{px - y} + \dfrac{k}{d} = -c$

30. Estimate: $\dfrac{(40{,}213 \times 10^5)(748{,}609 \times 10^{-30})}{(0.164289)(506{,}217 \times 10^2)}$

LESSON 67 *Radical denominators*

When a pair of two-part expressions are the same in all respects except that the signs in the middle are different, we say that each of the expressions is the **conjugate** of the other expression. Thus,

$-4 + \sqrt{3}$	is the conjugate of	$-4 - \sqrt{3}$
$-4 - \sqrt{3}$	is the conjugate of	$-4 + \sqrt{3}$
$-2 + bx$	is the conjugate of	$-2 - bx$
$-2 - bx$	is the conjugate of	$-2 + bx$
$a + b$	is the conjugate of	$a - b$
$a - b$	is the conjugate of	$a + b$

We have studied the products of conjugates and noted that the results are always the difference of two squares. Consider the following multiplications.

$$
\begin{array}{cc}
\text{(a)} & \text{(b)} \\
\end{array}
$$

$$
\begin{array}{rr}
-4 + \sqrt{3} & a + b \\
-4 - \sqrt{3} & a - b \\
\hline
16 - 4\sqrt{3} & a^2 + ab \\
\quad + 4\sqrt{3} - 3 & \quad - ab - b^2 \\
\hline
16 \qquad - 3 & a^2 \qquad - b^2 \\
\end{array}
$$

In this lesson we are interested in expressions that contain square root radicals, such as (a) above. We note that the product of the conjugates is $16 - 3$, which equals 13, a whole number. We can use this observation to help us simplify expressions such as the following:

$$\frac{1}{-4 + \sqrt{3}}$$

We can eliminate the radical in the denominator by multiplying above and below by the conjugate of the denominator. Again, we find another use for the most important rule in algebra, the denominator-numerator same-quantity rule, as we can use it to simplify this expression. **We say that an expression containing square root radicals is in simplified form when no radicand has a perfect square factor and when no radicals are in the denominator.**

example 67.1 Simplify: $\dfrac{1}{-4 + \sqrt{3}}$

solution We can eliminate the radical in the denominator if we multiply above and below by $-4 - \sqrt{3}$, which is the conjugate of $-4 + \sqrt{3}$.

$$\frac{1}{-4 + \sqrt{3}} \cdot \frac{\left(-4 - \sqrt{3}\right)}{\left(-4 - \sqrt{3}\right)} = \frac{-4 - \sqrt{3}}{16 + 4\sqrt{3} - 4\sqrt{3} - 3} = \frac{-4 - \sqrt{3}}{13}$$

Note that we were unable to get rid of the radical expression. All that we could do was to eliminate radical expressions in the denominator.

example 67.2 Simplify: $\dfrac{3}{2\sqrt{3} + \sqrt{2}}$

solution We simplify by multiplying above and below by the conjugate of the denominator.

$$\frac{3}{2\sqrt{3} + \sqrt{2}} \cdot \frac{\left(2\sqrt{3} - \sqrt{2}\right)}{\left(2\sqrt{3} - \sqrt{2}\right)} = \frac{6\sqrt{3} - 3\sqrt{2}}{12 - 2} = \frac{6\sqrt{3} - 3\sqrt{2}}{10}$$

practice   Simplify:

a. $\dfrac{1}{-3 + \sqrt{7}}$

b. $\dfrac{5}{3\sqrt{2} + \sqrt{3}}$

problem set 67

1. The number of Danish invaders who were repulsed varied directly as the number of Danes who attacked. If 12,400 attacked and 2000 were repulsed, how many attacked if 3000 were repulsed?

2. The temperature of a quantity of ideal gas was held constant at 700 K. If the pressure of 800 newtons per square meter was increased to 1200 newtons per square meter, what was the final volume if the initial volume was 300 cubic centimeters?

3. The original mixture weighed 400 pounds and was 20% fertilizer. How much of another mixture of 80% fertilizer should be added so that the result would be 32% fertilizer?

4. There were 75 more nurses than doctors. In fact, 8 times the number of nurses exceeded 10 times the number of doctors by 140. How many of each were there?

5. When the reaction was complete, the researchers found that 40 percent of the iridium had not reacted. If 240 grams had reacted, how much had not reacted?

Simplify:

6. $\dfrac{2}{-4 + \sqrt{5}}$

7. $\dfrac{1}{2\sqrt{2} + \sqrt{3}}$

8. $\dfrac{2}{3\sqrt{5} - 3}$

Add:

9. $\dfrac{4x + 2}{x - 2} - \dfrac{3}{2 - x}$

10. $\dfrac{4}{x^2 - 9} + \dfrac{2}{x - 3}$

11. Solve: $R_P T_P = 1062,\ R_T T_T = 295,\ T_P = T_T - 2,\ R_P = 6R_T$

Simplify:

12. $4x + \dfrac{a}{x + \dfrac{a}{b}}$

13. $a + \dfrac{x}{m + \dfrac{1}{x}}$

14. $(5 - i)(6 + 2i)$

15. $\sqrt{-4} - \sqrt{-9} + \sqrt{-2}\,\sqrt{2} - 4i^2$

16. Add: $10\underline{/217°} + 8\underline{/227°}$

17. Write $-4R - 3U$ in polar form.

18. Solve: $\begin{cases} \dfrac{2}{7}x - \dfrac{2}{5}y = 0 \\ 0.2x - 0.04y = 2.4 \end{cases}$

Solve by completing the square:

19. $-x = -1 - 4x^2$

20. $3x^2 + 5 = -2x$

21. Use similar triangles to find x and y.

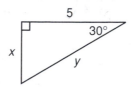

22. Use unit multipliers to convert 400 cubic centimeters per hour to cubic inches per minute.

23. The data points shown came from an experiment that involved silver (Ag) and gold (Au). Write the equation that expresses silver as a function of gold: $Ag = mAu + b$

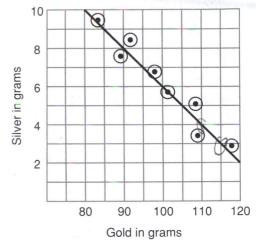

Simplify:

24. $\sqrt{2\sqrt[6]{4}}$

25. $\sqrt[5]{x^2yp}\sqrt[3]{xy^2}$

26. $4\sqrt{\dfrac{3}{11}} + 2\sqrt{\dfrac{11}{3}} - 2\sqrt{297}$

27. Find x: $\dfrac{a(b + c)}{x} - \dfrac{m}{y} = p$

28. Find c: $\dfrac{x}{a(b + c)} - \dfrac{y}{m} = p$

29. Find the equation of the line that passes through $(-2, 4)$ and $(5, 7)$.

30. Find the distance between $(-2, 4)$ and $(5, 7)$.

LESSON 68   *Scientific calculator • Scientific notation • Powers and roots*

68.A
scientific calculator

The calculator frees the user from mundane arithmetic chores and permits more emphasis to be placed on understanding. However, a calculator cannot be used to replace understanding itself. For instance, if the user does not know that the simplification of

$$-3^2$$

is -9, the calculator would be of no help in simplifying

$$-3.0165^2$$

because there would be no way of telling whether the calculator answer made sense or not.

Calculator instruction manuals often seem to concentrate on exotic manipulations that invariably lead to errors when tried by the beginner. The only sure way to avoid mistakes is to always estimate the answer before a calculator is used.

> **Always estimate the answer before using a calculator.**

There is almost never an excuse for accepting a wrong answer from a calculator because wrong answers usually differ greatly from any reasonable estimate of the correct answer.

The instructions in this book are for calculators that use algebraic operating systems.

68.B
scientific notation

Depressing the $\boxed{+/-}$ key on the calculator will change the sign of the number in the display. We use this key to enter negative values. To enter -55 we enter 55 and use the $\boxed{+/-}$ key to change it to -55.

ENTER	DISPLAY
55	55
+/−	−55

The exponent in scientific notation is entered by using the exponent key. This key is labeled exp on some calculators and EE on others. To enter -416.2×10^{-4}, we first enter the 416.2 and then change the sign.

ENTER	DISPLAY
416.2	416.2
+/−	−416.2

Now we enter the exponent.

ENTER	DISPLAY
exp	−416.2 00
4	−416.2 04

Lastly we change the sign of the exponent.

ENTER	DISPLAY
+/−	−416.2 −04

example 68.1   Simplify: $(-40,652 \times 10^{-8})(0.000324 \times 10^{15})$

solution   **We always estimate first.**

$$(-4.0652 \times 10^{-4})(3.24 \times 10^{11}) \approx -1.2 \times 10^8$$

Now we use the calculator. First we enter the negative number.

ENTER	DISPLAY
40652	40652
+/−	−40652

Then we enter the negative exponent.

ENTER	DISPLAY
exp	−40652 00
8	−40652 08
+/−	−40652 −08

Next we depress the multiply key. The use of the exp key in the previous step will switch some calculators into the scientific mode. In this case the display will remain in scientific notation. For other calculators the display will return to standard notation provided that the number is not too large or too small.

ENTER	DISPLAY
×	−4.0652 −04 (scientific mode)
	−0.00040652 (standard mode)

Now we enter the second number.

ENTER	DISPLAY
0.000324	0.000324
exp	0.000324 00
15	0.000324 15

Now we finish by depressing the equals key.

ENTER   DISPLAY

[=] $-131712480 \approx -1.32 \times 10^8$

We will accept an answer of -1.32×10^8 since it is close to our estimate of -1.2×10^8.

example 68.2   Simplify: $\dfrac{40,652 \times 10^{-8}}{0.000324 \times 10^{15}}$

solution   **We always estimate first.**

$$\frac{4.0 \times 10^{-4}}{3.0 \times 10^{11}} \approx 1.3 \times 10^{-15}$$

This time we will put the calculator in the scientific mode before we begin. Then all displays will be in scientific notation. We enter the numerator first.

ENTER	DISPLAY
40652	40652
[exp]	40652 00
8	40652 08
[+/−]	40652 −08

When we depress the divide key, the decimal point will shift but the display will be in scientific notation.

ENTER	DISPLAY
[÷]	4.0652 −4

Now we enter the bottom number and depress the equals key.

ENTER	DISPLAY
0.000324	0.000324
[exp]	0.000324 00
15	0.000324 15
[=]	1.254691358 −15

Our answer rounded to two places is **1.25×10^{-15}**, which agrees with our estimate of 1.3×10^{-15}.

68.C
powers and roots

We remember that a rational number is a number that can be written as a fraction of integers. Thus, the numbers

(a)   0.43   (b)   3.43   (c) $0.\overline{3}$ (d) $4.\overline{6}$

are rational numbers because they can be written as

(a) $\dfrac{43}{100}$ (b) $\dfrac{343}{100}$ (c) $\dfrac{1}{3}$ (d) $4\dfrac{2}{3} = \dfrac{14}{3}$

Every rational number can be written either as a decimal number that has a finite number of digits or as a decimal number whose digits repeat in a pattern. Conversely, decimal numbers that either have a finite number of digits or whose digits repeat in a pattern are rational numbers. Thus, the numbers

4.687254   and $4.672672\overline{672}$

are rational numbers. In Lesson 104 we will show how they can be written as fractions of integers.

Numbers such as π, e, $\sqrt{7}$, $\sqrt[5]{4}$, and $\sqrt[9]{3}$ are irrational numbers because they cannot be written as fractions of integers. The decimal representation of every irrational number has an infinite number of digits that do not have a repeating pattern. It is helpful to remember that if an integral root of an integer is not an integer, then the root is an irrational number. Thus, the numbers $4^{1/5}$, $\sqrt[32]{5}$, $75^{1/8}$, $\sqrt[15]{21}$, and $\sqrt[9]{23}$ are all irrational numbers.

The scientific calculator will give us an approximation of any rational or irrational root or power of any positive number. Thus we can raise 23 to the π power or take the πth root of 23. This procedure is explained in examples 68.3 and 68.4 below.

$$23^\pi \approx 18{,}966.80 \qquad \sqrt[\pi]{23} = 23^{1/\pi} \approx 2.71$$

The calculator will also raise negative numbers to either positive or negative integer powers

$$(-2.3)^{-3} \approx -0.082 \qquad (-2.3)^3 \approx -12.167$$

and will find odd roots of negative numbers.

$$\sqrt[3]{-4} = (-4)^{1/3} \approx -1.58 \qquad \sqrt[5]{-14} = (-14)^{1/5} \approx -1.69$$

We know that the square root of a negative number is an imaginary number. The discussion of the fourth, sixth, and higher even roots is a topic for advanced mathematics courses and will not be considered in this book. The calculator will not find even roots of negative numbers.

Scientific calculators are programmed to find only roots or powers of real numbers that are real numbers.

example 68.3   Evaluate: $\sqrt[3.28]{50.42}$

solution   **We always estimate first.** The third power of 3 is 27, and the third power of 4 is 64.

$$3^3 = 27$$

$$4^3 = 64$$

We guess that the answer will be some number between 3 and 4. Now we use the calculator.

ENTER	DISPLAY
50.42	50.42
inv y^x	50.42
3.28	3.28
=	**3.304**

This answer agrees with our estimate, so we accept it.

example 68.4   Evaluate: (a) $(9.26)^{4.58}$ (b) $(9.26)^{-4.58}$

solution   (a)   **We always estimate first.**

$$10^4 = 10{,}000$$

$$10^5 = 100{,}000$$

We guess that the answer will be some number between 10,000 and 100,000. Now we use the calculator.

ENTER	DISPLAY
9.26	9.26
y^x	9.26
4.58	4.58
=	**26,734.88**

This answer agrees with our estimate so we accept it.

(b) $(9.26)^{-4.58}$ is the reciprocal of $(9.26)^{4.58}$. One way to find the answer is to find the reciprocal of the answer to (a).

	ENTER	DISPLAY
	26,734.88	26,734.88
	1/x	3.74×10^{-5}

Another way is to raise 9.26 to the -4.58 power.

	ENTER	DISPLAY
	9.26	9.26
	y^x	9.26
	4.58	4.58
	+/–	-4.58
	=	3.74×10^{-5}

practice

Use a calculator to simplify. **Estimate first.**

a. $\dfrac{0.00042 \times 10^{-17}}{568,425 \times 10^{5}}$

b. $\sqrt[4.2]{156}$

c. $(0.00042 \times 10^{-17})(568,425 \times 10^{5})$

d. $(1.86)^{-4.86}$

problem set 68

1. The number of dastards varied directly as the number of poltroons. When there were 800 dastards, the poltroons numbered 9600. When there were 24,000 poltroons, how many dastards were there?

2. The temperature of a quantity of an ideal gas was held constant at 1400 K. If the pressure of 1200 newtons per square meter was increased to 1600 newtons per square meter, what was the final volume if the volume was initially 200 ml?

3. The original mixture weighed 800 pounds and was 40% slaked lime. How much of a 20% slaked lime mixture should be added to reduce the lime concentration to 36%?

4. The fugacious numbered 14 more than twice the number of ephemeral. Also, twice the number of fugacious was 100 less than 20 times the number of ephemeral. How many of each were there?

5. Twenty percent wanted to storm the fortress. If the rest totaled 6720 and just wanted to sleep in the shade, how many were in the advancing army?

First estimate the answer and then use a calculator to get a more exact answer.

6. (a) $\dfrac{0.000418 \times 10^{-14}}{501,635 \times 10^{6}}$ (b) $(0.00037 \times 10^{-13})(7231 \times 10^{4})$

7. (a) $\sqrt[3.8]{192}$ (b) $(1.76)^{-3.42}$

8. Use similar triangles to find x and y.

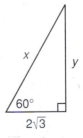

9. $ABCD$ is a parallelogram. N is the midpoint of $\overline{BC}$. M is the midpoint of $\overline{BN}$. Which of the three triangles whose base is $\overline{AD}$ has the largest area?

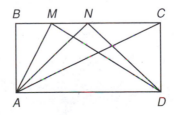

10. Find $m\widehat{MN}$.

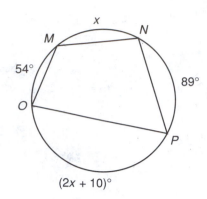

Simplify:

11. $\dfrac{1}{-2 - \sqrt{2}}$

12. $\dfrac{4}{3\sqrt{2} - 1}$

13. $\dfrac{2}{3\sqrt{3} - 5}$

Add:

14. $\dfrac{-7}{-x + 3} - \dfrac{2x}{x^2 - 9}$

15. $\dfrac{2}{4 - x} - \dfrac{3}{x^2 - 16}$

16. Solve: $R_G T_G = 140, R_B T_B = 140, R_B = 2R_G, T_B = T_G - 7$

Simplify:

17. $4r + \dfrac{m}{x + \dfrac{m}{x}}$

18. $3a + \dfrac{ax}{a + \dfrac{x}{a}}$

19. $(5i - 2)(3i + 4)$

20. $\sqrt{-2}\sqrt{2} - 3i^2 - \sqrt{-9} + 2i^4 + 4$

21. Add: $40\underline{/315°} + 10\underline{/24°}$

22. Solve: $\begin{cases} \dfrac{2}{5}x - \dfrac{2}{3}y = -2 \\ -0.06x - 0.4y = -7.2 \end{cases}$

Solve by completing the square:

23. $-2x = -1 - 4x^2$

24. $-3x^2 + 5 = -2x$

25. Use unit multipliers to convert 700 cubic centimeters per minute to cubic inches per hour.

26. $\sqrt[7]{3\sqrt[3]{3}}$

27. $\sqrt[3]{xy^5}\sqrt{x^5 y}$

28. $3\sqrt{\dfrac{2}{3}} - 5\sqrt{\dfrac{3}{2}} + 2\sqrt{24}$

29. Find b: $\dfrac{m}{x(a + b)} - \dfrac{p}{y} = c$

30. Find the equation of the line that passes through $(-8, 2)$ and $(5, -7)$.

LESSON 69 *Gas law problems*

In Lesson 57 we learned that the gas law for a fixed amount of an ideal gas can be written as

$$\frac{P_1 V_1}{T_1} = \frac{P_2 V_2}{T_2}$$

We have been substituting numbers in these equations and then solving. Many people believe it is a better procedure to solve the equation for the desired variable and then substitute. We will do that in the next two examples. Also, we will use values for the variables such as 0.0004×10^{16} so we can get practice in using scientific notation as well as in rearranging equations.

example 69.1   A quantity of an ideal gas had initial values of pressure, volume, and temperature of 40×10^4 atmospheres, 0.0003×10^{-6} cm^3, and 700×10^4 K, respectively. Find the final temperature if the final pressure was 0.0004×10^{15} atmospheres and the final volume was 0.015×10^{-14} cm^3. Begin by solving the equation for T_2.

solution   We need to solve the ideal gas law for T_2 in terms of the other variables. We begin by writing the ideal gas law.

$$\frac{P_1 V_1}{T_1} = \frac{P_2 V_2}{T_2}$$

Next we solve for T_2. **As always in a fractional equation, our first step is to eliminate the denominators. We do this by multiplying both sides by $T_1 T_2$.**

$$\cancel{T_1} T_2 \frac{P_1 V_1}{\cancel{T_1}} = \frac{P_2 V_2}{\cancel{T_2}} T_1 \cancel{T_2} \quad \rightarrow \quad T_2 P_1 V_1 = P_2 V_2 T_1$$

Now we complete our solution for T_2 by dividing both sides by $P_1 V_1$.

$$\frac{T_2 \cancel{P_1} \cancel{V_1}}{\cancel{P_1} \cancel{V_1}} = \frac{P_2 V_2 T_1}{P_1 V_1} \quad \rightarrow \quad T_2 = \frac{P_2 V_2 T_1}{P_1 V_1}$$

We finish by inserting the given values of P_2, V_2, T_1, P_1, and V_1 and simplifying.

$$T_2 = \frac{(0.0004 \times 10^{15})(0.015 \times 10^{-14})(700 \times 10^4)}{(40 \times 10^4)(0.0003 \times 10^{-6})}$$

$$= \frac{(4 \times 10^{11})(15 \times 10^{-17})(7 \times 10^6)}{(4 \times 10^5)(3 \times 10^{-10})}$$

$$= 35 \times 10^5$$

Thus, the final temperature is 3.5×10^6. **The temperature for T_1 was in kelvins, so the temperature for T_2 will also be in kelvins. The same units are always used throughout a problem.**

$$T_2 = 3.5 \times 10^6 \text{ K}$$

example 69.2   A quantity of ideal gas had initial values of 0.003×10^{14} lb/in.^{2}, 0.007 cm^3, and 7000 K. Find P_2 if the final volume was 3×10^2 cm^3 and the final temperature was 0.003×10^7 K. Solve the equation for P_2 as the first step.

solution   We write the ideal gas law equation.

$$\frac{P_1 V_1}{T_1} = \frac{P_2 V_2}{T_2}$$

Now, as always with fractional equations, we begin by eliminating the denominators.

$$\cancel{T_1} T_2 \frac{P_1 V_1}{\cancel{T_1}} = \frac{P_2 V_2}{\cancel{T_2}} T_1 \cancel{T_2} \quad \rightarrow \quad T_2 P_1 V_1 = P_2 V_2 T_1$$

Now we solve for P_2 by dividing by $V_2 T_1$.

$$\frac{T_2 P_1 V_1}{V_2 T_1} = \frac{P_2 \cancel{V_2} \cancel{T_1}}{\cancel{V_2} \cancel{T_1}} \quad \rightarrow \quad \frac{T_2 P_1 V_1}{V_2 T_1} = P_2$$

We finish by inserting the numbers and simplifying.

$$\frac{(0.003 \times 10^7)(0.003 \times 10^{14})(0.007)}{(3 \times 10^2)(7000)} = \frac{(3 \times 10^4)(3 \times 10^{11})(7 \times 10^{-3})}{(3 \times 10^2)(7 \times 10^3)}$$

$$= 3 \times 10^7$$

The units of P_2 are pounds per square inch because P_1 was in pounds per square inch. Thus,

$$P_2 = 3 \times 10^7 \text{ lb/in.}^2$$

practice   A quantity of an ideal gas had initial values of pressure, volume, and temperature of 50×10^3 atmospheres, 0.008×10^{-4} cm³, and 300×10^5 K. Find the final temperature if the final pressure was 0.00005×10^{14} atmospheres and the final volume was 0.000014×10^{-8} cm³. Begin by solving the ideal gas law equation for T_2.

problem set 69

1. A quantity of an ideal gas had initial values of pressure, volume, and temperature of 80×10^5 atmospheres, 0.0005×10^{-7} cm³, and 800×10^5 K. Find the final temperature if the final pressure was 0.0008×10^{16} atmospheres and the final volume was 0.013×10^{-12} cm³. Begin by solving the equation for T_2.

2. By using the Pythagorean theorem, we can see that the diagonal of any square is equal to the product of a side and $\sqrt{2}$ and that the length of a side equals the length of the hypotenuse divided by $\sqrt{2}$.

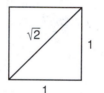 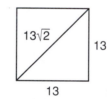 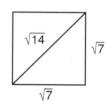

   The diagonal of a square is 7. What is the length of a side? What is the area of the square?

3. The borax varied inversely as the tungsten. If 400 tons of borax went with 5 tons of tungsten, how much borax went with 25 tons of tungsten?

4. Two solutions were available. One was 80% alcohol and the other was 40% alcohol. How much of each should be used to get 2000 ml of a solution that is 64% alcohol?

5. The bus headed north at noon at 50 miles per hour. At 2 p.m., the train headed north from the same station at 70 miles per hour. What time was it when the train got within 40 miles of the bus?

Add:

6. $\dfrac{1}{x - 5} - \dfrac{2x - 3}{5 - x}$

7. $\dfrac{2x + 3}{x - 2} + \dfrac{2x}{2 - x} - \dfrac{3}{-x + 2}$

8. $\dfrac{3x + 5}{x - 5} + \dfrac{2}{x^2 - 25}$

Simplify:

9. $\dfrac{4}{3 - \sqrt{2}}$

10. $\dfrac{2}{5 - 3\sqrt{2}}$

11. $\dfrac{2}{3 - 2\sqrt{8}}$

12. Solve: $R_G T_G = 171, R_R T_R = 171, R_R = 3R_G, T_R = T_G - 6$

Simplify:

13. $m + \dfrac{1}{m + \dfrac{1}{m}}$

14. $x + \dfrac{a}{x + \dfrac{1}{a}}$

15. $(4i - 2)(2i - 4)$

16. $-\sqrt{-2}\sqrt{-2} + 2i^3 - i^2$

17. Add: $20\underline{/30°} + 60\underline{/210°}$

18. Write $2R - 3U$ in polar form.

19. Solve: $\begin{cases} \dfrac{2}{3}x + \dfrac{2}{5}y = 28 \\ -0.05x - 0.2y = -5.5 \end{cases}$

Solve by completing the square:

20. $3x^2 + 8 = 5x$

21. $3x^2 + 8 = -5x$

22. Given: $m\widehat{AC} = 120°$
$\triangle ABC$ is equilateral.
$\overline{BC} = 10.$
Find: $m\widehat{BC}$, $m\widehat{AB}$, and the area of $\triangle ABC$.

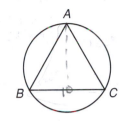

23. Use unit multipliers to convert 4 cubic feet per minute to cubic inches per hour.

24. The data points shown came from an experiment that involved chromium (Cr) and vanadium (V). Write the equation that expresses chromium as a function of vanadium:
$Cr = mV + b$

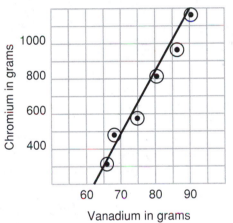

Simplify:

25. $\sqrt{27\sqrt[4]{3}}$

26. $\sqrt[3]{xm^5}\,\sqrt[4]{xm^2}$

27. $2\sqrt{\dfrac{9}{2}} + 5\sqrt{\dfrac{2}{9}} - 5\sqrt{50}$

28. Find x: $\dfrac{a(x + y)}{m} - \dfrac{c}{z} = k$

29. Solve: $\dfrac{x + 2}{5} - \dfrac{3x - 3}{4} = 2$

30. Use a calculator to evaluate. Estimate first.

(a) $\dfrac{472.2 \times 10^{-26}}{1658.27 \times 10^{10}}$

(b) $(1.24)^{-2.73}$

LESSON 70 *Advanced abstract equations*

Sometimes it is necessary to use the distributive property as the first step in solving a fractional equation.

example 70.1   Solve for b: $x = p\left(\dfrac{1}{a} + \dfrac{m}{b}\right)$

solution   We begin by multiplying to eliminate the parentheses. If we do this, we get

$$x = \frac{p}{a} + \frac{mp}{b}$$

Next we want to eliminate the denominators. We can do this if we multiply every numerator by the least common multiple of the denominators and cancel the denominators.

$$x \cdot ab = \frac{p}{\cancel{a}} \cdot \cancel{a}b + \frac{mp}{\cancel{b}} \cdot a\cancel{b} \quad \longrightarrow \quad xab = pb + mpa$$

Now we place all terms that have a factor of b on one side, then factor. Then we divide.

$$xab - pb = mpa \qquad \text{added } -pb$$

$$b(xa - p) = mpa \qquad \text{factored}$$

$$\boldsymbol{b = \frac{mpa}{xa - p}} \qquad \text{divided}$$

example 70.2   Find c: $xp = m\left(\dfrac{b}{1 + c} + \dfrac{a}{p}\right)$

solution   We begin by multiplying so we can eliminate the parentheses.

$$xp = \frac{mb}{1 + c} + \frac{ma}{p}$$

Now to eliminate the denominators, we multiply every numerator by the LCM of the denominators.

$$p(1 + c)xp = \frac{mbp(1 + c)}{1 + c} + \frac{ma}{p}p(1 + c)$$

Then we cancel the denominators and multiply to get

$$xp^2 + xp^2c = mbp + ma + mac$$

Next we place all terms with a c factor on one side, factor out the c, and divide.

$$xp^2c - mac = mbp + ma - xp^2 \qquad \text{added } -xp^2 - mac$$

$$c(xp^2 - ma) = mbp + ma - xp^2 \qquad \text{factored out } c$$

$$\boldsymbol{c = \frac{mbp + ma - xp^2}{xp^2 - ma}} \qquad \text{divided}$$

practice   Find k: $mx = p\left(\dfrac{z}{x + k} + \dfrac{2s}{y}\right)$

problem set 70

1. The initial values of pressure, volume, and temperature of a quantity of an ideal gas were 0.001×10^{-13} newton per square meter, 0.04×10^{14} liters, and 4×10^3 K, respectively. What was the final temperature if the final pressure and volume were 0.04×10^5 N/m^2 and 500 liters? Solve for T_2 as the first step.

2. The beaker contained 400 ml of a solution that was 20% alcohol. How many milliliters of a 50% alcohol solution must be added so that the result will be 26% alcohol?

3. Tourist tickets for the flight from Rhinelander to Hodag were $50 each, whereas first-class tickets were $100 each. If 60 people paid $5000, how many flew as tourists and how many flew first-class?

4. The container contained 680 grams of the compound $CaSO_4$. What was the weight of the sulfur (S) in the compound? (Ca, 40; S, 32; O, 16)

5. Find three consecutive even integers such that 4 times the product of the first and the third is 28 greater than the product of -10 and the sum of the second and the third.

6. Find b: $\quad m = x\left(\dfrac{1}{2p} + \dfrac{3z}{b}\right)$
                                                      7. Find s: $\quad ac = x\left(\dfrac{m}{r + s} + \dfrac{t}{z}\right)$

Add:

8. $\dfrac{3x - 2}{x - 2} - \dfrac{3x}{2 - x}$
                                                 9. $\dfrac{4}{x^2 + 8x + 12} + \dfrac{4x - 5}{x + 2}$

Simplify:

10. $\dfrac{2}{\sqrt{2} - 4}$
                 11. $\dfrac{2}{3\sqrt{12} - 2}$
                 12. $\dfrac{2}{2\sqrt{3} - 2}$

13. Solve: $R_B T_B = 65$, $R_X T_X = 104$, $R_X = 2R_B$, $T_X = T_B - 1$

Simplify:

14. $xy + \dfrac{a}{1 + \dfrac{a}{b}}$
                 15. $\dfrac{m}{y} + \dfrac{x}{a + \dfrac{1}{y}}$

16. $(2 - 3i)(5 - 6i)$
                 17. $-\sqrt{-4} - \sqrt{-2}\,\sqrt{-3} + \sqrt{-9}$

18. Add: $4\underline{/340°} + 6\underline{/320°}$
                 19. Write $-4R + 5U$ in polar form.

20. Solve: $\begin{cases} \dfrac{1}{5}x - \dfrac{1}{4}y = -2 \\[2mm] 0.07x + 0.3y = 5.5 \end{cases}$

Solve by completing the square:

21. $3x^2 + 1 = 4x$
                 22. $-4x + 7 = -3x^2$

23. Use similar triangles to find a and b.

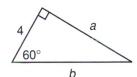

24. The data points shown came from an experiment that involved cobalt (Co) and nickel (Ni). Write the equation that expresses cobalt as a function of nickel: $Co = mNi + b$

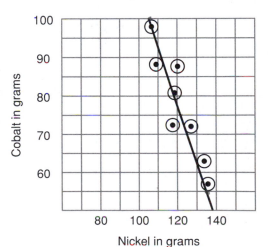

25. Use a calculator to simplify. Estimate first.

   (a) $\dfrac{46,831 \times 10^{-42}}{9140.26 \times 10^{-33}}$

   (b) $\sqrt[2.7]{146}$

Simplify:

26. $\sqrt[5]{x^4 y^3}\,\sqrt[4]{x^2 y}$

27. $\sqrt{\dfrac{7}{2}} + 2\sqrt{\dfrac{2}{7}} - 2\sqrt{126}$ **28.** Divide $4x^3 - 2$ by $x - 1$.

29. Find the equation of the line that passes through $(-2, 5)$ and that is perpendicular to the line $4x + 3y = 5$.

30. Solve: $\dfrac{4x - 2}{5} - \dfrac{3x - 2}{4} = 10$

LESSON 71 *Quadratic formula*

We have found that some quadratic equations can be solved by factoring and then using the zero factor theorem. We will use this procedure to solve $x^2 + 2x - 15 = 0$. First we factor and get

$$(x - 3)(x + 5) = 0$$

Now, from the zero factor theorem we know that if the product of two factors equals zero, one of the factors must be zero. So

If $x - 3 = 0$	If $x + 5 = 0$
$x = 3$	$x = -5$

We have also found that quadratic equations that cannot be factored, such as $x^2 + 3x - 3 = 0$, can be rearranged into the form

$$\left(x + \frac{3}{2}\right)^2 = \frac{21}{4}$$

This form of the equation can be solved by taking the square root of both sides of the equation.

$$\left(x + \frac{3}{2}\right)^2 = \frac{21}{4} \qquad \text{equation}$$

$$x + \frac{3}{2} = \pm\sqrt{\frac{21}{4}} \qquad \text{square root of both sides}$$

$$x = -\frac{3}{2} \pm \sqrt{\frac{21}{4}} \qquad \text{simplified}$$

As we know, this method is called *completing the square* and can be used to solve any quadratic equation. There is a quicker method, however, that we can use. We can complete the square on the general form of the quadratic equation and derive a formula whose use will give us the same answer. We begin this derivation by writing a general quadratic equation using the letters a, b, and c as the constants.

$$ax^2 + bx + c = 0$$

Next we give x^2 a unity coefficient by dividing every term by a, and we get

$$x^2 + \frac{b}{a}x + \frac{c}{a} = 0$$

Now we move $\frac{c}{a}$ to the right-hand side and use parentheses.

$$\left(x^2 + \frac{b}{a}x + \quad\right) = \quad -\frac{c}{a}$$

Note that we placed the $-\frac{c}{a}$ **well to the right of the equals sign.** Now we multiply $\frac{b}{a}$ by $\frac{1}{2}$ and square the result.

$$\left(\frac{b}{a} \cdot \frac{1}{2}\right)^2 = \frac{b^2}{4a^2}$$

Next we add $\frac{b^2}{4a^2}$ inside the parentheses and also to the other side of the equation. On the right we are careful to place $\frac{b^2}{4a^2}$ in front of $-\frac{c}{a}$.

$$\left(x^2 + \frac{b}{a}x + \frac{b^2}{4a^2}\right) = \frac{b^2}{4a^2} - \frac{c}{a}$$

Next we write the term in parentheses as a squared term and combine $\frac{b^2}{4a^2}$ and $-\frac{c}{a}$.

$$\left(x + \frac{b}{2a}\right)^2 = \frac{b^2 - 4ac}{4a^2}$$

Finally, we take the square root of both sides and solve for x.

$$x + \frac{b}{2a} = \pm\sqrt{\frac{b^2 - 4ac}{4a^2}} \qquad \text{took square roots}$$

$$x = -\frac{b}{2a} \pm \frac{\sqrt{b^2 - 4ac}}{2a} \qquad \text{solved for } x$$

$$x = \frac{-b \pm \sqrt{b^2 - 4ac}}{2a} \qquad \text{added}$$

This result is called the *quadratic formula* and should be memorized. It will be used in many higher mathematics courses.

The derivation of the quadratic formula will be required in future problem sets. This derivation requires only simple algebraic manipulations, and the requirement that a student be able to perform this derivation is not unreasonable.

example 71.1   Use the quadratic formula to find the roots of the equation $3x^2 - 2x + 5 = 0$.

solution   The formula is
$$x = \frac{-b \pm \sqrt{b^2 - 4ac}}{2a}$$

If we write the given equation just below the general quadratic equation,

$$ax^2 + bx + c = 0 \qquad \text{general equation}$$
$$3x^2 - 2x + 5 = 0 \qquad \text{given equation}$$

we note the following correspondences between the equations:

$$a = 3 \qquad b = -2 \qquad c = 5$$

If we use these numbers for a, b, and c in the quadratic formula, we get

$$x = \frac{-(-2) \pm \sqrt{(-2)^2 - 4(3)(5)}}{2(3)}$$

$$= \frac{2 \pm \sqrt{-56}}{6} = \frac{1}{3} \pm \frac{\sqrt{14}}{3}i$$

example 71.2   Solve $x^2 = 3x + 28$ by using the quadratic formula.

solution   The formula is
$$x = \frac{-b \pm \sqrt{b^2 - 4ac}}{2a}$$

We rearrange the given equation so that it is in standard form, and we write it just below the general quadratic equation.

$$ax^2 + bx + c = 0 \qquad \text{general equation}$$

$$x^2 - 3x - 28 = 0 \qquad \text{given equation}$$

We note the following correspondences between the coefficients.

$$a = 1 \qquad b = -3 \qquad c = -28$$

We use these values in the quadratic formula and simplify.

$$x = \frac{3 \pm \sqrt{9 - 4(1)(-28)}}{2(1)} \quad \longrightarrow \quad x = \frac{3 \pm \sqrt{9 + 112}}{2} \quad \longrightarrow \quad x = \frac{3 \pm \sqrt{121}}{2}$$

So

$$x = \frac{3 + 11}{2} \qquad \text{or} \qquad x = \frac{3 - 11}{2}$$

$$= \frac{14}{2} \qquad\qquad\qquad = -\frac{8}{2}$$

$$= 7 \qquad\qquad\qquad\qquad = -4$$

This means that the factors of the original equation are $(x - 7)(x + 4)$ and the equation could have been solved by factoring. **This shows that the quadratic formula can be used to solve any quadratic equation—even those that can be solved by factoring.**

practice   a.   Begin with $ax^2 + bx + c = 0$ and derive the quadratic formula.

b.   Use the quadratic formula to solve $2x^2 - 6x + 4 = 0$.

problem set 71

1.   The temperature of a quantity of an ideal gas was held constant at 740 kelvins. Find the final pressure if the initial pressure and volume were $40{,}000 \times 10^{-3}$ newtons per square meter and 4000×10^{-2} cm^3 and the final volume was 8000×10^{-2} cm^3. Begin by solving the equation for P_2.

2.   The number of folk dancers varied directly as the number of people who attended the festival. If 4800 attended the festival and 240 were folk dancers, how many attended the festival when the folk dancers totaled 600?

3.   One solution was 10% fluorine and the other was 30% fluorine. How much of each should be used to get 200 ml of a solution that is 17% fluorine?

4.   Sandy ran out at 8 mph and rode back to Ski Island in a jitney at 24 mph. If the total trip took 8 hours, how far did she run and how far did she ride?

5.   In the compound H_2CO_3, what percent of the weight of the compound is hydrogen (H)? (H, 1; C, 12; O, 16)

6.   Begin with $ax^2 + bx + c = 0$ and derive the quadratic formula.

Use the quadratic formula to solve:

7. $4x^2 - 2x - 6 = 0$ 8. $2x^2 = -x - 4$

9.   *MNOP* is a rhombus. Find x, y, and z.

10.   Find x: $r = m\left(\dfrac{1}{x + c} + \dfrac{3}{y}\right)$

11.   Add: $\dfrac{4x + 5}{x - 2} - \dfrac{4}{2 - x}$

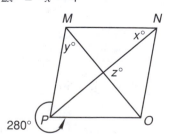

12. Simplify: $\dfrac{4}{2 - 3\sqrt{12}}$

13. Solve: $R_M T_M = 160,\ R_P T_P = 400,\ R_P = 2R_M,\ T_P = T_M + 1$

Simplify:

14. $x + \dfrac{x}{1 + \dfrac{1}{x}}$

15. $a + \dfrac{b}{a + \dfrac{a}{b}}$

16. $-3i^3 + 2\sqrt{-2}\sqrt{2} - \sqrt{-9}$

17. $(-i - 1)(-3i + 2)$

18. Add: $20\underline{/70°} + 10\underline{/40°}$

19. Write $4R - 4U$ in polar form.

20. Solve: $\begin{cases} \dfrac{3}{8}x - \dfrac{1}{2}y = 2 \\ 0.06x - 0.2y = -0.64 \end{cases}$

21. Solve $3x^2 - 2x + 5 = 0$ by completing the square.

22. Use similar triangles to find a and b.

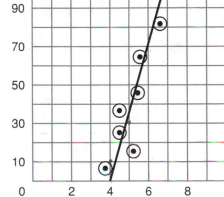

23. The data shown came from an experiment that involved lead (Pb) and boron (B). Write the equation that expresses lead as a function of boron: $Pb = mB + b$

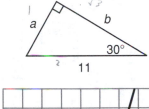

Simplify:

24. $\sqrt{81\sqrt{3}}$

25. $\sqrt[3]{x^5 y^6}\ \sqrt{xy^3}$

26. $3\sqrt{\dfrac{2}{5}} + 3\sqrt{\dfrac{5}{2}} - 6\sqrt{40}$

27. Expand: $(x - 2)^3$

28. Simplify: $\dfrac{x^2 y - \dfrac{1}{y}}{\dfrac{x^2}{y} - 6}$

29. Use a calculator to simplify. Estimate first.

(a) $\dfrac{-471,635 \times 10^5}{0.0071893 \times 10^{-14}}$

(b) $(2.4)^{-3.06}$

30. Use similar triangles to solve for b in terms of x, y, and z.

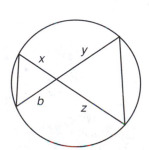

LESSON 72 *Lines from experimental data* • *Negative angles*

72.A

lines from experimental data

Thus far, in the problems dealing with experimental data points, the line indicated by these points has already been estimated and drawn. In science courses, it is necessary to do one's own estimate of the line indicated by the data points. In future problems of this type, the data points will be graphed, but the line will not be drawn. It will be necessary to estimate the location of the line indicated by the data points and draw the line.

example 72.1 Draw an estimate of the line indicated by the data points shown on the left. Then write the equation that expresses salt as a function of carbon.

$$S = mC + b$$

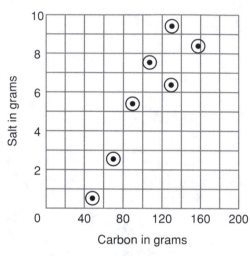

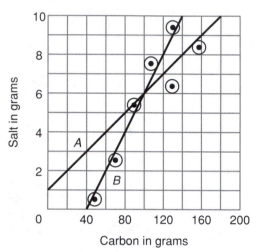

solution There is never an exact answer to these problems. The experimental data points are often scattered, and we can only estimate the position of the line. On the right we have drawn two lines, either of which could be said to represent the line indicated by the data. We have labeled the lines A and B. We'll exercise care in determining the slopes of these lines because each horizontal square has a value of 20, whereas each of the vertical squares has a value of 1. We drew triangles (not shown in the figure) to get the slopes.

$$\text{Slope of } A = \frac{3}{60} \text{ or } 0.05 \qquad \text{Slope of } B = \frac{6}{60} \text{ or } 0.1$$

$$\text{Equation: } S = 0.05C + b \qquad\qquad S = 0.1C + b$$

The intercept of line A appears to be 1, and we will use the coordinates of the point (100, 6) to calculate the intercept of line B.

$$S = 0.05C + b \qquad\qquad S = 0.1C + b$$
$$\qquad\qquad\qquad\qquad 6 = 0.1(100) + b$$
$$b = 1 \text{ (by inspection)} \qquad\qquad 6 = 10 + b$$
$$\qquad\qquad\qquad\qquad -4 = b$$

Equation A: So equation B is:

$$S = 0.05C + 1 \qquad S = 0.1C - 4$$

These equations appear to be very different, but the data points are very scattered. These equations are much closer to each other than they are to equations in which the numbers are greatly different such as

$$S = 475C - 23 \qquad \text{or} \qquad S = -578C - 460$$

Your answers to problems like these in the problem sets should be approximately the same as the answers given in the back of the book. **However, they will almost never be exactly the same as the answers given in the back of the book.**

72.B
negative angles

Rectangular coordinates designate the location of a point by giving the distance of the point to the left or right of the origin and the distance of the point above or below the origin. Thus, the coordinates

$$4R + 3U$$

tell us that the location of the point is 4 units to the right of the origin and 3 units above the origin. We can locate the same point by saying that it is 5 units from the origin at an angle of 36.87°.

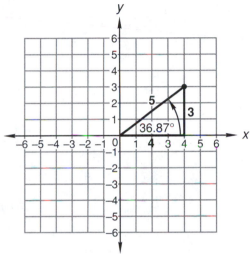

Remember to measure the angle counterclockwise from the positive *x* axis because this is the way mathematicians measure positive angles. **There is only one way to designate a point in rectangular coordinates, but we can use either positive or negative angles when we use polar coordinates. Negative angles are also measured from the positive *x* axis, but they are measured in the clockwise direction.**

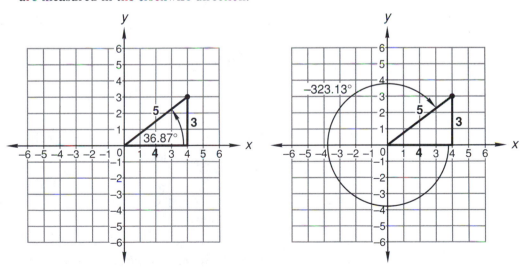

Since a full revolution is 360°, the negative angle is

$$-(360° - 36.87°) = -323.13°$$

Thus, we can say that these two notations designate the same point.

$$5\underline{/36.87°} \quad \text{equals} \quad 5\underline{/-323.13°}$$

example 72.2   Write $4\underline{/-210°}$ in rectangular coordinates.

solution   **We always measure the angle first and then measure the length, and then we draw the triangle.**

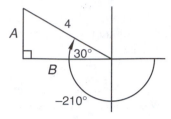

Now we solve for A and B.

$$\sin 30° = \frac{A}{4} \quad \rightarrow \quad A = 4 \sin 30° \quad \rightarrow \quad A = 4(0.5) \quad \rightarrow \quad A = 2$$

$$\cos 30° = \frac{B}{4} \quad \rightarrow \quad B = 4 \cos 30° \quad \rightarrow \quad B = 4(0.8660) \quad \rightarrow \quad B \approx 3.46$$

Thus, we can write

$$4\underline{/-210°} = -3.46R + 2U$$

practice   a.   Write $5\underline{/-230°}$ in rectangular coordinates.

b.   Estimate the location of the line indicated by the data points. Write the equation that gives hydrogen (H) as a function of carbon (C): $H = mC + b$

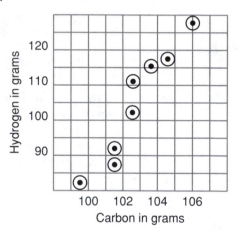

problem set 72

1.   A container contained 500 ml of a solution that was 52% water. How much water should be removed from the solution so that the remainder would be only 40% water?

2.   Pollyana felt that 31,314 things were felicific. If this was 3.07 times the number of things considered felicific by the average person, how many things did the average person think were felicific?

3.   Only 34 percent of the people in the mob carried a flambeau. If 5412 did not carry a flambeau, how many were in the mob?

4.   Flotsam and jetsam littered the beach. The pieces of flotsam numbered 160 more than the pieces of jetsam, and 6 times the number of pieces of jetsam outnumbered the number of pieces of flotsam by 40. How many pieces of each were there?

5.   A quantity of an ideal gas was confined in a container of fixed volume. If the initial pressure and temperature were 500×10^5 newtons per square meter and 0.0004×10^7 K, what was the pressure when the temperature was changed to 0.002×10^5 K? Begin by solving the equation for P_2.

6. Estimate the location of the line indicated by the data points shown. Then write the equation that expresses salt as a function of carbon:
$S = mC + b$

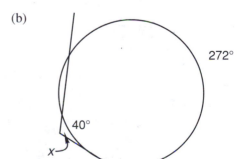

7. Write $3\angle{-250°}$ in rectangular coordinates.

8. Write $-60R - 20U$ in polar form.

9. Begin with $ax^2 + bx + c = 0$ and complete the square to derive the quadratic formula.

Use the quadratic formula to solve:

10. $5x^2 = -7 - 2x$

11. $3x^2 + 7x = -3$

12. Find m_1: $\quad a = x\left(\dfrac{1}{m_1} + \dfrac{y}{m_2}\right)$

13. Add: $\quad \dfrac{2x + 3}{x^2 - 2x - 8} + \dfrac{3x - 2}{x - 4}$

14. Simplify: $\quad \dfrac{3}{2 + 3\sqrt{20}}$

15. Solve: $R_P T_P = 693,\ R_C T_C = 165,\ R_P = 3R_C,\ T_P = T_C + 2$

Simplify:

16. $ab + \dfrac{b}{b + \dfrac{1}{b}}$

17. $x^2 + \dfrac{y}{y + \dfrac{1}{xy}}$

18. $(-3i - 5)(i + 5)$

19. $-4i^3 - 3i^2 + \sqrt{-9} - \sqrt{-3}\sqrt{-3}$

20. Solve: $\begin{cases} \dfrac{1}{4}x - \dfrac{1}{5}y = 2 \\ 0.03x - 0.4y = -1.64 \end{cases}$

21. Solve $2x^2 - x + 4 = 0$ by completing the square.

22. Find the measure of the angles labeled x.

(a)

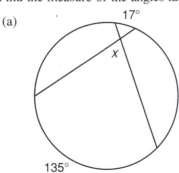

(b)

Simplify:

23. $\sqrt[5]{4\sqrt{2}}$

24. $\dfrac{-2^0}{-8^{-4/3}}$

25. $\sqrt[4]{a^5 y}\,\sqrt{ay^4}$

26. $2\sqrt{\dfrac{3}{5}} + 4\sqrt{\dfrac{5}{3}} - 2\sqrt{135}$

27. Find B.

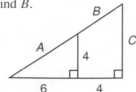

28. Solve: $(-2)^0 - 2^2 - 2 - 2^0 - |-2 - 2| - 2^3 = -2(-2x - 2)$

29. Find the equation of the line that passes through $(-7, 0)$ and is perpendicular to the line $4y - 3x = 1$.

30. Use a calculator to simplify. Estimate first.

(a) $\dfrac{-35,123 \times 10^4}{-798 \times 10^{-15}}$

(b) $\sqrt[3.8]{243}$

LESSON 73 *More on radical denominators*

In Lesson 67 we learned to eliminate the radical in the denominator of expressions such as

$$\frac{1}{2 - 3\sqrt{2}}$$

by multiplying above and below by the conjugate of the denominator.

$$\frac{1}{2 - 3\sqrt{2}} \cdot \frac{2 + 3\sqrt{2}}{2 + 3\sqrt{2}} = \frac{2 + 3\sqrt{2}}{4 - 6\sqrt{2} + 6\sqrt{2} - 18}$$

$$= \frac{2 + 3\sqrt{2}}{-14} = \frac{-2 - 3\sqrt{2}}{14}$$

If the original expression contains radicals above and below, such as

$$\frac{4 + \sqrt{3}}{2 - 3\sqrt{3}}$$

we still multiply above and below by the conjugate of the denominator. This will eliminate the radicals in the denominator but not in the numerator.

example 73.1 Simplify: $\dfrac{4 + \sqrt{3}}{2 - 3\sqrt{3}}$

solution We remember that an expression that contains square root radicals is in simplified form when no radicand has a perfect square as a factor and no radicals are in the denominator. We can rationalize the denominator if we multiply above and below by $2 + 3\sqrt{3}$.

$$\frac{4 + \sqrt{3}}{2 - 3\sqrt{3}} \cdot \frac{2 + 3\sqrt{3}}{2 + 3\sqrt{3}}$$

We have two multiplications to perform, one above and one below. Many people find it easier to do these multiplications separately and then write the answer. We will do this.

ABOVE	BELOW
$4 + \sqrt{3}$	$2 - 3\sqrt{3}$
$2 + 3\sqrt{3}$	$2 + 3\sqrt{3}$
$8 + 2\sqrt{3}$	$4 - 6\sqrt{3}$
$\phantom{8 + 2\sqrt{3}} + 12\sqrt{3} + 9$	$\phantom{4 - 6\sqrt{3}} + 6\sqrt{3} - 27$
$17 + 14\sqrt{3}$	$4 \phantom{- 6\sqrt{3}} - 27 = -23$

Thus, our simplification is

$$\frac{17 + 14\sqrt{3}}{-23} \qquad or \qquad \frac{-17 - 14\sqrt{3}}{23}$$

example 73.2 Simplify: $\dfrac{4 - \sqrt{2}}{4 + 3\sqrt{2}}$

solution We will multiply above and below by $4 - 3\sqrt{2}$.

$$\frac{4 - \sqrt{2}}{4 + 3\sqrt{2}} \cdot \frac{4 - 3\sqrt{2}}{4 - 3\sqrt{2}}$$

We have two multiplications to perform.

ABOVE	BELOW
$4 - \sqrt{2}$	$4 - 3\sqrt{2}$
$4 - 3\sqrt{2}$	$4 + 3\sqrt{2}$
$16 - 4\sqrt{2}$	$16 - 12\sqrt{2}$
$- 12\sqrt{2} + 6$	$+ 12\sqrt{2} - 18$
$22 - 16\sqrt{2}$	$16 - 18 = -2$

Thus, the simplification can be written as

$$\frac{22 - 16\sqrt{2}}{-2} \qquad \text{which simplifies to} \qquad -11 + 8\sqrt{2}$$

example 73.3 Simplify: $\dfrac{3\sqrt{12} - 2\sqrt{3}}{3\sqrt{3} - 2\sqrt{2}}$

solution We will multiply above and below by $3\sqrt{3} + 2\sqrt{2}$.

$$\frac{3\sqrt{12} - 2\sqrt{3}}{3\sqrt{3} - 2\sqrt{2}} \cdot \frac{3\sqrt{3} + 2\sqrt{2}}{3\sqrt{3} + 2\sqrt{2}}$$

Again we have two multiplications to perform.

ABOVE	BELOW
$3\sqrt{12} - 2\sqrt{3}$	$3\sqrt{3} - 2\sqrt{2}$
$3\sqrt{3} + 2\sqrt{2}$	$3\sqrt{3} + 2\sqrt{2}$
$9\sqrt{36} - 18$	$27 - 6\sqrt{6}$
$\phantom{9\sqrt{36}} + 6\sqrt{24} - 4\sqrt{6}$	$ + 6\sqrt{6} - 8$
$54 - 18 + 12\sqrt{6} - 4\sqrt{6} = 36 + 8\sqrt{6}$	$27 - 8 = 19$

Thus, our result is

$$\frac{36 + 8\sqrt{6}}{19}$$

practice Simplify:

a. $\dfrac{5 + \sqrt{3}}{2\sqrt{3}}$ 　　　　　　　　　　　　**b.** $\dfrac{2\sqrt{12} - 3\sqrt{3}}{2\sqrt{3} - 3\sqrt{2}}$

problem set 73

1. The chemist calculated that the carbon (C) in the compound C_2H_5Br weighed 48 grams. What was the total weight of the compound? (C, 12; H, 1; Br, 80)

2. Twice the number of pansies exceeded 4 times the number of daisies by 8. Also, 7 times the number of daisies was 4 less than 3 times the number of pansies. How many of each were there?

3. The wolf loped for a while at 16 mph and finished the journey by trotting at 12 mph. If the total trip was 256 miles, and he loped for 2 more hours than he trotted, how far did he trot?

4. The number of frangibles varied inversely with the strength of the clay. If 50 were frangible when the clay strength measured 50, how many were frangible when the clay strength dropped to 25?

5. The pressure of a quantity of ideal gas was held constant at 1100 newtons per square meter. The initial temperature and volume were 700×10^5 kelvins and 0.0004 liter. What was the final temperature if the final volume was 0.08 liter? Begin by solving the equation for T_2.

Simplify:

6. $\dfrac{3 + \sqrt{5}}{2 - 2\sqrt{5}}$

7. $\dfrac{\sqrt{8} - 2\sqrt{2}}{3\sqrt{2} - 2\sqrt{3}}$

8. The diagonal of a square is 7 meters. What is the length of one side of the square? What is the area of the square?

9. Estimate the location of the line indicated by the data points. Write the equation that gives hydrogen (H) as a function of carbon (C):
   $H = mC + b$

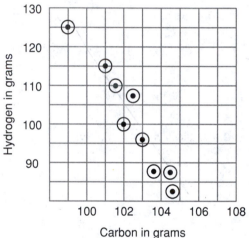

10. Add: $8/20° + 6/320°$

11. Write $4R + 9U$ in polar form.

12. Begin with $ax^2 + bx + c = 0$ and complete the square to derive the quadratic formula.

Use the quadratic formula to solve:

13. $2x^2 + 5 = -5x$

14. $2x^2 - 4 = -5x$

15. Find m_1: $xc = px\left(\dfrac{1}{km_1} - \dfrac{1}{m_2}\right)$

16. Add: $\dfrac{3x - 2}{x - 2} - \dfrac{4x - 3}{2 - x}$

17. Solve: $R_A T_A = 160,\ R_B T_B = 240,\ R_B = 2R_A,\ T_A + T_B = 7$

Simplify:

18. $ax^2 - \dfrac{a}{a - \dfrac{1}{ax}}$

19. $(3i + 2)(i - 4) - \sqrt{-9}$

20. Solve the system by graphing and then find an exact solution by using either substitution or elimination.
$$\begin{cases} 2x + 3y = 6 \\ x - 2y = 4 \end{cases}$$

21. Solve $x^2 + 6 = 3x$ by completing the square.

22. Find x, y, a, b, and c.

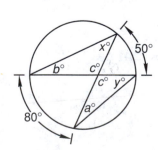

Simplify:

23. $3\sqrt{9\sqrt{3}}$

24. $\dfrac{-2^2}{-16^{-3/4}}$

25. $\sqrt{a^2 x^0 yx^{1/2} y^2}$

26. $2\sqrt{\dfrac{5}{8}} + 3\sqrt{\dfrac{8}{5}} - 2\sqrt{40}$

27. $\dfrac{x^3 - 16x - 6x^2}{x^2 - 8x - 20} \cdot \dfrac{-50 - 5x + x^2}{x^3 - 5x^2 - 24x}$

28. Find the equation of the line that passes through $(-5, 2)$ and has a slope of $-\frac{2}{5}$.

29. Use a calculator to simplify. Estimate first.

(a) $\dfrac{4{,}168{,}214 \times 10^{24}}{74.612 \times 10^{-5.34}}$

(b) $(4.01)^{-5.34}$

30. Solve: $\dfrac{3x - 5}{2} - \dfrac{x}{3} = 6$

LESSON 74   *Uniform motion with both distances given*

The distance diagrams and equations for the uniform motion problems that we have worked thus far have been similar to one of the following.

(a) $R_A T_A = R_B T_B$

(b) $R_A T_A + R_B T_B = 120$

(c) $R_A T_A + 40 = R_B T_B$

In (a), the distances traveled were equal. In (b), the sum of the distances traveled equaled 120, and in (c), the distance traveled by B was 40 greater than the distance traveled by A. **In some problems the distance by each object is given. In these problems, we get two distance equations.** Each of the equations will contain two different unknowns, so two more equations will be needed. One of these equations will be an equation about rates, and the other equation will be an equation about times.

example 74.1   Friar Tuck rode the 24 miles to the fair in Nottingham at a leisurely pace. He stayed too long at the fair and had to double his speed on the way back in order to get home in time. If his total traveling time was 9 hours, how fast did he travel in each direction? What were the times?

solution   **When both distances are given, we can write two distance equations.**

(a) $R_G T_G = 24$ (b) $R_B T_B = 24$

We have four unknowns but only two equations. We reread the problem to get the other two equations. One equation is a rate equation and one equation is a time equation.

(c) $R_B = 2R_G$ (d) $T_G + T_B = 9$

We will use equation (b) as our base equation.

$$R_B T_B = 24 \qquad \text{equation (b)}$$
$$(2R_G)(9 - T_G) = 24 \qquad \text{substituted}$$
$$18R_G - 2R_G T_G = 24 \qquad \text{multiplied}$$
$$18R_G - 2(24) = 24 \qquad \text{substituted}$$
$$R_G = 4 \text{ mph} \qquad \text{divided}$$

Since $R_B = 2R_G$, then $R_B = 8$ mph.

Now we will use these values in equations (a) and (b) to find T_G and T_B.

$R_G T_G = 24$	equation (a)		$R_B T_B = 24$	equation (b)
$4T_G = 24$	substituted		$8T_B = 24$	substituted
$T_G = 6$ hours	divided		$T_B = 3$ hours	divided

example 74.2   Atalanta could run 4 times as fast as her challenger could run. In fact, she could run 80 miles in 2 hours less than it took her challenger to run 28 miles. How fast could each of them run? How long did they run?

solution   **Both distances were given, so we will have two distance diagrams and two distance equations.**

$$D_A \qquad\qquad D_C$$
$$\overset{}{\underset{80}{\longmapsto}} \qquad\qquad \overset{}{\underset{28}{\longmapsto}}$$

$$\text{(a)} \quad R_A T_A = 80 \qquad\qquad \text{(b)} \quad R_C T_C = 28$$

We reread the problem and write the rate equation and time equation.

$$\text{(c)} \quad R_A = 4R_C \qquad\qquad \text{(d)} \quad T_A = T_C - 2$$

We will use equation (a) as our base equation.

$$R_A T_A = 80 \qquad \text{equation (a)}$$
$$(4R_C)(T_C - 2) = 80 \qquad \text{substituted}$$
$$4R_C T_C - 8R_C = 80 \qquad \text{multiplied}$$
$$4(28) - 8R_C = 80 \qquad \text{substituted}$$
$$32 = 8R_C \qquad \text{simplified}$$
$$4 \text{ mph} = R_C \qquad \text{divided}$$

Since Atalanta could run 4 times as fast as her competitor, her rate was **16 mph.** Now we use these rates in equations (a) and (b) to find the times.

$R_A T_A = 80$	equation (a)		$R_C T_C = 28$	equation (b)
$16T_A = 80$	substituted		$4T_C = 28$	substituted
$T_A = 5$ hours	divided		$T_C = 7$ hours	divided

practice   Bernard could race twice as fast as the local domestique. For this reason he could race 120 miles in 3 hours less than it took the domestique to race 105 miles. How fast could each of them race? What was the time of each racer?

problem set 74

1.   Little John rode the 28 miles to Sherwood Forest at a leisurely pace. He stayed too long at Robin's lair and had to double his speed on the way back in order to get home in time. If his total traveling time was 12 hours, how fast did he travel in each direction? What were his times?

2. Regina could cycle 4 times as fast as her challenger could cycle. In fact, she could cycle 80 miles in 2 hours less than it took her challenger to cycle 30 miles. How fast could each of them cycle? How long did they cycle?

3. The initial pressure, volume, and temperature of a quantity of an ideal gas were 0.004×10^5 newtons per square meter, 0.02×10^4 liters, and 0.06×10^6 kelvins, respectively. Find the final temperature if the final pressure and volume were 400×10^5 newtons per square meter and 500×10^4 liters. Begin by solving for T_2.

4. Two thousand liters of a solution was 92% alcohol. How much alcohol should be extracted to reduce the concentration to 80% alcohol?

5. Find three consecutive odd integers such that 4 times the product of the second and third is 12 greater than 20 times the sum of the first and second.

Simplify:

6. $\dfrac{2 - \sqrt{3}}{-\sqrt{3} - 2}$

7. $\dfrac{3\sqrt{2} - 4}{\sqrt{2} - 3}$

8. $\dfrac{4\sqrt{2} - 5}{2 - 3\sqrt{8}}$

9. Estimate the location of the line indicated by the data points and write the equation that gives nitrogen (N) as a function of fluorine (F):
$N = mF + b$

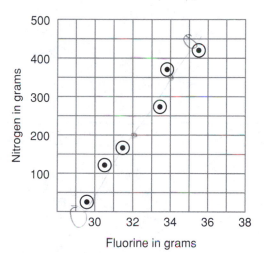

10. Add: $4\underline{/-135°} + 6\underline{/200°}$

11. Write $-5R - 5U$ in polar form.

12. Begin with $ax^2 + bx + c = 0$ and complete the square to derive the quadratic formula.

Use the quadratic formula to solve:

13. $-3x^2 - x = 4$

14. $-x - 3x^2 = -4$

15. Find x: $a = m\left(\dfrac{1}{pc} - \dfrac{k}{x}\right)$

16. Find x_1: $\dfrac{a}{m} = c\left(\dfrac{1}{x_1} + \dfrac{b}{x_2}\right)$

Simplify:

17. $ax - \dfrac{a}{x - \dfrac{x}{a}}$

18. $(2i - 3)(i - 3) + \sqrt{-9} + 3i^3$

19. Add: $\dfrac{3x - 2}{x^2 + 7x + 10} - \dfrac{1}{x + 5}$

20. Solve: $\begin{cases} \dfrac{3}{2}x - \dfrac{2}{5}y = 2 \\ 3x + 0.5y = 17 \end{cases}$

21. Solve $3x^2 - x = -7$ by completing the square.

22. Find x and y.

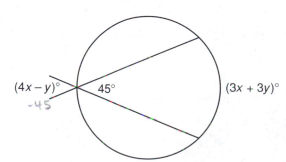

Simplify:

23. $2\sqrt{4\sqrt{2}}$

24. $\dfrac{-3^0(-3^0)^2}{-9^{-3/2}}$

25. $\sqrt{4x^2y^5}\ \sqrt[3]{8y^5x}$

26. $2\sqrt{\dfrac{6}{7}} + 3\sqrt{\dfrac{7}{6}} - 3\sqrt{42}$

27. Find the equation of the line that passes through $(-2, 0)$ and is parallel to the line $4x - y = 7$.

Solve:

28. $\sqrt{x - 5} - 2 = 7$

29. $\dfrac{4x - 3}{7} - \dfrac{x - 2}{3} = 5$

30. The diagonal of a square is $9\sqrt{2}$ meters, as shown. What is the length of one side of the square? What is the area of the square?

LESSON 75 *Factorable denominators and sign changes*

We have discussed how the addition of rational expressions can sometimes be facilitated by making sign changes in one or more of the expressions. For example, if we wish to add the expressions

$$\frac{3x + 2}{x - 5} + \frac{x - 3}{5 - x}$$

we can do so easily if we change the sign in front of the second expression and all the signs in the denominator of the second expression.

$$\frac{3x + 2}{x - 5} - \frac{x - 3}{x - 5}$$

Now the denominators are the same, and we can add the numerators and get

$$\frac{3x + 2 - x + 3}{x - 5} = \frac{2x + 5}{x - 5}$$

We have also learned to add algebraic expressions in which it is helpful to factor one or more denominators as the first step. In this lesson we will add expressions that require factoring in one term and sign changes in another term.

example 75.1   Add: $\dfrac{x + 3}{x^2 - x - 6} - \dfrac{3}{3 - x}$

solution   First we factor the first denominator.

$$\frac{x + 3}{(x - 3)(x + 2)} - \frac{3}{3 - x}$$

Now we need to change the signs in the second denominator. We do this and get

$$\frac{x + 3}{(x - 3)(x + 2)} + \frac{3}{x - 3}$$

We see that the second denominator needs an $x + 2$ factor to make the denominators the same, so we multiply above and below by $x + 2$.

$$\frac{x + 3}{(x - 3)(x + 2)} + \frac{3}{(x - 3)} \cdot \frac{(x + 2)}{(x + 2)}$$

Now both denominators are the same, and we can finish by adding the numerators.

$$\frac{x + 3 + 3x + 6}{(x - 3)(x + 2)} = \frac{4x + 9}{x^2 - x - 6}$$

We could have left the denominator in factored form, but we chose to multiply it out.

example 75.2   Add: $\dfrac{x + 7}{x^2 + 2x + 1} - \dfrac{3}{-1 - x}$

solution   This problem requires that we factor the denominator of the first expression and change the signs in the second expression. We factor first and get

$$\frac{x + 7}{(x + 1)(x + 1)} - \frac{3}{-1 - x}$$

To permit changing the signs in the second denominator, we must change either the sign in front or the sign above. We choose to change the sign in front, and now we have

$$\frac{x + 7}{(x + 1)(x + 1)} + \frac{3}{x + 1}$$

Now we need another $x + 1$ factor in the second denominator, so we multiply above and below by $x + 1$. Then we add the numerators.

$$\frac{x + 7}{(x + 1)(x + 1)} + \frac{3}{(x + 1)} \cdot \frac{(x + 1)}{(x + 1)} = \frac{x + 7 + 3x + 3}{(x + 1)(x + 1)} = \frac{4x + 10}{(x + 1)(x + 1)}$$

This time we left the denominator in factored form.

example 75.3   Add: $\dfrac{x - 3}{x^2 - 2x - 15} + \dfrac{x + 2}{5 - x}$

solution   Again we find that one denominator must be factored and the signs must be changed in the other denominator. We do this and get

$$\frac{x - 3}{(x - 5)(x + 3)} - \frac{x + 2}{x - 5}$$

Now we need another $x + 3$ factor in the second denominator.

$$\frac{x - 3}{(x - 5)(x + 3)} - \frac{x + 2}{(x - 5)} \cdot \frac{(x + 3)}{(x + 3)}$$

We finish by adding and simplifying, and get

$$\frac{x - 3 - x^2 - 5x - 6}{(x - 5)(x + 3)} = \frac{-x^2 - 4x - 9}{(x - 5)(x + 3)}$$

practice   Add: $\dfrac{x - 2}{x^2 - 3x - 4} + \dfrac{x + 5}{4 - x}$

problem set 75

1. David traveled 120 miles in 1 hour less than it took Emily to travel 360 miles. Emily could do this because she drove twice as fast as David. Find the rates and times of both.

2. A quantity of an ideal gas was confined in a container whose volume was fixed at 0.05 liter. The initial pressure and temperature were 0.0036×10^{-2} newton per square meter and 50×10^7 K. If the temperature was changed to 40×10^4 K, find the final pressure. Begin by solving for P_2.

3. The time to complete the job varied inversely as the number of men working. When 500 men worked, the job could be completed in 10 days. How long would 200 men take to complete the job?

4. The bus headed north 2 hours before the train headed north from the same town. The rate of the train was 60 mph and the rate of the bus was 40 mph. How long did it take the train to get 20 miles ahead of the bus?

5. Sandra bought lilies for $4 each and poinsettias for $6 each. She bought 2 fewer lilies than poinsettias and spent a total of $192. How many of each did she buy?

Add:

6. $\dfrac{x + 4}{x^2 + 2x - 3} - \dfrac{2}{1 - x}$

7. $\dfrac{x - 3}{x^2 - x - 12} - \dfrac{x + 2}{-3 - x}$

Simplify:

8. $\dfrac{2 - \sqrt{2}}{3 + \sqrt{2}}$

9. $\dfrac{2 - 3\sqrt{3}}{3 - 2\sqrt{12}}$

10. A square is inscribed in a circle whose area is 9π cm^2, as shown. What is the radius of the circle? What is the length of a diagonal of the square? What is the area of the square?

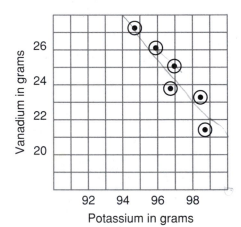

11. Estimate the location of the line indicated by the data points and write the equation that gives vanadium (V) as a function of potassium (K): $V = mK + b$

12. Begin with $ax^2 + bx + c = 0$ and complete the square to derive the quadratic formula.

13. Add: $10\underline{/30°} + 10\underline{/-380°}$

14. Write $-2R - 8U$ in polar form.

15. Use the quadratic formula to find the roots of $-2x + 4 = -5x^2$.

16. Find y: $\dfrac{m}{x} = cm\left(\dfrac{a}{x} + \dfrac{b}{y}\right)$

17. Use similar triangles to find x and y.

Simplify:

18. $a^2 y + \dfrac{a^2}{a + \dfrac{a}{y}}$

19. $(2 + i)(i - 4) - \sqrt{-16}$

20. Solve: $\begin{cases} \dfrac{1}{4}x + \dfrac{1}{3}y = 15 \\ 0.02x + 0.2y = 6.4 \end{cases}$

21. Solve $3x^2 + 2 = -x$ by completing the square.

22. The area of the triangle reduced by the area of the semicircle, as shown, is $(48 - 2\pi)$ m^2. Find H and r.

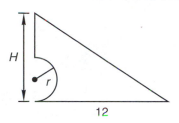

Simplify:

23. $4\sqrt{2\sqrt[3]{2}}$

24. $\dfrac{-2^0(-2^0)}{-4^{-3/2}}$

25. $\sqrt[4]{mp^5}\,\sqrt[3]{m^2p^4}$

26. $3\sqrt{\dfrac{7}{8}} + 2\sqrt{\dfrac{8}{7}} - 2\sqrt{56}$

27. Use a calculator to simplify. Estimate first.

    (a) $\dfrac{4813 \times 10^{-14}}{0.01903 \times 10^{-22}}$

    (b) $\sqrt[3.6]{198}$

28. Find the equation of the line through $(5, 7)$ which is perpendicular to the line passing through $(-2, 4)$ and $(-3, 5)$.

Solve:

29. $\sqrt{3x - 5} - 2 = 7$

30. $\dfrac{4x + 7}{2} - \dfrac{5x}{3} = 4$

LESSON 76 *Using both substitution and elimination • Negative vectors*

76.A
using both substitution and elimination

We have found that we can use either the substitution method or the elimination method to solve a system of two equations in two unknowns such as

$$\begin{cases} 3x + 2y = 8 \\ x + 3y = -2 \end{cases}$$

When we have to solve a system of three equations in three unknowns, it is sometimes helpful if we begin by using substitution and then finish by using elimination.

example 76.1 Use substitution and elimination as necessary to solve this system of equations.

$$\begin{cases} x = 2y & \text{(a)} \\ x + y + z = 9 & \text{(b)} \\ x - 3y - 2z = -8 & \text{(c)} \end{cases}$$

solution Equation (a) tells us that x is equal to $2y$. Thus, we will substitute $2y$ for x in equations (b) and (c) and then simplify.

    (b) $(2y) + y + z = 9$ → $3y + z = 9$ (d)

    (c) $(2y) - 3y - 2z = -8$ → $-y - 2z = -8$ (e)

Now we have two equations in y and z that can be solved by using either substitution or elimination. We will use elimination and will multiply the top equation by 2.

$$
\begin{array}{lllll}
\text{(d)} & 3y + z = 9 & \longrightarrow \; (2) \longrightarrow & 6y + 2z = 18 \\
\text{(e)} & -y - 2z = -8 & \longrightarrow \; (1) \longrightarrow & \underline{-y - 2z = -8} \\
& & & 5y \qquad\quad = 10 \\
& & & \qquad\quad\; \boldsymbol{y = 2}
\end{array}
$$

We can use 2 for y in either equation (d) or equation (e) to find z. This time we will use both equations to show that either one will give the desired result.

EQUATION (d)		EQUATION (e)	
$3y + z = 9$	equation	$-y - 2z = -8$	equation
$3(2) + z = 9$	substituted	$-(2) - 2z = -8$	substituted
$6 + z = 9$	multiplied	$-2z = -6$	simplified
$z = \boldsymbol{3}$	solved	$z = \boldsymbol{3}$	solved

Finally, we can use 2 for y and 3 for z in any of the first three equations to find x. This time we will use all three.

EQUATION (a)	EQUATION (b)	EQUATION (c)	
$x = 2y$	$x + y + z = 9$	$x - 3y - 2z = -8$	equation
$x = 2(2)$	$x + 2 + 3 = 9$	$x - 3(2) - 2(3) = -8$	substituted
$\boldsymbol{x = 4}$	$\boldsymbol{x = 4}$	$\boldsymbol{x = 4}$	solved

Thus we find that the solution to this system of three equations in three unknowns is the ordered triple $(\boldsymbol{4, 2, 3})$.

example 76.2   Use substitution and elimination as necessary to solve this system of equations.

$$
\begin{cases}
2x + 2y - z = 12 & \text{(a)} \\
3x - y + 2z = 21 & \text{(b)} \\
x - 3z = 0 & \text{(c)}
\end{cases}
$$

solution   If we solve equation (c) for x, we get

$$x = 3z$$

Next, in equations (a) and (b), we will replace x with $3z$ and then simplify.

$$
\begin{array}{lll}
\text{(a)} & 2(3z) + 2y - z = 12 \quad\longrightarrow\quad & 5z + 2y = 12 \quad \text{(d)} \\
\text{(b)} & 3(3z) - y + 2z = 21 \quad\longrightarrow\quad & 11z - y = 21 \quad \text{(e)}
\end{array}
$$

We can eliminate y if we multiply equation (e) by 2 and add.

$$
\begin{array}{llll}
\text{(d)} & 5z + 2y = 12 & \longrightarrow & 5z + 2y = 12 \\
\text{(e)} & 11z - y = 21 & \longrightarrow & \underline{22z - 2y = 42} \\
& & & 27z \qquad\;\; = 54 \\
& & & \qquad\quad\; \boldsymbol{z = 2}
\end{array}
$$

Now we can use 2 for z in either equation (d) or equation (c) to find y. We choose to use equation (d).

$5z + 2y = 12$	equation (d)
$5(2) + 2y = 12$	substituted
$2y = 2$	simplified
$\boldsymbol{y = 1}$	divided

Now we can use 2 for z and 1 for y in any of the original equations to find x. Equation (c) is the simplest so we will use it.

$$x - 3z = 0 \qquad \text{equation (c)}$$

$$x - 3(2) = 0 \qquad \text{substituted}$$

$$\mathbf{x = 6} \qquad \text{solved}$$

Thus, we find that the solution to this system of three equations in three unknowns is the ordered triple (**6, 1, 2**).

76.B
negative vectors

In Lesson 72, we noted that there is only one way to use rectangular coordinates to designate the location of a point, but more than one form of polar coordinates is possible because either positive angles or negative angles may be used.

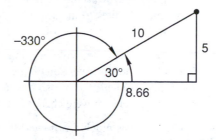

Since the point is 8.66 units to the right of the origin and 5 units above the origin, we can designate its location with rectangular coordinates by writing

$$8.66R + 5U$$

If we wish to use polar coordinates to name the same point, we can use either a positive angle or a negative angle, so we can write either

$$10\underline{/30°} \qquad \text{or} \qquad 10\underline{/-330°}$$

To make matters even more confusing, we note that it is also possible to use negative magnitudes to locate a point.

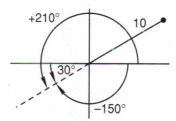

We see that if we begin by turning through an angle of +210° or −150°, we are pointing away from the point. Now if we back up 10 units, we are on the point. Thus, the point can also be designated by using negative magnitudes and writing

$$-10\underline{/210°} \qquad \text{or} \qquad -10\underline{/-150°}$$

example 76.3   Add: $-4\underline{/-20°} + 5\underline{/135°}$

solution   We begin by drawing the vectors. We lay out the angles first and then the magnitudes.

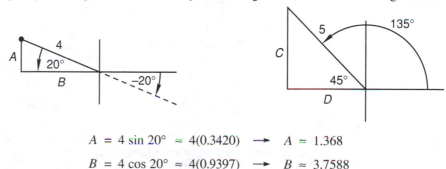

$$A = 4 \sin 20° \approx 4(0.3420) \quad \longrightarrow \quad A \approx 1.368$$

$$B = 4 \cos 20° \approx 4(0.9397) \quad \longrightarrow \quad B \approx 3.7588$$

$$C = 5 \sin 45° \approx 5(0.7071) \quad \longrightarrow \quad C \approx 3.5355$$

$$D = 5 \cos 45° \approx 5(0.7071) \quad \longrightarrow \quad D \approx 3.5355$$

Thus, we go $3.7588 + 3.5355 = 7.2943$ to the left and $1.368 + 3.5355 = 4.9035$ up. Therefore, our answer is

$$-7.29R + 4.90U$$

practice   a.   Solve: $\begin{cases} x = 3y \\ x + y + z = 56 \\ x - 2y - 3z = -25 \end{cases}$ b.   Add: $-2\underline{/-15°} + 5\underline{/110°}$

problem set 76

1.   Claudius could walk the 32 miles to Pompeii in 2 hours more than it took Tiberius to drive the 72 miles to the sea. Find the rate of each and the time of each if the rate of Tiberius was 3 times that of Claudius.

2.   The initial pressure, volume, and temperature of a quantity of an ideal gas were 700×10^5 newtons per square meter, 700×10^{-7} liter, and 56×10^4 K. What would be the final volume if the pressure were changed to 3500×10^4 newtons per square meter and the temperature were changed to 8000×10^5 K? Begin by solving the equation for V_2.

3.   The elixir was much too powerful, as 500 ml of it tested at 64% alcohol. How much alcohol had to be evaporated to reduce the alcohol concentration to 40%?

4.   The chemical formula for sodium hypochlorite is NaClO. Benjamin had 280 grams of chlorine (Cl) available. What would be the total weight of the oxygen (O) and sodium (Na) that he would need to make a batch of sodium hypochlorite? (Na, 23; O, 16; Cl, 35)

5.   When the chips were down, Chester found that 4 times the number of blue chips exceeded 3 times the number of red chips by 14. Also, 6 times the number of red chips exceeded the number of blue chips by 7. How many chips were down?

Solve:

6. $\begin{cases} x = 2y \\ x + y + z = -198 \\ x - 3y - 2z = 16 \end{cases}$ 7. $\begin{cases} 2x + 2y - z = 81 \\ 3x - y + 2z = 27 \\ x - 3z = 0 \end{cases}$

8.   Add: $-5\underline{/-20°} + 4\underline{/115°}$ 9.   Write $-4R - 1U$ in polar form.

10. Add: $\dfrac{5x + 2}{x^2 + 3x - 10} - \dfrac{2x}{2 - x}$

Simplify:

11. $\dfrac{2\sqrt{2} - 1}{1 - \sqrt{2}}$ 12. $\dfrac{3 - \sqrt{2}}{4 + 2\sqrt{8}}$

13. Estimate the location of the line indicated by the data points and write the equation that expresses aluminum (Al) as a function of boron (B): $Al = mB + b$

14. Begin with $ax^2 + bx + c = 0$ and derive the quadratic formula.

15. Find x: $\dfrac{a}{c + x} = m\left(\dfrac{1}{r} + \dfrac{1}{t}\right)$

16. Find t: $\dfrac{a}{c + x} = m\left(\dfrac{1}{r} + \dfrac{1}{t}\right)$

17. Solve $7x^2 - x - 1 = 0$ by completing the square.

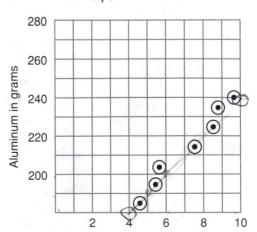

Simplify:

18. $x + \dfrac{a}{a + \dfrac{a}{x}}$

19. $(3 - i)(2 - i) - 2i^2 - \sqrt{-9}$

20.   Solve: $\begin{cases} \dfrac{1}{3}x + \dfrac{2}{3}y = 31 \\ 0.02x + 0.7y = 27.6 \end{cases}$

21.   Find the radius of the circle. The quadrilateral is a rectangle.

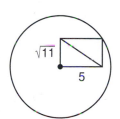

Simplify:

22. $5\sqrt{25\sqrt{5}}$

23. $\dfrac{-1^0(-1^0)}{-4^{-5/2}}$

24. $\sqrt[6]{my^3}\ \sqrt[4]{m^3y}$

25. $2\sqrt{\dfrac{3}{8}} + 3\sqrt{\dfrac{8}{3}} - 2\sqrt{216}$

26.   Use a calculator to simplify. Estimate first.

     (a) $\dfrac{0.00842 \times 10^{18}}{4{,}198{,}312 \times 10^{-13}}$

     (b) $(4.63)^{5.12}$

27.   Find the equation of the line that passes through $(-5, -7)$ and $(2, 4)$.

Solve:

28. $\sqrt{2x - 2} + 7 = 9$

29. $\dfrac{15x - 2}{3} - \dfrac{2x - 5}{4} = 3$

30.   Find the area of this figure. Dimensions are in inches.

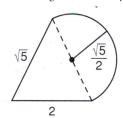

LESSON 77   *Advanced radical equations*   •   *Multiple radicals*

77.A

**advanced
radical
equations**

Thus far, we have restricted our investigation of radical equations to equations that contain one radical, such as

$$\sqrt{x - 1} = 4$$

We can solve this equation for x if we begin by squaring both sides of the equation.

$$\left(\sqrt{x - 1}\right)^2 = 4^2$$

Then we simplify and complete the solution.

$$x - 1 = 16$$

$$x = 17$$

As the final step, we must check the solution in the original equation because squaring both sides of an equation sometimes generates an equation that has more solutions than the original equation.

$$\sqrt{(17) - 1} = 4$$

$$\sqrt{16} = 4$$

$$4 = 4 \qquad \text{check}$$

It can be shown that if both sides of an equation are raised to the nth power ($n = 2, 3, 4, \ldots$), the solutions of the original equation (if any exist) are also solutions of the resulting equation. This permits us to solve many radical equations by isolating the radical and then raising both sides to the integral power that will eliminate the radical.

example 77.1   Solve: $\sqrt[3]{x^3 + 6x^2 - 4} - x - 2 = 0$

solution   We begin by isolating the radical on one side of the equation.

$$\sqrt[3]{x^3 + 6x^2 - 4} = x + 2$$

Now we raise both sides of the equation to the third power to eliminate the radical.

$$\left(\sqrt[3]{x^3 + 6x^2 - 4}\right)^3 = (x + 2)^3$$

$$x^3 + 6x^2 - 4 = x^3 + 6x^2 + 12x + 8 \qquad \text{multiplied}$$

$$-4 = 12x + 8 \qquad \text{simplified}$$

$$-12 = 12x \qquad \text{added} -8$$

$$\mathbf{-1} = x \qquad \text{divided}$$

Now we must check the answer in the original equation.

$$\sqrt[3]{(-1)^3 + 6(-1)^2 - 4} - (-1) - 2 = 0$$

$$\sqrt[3]{1} + 1 - 2 = 0$$

$$1 + 1 - 2 = 0 \qquad \text{check}$$

77.B

multiple radicals   When an equation contains more than one radical, it is sometimes necessary to square both sides of the equation more than once to eliminate all the radicals.

example 77.2   Solve: $\sqrt{k - 5} - \sqrt{k} + 1 = 0$

solution   We begin by rearranging the equation so that $\sqrt{k - 5}$ is isolated on the left-hand side.

$$\sqrt{k - 5} = \sqrt{k} - 1$$

Now we square both sides to eliminate the radical on the left.

$$\left(\sqrt{k - 5}\right)^2 = \left(\sqrt{k} - 1\right)^2$$

On the left we get $k - 5$, and on the right we must multiply $\sqrt{k} - 1$ by $\sqrt{k} - 1$ to get $k - 2\sqrt{k} + 1$.

$$k - 5 = k - 2\sqrt{k} + 1$$

Now we rearrange so that $\sqrt{k}$ is isolated.

$$-6 = -2\sqrt{k}$$

$$3 = \sqrt{k}$$

Next we square both sides again and get,

$$9 = k$$

Now we must check this value of k in the original equation.

$$\sqrt{(9) - 5} - \sqrt{(9)} + 1 = 0 \qquad \text{substituted}$$
$$\sqrt{4} - 3 + 1 = 0 \qquad \text{simplified}$$
$$0 = 0 \qquad \text{check}$$

example 77.3   Solve: $\sqrt{s - 8} + \sqrt{s} = 2$

solution   We begin by changing the equation so that $\sqrt{s - 8}$ is isolated on the left side.

$$\sqrt{s - 8} = 2 - \sqrt{s}$$

We will square both sides to eliminate the radical on the left side.

$$\left(\sqrt{s - 8}\right)^2 = \left(2 - \sqrt{s}\right)^2 \qquad \text{square both sides}$$
$$s - 8 = 4 - 4\sqrt{s} + s \qquad \text{multiplied}$$
$$-12 = -4\sqrt{s} \qquad \text{simplified}$$
$$3 = \sqrt{s} \qquad \text{divided by } -4$$

Now to finish, we square both sides again and get

$$9 = s$$

as a solution. Now we must check this value of s in the original equation.

$$\sqrt{(9) - 8} + \sqrt{(9)} = 2 \qquad \text{substituted}$$
$$\sqrt{1} + 3 = 2 \qquad \text{simplified}$$
$$4 = 2 \qquad \text{not true}$$

Our answer does not check, so the original equation has **no solution**.

practice   Solve:

a. $\sqrt[3]{x^3 + 3x^2 - 8} - x - 1 = 0$ **b.** $\sqrt{s - 16} + \sqrt{s} = 4$

problem set 77

1. Hahira could drive 320 miles in twice the time it took Sylvester to drive 240 miles. The speed of Sylvester was 20 mph greater than that of Hahira. What were the speeds and times of both?

2. Enjoyment for some varied inversely as the cost. If enjoyment measured 500 when the cost was $10, what did enjoyment measure when the cost was reduced to $1?

3. The drink was cloying because it was 30% sugar. If 50 liters were on hand, how much water should be added to reduce the sugar concentration to 3% sugar?

4. Rasputin searched for three consecutive multiples of 7 such that the sum of the first and 4 times the third exceeded 3 times the second by 133. What were the multiples?

5. Camilla ran and Quitman walked. Thus, Camilla could complete the trip in 6 hours while Quitman took 12 hours. What was Camilla's rate if she ran 4 miles per hour faster than Quitman walked?

Solve:

6. $\sqrt[3]{x^3 + 9x^2 - 27} - x - 3 = 0$ 7. $\sqrt{m - 12} - \sqrt{m + 2} = 0$

8. $\begin{cases} 2x + y - z = 7 \\ x - 2y + z = -2 \\ 2y + z = 0 \end{cases}$

9. $\begin{cases} x + y + z = 7 \\ 2x - y - z = -4 \\ z = 2y \end{cases}$

10. Use similar triangles to find r and t.

11. Add: $-4\underline{/-30°} + 6\underline{/-200°}$

12. Write $-2R - 7U$ in polar form.

13. Add: $\dfrac{7x + 2}{x^2 - 2x - 15} - \dfrac{2}{5 - x}$

Simplify:

14. $\dfrac{2 - \sqrt{2}}{2\sqrt{2} - 1}$

15. $\dfrac{3 + 2\sqrt{5}}{1 - \sqrt{5}}$

16. $a + \dfrac{a}{a + \dfrac{a}{x}}$

17. Find R_2: $\dfrac{a}{x} = m\left(\dfrac{a}{R_1} + \dfrac{b}{R_2}\right)$

18. Simplify: $-\sqrt{-9} - 3i^3 - 2i^4 + 2$

19. Solve: $\begin{cases} \dfrac{3}{7}x + \dfrac{2}{5}y = 10 \\ 0.03x - 0.2y = -1.58 \end{cases}$

20. Solve $-7x - 1 = 2x^2$ by completing the square.

21. Use the quadratic formula to solve: $-8x - 1 = 2x^2$

22. Given: $m\overparen{BC} = 30°$, $m\,\overline{OB} = 3$.
    Find: $m\angle AOC$, $m\angle OCA$, $m\angle OAC$,
    and the area of the $30°$ sector.

23. A circle is inscribed in a square whose area is 4 cm². How long is a side of the square? How long is a diagonal? What is the radius of the circle?

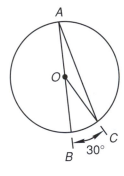

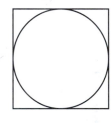

24. Use unit multipliers to convert 4000 cubic centimeters per second to cubic feet per minute.

Simplify:

25. $\sqrt[4]{4\sqrt[3]{2}}$

26. $\dfrac{(-3)^0(-3^0)}{-9^{-3/2}}$

27. $(2\sqrt{5} + 5)(5\sqrt{20} - 1)$

28. Find the equation of the line that passes through $(-2, 4)$ and is perpendicular to $5x + 4y = 3$.

29. Find the distance between $(-3, 8)$ and $(5, -2)$.

30. Solve: $\dfrac{3x - 1}{4} - \dfrac{x - 5}{7} = 1$

LESSON 78 *Force vectors at a point*

In the study of physical science much effort is devoted to understanding the relationships that force, mass, weight, velocity, and acceleration have to one another. In this study, it is helpful to use vectors to represent forces. If we do this, we can add the force vectors that act on a point to find the resultant force.

example 78.1   Two ropes are attached to a point and pulled on with the directions and magnitudes shown. What is the resultant force on the point?

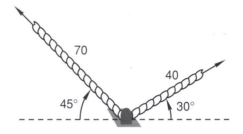

solution   If we complete the triangles, we get a picture of the problem. The left-hand vector pulls to the left and up, and the right-hand vector pulls to the right and up.

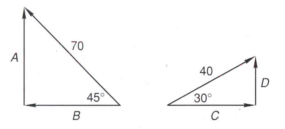

Next we solve for A, B, C, and D.

$A = 70 \sin 45°$ → $A \approx 70(0.7071) = 49.497$ $C = 40 \cos 30° \approx (40)(0.8660)$

$= 34.64$

$B = 70 \cos 45°$ → $B \approx 70(0.7071) = 49.497$ $D = 40 \sin 30° = 40(0.5) = 20$

Thus the total left-right force is $-49.497 + 34.64 = -14.857$ and the total up-down force is $+49.497 + 20 = 69.497$. Thus the resultant force is **$-14.86R + 69.50U$**.

To express the resultant force in polar coordinates, we need to find the angle and the hypotenuse of the triangle.

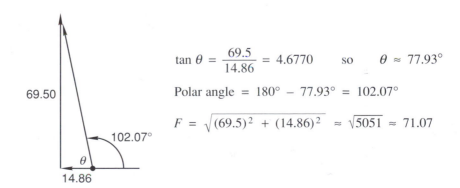

$\tan \theta = \dfrac{69.5}{14.86} = 4.6770$ so $\theta \approx 77.93°$

Polar angle $= 180° - 77.93° = 102.07°$

$F = \sqrt{(69.5)^2 + (14.86)^2} \approx \sqrt{5051} \approx 71.07$

Thus, the resultant force is **$71.07\underline{/102.07°}$**. This is the angle and pull that must be used with a single rope to get the same result obtained by using the two ropes.

example 78.2   Here a point is acted on by a push force and a pull force as shown. Find the resultant force on the point.

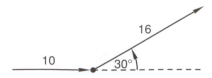

solution   The point can't tell the difference between a push of 10 from the left and a pull of 10 from the right. So we can redraw the problem, showing both vectors as pull vectors.

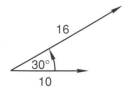

The horizontal vector is $10R + 0U$, and we find the horizontal and vertical components of the other vector to be $13.86R + 8U$.

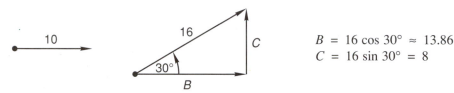

$$B = 16 \cos 30° \approx 13.86$$
$$C = 16 \sin 30° = 8$$

The resultant vector is the sum of the two vectors.

$$\begin{array}{r} 10.00R + 0U \\ 13.86R + 8U \\ \hline 23.86R + 8U \end{array}$$

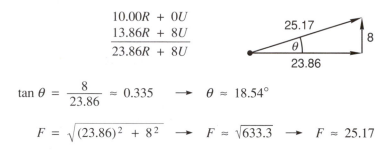

$$\tan \theta = \frac{8}{23.86} \approx 0.335 \quad \longrightarrow \quad \theta \approx 18.54°$$

$$F = \sqrt{(23.86)^2 + 8^2} \quad \longrightarrow \quad F \approx \sqrt{633.3} \quad \longrightarrow \quad F \approx 25.17$$

So the resultant force in polar form is

$$\mathbf{25.17\underline{/18.54°}}$$

practice   Two ropes are attached to a point and pulled on with the directions and magnitudes shown. What is the resultant force on the point?

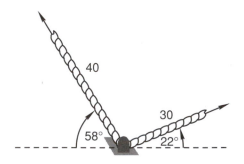

problem set
78

1.   Rita covered the 135 miles in 4 fewer hours than it took Jean to cover 945 miles. If Jean's speed was 3 times that of Rita, what was the speed of each and for how long did each one travel?

2.   Cathy and Ernie had four consecutive even integers. They found that the product of the second and fourth was 16 less than the product of −3 and the sum of the first and third. What were their integers?

3. Fourteen percent of the compound was pure chlorine. If 3440 grams was not chlorine, what was the total weight of the compound?

4. The assonance was uncanny because 42 percent of the chords sounded alike. If 232 chords had different sounds, how many chords were there in all?

5. The quotient of two numbers was 2 and their product was 200. What were the numbers?

6. The value of the fraction was 5. Twice the numerator was 6 greater than 7 times the denominator. What was the fraction?

Solve:

7. $\sqrt{x - 9} + \sqrt{x} = 3$ 8. $\sqrt{x - 8} + \sqrt{x} = 4$ 9. $\sqrt{k - 24} = 6 - \sqrt{k}$

10. Estimate the location of the line indicated by the data points and then write the equation that expresses steel (S) as a function of iron (I):
    $S = mI + b$

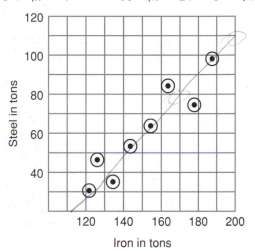

11. Solve: $\begin{cases} x - 2y - z = -9 \\ 2x - y + 2z = 7 \\ 3x - y = 0 \end{cases}$

$-y = 3x$

$y = 3x$

12. The two forces act on a point as shown. Find the resultant force.

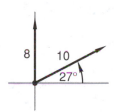

13. Add: $\dfrac{x - 3}{x^2 + 5x - 14} + \dfrac{3x}{2 - x}$

Simplify:

14. $\dfrac{3 - \sqrt{5}}{\sqrt{5} + 2}$ 15. $\dfrac{2 + 2\sqrt{2}}{3 - 3\sqrt{2}}$ 16. $\dfrac{4 - 3\sqrt{2}}{1 - \sqrt{2}}$

17. Use the quadratic formula to solve: $x^2 = -x - 1$

18. Find R_1: $mc = a\left(\dfrac{1}{x} + \dfrac{1}{R_1}\right)$ 19. Find c: $mc = a\left(\dfrac{1}{x} + \dfrac{1}{R_1}\right)$

Simplify:

20. $a + \dfrac{a}{a + \dfrac{x}{a}}$ 21. $-5i^3 - \sqrt{-9} + \sqrt{-3}\,\sqrt{-3}$

22. Use similar triangles to find a, b, and the area of this triangle.

23. Use unit multipliers to convert 400 cubic inches per second to cubic centimeters per minute.

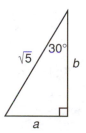

Simplify:

24. $2\sqrt[5]{4\sqrt[6]{2}}$

25. $\dfrac{-2^0(2^{-2})}{4^{-3/2}}$

26. $(2\sqrt{4} - 2)(3\sqrt{9} - 2)$

27. $4\sqrt{\dfrac{5}{8}} - 3\sqrt{\dfrac{8}{5}} + 2\sqrt{40}$

28. Use a calculator to simplify. Estimate first.

(a) $\dfrac{70{,}218 \times 10^{-4}}{5062 \times 10^5}$

(b) $\sqrt[5.4]{263}$

29. Find v if $\dfrac{\sqrt{mv}}{e} = p$ and if $m = 4 \times 10^7$, $e = 500$, and $p = 100 \times 10^{-14}$.

30. Find S in terms of T, U, and V.

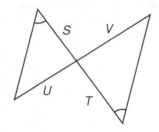

LESSON 79   *Metric volume* • *45-45-90 triangles*

79.A

metric volume

When we convert units of volume, it is necessary to use three unit multipliers for each step in the process. For example, to convert 7 cubic meters (m^3) to cubic centimeters (cm^3), we must use three unit multipliers, as shown here.

$$7\ m^3 \times \frac{100\ cm}{1\ m} \times \frac{100\ cm}{1\ m} \times \frac{100\ cm}{1\ m} = 7{,}000{,}000\ cm^3$$

The cubic meter is a rather large unit of volume, and the cubic centimeter is a rather small unit of volume. The **liter** is used to fill the requirement for an intermediate unit of volume and is about the size of a quart. **A liter is defined to equal the volume of 1000 cubic centimeters.**

1 liter = 1000 cubic centimeters

Because a cubic centimeter is one one-thousandth of a liter, a cubic centimeter is often called a *milliliter*, which is abbreviated ml.

1 cm^3 = 1 ml

We will use the terms *cubic centimeter* and *milliliter* interchangeably in this book because they are used interchangeably in chemistry and other science courses. **Because of the way liters are defined, only one unit multiplier is required to make volume unit conversions between cubic centimeters (milliliters) and liters.**

example 79.1   Use unit multipliers to convert 14,000 ml to liters.

solution   **We need only one unit multiplier.** We can use either

$$\text{(a)} \ \frac{1000 \text{ ml}}{1 \text{ liter}} \quad \text{or} \quad \text{(b)} \ \frac{1 \text{ liter}}{1000 \text{ ml}}$$

We will use form (b) so that the ml units will cancel.

$$14,000 \ \cancel{\text{ml}} \times \frac{1 \text{ liter}}{1000 \ \cancel{\text{ml}}} = 14 \text{ liters}$$

example 79.2   Use unit multipliers to convert 4 cubic feet to liters.

solution   We will go from cubic feet to cubic inches to cubic centimeters to liters. **Note that to go from cubic centimeters to liters, we need only one unit multiplier.**

$$4 \ \cancel{\text{ft}^3} \times \frac{12 \ \cancel{\text{in.}}}{1 \ \cancel{\text{ft}}} \times \frac{12 \ \cancel{\text{in.}}}{1 \ \cancel{\text{ft}}} \times \frac{12 \ \cancel{\text{in.}}}{1 \ \cancel{\text{ft}}} \times \frac{2.54 \ \cancel{\text{cm}}}{1 \ \cancel{\text{in.}}} \times \frac{2.54 \ \cancel{\text{cm}}}{1 \ \cancel{\text{in.}}} \times \frac{2.54 \ \cancel{\text{cm}}}{1 \ \cancel{\text{in.}}} \times \frac{1 \text{ liter}}{1000 \ \cancel{\text{cm}^3}}$$

$$= \frac{4(12)(12)(12)(2.54)(2.54)(2.54)}{1000} \text{ liters}$$

example 79.3   Use unit multipliers to convert 4.7 liters to cubic inches.

solution   We will go from liters to cubic centimeters to cubic inches.

$$4.7 \ \cancel{\text{liters}} \times \frac{1000 \ \cancel{\text{cm}^3}}{1 \ \cancel{\text{liter}}} \times \frac{1 \text{ in.}}{2.54 \ \cancel{\text{cm}}} \times \frac{1 \text{ in.}}{2.54 \ \cancel{\text{cm}}} \times \frac{1 \text{ in.}}{2.54 \ \cancel{\text{cm}}} = \frac{(4.7)(1000)}{(2.54)^3} \text{ in.}^3$$

79.B
45-45-90 triangles

We remember that if a right triangle has a 30° angle, the other angle is a 60° angle. If a right triangle has a 60° angle, the other angle is a 30° angle. Every 30-60-90 right triangle is similar to this right triangle.

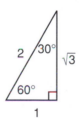

Every right isosceles triangle has two 45° angles. We call these triangles **45-45-90 triangles.** Every one of these triangles is similar to this triangle.

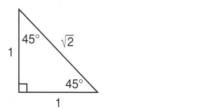

$$1^2 + 1^2 = H^2$$
$$2 = H^2$$
$$\sqrt{2} = H$$

example 79.4   Use similar triangles to find x and y.

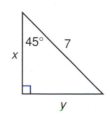

solution This is a 45-45-90 triangle. First we find the scale factor.

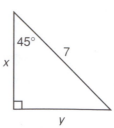

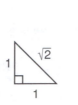

$$\sqrt{2}\ \overrightarrow{SF} = 7$$

$$\overrightarrow{SF} = \frac{7}{\sqrt{2}}$$

The scale factor has a radical in the denominator that we can rationalize by multiplying by $\sqrt{2}$ over $\sqrt{2}$.

$$\overrightarrow{SF} = \frac{7}{\sqrt{2}} \cdot \frac{\sqrt{2}}{\sqrt{2}} = \frac{7\sqrt{2}}{2}$$

Now we can find x and y.

$$1 \cdot \frac{7\sqrt{2}}{2} = x \qquad\qquad 1 \cdot \frac{7\sqrt{2}}{2} = y$$

$$\frac{7\sqrt{2}}{2} = x \qquad\qquad \frac{7\sqrt{2}}{2} = y$$

practice Use unit multipliers to convert:

    a. 12,000 liters to milliliters   **b.** 10 cubic feet to liters

    c. Use similar triangles to find y and z.

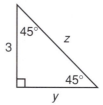

problem set 79

1. Profligate Pauline found that the total cost varied directly as the number purchased. The cost was \$2142 when she purchased 3. What would it cost her to purchase 10?

2. Darnley could travel 900 kilometers in 5 times the time it took Essex to travel 120 kilometers. Find the rates and times of both if Darnley's speed exceeded that of Essex by 10 kph.

3. What is the percent weight of the carbon (C) in a quantity of carbon dioxide, CO_2? (C, 12; O, 16)

4. The two jugs stood side by side on the shelf. One contained a solution that was 60% antiseptic, and the other contained a solution that was 90% antiseptic. How much of each should be used to get 50 milliliters of a solution that is 78% antiseptic?

5. When the hoplites collected their weapons, they found that the ratio of swords to spears was 2 to 7. Further, they found that 5 times the number of swords exceeded the number of spears by 120. How many of each did they have?

Use unit multipliers to convert:

6. 50,000 ml to liters   7. 20 cubic feet to liters

Solve:

8. $\sqrt{x^2 - 4x + 4} = x + 2$ 9. $\sqrt{s} = 4 - \sqrt{s + 8}$

10. $\begin{cases} 2x - y + 2z = 3 \\ x - y - 2z = -6 \\ 3x - y = 0 \end{cases}$

11. A 2-newton force and a 1-newton force act on a point as shown. Find the resultant force.

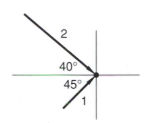

12. Add: $\dfrac{4x + 3}{x^2 - 9} - \dfrac{2x}{3 - x}$

Simplify:

13. $\dfrac{-2 - \sqrt{3}}{2\sqrt{3} + 2}$

14. $\dfrac{-1 - \sqrt{2}}{-5 - \sqrt{2}}$

15. Begin with $ax^2 + bx + c = 0$ and derive the quadratic formula by completing the square.

16. Find a: $\dfrac{a}{x} = m\left(\dfrac{a}{R_1} + \dfrac{b}{R_2}\right)$

17. Find b: $\dfrac{a}{x} = m\left(\dfrac{a}{R_1} + \dfrac{b}{R_2}\right)$

Simplify:

18. $ax - \dfrac{ax}{a - \dfrac{a}{x}}$

19. $\sqrt{-4} - 3i^2 - 2i^4 + 2 - \sqrt{-2}\sqrt{-2}$

20. Solve: $\begin{cases} \dfrac{1}{5}x - \dfrac{1}{4}y = -6 \\ 0.2x + 0.2y = 12 \end{cases}$

21. Use the quadratic formula to solve: $3x^2 - 1 = 2x$

22. Use similar triangles to find x and y.

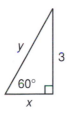

23. Use similar triangles to find a and b.

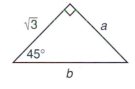

24. Use unit multipliers to convert 40 milliliters per second to cubic inches per hour.

Simplify:

25. $3\sqrt{9^2\sqrt{3}}$

26. $2\sqrt{\dfrac{7}{2}} + 3\sqrt{\dfrac{2}{7}} - 2\sqrt{126}$

27. $(4\sqrt{2} + 3)(5\sqrt{2} - 4)$

28. Use a calculator to simplify. Estimate first.

(a) $\dfrac{41,852 \times 10^{28}}{0.00492 \times 10^{-14}}$

(b) $(194)^{-1.09}$

29. Find the equation of the line that passes through $(5, 7)$ and has a slope of $\frac{1}{5}$.

30. Find the distance between $(4, -2)$ and $(4, -6)$.

LESSON 80 *Direct and inverse variation as ratios*

80.A

direct variation as a ratio

When a problem states that variable X varies directly as variable Y, we know that the relationship implied is

$$X = kY$$

and that the first step in solving the problem is to find the value of the constant of proportionality k.

example 80.1 A varies directly as B. If A is 50 when B is 5, what is the value of A when B is 7?

solution The equation implied is

$$A = kB$$

If we use 50 for A and 5 for B, we find that the constant k equals 10.

$$50 = k(5) \quad \longrightarrow \quad k = 10$$

Now we can completely define the relationship by writing

$$A = 10B$$

We are asked to find the value of A when B is 7, so we replace B with 7 and get

$$A = 10(7) \quad \longrightarrow \quad A = 70$$

There is another way to work this problem because the statement that A varies directly as B also implies the equation

$$\frac{A_1}{A_2} = \frac{B_1}{B_2}$$

Note that both A's are on the same side, and that A_1 and B_1 are both on top. We have been given initial values of A_1 and B_1 of 50 and 5, respectively, and a value of 7 for B_2. If we insert these values, we get

$$\frac{50}{A_2} = \frac{5}{7}$$

We solve this equation by first cross multiplying and then dividing both sides by 5.

$$7 \cdot 50 = 5A_2 \quad \longrightarrow \quad \frac{7 \cdot 50}{5} = \frac{5A_2}{5} \quad \longrightarrow \quad 70 = A_2$$

It is important to understand that there are two ways to work these problems; some authors of upper-level science books use the first approach, whereas others use the second approach.

example 80.2 Cost varies directly as the number purchased. If 12 can be purchased for $78, how much would 42 cost?

solution The two equations that we can use are

$$\text{(a)} \quad C = kN \qquad \text{and} \qquad \text{(b)} \quad \frac{C_1}{C_2} = \frac{N_1}{N_2}$$

We will first use equation (a) and three steps.

$$\text{(a)} \quad \text{Step 1} \qquad C = kN \quad \longrightarrow \quad 78 = k12 \quad \longrightarrow \quad k = 6.5$$

$$\text{Step 2} \qquad C = 6.5N$$

$$\text{Step 3} \qquad C = 6.5(42) \quad \longrightarrow \quad C = \$273$$

Now we will use equation (b).

$$(b) \quad \frac{C_1}{C_2} = \frac{N_1}{N_2} \quad \rightarrow \quad \frac{78}{C_2} = \frac{12}{42} \quad \rightarrow \quad 78 \cdot 42 = 12C_2$$

$$\frac{78 \cdot 42}{12} = \frac{12C_2}{12} \quad \rightarrow \quad \mathbf{\$273} = C_2$$

80.B
inverse variation as a ratio

If A varies inversely as B, the following equation is implied.

$$A = \frac{k}{B}$$

This statement also implies the inverted ratio

$$\frac{A_1}{A_2} = \frac{B_2}{B_1}$$

Note that both A's are on the same side, and both B's are on the other side. However, the B's are in the inverted form. We will work the next problem using both formats.

example 80.3 Blues vary inversely as yellows squared. If 100 blues go with 2 yellows, how many blues go with 10 yellows?

solution The two relationships implied are

$$(a) \quad B = \frac{k}{Y^2} \qquad (b) \quad \frac{B_1}{B_2} = \frac{Y_2^2}{Y_1^2}$$

We will use equation (a) first. We begin by finding k.

$$\text{Step 1} \quad B = \frac{k}{Y^2} \quad \rightarrow \quad 100 = \frac{k}{(2)^2} \quad \rightarrow \quad k = 400$$

$$\text{Step 2} \quad B = \frac{400}{Y^2}$$

$$\text{Step 3} \quad B = \frac{400}{(10)^2} \quad \rightarrow \quad B = 4$$

Thus, 4 blues go with 10 yellows. Now we will rework the problem using equation (b).

$$\frac{B_1}{B_2} = \frac{Y_2^2}{Y_1^2} \quad \rightarrow \quad \frac{100}{B_2} = \frac{(10)^2}{(2)^2} \quad \rightarrow \quad 400 = 100B_2 \quad \rightarrow \quad 4 = B_2$$

We get the same answer with either approach.

practice Use the equal ratio method to work each problem:

a. Cost varies inversely as the number purchased. If 15 can be purchased for $225, how much would 42 cost?

b. Blues vary inversely as yellows squared. If 100 blues go with 3 yellows, how many blues go with 10 yellows?

problem set 80

1. Cost varies directly as the number purchased. If 14 can be purchased for $119, how much would 32 cost? Work the problem using the direct variation method.

2. Blues vary inversely as yellows squared. If 50 blues go with 5 yellows, how many blues go with 10 yellows? Use the equal ratio method to work the problem.

3. The ratio of acrobats in blue to acrobats in pink was 4 to 5. Those who wore blue numbered 1200 fewer than twice the number who wore pink. How many wore blue, and how many wore pink?

4. The place was crowded because $3\frac{1}{4}$ times as many people had come as the fire marshal would permit. If 650 had come, how many would the fire marshal permit?

5. Dogs were \$40 each, and cats cost only \$2 each. If Doerun bought 30 animals and paid a total of \$820, how many animals of each kind did he buy?

Solve:

6. $\sqrt{x^2 - x - 2} - x + 2 = 0$

7. $\sqrt{p + 20} + \sqrt{p} = 10$

8. $\sqrt{s - 18} + \sqrt{s - 36} = 0$

9. $\begin{cases} x + y + z = 8 \\ 2x - 3y - z = -6 \\ 2x - z = 0 \end{cases}$

$z = -2x$

10. Add: $4\underline{/60°} - 6\underline{/-200°}$

11. Write $-2R + 6U$ in polar form.

12. Add: $\dfrac{4x + 2}{x^2 - 6x - 16} - \dfrac{3}{x - 8}$

Simplify:

13. $\dfrac{3\sqrt{2} - 1}{1 + \sqrt{2}}$

14. $\dfrac{2 - 3\sqrt{2}}{3 - 2\sqrt{2}}$

15. Find B.

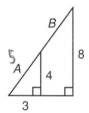

16. Solve $-3x^2 + 2 = -3x$ by completing the square.

17. Find y: $a = xm\left(\dfrac{p}{y} + \dfrac{q}{c}\right)$

18. Simplify: $3a - \dfrac{3}{a - \dfrac{3}{a}}$

19. Use similar triangles to find:

(a) m and n

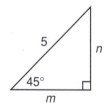

(b) c and d

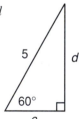

20. $-\sqrt{-4} + \sqrt{-9} - i^3 + \sqrt{-2}\sqrt{-2} - 4i^4$

21. Solve: $\begin{cases} \dfrac{2}{3}x - \dfrac{1}{4}y = 6 + 2 \\ \qquad\qquad\quad 8 \\ 0.07x + 0.06y = 1.32 \end{cases}$

22. Use the quadratic formula to solve $-x^2 = -x - 5$.

23. Use unit multipliers to convert 600 cubic centimeters per minute to cubic feet per hour.

24. The figure on the left is the base of a right solid that is 6 ft high. How many 1-inch sugar cubes will the solid hold? Dimensions are in inches.

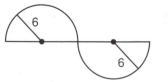
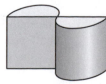

Simplify:

25. $\sqrt[4]{x^5 y} \, \sqrt{xy^3}$

26. $\dfrac{-2.^0(-2)^0}{-(4)^{-3/2}}$

27. Solve by graphing and then use either substitution or elimination to get an exact solution:

$$\begin{cases} x - 3y = 6 \\ 2x + y = -1 \end{cases}$$

28. Add: $\dfrac{3}{ax} + \dfrac{3x}{a^2 x} + \dfrac{7x}{x + a}$

29. Solve: $x^3 = 4x^2 + 32x$

30. Find the equation of the line that passes through $(-2, -3)$ and is perpendicular to $-x - y - 1 = 0$.

LESSON 81   *Complex numbers*

81.A
complex numbers and real numbers

A complex number is a number that has a real part and an imaginary part. When the real part is written first, we say that we have written the complex number in **standard form.** Thus, the general expression for a complex number in standard form is

$$a + bi$$

The letter a can be any real number, and the letter b can be any real number. All of these numbers

(a) $-\sqrt{2} + 3i$ (b) $-\dfrac{4\sqrt{3}}{5} + 2\sqrt{3}\, i$ (c) $3 - \dfrac{23}{\sqrt{2}} i$

are complex numbers in standard form because all the replacements for a and b are real numbers. If a equals zero, the number does not have a real part. Thus, the following numbers are complex numbers whose real parts equal zero.

(d) $+3i$ (e) $+2\sqrt{3}i$ (f) $-\dfrac{23}{\sqrt{2}} i$

If the coefficient of the imaginary part of a complex number is zero, we get a complex number that has only a real part, such as the following:

(g) $-\sqrt{2}$ (h) $-\dfrac{4\sqrt{3}}{2}$ (i) 3

Thus, we see that every real number is a complex number whose imaginary part is zero, and every imaginary number is a complex number whose real part is zero. **Thus, the set of real numbers is a subset of the set of complex numbers, and the set of imaginary numbers is also a subset of the set of complex numbers.**

The complex number

$$\dfrac{4 - 3i}{5}$$

is not in standard form because it is not in the form $a + bi$. However, it takes only a slight change to write it in standard form as

$$\dfrac{4}{5} - \dfrac{3}{5} i$$

81.B
products of complex conjugates

We have noted that the product of a two-part number and its conjugate has the form $a^2 - b^2$.

$$
\begin{array}{r}
a + b \\
a - b \\
\hline
a^2 + ab \\
- ab - b^2 \\
\hline
a^2 \quad - b^2
\end{array}
\qquad
\begin{array}{r}
3 + 5\sqrt{2} \\
3 - 5\sqrt{2} \\
\hline
9 + 15\sqrt{2} \\
- 15\sqrt{2} - 50 \\
\hline
9 \qquad - 50
\end{array}
$$

If we multiply a complex number in standard form by its conjugate, we get an answer in the form $a^2 + b^2$. The sign change between a^2 and b^2 is caused by the presence of an i^2 factor in the second part of the product.

$$
\begin{array}{r}
a + bi \\
a - bi \\
\hline
a^2 + abi \\
- abi - b^2 i^2 \\
\hline
a^2 \qquad + b^2
\end{array}
\qquad
\begin{array}{r}
3 + 4i \\
3 - 4i \\
\hline
9 + 12i \\
- 12i - 16i^2 \\
\hline
9 \qquad + 16
\end{array}
$$

We note that neither product has an imaginary part. We will find that we can use this fact to eliminate the i factor in the denominator of a fraction of complex numbers.

81.C
division of complex numbers

The notation

$$
\frac{x^2 - 2x + 7}{x + 3}
$$

indicates that $x^2 - 2x + 7$ is to be divided by $x + 3$. There is a format and a procedure we can use to perform this division.

$$
\begin{array}{r}
x - 5 \\
x + 3 \overline{) x^2 - 2x + 7} \\
\underline{x^2 + 3x} \\
-5x + 7 \\
\underline{-5x - 15} \\
22
\end{array}
$$

By using this procedure, we find that $\dfrac{x^2 - 2x + 7}{x + 3}$ equals $x - 5 + \dfrac{22}{x + 3}$.

If we encounter the expression

$$
\frac{2 + 3i}{4 - 2i}
$$

we see that this notation indicates that $2 + 3i$ is to be divided by $4 - 2i$. **Unfortunately, there is no simple format or procedure that we can use to perform this division. However, we can change the form of the expression by multiplying above and below by the conjugate of the denominator. The resulting expression will have a denominator that is a real number.**

example 81.1   Simplify: $\dfrac{2 + 3i}{4 - 2i}$

solution   We can change the denominator to a rational number if we multiply above and below by $4 + 2i$, which is the **conjugate of the denominator.**

$$
\frac{2 + 3i}{4 - 2i} \cdot \frac{4 + 2i}{4 + 2i}
$$

We have two multiplications indicated, one above and one below. We will use the vertical format for each multiplication.

ABOVE	BELOW
$2 + 3i$	$4 - 2i$
$4 + 2i$	$4 + 2i$
$\overline{8 + 12i}$	$\overline{16 - 8i}$
$\quad + 4i + 6i^2$	$\quad + 8i - 4i^2$
$\overline{8 + 16i - 6} = 2 + 16i$	$\overline{16 \quad + 4} = 20$

Thus, we can write our answer as

$$\frac{2 + 16i}{20} = \frac{1 + 8i}{10}$$

This answer is not in the preferred form of $a + bi$. We can write this complex number in standard form if we write

$$\frac{1}{10} + \frac{4}{5}i$$

example 81.2   Simplify: $\dfrac{4 - 2i}{2i - 3}$

solution   Although it is not necessary, we will begin by writing the denominator in standard form as

$$\frac{4 - 2i}{-3 + 2i}$$

We can change the denominator to a real number if we multiply above and below by $-3 - 2i$.

$$\frac{4 - 2i}{-3 + 2i} \cdot \frac{-3 - 2i}{-3 - 2i}$$

We have two multiplications to perform.

ABOVE	BELOW
$4 - 2i$	$-3 + 2i$
$-3 - 2i$	$-3 - 2i$
$\overline{-12 + 6i}$	$\overline{9 - 6i}$
$\quad - 8i + 4i^2$	$\quad + 6i - 4i^2$
$\overline{-12 - 2i - 4} = -16 - 2i$	$\overline{9 \quad + 4} = 13$

Thus, the new form of the expression is

$$\frac{-16 - 2i}{13}$$

which can be written in standard form as follows

$$-\frac{16}{13} - \frac{2}{13}i$$

practice   Simplify:

a. $\dfrac{2 + 3i}{3 - 3i}$ b. $\dfrac{2 - 2i}{2i - 2}$

**problem set
81**

1. Monkeys varied directly as turtles squared. When there were 2 turtles, there were 100 monkeys. How many monkeys were there when there were 5 turtles? Work the problem once using the direct variation method and again using the equal ratio method.

2. The number of macaws varied inversely as the number of apes squared. When there were 4 macaws, there were 10 apes. How many macaws were there when there were only 2 apes?

3. Roger made the 375-mile trip in 10 hours less than it took Judy. This was because he traveled 3 times as fast as Judy traveled. How fast did each travel, and for how long did each travel?

4. Wittlocoodee had one solution that was 10% glycol and another that was 40% glycol. How much of each should she use to get 200 liters of solution that was 19% glycol?

5. The curmudgeon chortled with glee when the results were announced because only 60% had made it. If 1120 had not made it, how many had tried?

Simplify:

6. $\dfrac{3 - i}{2 + 5i}$

7. $\dfrac{3 - 2i}{2i - 4}$

8. Find an angle whose supplement is 30° greater than 4 times its complement.

Solve:

9. $\sqrt{x^2 - x + 30} - 3 = x$

10. $\sqrt{p - 48} = 12 - \sqrt{p}$

11. $\begin{cases} x + 2y - 3z = 5 \\ 2x - y - z = 0 \\ y - 3z = 0 \end{cases}$

12. The two forces act on a point as shown. Find the resultant force.

13. Add: $\dfrac{7x - 2}{x^2 - 9} + \dfrac{3x}{3 - x}$

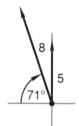

Simplify:

14. $\dfrac{\sqrt{2} - 5}{\sqrt{2} - 2}$

15. $\dfrac{2\sqrt{3} - 1}{1 - 3\sqrt{3}}$

16. $\dfrac{1 + \sqrt{2}}{3 - \sqrt{2}}$

17. Find x and y.

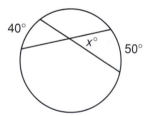
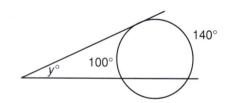

18. Begin with $ax^2 + bx + c$ and derive the quadratic formula by completing the square.

Simplify:

19. $2a^2 - \dfrac{3a}{a + \dfrac{1}{a}}$

20. $\sqrt{-9} - \sqrt{-2}\sqrt{-2} + \sqrt{-2}\sqrt{2} - 3i^3 - 2i^2$

21. Solve: $\begin{cases} \dfrac{2}{3}x - \dfrac{1}{3}y = 6 \\ 0.15x + 0.01y = 0.84 \end{cases}$

22. Solve $2 = -2x^2 - 3x$ by using the quadratic formula.

23. Divide $4x^3 - x + 2$ by $x - 4$.

24. Find x and y.

Simplify:

25. $3\sqrt{9\sqrt[4]{3}}$

26. $2\sqrt{\dfrac{1}{5}} - 3\sqrt{5} + 3\sqrt{20}$

27. Add: $\dfrac{x}{x + y} + \dfrac{3}{x^2 y} + \dfrac{2}{xy}$

28. Solve: $-4^2 - 3^0 - 2^0(x - x^0) - 3^0(-2x - 5) = 7$

29. Find the distance between $(-2, 8)$ and $(5, -3)$.

30. Use a calculator to simplify. Estimate first.

(a) $\dfrac{0.5061 \times 10^5}{0.0071643 \times 10^{-18}}$

(b) $\sqrt[6.2]{594}$

LESSON 82 *Algebraic simplifications*

There are three rules of algebra that stand above all the rest. Two of the rules cannot be used unless an equals sign is present, for these rules apply only to equations. We state them informally by saying the following.

1. The same quantity can be added to both sides of an equation without changing the answer(s) to the equation.
2. Every term on both sides of an equation can be multiplied by the same quantity (except zero) without changing the answer(s) to the equation.

The third rule can be used on the individual terms of an equation and can also be used on expressions that are not terms of an equation. With few exceptions, if no equals sign is present, this rule is the only rule that can be used.

3. The denominator and the numerator of an algebraic expression can be multiplied by the same quantity (except zero) without changing the value of the expression. We call this rule the denominator-numerator same-quantity rule.

In Lesson 64, we found that an expression such as

$$a + \dfrac{1}{1 + \dfrac{1}{x}}$$

can be written as a simple fraction by using the denominator-numerator same-quantity rule several times. We review this procedure by first simplifying $1 + \dfrac{1}{x}$, and we get

$$a + \dfrac{1}{\dfrac{x + 1}{x}}$$

Next we use the same rule again to simplify the second term and get

$$a + \dfrac{x}{x + 1}$$

We finish by using the rule again to change the form of the first term, and then we add the two terms.

$$\dfrac{a(x + 1)}{x + 1} + \dfrac{x}{x + 1} = \dfrac{ax + a + x}{x + 1}$$

In this lesson we will discuss the repeated use of the denominator-numerator same-quantity rule to simplify expressions that are just a little more complicated.

example 82.1   Simplify: $\dfrac{a}{1 + \dfrac{a}{1 + \dfrac{1}{x}}}$

solution   We begin by adding $1 + \frac{1}{x}$, and we get

$$\dfrac{a}{1 + \dfrac{a}{\dfrac{x + 1}{x}}}$$

Next, we simplify the second term of the denominator and get

$$\dfrac{a}{1 + \dfrac{xa}{x + 1}}$$

Now we add the two terms in the denominator and get

$$\dfrac{a}{\dfrac{x + 1 + xa}{x + 1}}$$

We finish by multiplying above and below by the reciprocal of the denominator.

$$\dfrac{a(x + 1)}{x + 1 + xa}$$

example 82.2   Simplify: $\dfrac{b}{a + \dfrac{c}{x + \dfrac{1}{y}}}$

solution   We will use the same procedure we used in the preceding example. We begin by adding $x + \frac{1}{y}$ and get

$$\dfrac{b}{a + \dfrac{c}{\dfrac{xy + 1}{y}}}$$

Next, we simplify the second term in the denominator and get

$$\dfrac{b}{a + \dfrac{cy}{xy + 1}}$$

Now we add the two terms in the denominator and get

$$\dfrac{b}{\dfrac{axy + a + cy}{xy + 1}}$$

We finish by multiplying above and below by the reciprocal of the denominator.

$$\dfrac{b(xy + 1)}{axy + a + cy}$$

practice   Simplify:

a. $\dfrac{1}{3 + \dfrac{x}{3 + \dfrac{3}{a}}}$

b. $\dfrac{m}{x + \dfrac{p}{q + \dfrac{1}{z}}}$

problem set 82

1.   The discipline quotient varied inversely as the square of the number of unruly students. If the discipline quotient was 300 when the number of unruly students totaled 5, what was the discipline quotient when the number of unruly students totaled 10? Use the equal ratio method to solve the problem.

2.   The plane could fly 1920 miles in 4 more hours than it took the racer to drive 320 miles. The speed of the plane was 3 times the speed of the racer. Find the times of both and the speeds of both.

3.   The initial pressure, temperature, and volume of a quantity of an ideal gas were 5 atmospheres, 540 K, and 250 cubic meters, respectively. What would the temperature be if the pressure were increased to 50 atmospheres and the volume reduced to 200 cubic meters?

4.   After running for a while at 10 mph, Mingo slowed to a walk at 5 mph. If he traveled 60 miles in 8 hours, how far did he run and how far did he walk?

5.   The total weight of the sodium monohydrogen phosphate, Na_2HPO_4, was 852 grams. What was the weight of the sodium (Na) in this amount of the compound? What percent by weight of the compound was sodium? (Na, 23; H, 1; P, 31; O, 16)

Simplify:

6. $\dfrac{m}{2 + \dfrac{m}{2 + \dfrac{2}{p}}}$

7. $\dfrac{p}{a + \dfrac{b}{m + \dfrac{3}{y}}}$

8.   A circular cone whose radius is 3 meters has a volume of $12\pi \text{ m}^3$. Find the height of the cone.

Simplify:

9. $\dfrac{2 - 3i}{4 + i}$

10. $\dfrac{5 - i}{2 - 3i}$

Solve:

11. $\sqrt{x^2 - x + 47} - 5 = x$

12. $\sqrt{x + 24} + \sqrt{x} = 12$

13. $\begin{cases} 2x - 2y - z = 16 \\ 3x - y + 2z = 5 \\ -y + 3z = 0 \end{cases}$

14.   Add: $5\underline{/70°} - 30\underline{/-20°}$

15.   Write $-4R + 7U$ in polar form.

16.   Add: $\dfrac{3x - 2}{x^2 - 9} + \dfrac{x}{3 - x}$

Simplify:

17. $\dfrac{3\sqrt{2} - 1}{1 + \sqrt{2}}$

18. $\dfrac{-2 - \sqrt{5}}{2 + 2\sqrt{5}}$

19.   Find B.

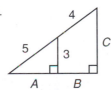

20. Find a: $\quad \dfrac{z}{m^2} = \dfrac{p}{m}\left(\dfrac{x}{a} + y\right)$ **21.** Find p: $\quad \dfrac{z}{m^2} = \dfrac{p}{m}\left(\dfrac{x}{a} + y\right)$

22. The diagonal of a square is 6 meters long. How long is a side of the square? What is the area of the square?

23. Solve: $\begin{cases} \dfrac{2}{5}x - \dfrac{1}{3}y = -1 \\ 0.07x + 0.2y = 2.15 \end{cases}$

24. Begin with $ax^2 + bx + c = 0$ and complete the square to derive the quadratic formula.

25. Solve $-3x^2 - 2 = 5x$ by using the quadratic formula.

Simplify:

26. $\sqrt{x^5 y}\ \sqrt[3]{x^2 y^5}$ **27.** $3\sqrt{\dfrac{2}{9}} + 3\sqrt{\dfrac{9}{2}} - 4\sqrt{50}$ **28.** $\dfrac{-2^{0}}{-4^{-5/2}}$

29. Solve by graphing and then get an exact solution by using either substitution or elimination:

$$\begin{cases} 4x - 3y = -3 \\ 4x + 3y = 6 \end{cases}$$

30. Use unit multipliers to convert 400 cubic centimeters per second to cubic feet per minute.

LESSON 83 *Variable exponents*

83.A
product rule with variables

The two major rules for exponents are the product rule and the power rule. The product rule tells us that to multiply exponential expressions whose bases are equal, we add the exponents.

$$x^2 \cdot x^5 = x^7$$

We can see the reason for this rule if we break down the factors and get

$$xx \cdot xxxxx$$

for there are seven x factors, and we can indicate seven x factors by writing

$$x^7$$

Since variables represent unspecified numbers, we use the same rule when the exponents are variables. If the bases are the same, the exponential expressions are multiplied by adding the exponents. Thus,

$$x^a \cdot x^b = x^{a+b}$$

example 83.1 Simplify: $\quad x^a y^{2x} x^{a/2} y^{3x/4}$

solution We simplify by adding the exponents of like bases.

$$x^{a+a/2} y^{2x+3x/4}$$

Now since

$$a + \frac{a}{2} = \frac{3a}{2} \qquad \text{and} \qquad 2x + \frac{3x}{4} = \frac{11x}{4}$$

we finish by writing

$$x^{3a/2}y^{11x/4}$$

example 83.2 Simplify: $\dfrac{x^{a-2}\,y^{a+4}}{x^{a/2}\,y^{2a}}$

solution We begin by writing all exponential expressions in the numerator.

$$x^{a-2}x^{-a/2}y^{a+4}y^{-2a}$$

Now we finish by adding the exponents of x and the exponents of y and get

$$x^{a/2-2}y^{-a+4}$$

83.B
power rule with variables

The power rule for exponents tells us that when we have a notation such as

$$(x^3)^2$$

we simplify by multiplying the exponents and get

$$x^6$$

We can see the reason for this rule if we remember that if a quantity is squared we multiply the quantity by itself. Thus,

$$(x^3)^2 \qquad \text{means} \qquad (x^3)(x^3) \qquad \text{which is } x^6$$

The rule is the same when the exponents contain variables.

example 83.3 Simplify: $(x^{a+2})^2$

solution We multiply $a + 2$ by 2 and get

$$x^{2a+4}$$

example 83.4 Simplify: $\dfrac{(x^a)^b(y^a)^{b+2}}{x^{-a}}$

solution First we use the power rule twice in the numerator and get

$$\dfrac{x^{ab}\,y^{ab+2a}}{x^{-a}}$$

We finish by bringing up the x^{-a} from below, and we get

$$x^{ab+a}y^{ab+2a}$$

practice Simplify:

a. $x^m a^{3x} x^{m/5} a^{x/2}$

b. $\dfrac{m^{a-2}\,p^{a+1}}{m^{a/3}\,p^{2a}}$

c. $\dfrac{(m^x)^z(z)^{m+3}}{m^{-x}}$

problem set 83

1. The number who were resentful varied directly as the number of invidious comparisons that were made. When 1200 were resentful, 300 invidious comparisons had been made. When 100 invidious comparisons were made, how many were resentful? Work once using the equal ratio method and again using the direct variation method.

2. It took Raul twice as long to drive the 420 kilometers to Merida as it took to drive the 270 kilometers to Tulum. The disparity occurred because Raul drove 20 kilometers per hour faster when he went to Tulum. What were his times and rates to both destinations?

3. Americus won the race with Ashburn by 400 yards. If Americus' speed was 5 yards per second and her time was 400 seconds, how fast did Ashburn run?

4. Millsap observed that 4 times the number of igneous rocks exceeded 8 times the number of sedimentary rocks by 80. He also noted that 10 times the number of sedimentary rocks exceeded the number of igneous rocks by 140. How many rocks were igneous, and how many were sedimentary?

5. Lynn and Laws knew that the confection should be exactly 8% sugar. It was to be made using one component that was 5% sugar and another that was 20% sugar. How much of each should they use to get 800 ml of the confection?

6. Find x and y.

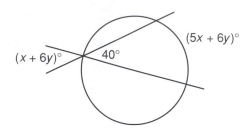

Simplify:

7. $\dfrac{a^{x-3}\,m^{x+2}}{a^{x/3}\,m^{3x}}$

8. $\dfrac{(a^x)^m\,(b^x)^{m+3}}{a^{-m}}$

9. $\dfrac{a}{x + \dfrac{1}{a + \dfrac{1}{x}}}$

10. $\dfrac{b}{a + \dfrac{b}{a + \dfrac{a}{b}}}$

11. $\dfrac{2 - 4i}{1 + i}$

12. $\dfrac{3 + 5i}{2 - 2i}$

Solve:

13. $\begin{cases} 2x + 3y - z = -3 \\ x + 2y = 0 \quad X=-2y \\ x - 2y + z = -2 \end{cases}$

14. $\sqrt{k} + \sqrt{k + 32} = 8$

15. Two forces act on a point as shown. Find the resultant force.

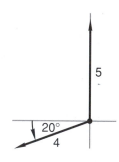

16. Add: $\dfrac{5}{x} + \dfrac{6}{x^2 - 4} - \dfrac{3x}{2 - x}$

Simplify:

17. $\dfrac{-2\sqrt{2} - 2}{4 + \sqrt{2}}$

18. $\dfrac{-\sqrt{3} - 3}{2 - \sqrt{3}}$

19. Find m: $p = \dfrac{a}{x} - c\left(\dfrac{a}{m} - y\right)$

20. Find c: $p = \dfrac{a}{x} - c\left(\dfrac{a}{m} - y\right)$

21. Use similar triangles to find x and y.

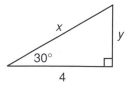

22. Use similar triangles to find m and p.

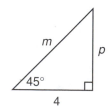

23. Solve: $\begin{cases} x - \dfrac{2}{5}y = 11 \\ -0.05x - 0.2y = 1.65 \end{cases}$

24. Solve $2x^2 - 5x = 5$ by completing the square.

Simplify:

25. $\sqrt{\dfrac{7}{4}} + 2\sqrt{\dfrac{4}{7}} - 5\sqrt{63}$ **26.** $\dfrac{-2^0(-3^0)}{-27^{-2/3}}$ **27.** $\left(\sqrt[3]{x^2\,y}\right)^4$

28. Use unit multipliers to convert 4 liters per second to cubic centimeters per hour.

29. Find the distance between $(-3, 2)$ and $(5, -7)$.

30. Find the equation of the line that passes through $(-3, 2)$ and $(5, -7)$.

LESSON 84 *Solutions of equations*

The degree of a term of a polynomial is the sum of the exponents of the variables of the term. Thus

$$x^3 \qquad \text{is a third-degree term}$$

$$x^2ymp \qquad \text{is a fifth-degree term}$$

$$47x \qquad \text{is a first-degree term}$$

The degree of a polynomial equation is the same as the degree of the highest-degree term of the equation. Thus,

$$3x^2 + 2x + 5 = 0 \qquad \text{is a second-degree equation}$$

$$3x^3 + 2x^2 + 3 = 0 \qquad \text{is a third-degree equation}$$

$$xy + 5 = 0 \qquad \text{is a second-degree equation}$$

$$y = 3x + 2 \qquad \text{is a first-degree equation}$$

Thus far, we have restricted our equation-solving efforts to first- and second-degree polynomial equations. First-degree equations in two or more unknowns are called **linear equations** because the graph of a first-degree equation in two unknowns is a straight line. The equation

$$3y + 2x = 4$$

is a first-degree polynomial equation in two unknowns, and thus the graph of this equation is a straight line. There is an infinite number of ordered pairs of x and y that will satisfy this equation, and the graph of these points is the line.

When we have a system of two linear equations in two unknowns, there are three possibilities. The first is that the equations are equations of two lines that intersect, as shown in (a). Each of these lines has an infinite number of ordered pairs of x and y that satisfy its equation, but only one ordered pair, $(-1, 1)$, satisfies both equations and thus lies on both lines. A system of linear equations whose graphs intersect in one point is called a **consistent system** of equations.

In (b) we have graphed both equations and find they have the same graph.

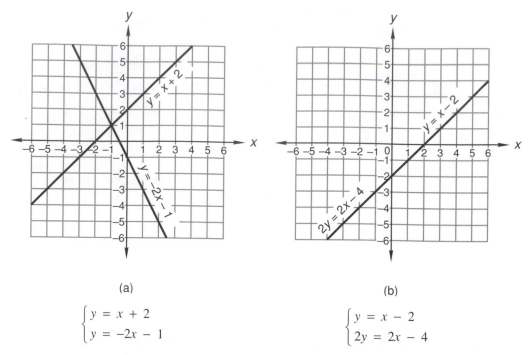

(a)

$$\begin{cases} y = x + 2 \\ y = -2x - 1 \end{cases}$$

(b)

$$\begin{cases} y = x - 2 \\ 2y = 2x - 4 \end{cases}$$

Thus, the coordinates of any point that satisfy one of these equations satisfy the other equation. As we see, there is an infinite number of points whose coordinates satisfy both equations. We say that this system of equations is a **dependent system** of equations.

The third possibility is that the lines have the same slopes but different intercepts and thus are parallel lines. Both of the lines in (c) have a slope of 2 and thus are parallel. These lines never cross and have no point in common. Systems of linear equations that have no solutions are called **inconsistent systems** of equations.

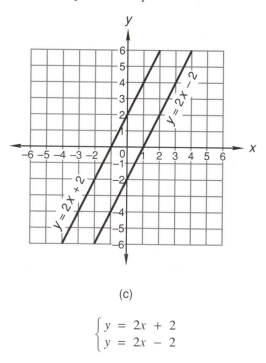

(c)

$$\begin{cases} y = 2x + 2 \\ y = 2x - 2 \end{cases}$$

The words "consistent," "inconsistent," and "dependent" are not important. The important thing to remember is that given a system of two linear equations in two variables there are three possibilities.

1. The lines cross and thus have one point in common.
2. The equations are two forms of the equation of the same line.
3. The lines are parallel and thus have no point(s) in common.

If a system has one solution, the lines will cross and the system can be solved by graphing or by using either the substitution method or the elimination method. If the lines do not cross, an attempt to find a solution by using either substitution or elimination must result in failure because there is no solution.

An attempt to solve a dependent system will cause the elimination of all variables, and we will end up with a true numerical statement such as

$$0 = 0 \quad \text{or} \quad 5 = 5$$

We will demonstrate by trying to solve system (b) by using both substitution and elimination.

SUBSTITUTION   ELIMINATION

$$2(x - 2) = 2x - 4$$
$$\underline{2x - 4 = 2x - 4}$$
$$0 = 0 \quad \text{True}$$

$$y = x - 2 \quad \rightarrow \quad (2) \quad \rightarrow \quad 2y = 2x - 4$$
$$2y = 2x - 4 \quad \rightarrow \quad (-1) \quad \rightarrow \quad \underline{-2y = -2x + 4}$$
$$0 = 0 \quad \text{True}$$

A system of two equations whose graphs are parallel has no solution, so an attempt to solve one of these systems will also result in failure. Again, the variables will be eliminated, but this time the final result will be a false numerical statement such as

$$0 = 5 \quad \text{or} \quad -3 = 7$$

We will demonstrate by trying to find a solution for system (c).

SUBSTITUTION   ELIMINATION

$$2x - 2 = 2x + 2$$
$$\underline{-2x \quad\quad\quad -2x}$$
$$-2 = +2 \quad \text{False}$$

$$y = 2x + 2 \quad \rightarrow \quad (-1) \quad \rightarrow \quad -y = -2x - 2$$
$$y = 2x - 2 \quad \rightarrow \quad (1) \quad \rightarrow \quad \underline{y = 2x - 2}$$
$$0 = -4 \quad \text{False}$$

The graph of a system of three linear equations in three unknowns is a graph of three planes. If the three planes intersect in a common point as in the following figure, an **ordered triple** (x, y, z) exists that will satisfy all three equations. We say that the equations are **independent equations.** We also say that the equations are **consistent equations.** If no ordered triple will satisfy all three equations, we say that the equations are not independent or we say they are not consistent.

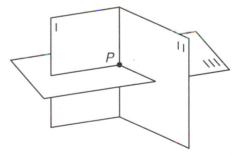

The equations of the three parallel planes of the figure on the left cannot be satisfied by one ordered triple (x, y, z), for there is no point that is common to all three planes. Thus, these equations are inconsistent equations. Some equations are neither consistent nor inconsistent. The equations of the three planes of the figure on the right are neither consistent nor inconsistent equations. These planes intersect to form a line, and thus there is an infinite number of ordered triples (x, y, z) that satisfies all three equations.

practice   Is the following system of equations consistent, inconsistent, or dependent?

$$\begin{cases} 3x - y = -2 \\ 2y = 6x + 5 \end{cases}$$

**problem set
84**

1. The number who were admired varied inversely with the number who had a proclivity for bragging. When 80 bragged, only 20 were admired. How many were admired when 10 bragged? Work once using the equal ratio method and again using the direct variation method.

2. **The temperature in the gas law equations must be in kelvins, and a temperature in kelvins is a number 273 greater than the number that designates the same temperature in degrees Celsius. Thus, 50 degrees Celsius (50°C) equals 273 + 50 = 323 kelvins.** If in an experiment the volume was held constant and the initial pressure and temperature were 4 atmospheres and 50°C, what would be the final temperature in degrees Celsius if the pressure were increased to 8 atmospheres?

3. When the compound was analyzed, its chemical formula was determined to be $Li_2Ca_2O_7$. If the weight of the lithium (Li) was 56 grams, what was the total weight of the compound? (Li, 7; Ca, 40; O, 16) What percent by weight of the compound was lithium?

4. The chemist pondered. She had 160 ml of a solution that was 10% glycol. The other solution available was 30% glycol. How much of the other solution should she use to get a solution that was 22% glycol?

5. It really took 150 percent more borax than had been agreed on. If the agreement was for 140,000 tons of borax, how many tons did it really take?

Simplify:

6. $y^b x^{c+2} y^{b/3} x^{-2-p}$

7. $\dfrac{y^b x^{c+3}}{y^{b/3} x^{2-p}}$

8. $(x^{a+3})^b x^{-2b+4}$

9. $\dfrac{(y^a)^{b+2} y^{-ab}}{y^{-2+a}}$

10. $\dfrac{p}{a + \dfrac{m}{1 + \dfrac{1}{am}}}$

11. $\dfrac{1}{p - \dfrac{b}{b - \dfrac{1}{x}}}$

12. $\dfrac{2 - 3i}{1 + i}$

13. $\dfrac{3 + 4i}{3 - 3i}$

Solve:

14. $\sqrt{s - 48} + \sqrt{s} = 8$

15. $\begin{cases} 3x - y - 2z = -6 \\ 2x - y + z = 2 \\ -y + z = 0 \end{cases}$

16. Add: $-6\underline{/-150°} + 4\underline{/20°}$

17. Write $3R + 8U$ in polar form.

18. Add: $\dfrac{4}{x^2} - \dfrac{2}{-x(x - 3)} + \dfrac{5x}{3 - x}$

19. Simplify: $\dfrac{4\sqrt{2} - 5}{3\sqrt{2} + 2}$

20. Multiply: $(2 + 3\sqrt{20})(4 - 5\sqrt{45})$

21. Find c: $\dfrac{m}{c} + x = p\left(\dfrac{1}{x} + \dfrac{b}{y}\right)$

22. Simplify: $\sqrt{-3}\sqrt{3} - \sqrt{2}\sqrt{2} - \sqrt{-4} + 3i^2 - 2i^5$

23. Solve $-2x^2 - x = 3$ by using the quadratic formula.

24. A circle is inscribed in a square whose area is 36 m². What is the length of a side of the square? What is the radius of the circle? What is the area of the circle?

25. Estimate the location of the line indicated by the data points and then write the equation that expresses yttrium as a function of boron: $Y = mB + b$

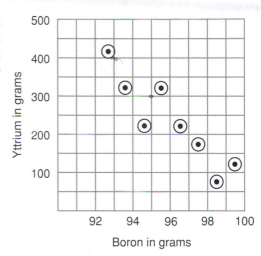

26. Use a calculator to simplify. Estimate first.

   (a) $\dfrac{-2.065 \times 10^4}{-500 \times 10^6}$

   (b) $(84.9)^{-4.91}$

27. Simplify: $3\sqrt{\dfrac{5}{3}} + 2\sqrt{\dfrac{3}{5}} - 4\sqrt{60}$

28. Use unit multipliers to convert 40 centimeters per second to miles per hour.

29. Solve: $-2^0[-2 - 3(x - 2^2)] - 3(2x - 5^0) = 7 - 2(2x + 2)$

30. Simplify: $-3\{[(-2 - 3) - 2] - 2[-4(-3 - 2^0)]\} + 2^0(-3)$

LESSON 85   *Systems of nonlinear equations*

We know that the graph of a first-degree polynomial equation in two unknowns is a straight line. Each of the variables in the equation

$$y = 2x - 4$$

has an understood exponent of 1, so this equation is a first-degree equation. Also, the equation has two unknowns, and thus the equation has a straight line as its graph. The equation

$$3x - 2y + z = 17$$

is also a first-degree polynomial equation, but it has three unknowns. The graph of this equation is not a line but a plane because the presence of three unknowns requires that we use three dimensions to picture it. Nonetheless, we still call this equation a linear equation, for we give this name to any first-degree polynomial equation that has two or more unknowns. This is the reason that the equations

$$4x + 2y - 3z - 6k = 14$$

and $$3x + 7y - 2z - 6p + 3m - n = 21$$

can be called linear equations even though each has more than three unknowns.

Equations that have variables at least one of whose exponents is not 1, such as

$$x^2 + y = 5 \quad \text{or} \quad x^2 + y^2 = 7$$

and equations in which some terms are products of variables, such as

$$xy = 4 \quad \text{and} \quad 4x - xy = 7$$

have graphs that are not straight lines, and thus we call these equations **nonlinear equations.**
The most important nonlinear figures are called **conic sections** because they can be formed by
a plane cutting a single right circular cone or a double right circular cone. Four important
conic sections are the **circle**, the **ellipse**, the **parabola**, and the **hyperbola.** The following
figures show how a plane can cut a right circular cone to form conic sections, and two typical
equations for each section are given.

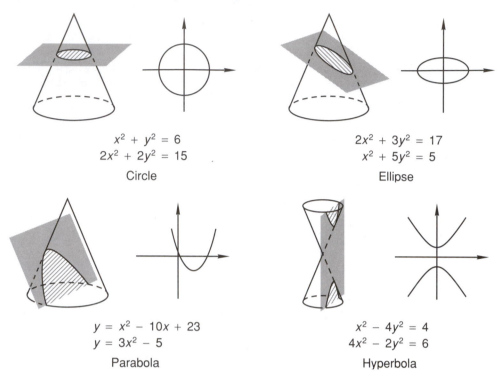

$$x^2 + y^2 = 6$$
$$2x^2 + 2y^2 = 15$$
Circle

$$2x^2 + 3y^2 = 17$$
$$x^2 + 5y^2 = 5$$
Ellipse

$$y = x^2 - 10x + 23$$
$$y = 3x^2 - 5$$
Parabola

$$x^2 - 4y^2 = 4$$
$$4x^2 - 2y^2 = 6$$
Hyperbola

It is interesting to note that if the plane is tangent to the cone, the set of points where the plane
and the cone touch would form a straight line; for this reason, the straight line can also be
considered to be a conic section.

   **A nonlinear system of equations is a system in which at least one of the equations
is a nonlinear equation.** These systems often have more than one solution because the graphs
intersect in more than one point.

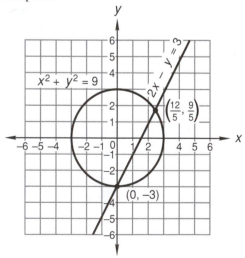

   In this figure, we show the two points of intersection of the graphs of a circle and a
straight line. The coordinates of either point will satisfy both the equation of the line and the
equation of the circle. The study of the graphs of conic sections is a topic for a more advanced
course, but the solutions of systems of nonlinear equations is a topic suitable for study at this
level.

To solve systems of nonlinear equations, we use the substitution method and the elimination method just as we have done to solve systems of linear equations. Some nonlinear systems lend themselves more readily to the use of one of these methods than to the other, as we will see. Also we will find that it is important to remember that these systems often have more than one solution. We will begin solving for the points of intersection of the circle and the line shown in the preceding figure.

example 85.1   Solve the system: $\begin{cases} x^2 + y^2 = 9 & \text{(a)} \\ 2x - y = 3 & \text{(b)} \end{cases}$

solution   We cannot use elimination because the variables in (a) are squared and the variables in (b) are not. Thus, we will solve equation (b) for y and then square both sides.

$$\text{(b)} \qquad 2x - 3 = y \qquad \text{solved for } y$$
$$4x^2 - 12x + 9 = y^2 \qquad \text{squared both sides}$$

Now we can use $4x^2 - 12x + 9$ instead of y^2 in equation (a).

$$x^2 + y^2 = 9 \qquad \text{equation (a)}$$
$$x^2 + (4x^2 - 12x + 9) = 9 \qquad \text{substituted}$$
$$5x^2 - 12x = 0 \qquad \text{simplified}$$
$$x(5x - 12) = 0 \qquad \text{factored}$$

$$\text{If } x = 0 \qquad\qquad \text{If } 5x - 12 = 0$$
$$x = 0 \qquad\qquad\qquad 5x = 12$$
$$x = \frac{12}{5}$$

We have found two values of x. We will use the linear equation (b) to find a value for y for both values of x.

$$2x - 3 = y \qquad \text{equation (b)} \qquad\qquad 2x - 3 = y \qquad \text{equation (b)}$$
$$2(0) - 3 = y \qquad \text{substituted} \qquad\qquad 2\left(\frac{12}{5}\right) - 3 = y \qquad \text{substituted}$$
$$-3 = y \qquad \text{solved} \qquad\qquad\qquad \frac{9}{5} = y \qquad \text{solved}$$

Thus, the ordered pair $(0, -3)$ is a solution, and $\left(\frac{12}{5}, \frac{9}{5}\right)$ is also a solution.

example 85.2   Solve the system: $\begin{cases} BT_D + 5T_D = 25 & \text{(a)} \\ BT_D - 5T_D = 15 & \text{(b)} \end{cases}$

solution   These same equations will be used in a later lesson to solve a problem about a boat in a river. B stands for the speed of the boat, and T_D stands for the time downstream. We will solve this system twice. The first time we add the equations just as they are and eliminate the isolated variable T_D.

$$\begin{array}{lll} \text{(a)} & BT_D + 5T_D = 25 & \\ \text{(b)} & \underline{BT_D - 5T_D = 15} & \\ & 2BT_D \qquad\quad = 40 & \text{added the equations} \\ \text{(c)} & BT_D \qquad\quad = 20 & \text{divided by 2} \end{array}$$

We do not yet have a solution, for BT_D has two variables. The next step is to use 20 as a replacement for BT_D in either equation (a) or equation (b). We decide to use equation (a), so we get

$$20 + 5T_D = 25 \qquad \text{equation (a)}$$
$$5T_D = 5 \qquad \text{added } -20$$
$$T_D = 1 \qquad \text{divided}$$

Now we can use 1 for T_D in either equation (a), (b), or (c) to solve for B. We decide to use equation (c).

$$BT_D = 20 \qquad \text{equation (c)}$$
$$B(1) = 20 \qquad \text{substituted}$$
$$\boldsymbol{B = 20} \qquad \text{solved}$$

We could use a less involved procedure if we notice that we can eliminate the double variable BT_D in the beginning if we multiply either equation (a) or equation (b) by -1. We choose to multiply (b) by -1.

$$
\begin{array}{lllll}
\text{(a)} & BT_D + 5T_D = 25 & \longrightarrow & \text{(1)} & \longrightarrow \\
\text{(b)} & BT_D - 5T_D = 15 & \longrightarrow & \text{(-1)} & \longrightarrow \\
\end{array}
$$

$$
\begin{array}{r}
BT_D + 5T_D = \ \ 25 \\
-BT_D + 5T_D = -15 \\
\hline
10T_D = \ \ 10 \\
\boldsymbol{T_D = 1}
\end{array}
$$

Now we can use 1 for T_D in either (a) or (b). We will use (a).

$$BT_D + 5T_D = 25 \qquad \text{equation (a)}$$
$$B(1) + 5(1) = 25 \qquad \text{substituted}$$
$$B + 5 = 25 \qquad \text{simplified}$$
$$\boldsymbol{B = 20} \qquad \text{solved}$$

practice Solve each system:

a. $\begin{cases} BT_D + 9T_D = 36 \\ BT_D - 9T_D = 18 \end{cases}$

b. $\begin{cases} x^2 + y^2 = 25 \\ 2x - y = 5 \end{cases}$

problem set 85

1. The number that were improvident varied directly as the number that were thoughtless. If 800 were improvident when 2400 were thoughtless, how many were improvident when only 9 were thoughtless? Work once using the equal ratio method and again using the direct variation method.

2. Pippin could cover 640 miles in twice the time it took le Bref to cover 280 miles. If Pippin's rate exceeded that of le Bref by 20 miles per hour, find the rates and times of both men.

3. More were hoydens than were demure. In fact, 5 times the number that were hoydens exceeded the number that were demure by 90. Also, 3 times the number of demure was only 10 greater than the number of hoydens. How many of each were there?

4. The final mixture had to be exactly 34% gravel. Two piles were available. One was 10% gravel and the other was 50% gravel. How much of each should be used to get 400 cubic feet of the desired composition?

5. Lothario looked for consecutive odd integers. He wanted three such that the product of the first and the third was 25 less than the product of 10 and the opposite of the second. What integers did he want?

Solve:

6. $\begin{cases} BT_D + 6T_D = 24 \\ BT_D - 6T_D = 12 \end{cases}$

7. $\begin{cases} x^2 + y^2 = 16 \\ 2x - y = 4 \end{cases}$

Simplify:

8. $\dfrac{a^x(b^{y-2})^x}{a^{2x}(b^{-2})^x}$

9. $\dfrac{a^x b^{x/3} b^{-2}}{a^{x/2}}$

10. $\dfrac{x}{a + \dfrac{b}{c + \dfrac{x}{m}}}$

11. $\dfrac{a}{2 + \dfrac{c}{c + \dfrac{b}{c}}}$

12. $\dfrac{2 - 2i}{3 - 5i}$

13. $\dfrac{4 - \sqrt{2}\,i}{3 + \sqrt{2}\,i}$

Solve:

14. $\sqrt{k - 32} + \sqrt{k} = 8$

15. $\begin{cases} x + 2y + 2z = 6 \\ 2x - y + 3z = 6 \\ y - z = 0 \end{cases}$

16. The two forces act on the point as shown. Find the resultant force.

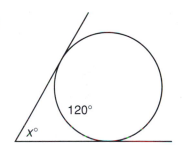

17. Add: $\dfrac{2x + 3}{x - a} - \dfrac{4}{a - x}$

Simplify:

18. $\dfrac{3 - 2\sqrt{2}}{5 - \sqrt{2}}$

19. $\dfrac{4 + \sqrt{3}}{2 - 2\sqrt{3}}$

20. Find p: $c = m\left(\dfrac{d}{c} - p\right)$

21. Simplify: $-\sqrt{-2}\sqrt{2} - 3i^3 + 2i + \sqrt{-2}\sqrt{-2} - \sqrt{-9}$

22. Find x.

(a)

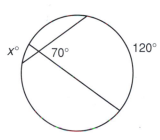

(b)

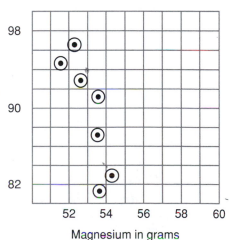

23. Solve: $\begin{cases} \dfrac{1}{9}x + \dfrac{1}{3}y = 3 \\ 0.3x - 0.04y = 2.46 \end{cases}$

24. Begin with $ax^2 + bx + c = 0$ and derive the quadratic formula by completing the square.

25. Estimate the location of the line indicated by the data points and then write the equation that expresses sodium (Na) as a function of magnesium (Mg): $Na = mMg + b$

26. Solve by graphing and then find an exact solution by using substitution or elimination:

$\begin{cases} x - 4y = -8 \\ 3x + y = 6 \end{cases}$

27. Write equal ratios and find x in terms of r, s, and t.

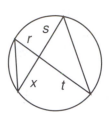

28. Use unit multipliers to convert 100 cubic feet per second to cubic centimeters per minute.

Simplify:

29. $3\sqrt{\dfrac{5}{12}} + 3\sqrt{\dfrac{12}{5}} + 3\sqrt{240}$

30. $\dfrac{-2^{0}}{-4^{-5/2}}$

LESSON 86   *Greater than • Trichotomy and transitive axioms • Irrational roots*

86.A

greater than

One number is said to be *greater than* another number if its graph on the number line **lies to the right of the graph of the other number.** Thus, we see that -1 is greater than -4 because the graph of -1 lies to the right of the graph of -4.

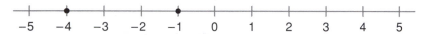

The so-called new mathematics of the 1960s introduced the number line at the elementary algebra level and used it extensively to give students a visual representation of the relationships between real numbers. Before this time, the concept of greater than was explained using only words and symbols. These explanations were abstract, and many students had difficulty in understanding why -1 is greater than -4. When we use the number line definition of greater than, we can look at the number line and see that -1 is greater than -4.

Graphing inequalities on a number line affords practice with the concept of greater than and also allows us to remember the definitions of the domain and of the subsets of the set of real numbers. Recall that the domain of an equation or inequality is the set of permissible replacement values of the variable.

example 86.1   Graph: $x < 3$; $D = \{\text{Positive integers}\}$

solution   We are asked to graph the numbers that are less than 3, but are allowed to consider only positive integers as replacements for the variable. The solution is

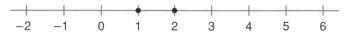

because there are only two positive integers that are less than 3.

example 86.2   Graph: $x > -3$; $D = \{\text{Reals}\}$

solution   We are asked to designate all real numbers that are greater than -3.

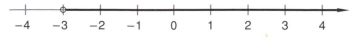

The open circle at -3 indicates that -3 is not a member of the solution set of this inequality.

86.B

trichotomy and transitive axioms

Some fundamental concepts of mathematics are difficult to remember because they are so self-evident. The trichotomy axiom and the transitive axiom are good examples. The trichotomy axiom can be demonstrated by having someone write a number on a piece of paper. Then let that person turn the paper over and write a number on the other side. There are then exactly three possibilities:

1. The second number is the same number as the first number.
2. The second number is greater than the first number.
3. The second number is less than the first number.

These statements are self-evident but are not trivial. They tell us that the real numbers are arranged in order, and thus we can say that the real numbers constitute an **ordered set**. We say that these three statements form the **trichotomy axiom**. Trichotomy comes from the Greek word *trikha*, which means "in three parts."

The transitive axiom also has three parts. It is also self-evident but not trivial. If Arthur is larger than Billy and Billy is larger than Susan, then Arthur is larger than Susan. The same statement can be made using *smaller than* or *the same size as* instead of *larger than*. Real numbers are just like people in this respect because this thought also applies to real numbers. We state both of these axioms formally in the box below.

Axioms	
For any real numbers, a, b, and c:	
Trichotomy Axiom	Exactly one of the following is true: $\quad a < b, \qquad a = b, \qquad$ or $\qquad a > b$
Transitive Axiom	If $a > b$ and $b > c$, then $a > c$. If $a < b$ and $b < c$, then $a < c$. If $a = b$ and $b = c$, then $a = c$.

example 86.3   Graph: $x \not\leq -4$; $D = \{\text{Negative integers}\}$

solution   We are asked to graph the negative integers that are not less than or equal to -4. **By the trichotomy axiom, if the numbers are not less than or equal to -4, they must be greater than -4.** Thus, we can say the same thing by writing

$$x > -4$$

In the graph we show the negative integers that are greater than -4.

Note that we do not use an open circle at -4.

example 86.4   Graph: $-x + 4 \not> 2$; $D = \{\text{Reals}\}$

solution   Not greater than means equal to or less than, so we can replace the given symbol with $\leq$.

$$-x + 4 \leq 2$$

Now we isolate x by adding -4 to both sides. We do not reverse the inequality symbol when we add a negative number to both sides.

$$
\begin{array}{r}
-x + 4 \leq 2 \\
\underline{-4 \quad -4} \\
-x \leq -2
\end{array}
$$

We need to solve for x, not for $-x$. Thus, we multiply both sides by -1 and reverse the inequality symbol.

$$x \geq 2$$

We finish by graphing this inequality, and we indicate all real numbers that equal 2 or are greater than 2.

example 86.5   Graph: $-x - 4 \ngeq -2$; $D = \{\text{Negative integers}\}$

solution   We begin by replacing the symbol for not greater than or equal to with the symbol for less than.

$$-x - 4 < -2$$

Now we add $+4$ to both sides and get

$$-x < +2$$

Next we multiply both sides by -1, reverse the inequality symbol, and graph the result.

$$x > -2$$

```
   +----+----●----+----+----+----+----+--
  -3   -2   -1    0    1    2    3
```

The solution is -1 because -1 is the only negative integer that is greater than -2.

86.C

irrational roots   The solution for the points of intersection of the circle and the line shown below was discussed in the preceding lesson. Finding these points required that we find the solutions to the quadratic equation

$$5x^2 - 12x = 0$$

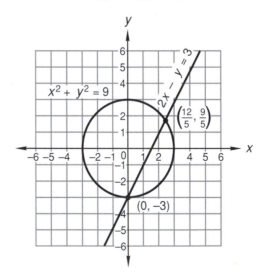

This equation can be factored, and thus the solutions to this equation are rational numbers. Some equations that describe the points of intersection of a line and a circle cannot be factored because the solutions to these equations are irrational numbers. These solutions can be found quickly and easily by using the quadratic formula. Many people always use this formula when solving real-life quadratic equations because so many of these equations cannot be solved by factoring.

example 86.6   Solve: $\begin{cases} x^2 + y^2 = 9 & \text{(a)} \\ y - x = 1 & \text{(b)} \end{cases}$

solution   We will begin by solving equation (b) for y, and then we will square both sides.

$$y - x = 1 \qquad\qquad \text{equation (b)}$$

$$y = x + 1 \qquad\qquad \text{solved for } y$$

$$y^2 = x^2 + 2x + 1 \qquad \text{squared both sides}$$

Now we will replace y^2 in equation (a) with $x^2 + 2x + 1$.

$$x^2 + (x^2 + 2x + 1) = 9 \qquad \text{substituted}$$

$$2x^2 + 2x - 8 = 0 \qquad \text{simplified}$$

$$x^2 + x - 4 = 0 \qquad \text{divided by 2}$$

This equation cannot be solved by factoring so we will use the quadratic formula.

$$x = \frac{-b \pm \sqrt{b^2 - 4ac}}{2a} \quad\longrightarrow\quad x = \frac{-1 \pm \sqrt{1 - 4(1)(-4)}}{2} \quad\longrightarrow\quad x = -\frac{1}{2} \pm \frac{\sqrt{17}}{2}$$

$$\longrightarrow\quad x = -\frac{1}{2} + \frac{\sqrt{17}}{2} \qquad \text{and} \qquad x = -\frac{1}{2} - \frac{\sqrt{17}}{2}$$

Now we could use either equation (a) or equation (b) to find the values of y. We will use the equation of the line to find y because it has no squared terms and is easier to use.

$$y = x + 1 \qquad\qquad y = x + 1 \qquad\qquad \text{equation (b)}$$

$$y = \left(-\frac{1}{2} + \frac{\sqrt{17}}{2}\right) + 1 \qquad y = \left(-\frac{1}{2} - \frac{\sqrt{17}}{2}\right) + 1 \qquad \text{substituted}$$

$$y = \frac{1}{2} + \frac{\sqrt{17}}{2} \qquad\qquad y = \frac{1}{2} - \frac{\sqrt{17}}{2} \qquad\qquad \text{simplified}$$

Thus, the ordered pairs of x and y that satisfy the given system are

$$\left(-\frac{1}{2} + \frac{\sqrt{17}}{2}, \frac{1}{2} + \frac{\sqrt{17}}{2}\right) \qquad \text{and} \qquad \left(-\frac{1}{2} - \frac{\sqrt{17}}{2}, \frac{1}{2} - \frac{\sqrt{17}}{2}\right)$$

practice   **a.**   Graph: $-x \nleq 4$; $D = \{\text{Integers}\}$ **b.**   Solve: $\begin{cases} x^2 + y^2 = 11 \\ y - x = 1 \end{cases}$

problem set 86

1. The pressure of an ideal gas was held constant at 450 millimeters of mercury. The volume was 400 liters and the temperature was 1000°C. What was the volume when the temperature was increased to 2000°C? Begin by adding 273 to convert degrees Celsius to kelvins.

2. Cuthbert rode jauntily out of town on his scooter at 20 miles per hour. Halfway to the swamp his scooter broke down, and he had to push it at 2 miles per hour all the way home. If he was gone for 11 hours, how far was it to the swamp?

3. The large jug contained 100 liters of a solution that was only 50% alcohol. How many liters of a 20% solution should be added to get a solution that is 23% alcohol?

4. The total weight of the carbonic acid, H_2CO_3, was 372 grams. What was the weight of the carbon (C) in this amount of acid? (H, 1; C, 12; O, 16)

5. After the count was completed and the tally made, the observers were amazed to find that only 0.14 of the ducks were ring-necked. If 120,000 ducks were tallied, how many were ring-necked?

6. A square is inscribed in a circle whose area is 25π m². What is the radius of the circle? What is the length of a diagonal of the square? What is the length of a side of the square? What is the area of the square?

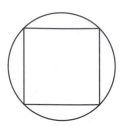

Graph:

7. $x \not\leq -2$; D = {Negative integers}

8. $-x + 3 \not> 2$; D = {Reals}

9. $-x - 6 \not\geq -3$; D = {Negative integers}

Solve:

10. $\begin{cases} BT_D + 3T_D = 60 \\ BT_D - 3T_D = 36 \end{cases}$

11. $\begin{cases} x^2 + y^2 = 4 \\ y - x = 1 \end{cases}$

Simplify:

12. $\dfrac{x^{2a}(y^b)^{2a} x^{a/3}}{y^{ba/3}}$

13. $\dfrac{(x^{a+2})^2}{x^{2-a}}$

14. $\dfrac{1}{x + \dfrac{a}{x + \dfrac{1}{a}}}$

15. $\dfrac{2 - 3i}{-5 + i}$

16. $\dfrac{3 + 2i}{5 - i}$

17. The average of five numbers is 9. Four of the numbers are 2, 1, 6, and 13. What is the fifth number?

Solve:

18. $\sqrt{p + 48} = 8 - \sqrt{p}$

19. $\begin{cases} x + 2y - z = 0 \\ 3x + y - 2z = 3 \\ 2x - z = 0 \end{cases}$

20. Add: $-10\underline{/-40°} + 10\underline{/-220°}$

21. Write $-3R - 10U$ in polar form.

22. Simplify: $\dfrac{-3 - 2\sqrt{3}}{1 - 3\sqrt{3}}$

23. Find b: $\dfrac{x + 2}{y} - c = m\left(\dfrac{a}{b} + x\right)$

Simplify:

24. $-\sqrt{-16} - \sqrt{3}\sqrt{-3} + \sqrt{-3}\sqrt{-3}$

25. $5\sqrt{\dfrac{2}{7}} + 3\sqrt{\dfrac{7}{2}} - 2\sqrt{56}$

26. Solve $-3x^2 - 1 = 6x$ by completing the square.

27. Use unit multipliers to convert 40 square inches per minute to square yards per hour.

28. Solve $-2x^2 - 1 = 6x$ by using the quadratic formula.

29. Solve by graphing and then find an exact solution by using either substitution or elimination:

$$\begin{cases} 2x - 5y = -15 \\ 3x + 4y = -4 \end{cases}$$

30. Find the equation of the line that passes through $(7, 2)$ and is perpendicular to the line $4x - 6y = 25$.

LESSON 87 *Slope formula*

The *slope* of a line is defined to be the change in the *y* coordinate divided by the change in the *x* coordinate as we move from one point on the line to another point on the line. We will use the points $(-3, 4)$ and $(5, -1)$ on the line graphed below to investigate. When we graph the points and draw the triangle, we see that the slope is negative and that the rise over the run is 5 over 8, so the slope of this line is $-\frac{5}{8}$.

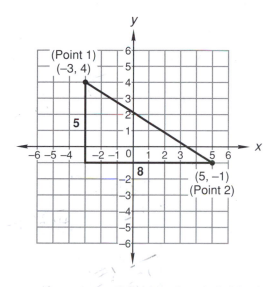

We can get the same answer if we use the definition given in bold print above. We will find the slope twice to demonstrate that we get the same answer when we move from point 1 to point 2 as we do if we move from point 2 to point 1. The definition defines the slope as

$$m = \frac{\text{change in the } y \text{ coordinate}}{\text{change in the } x \text{ coordinate}}$$

First, we will move from point 1 to point 2. When we do, the *y* coordinate changes from 4 to -1, a change of -5.

$$\text{Change in } y = -5$$

The *x* coordinate changes from -3 to $+5$, a change of $+8$.

$$\text{Change in } x = +8$$

Thus, when we move from point 1 to point 2, we find

$$m = \frac{\text{change in the } y \text{ coordinate}}{\text{change in the } x \text{ coordinate}} = \frac{-5}{+8} = -\frac{5}{8}$$

If we move from point 2 to point 1, the *y* coordinate changes from -1 to $+4$, a change of $+5$; and the *x* coordinate changes from 5 to -3, a change of -8.

$$m = \frac{\text{change in the } y \text{ coordinate}}{\text{change in the } x \text{ coordinate}} = \frac{+5}{-8} = -\frac{5}{8}$$

We get the same answer both ways, because $+5$ divided by -8 is the same number as -5 divided by $+8$.

If we call the points point 1 and point 2 with coordinates (x_1, y_1) and (x_2, y_2),

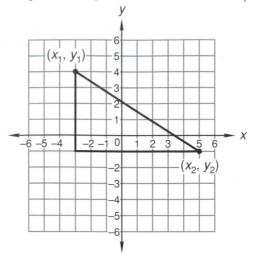

we see that we can find the slope by using either of these formulas:

$$m = \frac{y_2 - y_1}{x_2 - x_1} \quad \text{or} \quad m = \frac{y_1 - y_2}{x_1 - x_2}$$

It makes no difference which subscript comes first as long as the subscripts are in the same order above and below. It also makes no difference which point is called point 1 and which point is called point 2. Of course, the y's must always be in the numerator.

It is very easy to make a mistake in signs when using the slope formula, in which case the slope found will be incorrect.

example 87.1 Use the slope formula to find the slope of the line that passes through $(2, -5)$ and $(-3, 7)$.

solution We will work the problem twice. The first time $(2, -5)$ will be point 1. We can prevent some mistakes if we first identify the coordinates.

$$x_1 = 2 \qquad y_1 = -5 \qquad x_2 = -3 \qquad y_2 = 7$$

Now we use the formula.

$$m = \frac{y_2 - y_1}{x_2 - x_1} \quad \rightarrow \quad m = \frac{7 - (-5)}{-3 - 2} = \frac{7 + 5}{-5} = -\frac{12}{5}$$

Now we will find the slope again, but this time we will let $(-3, 7)$ be point 1.

$$x_1 = -3 \qquad y_1 = 7 \qquad x_2 = 2 \qquad y_2 = -5$$

We will get the same answer as we did previously.

$$m = \frac{y_2 - y_1}{x_2 - x_1} \quad \rightarrow \quad m = \frac{-5 - 7}{2 - (-3)} = \frac{-12}{2 + 3} = -\frac{12}{5}$$

example 87.2 Use the slope formula to find the slope of the line that passes through $(-5, 100)$ and $(-33, 57)$.

solution We decide to let $(-5, 100)$ be point 1. Thus, we get

$$x_1 = -5 \qquad y_1 = 100 \qquad x_2 = -33 \qquad y_2 = 57$$

Now we use the formula to find the slope.

$$m = \frac{y_1 - y_2}{x_1 - x_2} \quad \rightarrow \quad m = \frac{100 - 57}{-5 - (-33)} = \frac{43}{-5 + 33} = \frac{43}{28}$$

We see that the slope formula will yield the slope, but it will yield the incorrect slope unless care is used in handling the signs. Thus, many people believe that the graphical method of finding the slope is the most reliable.

practice Use the slope formula to find the slope of the line that passes through $(-3, 96)$ and $(-11, 49)$.

problem set 87

1. Juarez's speed was 4 times as great as that of Benito. Thus, Juarez could travel 1440 miles in only 4 hours more than it took Benito to travel 120 miles. Find the speed of both and the times of both.

2. The hours slept varied inversely as the intensity of the hypnophobia. When hypnophobia measured 400 on the H scale, Trishnutt could sleep 10 hours. How long could she sleep when the reading was 200? Work once using the equal ratio method and again using the direct variation method.

3. The mixture in the brown barrel was 80% sand, and the mixture in the blue barrel was 30% sand. How many cubic inches of each should be used to get 600 cubic inches of a mixture that is 40% sand?

4. A few were holographs but most had been printed on a press. Ten times the number of holographs was 102 less than the number of printed ones. Also, 100 times the number of holographs exceeded the number of printed ones by 168. How many of each kind were there?

5. To bottle oxygen, the manufacturer separated the oxygen from $KClO_3$. What percent by weight of this compound is oxygen (O)? If 576 grams of oxygen were needed, how many grams of $KClO_3$ were required? (K, 39; Cl, 35; O, 16)

6. Write a short explanation of why the slope formula works, and include an explanation of why either point can be point 1 or point 2.

7. Use the slope formula to find the slope of the line that passes through $(-2, 108)$ and $(-21, 47)$.

Graph:

8. $x \not< -4$; $D = \{\text{Integers}\}$

9. $-x + 2 \leq -3$; $D = \{\text{Positive integers}\}$

Solve:

10. $\begin{cases} BT_D + 4T_D = 36 \\ BT_D - 4T_D = 12 \end{cases}$

11. $\begin{cases} x^2 + y^2 = 4 \\ x - 2y = 1 \end{cases}$

Simplify:

12. $\dfrac{m^a m^{2a+2} y^{-b}}{y^{2-b}}$

13. $\dfrac{(m^{a-3})^2 y}{m^a y^{b+1}}$

14. $\dfrac{a}{a + \dfrac{a}{a + \dfrac{b}{a}}}$

15. $\dfrac{5i - 2}{-1 - i}$

16. $\dfrac{-3 + 2i}{-2 - i}$

17. Use a calculator to simplify. Estimate first.
    (a) $(9315 \times 10^3)(-2.065 \times 10^4)$
    (b) $\sqrt[2.7]{1001.94}$

Solve:

18. $\begin{cases} x + y - 2z = 7 \\ 3x - y - z = 3 \\ 2x + z = 0 \end{cases}$

19. $\sqrt{z} - \sqrt{z - 45} = 5$

20. The two forces act on the point as shown. Find the resultant force.

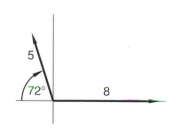

Simplify:

21. $\dfrac{2 - \sqrt{3}}{\sqrt{3} + 2}$

22. $-3i^3 - \sqrt{-2}\,\sqrt{-3}$

23. $3\sqrt{\dfrac{7}{3}} + 3\sqrt{\dfrac{3}{7}} - 4\sqrt{189}$

24. Find c: $\dfrac{y + 4}{m} = p\left(\dfrac{a}{b} + \dfrac{1}{c}\right)$

25. Find x and y.

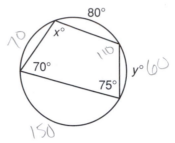

26. Solve $-3x^2 - 4 = 2x$ by using the quadratic formula.

27. Use unit multipliers to convert 10 kilometers per hour to inches per second.

28. Begin with $ax^2 + bx + c = 0$ and derive the quadratic formula by completing the square.

Simplify:

29. $3\sqrt{9\sqrt[4]{3}}$

30. $\dfrac{+9^{-3/2}}{+(-27)^{-2/3}}$

LESSON 88   *The distance formula • The relationship $PV = nRT$*

88.A
the distance formula

In Lesson 10, we discussed the fact that the square drawn on the hypotenuse of a right triangle has the same area as the sum of the areas of the squares drawn on the other two sides.

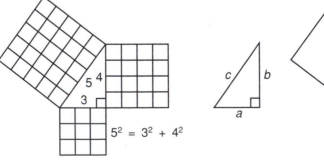

From this geometric approach, we can deduce the algebraic formula

$$c^2 = a^2 + b^2$$

where c is the length of the hypotenuse and a and b are the lengths of the other two sides. We have been using this formula to find the length of the missing side in a right triangle when the other two sides are given. If the distance between two points is required, we know that we can find the answer by graphing the points, drawing the triangle, and then using the algebraic form of this relationship, which is called the Pythagorean theorem.

To find the distance between $(4, -4)$ and $(-2, 3)$, we first graph the points. Then we draw the triangle and find the lengths of the vertical and horizontal sides. Now we use 6 for a and 7 for b and solve for c.

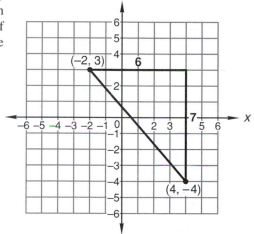

$$c^2 = 6^2 + 7^2$$

$$c = \sqrt{6^2 + 7^2} \quad \longrightarrow \quad c = \sqrt{85}$$

We note that the length of the vertical side of the triangle is 7, which is the difference in the y coordinates of the points,

$$+3 \text{ to } -4 \text{ is a distance of } 7$$

and that the length of the horizontal side of the triangle is 6, which is the difference in the x coordinates of the points.

$$-2 \text{ to } +4 \text{ is a distance of } 6$$

To develop a general formula for distance we will call the points point 1 with coordinates (x_1, y_1) and point 2 with coordinates (x_2, y_2). Either point can be point 1.

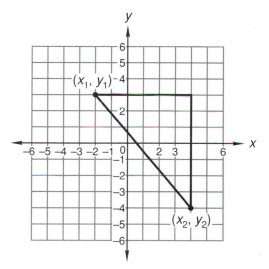

We see that the lengths of the sides are the absolute values of the differences in the coordinates. Since any number squared is positive, the absolute value notation is not required. Thus, the distance between the points can be represented by

$$D = \sqrt{(x_1 - x_2)^2 + (y_1 - y_2)^2}$$

This is called the **distance formula**. It is simply an algebraic expression that says that the distance is the square root of the sum of the squares of the sides of the triangle. In this formula either point can be used as point 1, and thus the following formula will produce the same result as the formula above.

$$D = \sqrt{(x_2 - x_1)^2 + (y_2 - y_1)^2}$$

example 88.1 Use the distance formula to find the distance between $(4, -2)$ and $(-5, 3)$.

solution It is easy to make mistakes in sign. Thus, we will be careful and begin by writing down the values of x_1, y_1, x_2, and y_2.

$$x_1 = 4 \qquad y_1 = -2 \qquad x_2 = -5 \qquad y_2 = 3$$

Now we place these values in one of the two forms of the distance formula and simplify.

$$D = \sqrt{(x_1 - x_2)^2 + (y_1 - y_2)^2} \quad \longrightarrow \quad D = \sqrt{[4 - (-5)]^2 + [-2 - (3)]^2}$$

$$D = \sqrt{81 + 25} \quad \longrightarrow \quad D = \sqrt{106}$$

88.B
the relationship $PV = nRT$

Some people say that unfamiliar concepts are difficult concepts and that familiar concepts are easy concepts. These people have confused the words **difficult** and **different.** We have found that algebraic concepts that seem difficult at first are really just different, and they become familiar concepts after we have worked with them for a while. **Thus we can correctly say that algebra is not difficult—it is just different.**

The same is true for chemistry, physics, and other sciences. This book was written to permit the student to develop a firm understanding of the foundations of algebra and geometry and to prepare for the concepts of advanced courses in mathematics and science. We have introduced some problems that will be encountered in chemistry. These problems are part mathematics and part chemistry. Their introduction at this level permits the mathematical part to be learned and allows some of the newness of these problems to rub off. Another kind of problem that will be encountered in both physics and chemistry uses the relationship

$$PV = nRT$$

To use this formula, values for four of the five variables must be given; the formula is then used to find the value of the variable that is unknown. The letter R stands for a constant whose value depends on the units used for the other variables. We will always use the units listed below for the other variables, and thus R will always be 0.0821. The symbols represent the following:

$$P = \text{pressure in atmospheres}$$
$$V = \text{volume in liters}$$
$$n = \text{number of moles of gas}$$
$$R = \text{a constant } (0.0821)$$
$$T = \text{temperature in kelvins}$$

We will not worry about the units. We will insert the numbers for the known values and then simplify.

example 88.2 Use the relationship $PV = nRT$ to find the number of moles in an amount of gas when the temperature is 273 K, the pressure is 1 atmosphere, and the volume is 8 liters ($R = 0.0821$).

solution Since we are finding n, we will begin by solving the abstract equation for n. We do this by dividing both sides by RT.

$$\frac{PV}{RT} = \frac{nRT}{RT} \quad \longrightarrow \quad \frac{PV}{RT} = n$$

Now we make the indicated substitutions and simplify.

$$\frac{(1)(8)}{(0.0821)(273)} = n$$

If a calculator is handy, this reduces to 0.357 mole. A mole is a measure of how many atoms of gas are present. This concept will be discussed in great detail in an elementary course in chemistry.

example 88.3   Use the formula $PV = nRT$ to find the volume of 0.832 mole of a gas at a pressure of 3 atmospheres and a temperature of 400 K ($R = 0.0821$).

solution   We begin by solving the equation for V. We do this by dividing both sides by P.

$$\frac{PV}{P} = \frac{nRT}{P} \quad \longrightarrow \quad V = \frac{nRT}{P}$$

Now we make the indicated substitutions and simplify.

$$V = \frac{(0.832)(0.0821)(400)}{3}$$

And if a calculator is handy, we find the volume is

$$V = \textbf{9.11 liters}$$

practice   **a.** Use the distance formula to find the distance between $(4, -2)$ and $(-3, 1)$.

b. Use the relationship $PV = nRT$ to find the number of moles of a quantity of gas when the temperature is 159 K, the pressure is 2 atmospheres, and the volume is 4 liters ($R = 0.0821$). Begin by solving the abstract equation for n.

problem set 88

1. Find three consecutive even integers such that the product of the first and third is 12 greater than the product of 6 and the opposite of the second.

2. The tin (Sn) in the tin II chromate, $SnCrO_4$, weighed 595 grams. What was the total weight of the tin II chromate? (Sn, 119; Cr, 52; O, 16)

3. David's speed was 20 miles per hour less than Gretchen's speed. Thus, David could travel 400 miles in one-half the time it took Gretchen to travel 1120 miles. Find the speeds of both and the times of both.

4. The initial pressure, volume, and temperature of a quantity of an ideal gas were recorded as 740 millimeters of mercury, 10 liters, and 300°C. Find the final volume if the final pressure was 1480 millimeters of mercury and the temperature was 1200°C. Begin by solving for V_2. Remember to add 273 to convert degrees Celsius to kelvins.

5. Use the relationship $PV = nRT$ to find the number of moles in a quantity of gas when the temperature is 251 K, the pressure is 1 atmosphere, and the volume is 5 liters ($R = 0.0821$). Begin by solving the abstract equation for n.

6. Write a short explanation of why the distance formula works, and include a discussion of why either point can be point 1 or point 2.

7. Use the distance formula to find the distance between $(5, -2)$ and $(-3, 3)$.

Graph:

8. $-x - 2 \leq 4; D = \{$Negative integers$\}$

9. $-x + 2 \not> 3; D = \{$Negative integers$\}$

Solve:

10. $\begin{cases} BT_D + 6T_D = 22 \\ BT_D - 6T_D = 10 \end{cases}$

11. $\begin{cases} x^2 + y^2 = 2 \\ x - y = 1 \end{cases}$

Simplify:

12. $\dfrac{(x^2)^{a+b} \, x^{-2a+b} \, y^a}{y^{a/4}}$

13. $\dfrac{(y^{2a+2})^2 \, y^{a/2} b}{y^a \, b^{2a}}$

14. $\dfrac{r}{m + \dfrac{r}{\dfrac{1}{r} + m}}$

15. $\dfrac{pc}{p - \dfrac{c^2}{p - \dfrac{1}{pc}}}$

16. $\dfrac{4i - 1}{3i - 2}$

17. Use a calculator to simplify. Estimate first.

   (a) $\dfrac{5712 \times 10^{-2}}{0.0416 \times 10^{3}}$ (b) $(184.3)^{-1.62}$

Solve:

18. $\sqrt{z - 35} + \sqrt{z} = 7$ 19. $\begin{cases} 2x + 2y - z = 14 \\ 3x + 3y + z = 16 \\ x - 2y = 0 \end{cases}$

20. Add: $-20\underline{/-200°} + 30\underline{/-30°}$ 21. Write $4R - 12U$ in polar form.

Simplify:

22. $4i^2 - \sqrt{-9}$ 23. $4\sqrt{\dfrac{9}{3}} + 3\sqrt{\dfrac{3}{9}} - 5\sqrt{27}$

24. $\dfrac{1 - \sqrt{2}}{3 - 2\sqrt{2}}$ 25. $\dfrac{4 + \sqrt{3}}{1 - \sqrt{3}}$

26. Find r: $\dfrac{x + 2y}{c} = y\left(\dfrac{1}{x} - \dfrac{1}{r}\right)$

27. Solve $-x^2 - 2x - 2 = 0$ by completing the square.

28. Use unit multipliers to convert 10 milliliters per second to cubic inches per minute.

29. Find x and y.

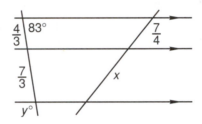

30. Solve: $\dfrac{-3x - 2}{2} - \dfrac{2x - 4}{3} = 7$

LESSON 89 *Conjunctions • Disjunctions • Products of chords and secants*

89.A
conjunctions

The English prefixes *com-* and *con-* come from the Latin word *cum*, which means "with" or "together." For example, the English word *compose* means "to make by placing together," and the *confluence* of two rivers means "the place where the two rivers flow together." We use the prefix *con-* in algebra in the word *conjunction* to describe a situation where two or more restrictions on the variable must apply (join together) at the same time. We use the word **and** to denote conjunctions by writing **and** between the two symbolic statements of the restrictions on the variable.

example 89.1   Graph: $-x - 3 \le -2$ and $x - 2 < 1$; $D = \{\text{Integers}\}$

solution   **The word *and* means that both conditions must be met.** The additive property of inequality permits us to add the same quantity to both sides as required to find x.

$$-x - 3 \leq -2 \quad \text{and} \quad x - 2 < 1$$
$$\underline{+ 3 \quad +3} \qquad \qquad \underline{+ 2 \quad +2}$$
$$-x \quad \leq 1 \quad \text{and} \quad x \quad < 3$$

Now on the left we must mentally multiply both sides by -1 and reverse the inequality symbol to solve for x.

$$x \geq -1 \quad \text{and} \quad x < 3$$

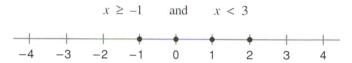

In the graph we have indicated the integers that are less than 3 **and** are greater than or equal to -1.

example 89.2   Graph: $-2 < x - 2 \leq 4$; $D = \{\text{Reals}\}$

solution   **This notation is an alternative way to write a conjunction.** To find the two conditions, we first cover up the ≤ 4 on the right and read the first condition as

$$-2 < x - 2$$

Then we cover up the $-2 <$ on the left and read the second condition as

$$x - 2 \leq 4$$

We now add 2 to both sides of these inequalities to solve for x.

$$-2 < x - 2 \quad \text{and} \quad x - 2 \leq 4$$
$$\underline{+2 < \quad + 2} \qquad \qquad \underline{+ 2 \quad +2}$$
$$0 < x \quad \text{and} \quad x \quad \leq 6$$

Thus, the solution is the graph of the real numbers that are greater than 0 **and** are less than or equal to 6.

89.B

disjunctions   The Latin prefix *dis-* indicates negation or reversal. For example, *disapprove* means "not to approve" and *disagree* means "not to agree." In algebra, we use the prefix *dis-* in the word *disjunction*, which describes a situation where it is not necessary to satisfy both restrictions on a variable. A number satisfies a disjunction if it satisfies either of the two restrictions stated in the disjunction. We use the word **or** to designate a disjunction.

example 89.3   Graph: $-x \geq 3$ or $-x < -1$; $D = \{\text{Reals}\}$

solution   The word **or** indicates that a number satisfies this disjunction if it satisfies the left-hand inequality or if it satisfies the right-hand inequality. We solve the inequalities for x by multiplying both sides by -1 and reversing the inequality symbols.

$$x \leq -3 \quad \text{or} \quad x > 1$$

Any real number that satisfies either of the restrictions is a solution to the disjunction.

example 89.4   Graph: $-x - 2 \geq 0$ or $x + 3 > 6$; $D = \{\text{Integers}\}$

solution   We begin by solving both inequalities for x. As the last step on the left, we mentally multiply both sides by -1 and reverse the inequality symbol.

$$-x - 2 \geq 0 \quad \text{or} \quad x + 3 > 6$$
$$\underline{\quad + 2 \quad +2 \quad} \quad \quad \underline{\quad - 3 \quad -3 \quad}$$
$$-x \quad \quad \geq \; 2 \quad \text{or} \quad x \quad \quad > \; 3$$
$$x \leq -2 \quad \text{or} \quad x > 3$$

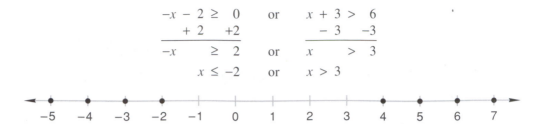

This disjunction is satisfied by all integers that are greater than 3 **or** are less than or equal to −2. A number qualifies if it satisfies either one of these conditions. Unfortunately (or fortunately), there is no concise notation for a disjunction. Both conditions must be written out, and the word **or** must be written between the conditions.

89.C
products of chords and secants

We review by remembering that the measures of the vertical angles formed when two chords intersect equal half the sum of the measures of the intercepted arcs. When secants or tangents intersect outside a circle, the measure of the angle formed equals half the difference of the measures of the intercepted arcs.

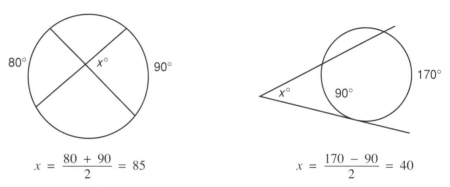

$$x = \frac{80 + 90}{2} = 85 \qquad\qquad\qquad x = \frac{170 - 90}{2} = 40$$

In this lesson we will prove relationships about the length of the segments. **The first is that when the chords intersect inside the circle, the products of the chord segments are equal.**

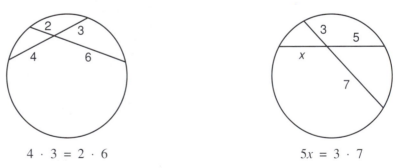

$$4 \cdot 3 = 2 \cdot 6 \qquad\qquad\qquad 5x = 3 \cdot 7$$

We will also prove that **when secants intersect outside the circle the products of the external segments and the whole segments are equal.**

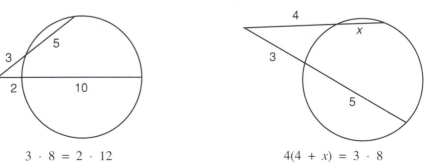

$$3 \cdot 8 = 2 \cdot 12 \qquad\qquad\qquad 4(4 + x) = 3 \cdot 8$$

We can also prove that **if one of the lines is a tangent, one product is the square of the tangent segment.**

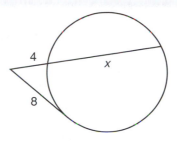

$$4(x + 4) = 8 \cdot 8$$

To prove that the products of the chords are equal, we connect the ends of the chords to form two similar triangles.

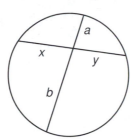
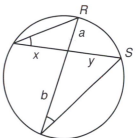

The two angles marked in the right-hand figure are equal because they both intercept arc *RS*. The vertical angles are equal so the triangles are similar by AAA. If we write the ratios and cross multiply, we get

$$\frac{a}{x} = \frac{y}{b} \quad \longrightarrow \quad ab = xy \qquad \text{QED}$$

To prove the secant-segment rule, we begin with the figure on the left. Then we draw dotted lines to connect both *R* and *T* to the ends of arc *SU* so the angles at *R* and *T* are equal. Then we look at the two triangles separately on the right.

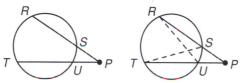
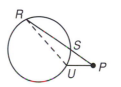
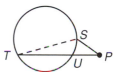

Since the angles at *R* and *T* are equal and both triangles contain angle *P*, the triangles are similar by AA. If we write the ratios of the corresponding sides, we get

$$\frac{RP}{TP} = \frac{UP}{SP} \qquad \text{corresponding sides}$$

and if we cross multiply, we get

$$RP \cdot SP = TP \cdot UP$$

Each of these products has two factors. One factor is the length of the whole secant segment and the other factor is the length of the external segment of the same secant.

We can develop the tangent rule by increasing the angle at *P*. As we do this, the length of the internal segment gets shorter and shorter until *T* and *U* designate the same point and the product becomes $(TP)^2$.

In the left-hand figure below, the product of the length of one secant segment times the length of its external segment equals the product of the length of the other secant segment times the length of its external segment.

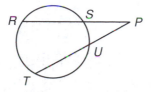
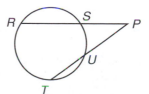
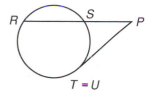

In the center figure, T and U are closer together, and the equation is the same. In the right-hand figure, the length of the secant segment is the same as the length of the external segment. **So in the case of a tangent and a secant drawn from a point outside the circle, the product of the length of the secant segment and the length of its external segment equals the square of the length of the tangent segment from the point.**

practice   Graph on a number line:

   a. $-x - 4 \le -1$ and $x - 1 < 1$; $D = \{\text{Integers}\}$

   b. $-1 < x - 1 \le 3$; $D = \{\text{Reals}\}$

   c.   Find x.   **d.**   Find y.

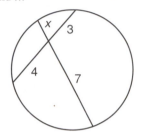   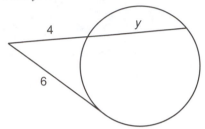

problem set   1.   The number of altercations varied directly as the number who were belligerent by
89   nature. If there were 500 altercations when 10 were belligerent, how many altercations
                        would there be if 42 were belligerent? Work once using the ratio format and once using
                        the variation format.

   2.   The road was rocky, so after running for a while at 10 kilometers per hour, Clytemnestra
            slowed to a 5-kilometer-per-hour walk. If the total trip was 65 kilometers and she made
            it in 9 hours, how far did she walk and how far did she run?

   3.   Eight hundred liters of a 79% glycol solution was available. How much pure glycol
            should be extracted so that the remainder would be only 30% glycol?

   4.   Some were fast, and the rest were slow. Ten times the number of fast was 140 less than
            twice the number of slow. Also, one-half the number of slow exceeded 3 times the
            number of fast by 10. How many were fast and how many were slow?

   5.   When the Grinch peered through the crack, he could see that the number of spotted ones
            had increased 640 percent. If he could now identify 592 spotted ones, how many spotted
            ones had he seen previously?

Graph on a number line:

   6. $-x - 5 \le -3$ and $x - 3 < 1$; $D = \{\text{Integers}\}$

   7. $-4 < x - 4 \le 1$; $D = \{\text{Reals}\}$

   8. $-x \not\ge 2$ or $-x < 1$; $D = \{\text{Reals}\}$

   9.   Find x.   **10.**   Find y.

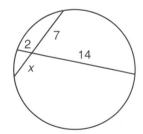   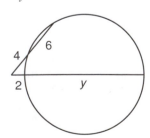

Solve:

   11. $\begin{cases} BT_D + 5T_D = 57 \\ BT_D - 5T_D = 27 \end{cases}$ **12.** $\begin{cases} x^2 + y^2 = 3 \\ x - y = 2 \end{cases}$

Simplify:

13. $\dfrac{x^a y^{2b}(x^{a+2})^{1/2}}{y^{3b}}$

14. $\dfrac{(y^{a+2})^a(y^a)^a}{y^{2+a}}$

15. $\dfrac{k}{m + \dfrac{m}{a + \dfrac{1}{m}}}$

16. $\dfrac{m}{a + \dfrac{x}{b + \dfrac{d}{c}}}$

17. $\dfrac{3 - 2i}{i - 4}$

18. $\dfrac{2 - 3i}{4i - 1}$

Solve:

19. $\begin{cases} 3x + 2y + z = 9 \\ x - 2y - 2z = -3 \\ 2x + z = 0 \end{cases}$

20. $\sqrt{s} = 3 + \sqrt{s - 21}$

21. The two forces are applied to an object as indicated. Find the resultant force on the object.

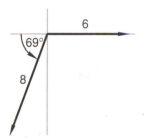

Simplify:

22. $\dfrac{4 + 2\sqrt{3}}{3\sqrt{3} - 2}$

23. Find y: $\dfrac{x}{a + c} = m\left(\dfrac{x}{y} + c\right)$

24. Find a: $\dfrac{x}{a + c} = m\left(\dfrac{x}{y} + c\right)$

Simplify:

25. $3i^3 + 5i - \sqrt{-2}\,\sqrt{2}$

26. $3\sqrt{\dfrac{3}{8}} + 4\sqrt{\dfrac{8}{3}} - 2\sqrt{24}$

27. Solve $-7x^2 = -x - 5$ by using the quadratic formula.

28. Use unit multipliers to convert 10 cubic feet per second to cubic inches per minute.

29. Find x, A, and B.

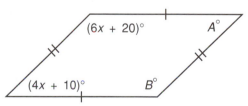

30. Find the distance between $(-2, 3)$ and $(8, 4)$.

LESSON 90 *Systems of three equations*

The solution to a system of two linear equations in two unknowns, such as

$$\begin{cases} 3x + 2y = -4 & \text{(a)} \\ 2x - 4y = -8 & \text{(b)} \end{cases}$$

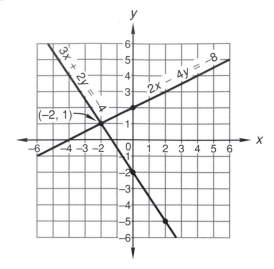

is the value of x and the value of y that will satisfy both equations. A solution exists only if the graphs of the equations intersect. The graphs of the two equations intersect at the point whose coordinates are $x = -2$ and $y = 1$, so we say that the solution of this system is the **ordered pair** $(-2, 1)$. If we substitute these values for x and y in equations (a) and (b), both equations will become true numerical equations.

(a) $3(-2) + 2(1) = -4$ (b) $2(-2) - 4(1) = -8$

$-6 + 2 = -4$ $-4 - 4 = -8$

$-4 = -4$ True $-8 = -8$ True

The solution to a system of three linear equations in three unknowns, such as

$$\begin{cases} x + 2y + z = 4 & \text{(a)} \\ 2x - y - z = 0 & \text{(b)} \\ 2x - 2y + z = 1 & \text{(c)} \end{cases}$$

is the value of x, the value of y, and the value of z that will satisfy all three equations. The graph of a linear equation in three unknowns is a plane, and thus the three equations shown describe three planes. If the three planes meet in one point, the three coordinates of this point are the **ordered triple** (x, y, z) that will satisfy all three equations.

A graphical solution of this system is not feasible because of the difficulty in drawing accurate graphs in three dimensions. However, we may use either substitution or elimination to find the answer. **The most common procedure is to add two equations so that one variable is eliminated. Then the remaining equation is added to one of those already used so that the same variable is eliminated. The result will be two equations in two unknowns that can be solved by using either substitution or elimination.**

example 90.1 Solve: $\begin{cases} x + 2y + z = 4 & \text{(a)} \\ 2x - y - z = 0 & \text{(b)} \\ 2x - 2y + z = 1 & \text{(c)} \end{cases}$

solution We choose to eliminate z. As the first step we will add (a) to (b) so that z is eliminated. Then we can add (c) to either (a) or (b) and again eliminate z. We decide to add (c) to (b).

(a) $x + 2y + z = 4$ (b) $2x - y - z = 0$
(b) $\underline{2x - y - z = 0}$ (c) $\underline{2x - 2y + z = 1}$
(d) $3x + y \qquad = 4$ (e) $4x - 3y \qquad = 1$

Now we have the two equations, (d) and (e), in the two unknowns, x and y. We will use elimination and add (e) to the product of (d) and 3.

$$
\begin{array}{rl}
3(d) & 9x + 3y = 12 \\
(e) & \underline{4x - 3y = 1} \\
& 13x = 13
\end{array}
$$

$$x = 1$$

Now we can use $x = 1$ in either (d) or (e) to find y. We will use both to show that both will yield the same answer.

$$
\begin{array}{ll}
\text{(d)} \quad 3(1) + y = 4 & \text{(e)} \quad 4(1) - 3y = 1 \\
\phantom{\text{(d)} \quad} 3 + y = 4 & \phantom{\text{(e)} \quad} 4 - 3y = 1 \\
\phantom{\text{(d)} \quad 3 + {}} y = 1 & \phantom{\text{(e)} \quad 4 - {}} -3y = -3 \\
 & \phantom{\text{(e)} \quad 4 - 3} y = 1
\end{array}
$$

Now we can use $x = 1$ and $y = 1$ in either (a), (b), or (c) to find z. This time we will use all three equations to show that all three will produce the same result.

$$
\begin{array}{lll}
\text{(a)} \quad (1) + 2(1) + z = 4 & \text{(b)} \quad 2(1) - (1) - z = 0 & \text{(c)} \quad 2(1) - 2(1) + z = 1 \\
\phantom{\text{(a)} \quad} 1 + 2 + z = 4 & \phantom{\text{(b)} \quad} 2 - 1 - z = 0 & \phantom{\text{(c)} \quad} 2 - 2 + z = 1 \\
\phantom{\text{(a)} \quad} 3 + z = 4 & \phantom{\text{(b)} \quad} -z = -1 & \phantom{\text{(c)} \quad} 0 + z = 1 \\
\phantom{\text{(a)} \quad 3 + {}} z = 1 & \phantom{\text{(b)} \quad -{}} z = 1 & \phantom{\text{(c)} \quad 0 + {}} z = 1
\end{array}
$$

Thus, the solution is the ordered triple **(1, 1, 1)**.

example 90.2   Solve: $\begin{cases} 2x - y + 3z = 9 & \text{(a)} \\ x + 2y + z = 8 & \text{(b)} \\ x - 2y + z = 0 & \text{(c)} \end{cases}$

solution   This time we decide to eliminate y. Thus, on the left, we multiply (a) by 2 and add this product to (b). On the right, we add (b) and (c) in their present form.

$$
\begin{array}{ll}
\begin{array}{rl}
2(a) & 4x - 2y + 6z = 18 \\
(b) & \underline{x + 2y + z = 8} \\
(d) & 5x + 7z = 26
\end{array}
&
\begin{array}{rl}
(b) & x + 2y + z = 8 \\
(c) & \underline{x - 2y + z = 0} \\
(e) & 2x + 2z = 8
\end{array}
\end{array}
$$

Now to eliminate x, we will multiply (d) by 2 and (e) by -5.

$$
\begin{array}{rl}
2(d) & 10x + 14z = 52 \\
-5(e) & \underline{-10x - 10z = -40} \\
& 4z = 12
\end{array}
$$

$$z = 3$$

Now we could use $z = 3$ in either (d) or (e) to find x. We will use (e).

$$
\begin{array}{l}
\text{(e)} \quad 2x + 2(3) = 8 \\
\phantom{\text{(e)} \quad} 2x + 6 = 8 \\
\phantom{\text{(e)} \quad} 2x = 2 \\
\phantom{\text{(e)} \quad} x = 1
\end{array}
$$

Now we could use $x = 1$ and $z = 3$ in either (a), (b), or (c) to find y. We decide to use (b).

$$
\begin{array}{l}
\text{(b)} \quad (1) + 2y + (3) = 8 \\
\phantom{\text{(b)} \quad} 2y = 4 \\
\phantom{\text{(b)} \quad} y = 2
\end{array}
$$

Thus, the solution to this system of three linear equations is the ordered triple **(1, 2, 3)**.

practice   Solve: $\begin{cases} x - 2y + 2z = 2 \\ x + y - z = 3 \\ 3x - 2y - 2z = 2 \end{cases}$

problem set 90

1. It took Pedro 5 times as long to travel 300 miles as it took Roberto to travel 160 miles. If Roberto's speed was 50 mph greater than Pedro's speed, find the speeds of both and the times that both were traveling.

2. Use the formula $PV = nRT$ to find the volume of 0.832 mole of gas at 3 atmospheres of pressure and a temperature of 400 K. Begin by solving for V ($R = 0.0821$).

3. Sergio painted some boats red and painted the rest blue. The number of red boats was 5 less than 3 times the number of blue boats. Also, 6 times the number of blue boats was 70 less than 10 times the number of red boats. How many were red and how many were blue?

4. Only 10% of the first solution was glycerine, but 40% of the second solution was glycerine. How many liters of each should Lolita use to get 800 liters of a solution that is 13% glycerine?

5. As the Grinch watched in horror, $\frac{7}{16}$ of the spotted ones metamorphosed into striped ones. If the number that metamorphosed totaled 672, how many were spotted when the Grinch began to watch?

Solve:

6. $\begin{cases} x + y - z = 3 \\ -x - 2y - 2z = 0 \\ x - 2y - 2z = 4 \end{cases}$

7. $\begin{cases} 2x - y + z = 2 \\ x + 2y + 2z = 3 \\ 2x - 2y + z = 0 \end{cases}$

Graph on a number line:

8. $-1 \le x - 1 < 4;\ D = \{\text{Integers}\}$

9. $x - 2 \ngeq 0$ or $x - 2 > 2;\ D = \{\text{Reals}\}$

Solve:

10. $\begin{cases} BT_D + 3T_D = 28 \\ BT_D - 3T_D = 16 \end{cases}$

11. $\begin{cases} x^2 + y^2 = 18 \\ y - x = 4 \end{cases}$

Simplify:

12. $\dfrac{a^2 a^{x/2}(a^2)^x}{(a^{-3})^{-x}}$

13. $\dfrac{y^c(m^{-b})^2}{m^{-b/2}}$

14. $\dfrac{x}{xm - \dfrac{m}{m - \dfrac{1}{x}}}$

15. $\dfrac{p}{x - \dfrac{xp}{1 - \dfrac{p}{x}}}$

16. $\dfrac{1 + 2i}{-5 - i}$

17. The radius of the circle is 5 cm. Find the length of $\overset{\frown}{ABC}$.

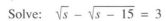

18. Solve: $\sqrt{s} - \sqrt{s - 15} = 3$

19. Add: $4\underline{/40°} - 6\underline{/-120°}$

20. Write $-4R - 10U$ in polar form.

Simplify:

21. $\dfrac{-2 - \sqrt{3}}{2 - 2\sqrt{3}}$

22. $\dfrac{4 + \sqrt{3}}{3 - 2\sqrt{3}}$

23. $-2i^2 + \sqrt{-4}\sqrt{4} - \sqrt{-3}\sqrt{-3} - 2i^5$

24. Find c: $\dfrac{x}{y} - m = p\left(\dfrac{r}{c} - \dfrac{1}{b}\right)$ **25.** Simplify: $\sqrt[3]{9\sqrt[4]{3}}$

26. Find x.

(a)

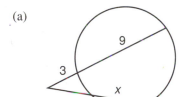

(b)

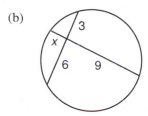

27. Solve $-6x^2 - x - 5 = 0$ by completing the square.

28. Use unit multipliers to convert 42 cubic feet to cubic centimeters.

29. Find B.

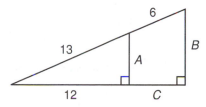

30. Find the equation of the line that passes through $(-2, 7)$ and is perpendicular to $x + 5y = 7$.

LESSON 91 *Linear inequalities • Greater than or equal to; less than or equal to • Systems of linear inequalities*

91.A

linear inequalities

A line divides the set of all the points in a plane into three disjoint subsets. These subsets are the set of points that lie on the line and the two sets of points that lie on either side of the line.

In the figure, we have graphed the equation $y = -\frac{1}{2}x + 1$. The coordinates of all the points that lie on the line will satisfy this equation. The coordinates of any point not on the line will satisfy one of the following inequalities.

(a) $y > -\dfrac{1}{2}x + 1$ (b) $y < -\dfrac{1}{2}x + 1$

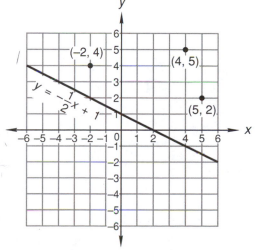

If the coordinates of a point on one side of the line will satisfy one of these inequalities, the coordinates of all the points on the same side of the line will satisfy the same inequality. We will demonstrate this by using in the left-hand equation the coordinates of the three points graphed above. All the resulting inequalities will be true inequalities.

USING (−2, 4)

$$4 > -\frac{1}{2}(-2) + 1$$

$$4 > 1 + 1$$

$$4 > 2 \quad \text{True}$$

USING (4, 5)

$$5 > -\frac{1}{2}(4) + 1$$

$$5 > -2 + 1$$

$$5 > -1 \quad \text{True}$$

USING (5, 2)

$$2 > -\frac{1}{2}(5) + 1$$

$$2 > -\frac{5}{2} + 1$$

$$2 > -\frac{3}{2} \quad \text{True}$$

Thus, the coordinates of any point above the line will satisfy this inequality and will not satisfy inequality (b),

$$\text{(b)} \quad y < -\frac{1}{2}x + 1$$

because this inequality is satisfied by the coordinates of all points below the line. We can always choose a point blindly and try both inequalities to see which one the coordinates of the point will satisfy. But if we remember that *y* is *greater than* as we go up and that *y* is *less than* as we go down, beginning with a test point is unnecessary. We read

$$y > -\frac{1}{2}x + 1$$

as "*y* is greater than" and remember that "*y* is greater than" means "above," so this inequality is satisfied by all points above the line. The other inequality is read as "*y* is less than" and is satisfied by all points below the line.

y IS GREATER THAN (POINTS ABOVE)

$$y > -\frac{1}{2}x + 1$$

y IS LESS THAN (POINTS BELOW)

$$y < -\frac{1}{2}x + 1$$

Our surmise can be checked by using a test point if we wish.

91.B
greater than or equal to; less than or equal to

The two inequalities

$$\text{(a)} \quad y \leq \frac{1}{3}x + 2 \quad \text{and} \quad \text{(b)} \quad y < \frac{1}{3}x + 2$$

both designate the points that lie below the same line, for the coordinates of any point below this line will satisfy both inequalities. In addition, the coordinates of any point that lies on the line will satisfy inequality (a), for this inequality is read "*y* is less than or equal to one-third *x* plus two."

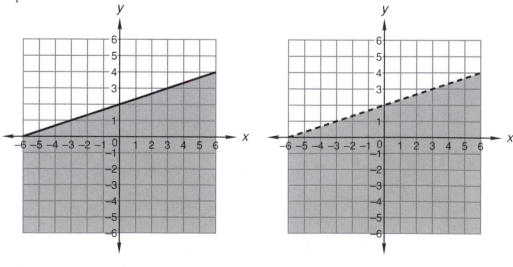

$$\text{(a)} \quad y \leq \frac{1}{3}x + 2$$

$$\text{(b)} \quad y < \frac{1}{3}x + 2$$

When we graph inequalities, we shade the regions whose coordinates satisfy the inequality. In both figures, the area below the line is shaded. **In the left-hand figure, the line is drawn as a solid line to indicate that the coordinates of the points of the line also satisfy the inequality. In the right-hand figure, the line is drawn as a dashed line to indicate that the coordinates of the points on the line do not satisfy the inequality.**

91.C

systems of linear inequalities

If two linear equations are not equivalent equations, then the lines designated by them either intersect or the lines are parallel.

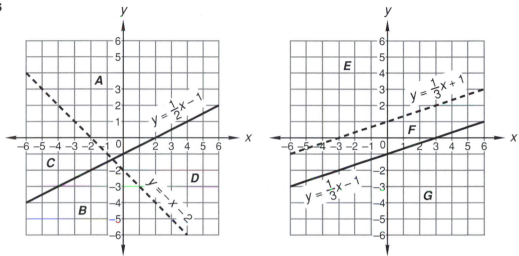

The lines in the left-hand figure intersect and divide the figure into four distinct regions.

Region A is above the dashed line and on or above the solid line.

Region B is below the dashed line and on or below the solid line.

Region C is below the dashed line and on or above the solid line.

Region D is above the dashed line and on or below the solid line.

The systems of linear inequalities that define these regions are

$$A \begin{cases} y \geq \dfrac{1}{2}x - 1 \\ y > -x - 2 \end{cases} \qquad B \begin{cases} y \leq \dfrac{1}{2}x - 1 \\ y < -x - 2 \end{cases}$$

$$C \begin{cases} y \geq \dfrac{1}{2}x - 1 \\ y < -x - 2 \end{cases} \qquad D \begin{cases} y \leq \dfrac{1}{2}x - 1 \\ y > -x - 2 \end{cases}$$

In the right-hand figure,

Region E is above both lines.

Region F is on or above the solid line and below the dashed line.

Region G is below the dashed line and on or below the solid line.

The systems of inequalities that define these regions are

$$E \begin{cases} y > \dfrac{1}{3}x + 1 \\ y \geq \dfrac{1}{3}x - 1 \end{cases} \qquad F \begin{cases} y < \dfrac{1}{3}x + 1 \\ y \geq \dfrac{1}{3}x - 1 \end{cases} \qquad G \begin{cases} y < \dfrac{1}{3}x + 1 \\ y \leq \dfrac{1}{3}x - 1 \end{cases}$$

example 91.1   Graph the solution to: $\begin{cases} y < \dfrac{1}{2}x + 2 \\ y \geq -x - 3 \end{cases}$

solution   The first step is to draw the lines. We will draw $y = -x - 3$ as a solid line and $y = \dfrac{1}{2}x + 2$ as a dashed line.

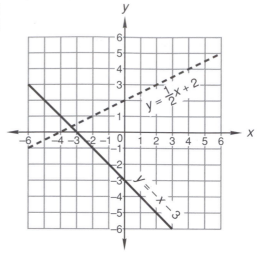

The region we wish to find is above or on the solid line and below the dashed line. We shade this region in the following figure.

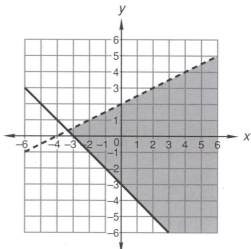

example 91.2   Graph the solution to: $\begin{cases} y > x - 2 \\ y \leq x + 1 \end{cases}$

solution   In the left-hand figure, we draw the lines; and in the right-hand figure we shade the region between the lines because the inequalities specify the points that lie above the bottom line that are also on or below the top line.

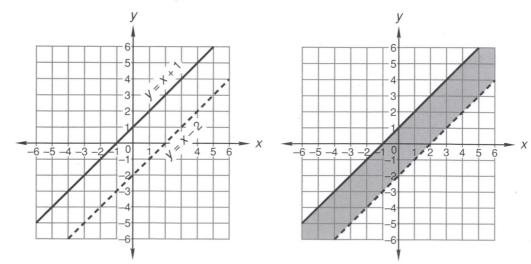

practice   Graph:

a. $\begin{cases} y \geq -x - 2 \\ y \leq \dfrac{1}{3}x - 2 \end{cases}$

b. $\begin{cases} y > x - 3 \\ y \leq x + 3 \end{cases}$

problem set 91

1.   The pressure of a quantity of an ideal gas at a temperature of 127°C was 740 millimeters of mercury. If the volume was not changed, what would the pressure be if the temperature were raised to 1327°C? Begin by adding 273 to convert degrees Celsius to kelvins.

2.   The solution in the large container was 60% alcohol, and the solution in the small container was only 30% alcohol. How much of each should be used to get 300 ml that is 42% alcohol?

3.   The container held the compound As_4O_6. What was the percentage by weight of the arsenic (As) in this compound? (As, 75; O, 16)

4.   The large plane could travel 3000 miles in 1 more hour than it took the small plane to cover 800 miles. If the rate of the large plane was 3 times the rate of the small plane, find the rates and the times of both.

5.   The number that were glabrous varied inversely as the average age of those present. If 35 were glabrous when the average age was 70, how many were glabrous when the average age was only 50? Work once using the direct variation method and again using the equal ratio method.

Graph:

6. $\begin{cases} y \geq -x - 1 \\ y < \dfrac{1}{3}x + 1 \end{cases}$

7. $\begin{cases} y > x - 1 \\ y \leq x + 2 \end{cases}$

8.   Solve: $\begin{cases} x + 2y - z = 1 \\ 2x - y + 2z = 9 \\ x - 2y - 3z = -9 \end{cases}$

Graph on a number line:

9. $4 < x + 4 < 6$; $D = \{\text{Integers}\}$

10. $x + 2 \not\leq 5$ or $x + 5 < 6$; $D = \{\text{Reals}\}$

Solve:

11. $\begin{cases} BT_D + 2T_D = 51 \\ BT_D - 2T_D = 39 \end{cases}$

12. $\begin{cases} x^2 + y^2 = 12 \\ x + y = 4 \end{cases}$ $X = 4-y$

Simplify:

13. $\dfrac{(x^b)^{2-a}x^{ab}}{x^{ab/2}}$

14. $\dfrac{a}{b + \dfrac{1}{cx + \dfrac{1}{x}}}$

15. $\dfrac{m}{a + \dfrac{ma}{a + \dfrac{m}{a}}}$

16. $\dfrac{2i + 7}{i + 2}$

17. $\dfrac{3i - 6}{-2i + 1}$

18. $\dfrac{1 - \sqrt{2}}{4 - 5\sqrt{2}}$

19. $\dfrac{4 + 3\sqrt{2}}{-\sqrt{2}}$

20.   Find y: $\dfrac{x}{my} = d\left(\dfrac{r}{a} + b\right)$

21.   Find a: $\dfrac{x}{my} = d\left(\dfrac{r}{a} + b\right)$

22.   The two forces are applied on the object as indicated. Find the resultant force on the object.

23. Draw the line suggested by the data points, and write the equation that expresses output as a function of input.

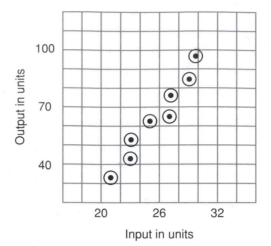

Simplify:

24. $\dfrac{-9^{3/2}}{27^{2/3}}$

25. $-2i^5 + 3i^3 - \sqrt{-9} + \sqrt{-4} - \sqrt{-2}\,\sqrt{-2}$

26. $2\sqrt{\dfrac{9}{5}} + 3\sqrt{\dfrac{5}{9}} + 3\sqrt{45}$

27. Begin with $ax^2 + bx + c = 0$ and derive the quadratic formula by completing the square.

28. Use unit multipliers to convert 400 milliliters per second to cubic inches per minute.

29. Divide $3x^3 - 2x + 2$ by $x + 2$.

30. Find x.

(a)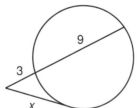

(b)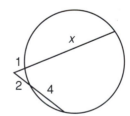

LESSON 92 *Boat-in-the-river problems*

Robert has a boat that has a speed of 11 mph in still water. If the boat, with the motor turned off, is placed in a river in which the water flows at 3 mph, the boat will drift downstream at 3 mph. If the boat is headed downstream with the motor turned on, its speed (rate) downstream would be 14 mph, which is 3 mph plus 11 mph. If the boat is turned around and is headed upstream, it must go against the current; and its speed upstream would be only 8 mph, which is 11 mph minus 3 mph. Thus, in general, the downstream rate is the speed of the boat plus the speed of the water, and the upstream rate is the speed of the boat minus the speed of the water.

$$\text{Downstream rate} = B + W$$

$$\text{Upstream rate} = B - W$$

The distance downstream equals the rate downstream times the time downstream, and the distance upstream equals the rate upstream times the time upstream. These two statements lead to the following two equations, which can be used to solve almost all boat-in-the-river problems.

DOWNSTREAM EQUATION	UPSTREAM EQUATION
$(B + W)T_D = D_D$	$(B - W)T_U = D_U$

These two equations contain six unknowns: B, W, T_D, D_D, T_U, and D_U. Thus, boat-in-the-river problems must contain six statements of equality because six equations are required.

example 92.1   Robert and Clay could go 60 miles downstream in the same time it took them to go 20 miles upstream. If the speed of their boat was 8 mph in still water, what was the speed of the current and what were their times?

solution   We begin by recording the downstream equation and the upstream equation.

DOWNSTREAM EQUATION	UPSTREAM EQUATION
(a) $(B + W)T_D = D_D$	(b) $(B - W)T_U = D_U$

Rather than write the other six equations, we will just make the necessary substitutions. The times were equal, so we will use T for both T_D and T_U. The rate of the boat was 8, the distance downstream was 60, and the distance upstream was 20.

$$\text{(a)} (8 + W)T = 60 \qquad \text{(b)} (8 - W)T = 20$$

Next we multiply as required in both equations and get

$$\text{(a)} 8T + WT = 60 \qquad \text{(b)} 8T - WT = 20$$

We will use elimination to solve.

$$
\begin{array}{l}
\text{(a)} \quad 8T + WT = 60 \\
\text{(b)} \quad \underline{8T - WT = 20} \\
\phantom{\text{(b)} \quad} 16T = 80 \quad \rightarrow \quad \mathbf{T = 5 \text{ hours}}
\end{array}
$$

We will now use equation (a) to solve for W.

$$8(5) + W(5) = 60 \qquad \text{substituted}$$
$$40 + 5W = 60 \qquad \text{multiplied}$$
$$5W = 20 \qquad \text{added} - 40$$
$$\mathbf{W = 4 \text{ mph}} \qquad \text{divided}$$

Thus, the time for both trips was 5 hours, and the speed of the current was 4 mph.

example 92.2   The steamboat *Juby Fountain* could go 70 miles downstream in 5 hours but required 6 hours to go 48 miles upstream. What was the speed of the boat in still water and what was the speed of the current?

solution   We begin by writing both equations.

DOWNSTREAM EQUATION	UPSTREAM EQUATION
(a) $(B + W)T_D = D_D$	(b) $(B - W)T_U = D_U$

Next we replace D_D, T_D, D_U, and T_U with 70, 5, 48, and 6, respectively.

$$\text{(a)} (B + W)5 = 70 \qquad \text{(b)} (B - W)6 = 48$$

Now we multiply in both equations and get

$$\text{(a)} 5B + 5W = 70 \qquad \text{(b)} 6B - 6W = 48$$

We choose to multiply equation (a) by 6 and to multiply equation (b) by 5 and then use elimination.

$$\begin{array}{llll}
\text{(a)} & 5B + 5W = 70 & \rightarrow \;(6)\; \rightarrow & \text{(a}') \quad 30B + 30W = 420 \\
\text{(b)} & 6B - 6W = 48 & \rightarrow \;(5)\; \rightarrow & \text{(b}') \quad \underline{30B - 30W = 240} \\
& & & \qquad\quad 60B \qquad\quad = 660
\end{array}$$

$$B = 11 \text{ mph}$$

Now we will use 11 for B in equation (a) and solve for W.

$$\begin{array}{lll}
5(11) + 5W = 70 & \qquad & \text{substituted} \\
55 + 5W = 70 & & \text{multiplied} \\
5W = 15 & & \text{added } -55 \\
W = 3 \text{ mph} & & \text{divided}
\end{array}$$

Thus, the speed of the boat in still water was 11 mph, and the speed of the current was 3 mph.

example 92.3 The water in the Flint River flows at 5 kilometers per hour. A speedboat can go 15 kilometers upstream in the same time it takes to go 25 kilometers downstream. How fast can the boat go in still water?

solution We begin by writing both equations.

$$\qquad\qquad \text{D\scriptsize OWNSTREAM} \text{ E\scriptsize QUATION} \qquad\qquad\qquad\qquad \text{U\scriptsize PSTREAM} \text{ E\scriptsize QUATION}$$

$$\text{(a)} \quad (B + W)T_D = D_D \qquad\qquad\qquad \text{(b)} \quad (B - W)T_U = D_U$$

Next we reread the equations and make the required substitutions. We use T for both T_D and T_U since these times are equal.

$$\text{(a)} \quad (B + 5)T = 25 \qquad\qquad \text{(b)} \quad (B - 5)T = 15$$

Now we multiply.

$$\text{(a)} \quad BT + 5T = 25 \qquad\qquad \text{(b)} \quad BT - 5T = 15$$

We can eliminate the double variable BT if we multiply equation (b) by (-1) and add it to equation (a).

$$\begin{array}{lrl}
\text{(a)} & BT + 5T = & 25 \\
(-1)\text{(b)} & \underline{-BT + 5T =} & \underline{-15} \\
& 10T = & 10
\end{array}$$

$$T = 1 \text{ hour}$$

Now we use 1 for T in equation (a) and solve for the speed of the boat.

$$\begin{array}{lll}
B(1) + 5 = 25 & \qquad & \text{substituted} \\
B = 20 \text{ kph} & & \text{added } -5
\end{array}$$

Thus, we find that the speed of the boat in still water is 20 kph.

practice **a.** Annie and Patrice could go 50 miles downstream in the same time it took them to go 30 miles upstream. If the speed of their boat was 4 mph in still water, what was the speed of the current and what were their times?

b. The sloop *Zollie* could go 33 miles downstream in 3 hours but required 4 hours to go 12 miles upstream. What was the speed of the boat in still water and what was the speed of the current?

problem set 92 **1.** Alonzo and Rupert could go 70 miles downstream in the same time it took them to go 30 miles upstream. If the speed of their boat was 10 mph in still water, what was the speed of the current and what were their times?

2. The motorboat *Alexis* could go 60 miles downstream in 4 hours but required 5 hours to go 55 miles upstream. What was the speed of the boat in still water, and what was the speed of the current?

3. The water in the Kern River flows at 7 kilometers per hour. A speedboat can go 21 kilometers upstream in the same time it takes to go 35 kilometers downstream. How fast can the boat go in still water?

4. The gustatory score varied directly as the number of delicious comestibles offered. If the gustatory score was 500 when 20 delicious comestibles were offered, what offering was necessary for a gustatory score of 1750? Work once using the equal ratio method and again using the direct variation method.

5. Harriet gave Wilbur a 50-yard head start. How long did it take her to catch Wilbur if her speed was 8 yards per second and his was only 6 yards per second?

Graph:

6. $\begin{cases} y \ge x + 1 \\ y > -\dfrac{4}{5}x - 1 \end{cases}$

7. $\begin{cases} y \le \dfrac{1}{2}x + 2 \\ x < 2 \end{cases}$

Solve:

8. $\begin{cases} x + y + z = 6 \\ 3x - y + z = 8 \\ x - 2y + z = 0 \end{cases}$

9. $\begin{cases} x^2 + y^2 = 10 \\ x + y = 4 \end{cases}$ $x = 4 - y$

Graph on a number line:

10. $-4 \le x - 2 < 2$; $D = \{\text{Reals}\}$

11. $x - 1 \not< 2$ or $x + 2 \not\ge 2$; $D = \{\text{Integers}\}$

Simplify:

12. $\dfrac{(y^{a+2})^2 x^{2b/3}}{x^b y^{-a}}$

13. $\dfrac{x}{y + \dfrac{y}{\dfrac{1}{x} + y}}$

14. $\dfrac{m}{a + \dfrac{a}{1 + \dfrac{a}{m}}}$

15. $\dfrac{i - 5}{7 - i}$

16. $\dfrac{3i + 2}{2i - 3}$

17. Solve: $\sqrt{z} - 3 = \sqrt{z - 27}$

18. Add: $-5\underline{/20°} + 8\underline{/-150°}$

19. Write $-5R + 20U$ in polar form.

Simplify:

20. $\dfrac{3 - 5\sqrt{2}}{2 - \sqrt{2}}$

21. $\dfrac{8 - \sqrt{2}}{4 - \sqrt{8}}$

22. $\sqrt[5]{x^4 y^3} \sqrt[3]{xy^2}$

23. Find R_1: $\dfrac{a}{by} = x\left(\dfrac{1}{R_1} + \dfrac{1}{R_2}\right)$

24. Find y: $\dfrac{a}{by} = x\left(\dfrac{1}{R_1} + \dfrac{1}{R_2}\right)$

Simplify:

25. $i^3 - 2i^4 + \sqrt{-9} - \sqrt{-2}\sqrt{-2}$

26. $\sqrt{\dfrac{5}{12}} - 3\sqrt{\dfrac{12}{5}} + 2\sqrt{60}$

27. Solve $-3x^2 - x = 5$ by using the quadratic formula.

28. Use unit multipliers to convert 15 centimeters per second to yards per minute.

29. Add: $\dfrac{4}{x^2 - 9} - \dfrac{3x}{-3 + x}$

30. Find x and y.

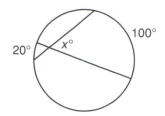
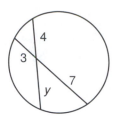

LESSON 93   *The discriminant*

In Lesson 70, we completed the square on the general form of the quadratic equation $ax^2 + bx + c = 0$ to derive the quadratic formula. This formula can be used to solve any quadratic equation.

$$x = \frac{-b \pm \sqrt{b^2 - 4ac}}{2a}$$

We remember that quadratic equations can have three different types of solutions.

(a) A single real number
(b) Two different real numbers
(c) Two different complex numbers

Which type of solution an equation will have is determined by the value of the expression

$$b^2 - 4ac$$

in the quadratic formula. Because this expression determines the type of answer we will get, we say that this expression **discriminates** between the possible types of answers. This is the reason we call this part of the quadratic formula the **discriminant**. We can see how this expression discriminates if we use the quadratic formula to solve the following quadratic equations.

(a) $x^2 + 4x + 4 = 0$ (b) $x^2 + 5x - 1 = 0$ (c) $x^2 + 5x + 7 = 0$

SOLUTION	SOLUTION	SOLUTION
$a = 1, b = 4, c = 4$	$a = 1, b = 5, c = -1$	$a = 1, b = 5, c = 7$
$x = \dfrac{-4 \pm \sqrt{4^2 - 4(1)(4)}}{2}$	$x = \dfrac{-5 \pm \sqrt{5^2 - 4(1)(-1)}}{2}$	$x = \dfrac{-5 \pm \sqrt{5^2 - 4(1)(7)}}{2}$
$= -2 \pm \dfrac{\sqrt{0}}{2}$	$= \dfrac{-5 \pm \sqrt{25 + 4}}{2}$	$= \dfrac{-5 \pm \sqrt{25 - 28}}{2}$
$x = -2$	$= \dfrac{-5 \pm \sqrt{29}}{2}$	$= \dfrac{-5 \pm \sqrt{-3}}{2}$
	$x = -\dfrac{5}{2} \pm \dfrac{\sqrt{29}}{2}$	$x = -\dfrac{5}{2} \pm \dfrac{\sqrt{3}}{2} i$

In (a) the value of $b^2 - 4ac$ is *zero*, so we get the answer

$$-2 \pm 0$$

for a solution. This has only one value, which is -2. Thus, we see that

(a) When $b^2 - 4ac$ equals zero, the solution is one real number.

In (b) the value of $b^2 - 4ac$ is 29, **a positive number,** so we get the two solutions to this equation,

$$-\frac{5}{2} + \frac{\sqrt{29}}{2} \quad \text{and} \quad -\frac{5}{2} - \frac{\sqrt{29}}{2}$$

From this we see that

   (b) **When $b^2 - 4ac$ is a positive number, there are two real number solutions.**

In (c) the value of $b^2 - 4ac$ is -3, **a negative number.** In this case, we get the following complex numbers as solutions.

$$-\frac{5}{2} + \frac{\sqrt{3}}{2}i \quad \text{and} \quad -\frac{5}{2} - \frac{\sqrt{3}}{2}i$$

From this we see that

   (c) **When $b^2 - 4ac$ is a negative number, the equation has two complex solutions which are conjugates.**

example 93.1   What kind of solutions does the equation $x^2 = -4x + 2$ have? Do not solve.

solution   We rearrange the equation into standard form and find the values of a, b, and c.

$$x^2 + 4x - 2 = 0$$

$$a = 1, b = 4, c = -2$$

Now we find the value of $b^2 - 4ac$:

$b^2 - 4ac$	discriminant
$(4)^2 - 4(1)(-2)$	substituted
$16 + 8$	multiplied
24	added

In this equation the discriminant is a positive number, so the equation has two real, unequal solutions.

example 93.2   What kind of roots does the equation $-2x = -3x^2 - 8$ have? Do not solve.

solution   We first write the equation in standard form. Then we identify a, b, and c and find the value of the discriminant $b^2 - 4ac$.

$3x^2 - 2x + 8 = 0$	standard form
$a = 3, b = -2, c = 8$	values of a, b, and c
$b^2 - 4ac$	discriminant
$(-2)^2 - 4(3)(8)$	substituted
$4 - 96$	multiplied
-92	added

The discriminant in this equation has a value of -92, which is a negative number. Thus, the solution to this equation is a pair of complex numbers that are conjugates.

practice   Use the discriminant to determine the type of solution for each of the following equations. Do not solve.

   a. $x^2 = -3x + 2$ **b.** $-4x = -3x^2 - 3$

problem set 93

1. The *Robert E. Lee* could go 45 miles down the Old Muddy in the same time it took it to go 15 miles up the same stream. If the current in the Old Muddy was 5 miles per hour, what was the speed of the *Robert E. Lee* in still water?

2. The *Memphis Belle* could go 48 miles downstream in 4 hours but required 8 hours to go 64 miles upstream. What was the speed of the *Memphis Belle* in still water and what was the speed of the current?

3. The hydro could go 40 miles per hour on a lake. The same boat could go 210 miles down the Echeconnee River in one-half the time that it took to go 380 miles up the Echeconnee River. How fast did the Echeconnee flow?

4. Charlemagne trudged the 40 miles to the battle in 4 hours longer than it took Roland to travel 48 miles to the same battle. Roland rode a horse and traveled at twice the speed of Charlemagne. Find the rates and times of both.

5. The volume of a quantity of an ideal gas was held constant at 3 liters. The initial temperature and pressure were 3727°C and 5 newtons per square meter. What would the new temperature be in kelvins if the pressure were reduced to 1 newton per square meter? Begin by adding 273 to convert degrees Celsius to kelvins.

Use the discriminant to determine the type of solution that each equation has. Do not solve.

6. $x^2 = -5x + 1$

7. $-3x = -2x^2 - 5$

8. Graph: $\begin{cases} x + 2y < 2 \\ y \geq -1 \end{cases}$

9. Graph on a number line: $-3 < x - 4 < 2$; $D = \{\text{Integers}\}$

Solve:

10. $\begin{cases} x + y - 2z = -3 \\ 2x + y + z = 7 \\ 3x - y - z = 13 \end{cases}$

11. $\begin{cases} x^2 + y^2 = 1 \\ x - 2y = 2 \end{cases}$

Simplify:

12. $(y^{a+2})^2 y^{a/3} y^2$

13. $\dfrac{x}{1 + \dfrac{a}{b + \dfrac{1}{c}}}$

14. $\dfrac{m}{2x + \dfrac{3}{3 + \dfrac{1}{m}}}$

15. $\dfrac{3i - 5}{i - 7}$

16. $\dfrac{2i + 4}{3i + 2}$

17. Solve: $\sqrt{p} = 5 + \sqrt{p - 35}$

18. The two forces are applied to the point as indicated. Find the resultant force.

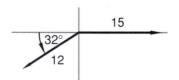

Simplify:

19. $\dfrac{3 + 4\sqrt{2}}{3\sqrt{2} - 4}$

20. $\dfrac{5 + \sqrt{5}}{5 + 2\sqrt{5}}$

21. $\sqrt[3]{4\sqrt[5]{2}}$

22. $\sqrt[4]{xy^6} \sqrt{x^3 y^5}$

23. Find a: $\dfrac{a + c}{m} = m\left(\dfrac{1}{x} + \dfrac{b}{d}\right)$

24. Find x: $\dfrac{a + c}{m} = m\left(\dfrac{1}{x} + \dfrac{b}{d}\right)$

Simplify:

25. $i^3 - 2i^2 - \sqrt{-9} + \sqrt{-4}\sqrt{-4} + 2$

26. $3\sqrt{\dfrac{2}{9}} + \sqrt{\dfrac{9}{2}} - 3\sqrt{18}$

27.   Solve $3x^2 = -x - 2$ by completing the square.

28.   Use unit multipliers to convert 10 meters per second to feet per minute.

29.   Divide $x^3 - 3x^2 - 2$ by $x + 3$.

30.   Find x and y.

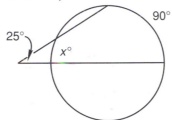   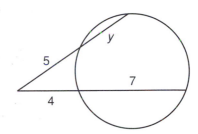

LESSON 94   *Dependent and independent variables • Functions • Functional notation*

94.A
dependent and independent variables

If we look at the equation

$$y = 2x + 4$$

we see that if we assign a value to x, then the equation will give us a value for y. For instance, if we let x equal 5, then we get

$$y = 2(5) + 4 \qquad \text{substituted 5 for } x$$
$$y = 10 + 4 \qquad \text{multiplied}$$
$$y = \mathbf{14} \qquad \text{added}$$

We find that y equals 14. It was not necessary to assign the value to x. We could have assigned a value to y and solved the equation to find the matching value of x. If we let y equal 14,

$$14 = 2x + 4 \qquad \text{substituted 14 for } y$$
$$10 = 2x \qquad \text{added} -4$$
$$\mathbf{5} = \mathbf{x} \qquad \text{divided}$$

we find that the matching value of x is 5.

   Thus we see that when we have an equation in two variables, we can assign a value to either variable and use the equation to find the value of the other variable. **We call the variable to which we assign the value the** *independent variable*, **and we call the other variable the** *dependent variable* **because its value** *depends* **on the value assigned to the independent variable.** To reduce confusion as to which variable is the dependent variable, we normally use the letters x and y to represent the variables, and we let x be the variable to which we assign values. **Thus, x will almost always be the independent variable, and the value of y will depend on the value of x.** This is the reason that we graph a line by solving the equation for y.

$$y = 2x - 3$$

x	0	2	-2
y			

Then we assign values to x and see what values of y the equation pairs with the chosen values of x.

94.B
functions

Some equations have only one answer for y for any chosen value of x. Both of the following equations

$$\text{(a)} \quad y = 2x + 4 \qquad \text{(b)} \quad y = x^3 + 4x + 3$$

are equations of this kind. For instance, if we give x a value of -3, each equation will give us one answer for y.

$$\begin{aligned}
\text{(a)} \quad y &= 2(-3) + 4 & \text{(b)} \quad y &= (-3)^3 + 4(-3) + 3 \\
y &= -6 + 4 & y &= -27 - 12 + 3 \\
\boldsymbol{y} &= \boldsymbol{-2} & \boldsymbol{y} &= \boldsymbol{-36}
\end{aligned}$$

We see that when we let x equal -3, equation (a) gives us an answer of -2 for y and equation (b) gives us an answer of -36 for y. Some people prefer not to use the word **answer**, and they would say that equation (a) **pairs** or **matches** the y value of -2 with the x value of -3. They would also say that equation (b) **pairs** or **matches** the y value of -36 with the x value of -3.

Not all equations have just one answer for y for every value of x. For example, in the equation

$$y^2 = x$$

if we let x equal 4

$$y^2 = 4$$

then both $+2$ and -2 are paired values of y, for both $(+2)^2$ and $(-2)^2$ equal 4.

$$(+2)^2 = 4 \qquad (-2)^2 = 4$$

Mathematicians have found it useful to have a special name for equations that have only one answer for y for every value of x. They call these equations functional relationships and use the word **function** when discussing these relationships. Unfortunately (or fortunately), the definition of a function has been extended to cover any situation where each member of a given set of numbers or letters has only one answer.

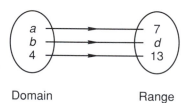

Domain   Range

Here we see that the answer for a is 7, the answer for b is d, and the answer for 4 is 13. This figure falls under the definition of a *function*, and we discover that an equation is not necessary. All we need is two sets. The first set is the values of x that we can use, and the second set is the collection of all the answers for these values of x. We remember that the **domain** is the word we use for the permissible values of x, so we call the first set the *domain of the function*. Instead of calling the answers *answers*, we call them **images. Thus, in the above diagrams, the image of a is 7, and the images of b and 4 are d and 13, respectively. We call the collection of all the images the *range* of the function.**

There are two accepted definitions for a function. One definition says that **the pairing or the matching is the function,** while the second definition says that the **ordered pairs themselves are the function.**

> A **function** is a **mapping** between two sets that associates with each element of the first set a **unique** (one and only one) element of the second set. The first set is called the **domain** of the function. For each element x of the domain, the corresponding element y of the second set is called the **image** of x under the function. The set of all images of the elements of the domain is called the **range** of the function.

> A **function** is a **set of ordered pairs** in which no two pairs have the same first element and different second elements.

Now we need to know what to think when we see the word function. We will be correct if we always think that

> A function is something that has for every value of *x* exactly one answer.

If the relationship has one or more answers, it is called a **relation.** Thus, we see that every function can also be called a relation; but every relation is not a function, for many relations have more than one answer.

We will remember that for a relationship to be called a function:

1. **The domain must be specified or implied.**
2. **A way must be designated to find every image (answer).**
3. **There is exactly one answer for every member of the domain.**

example 94.1   Which of the following depict functions?

(a) 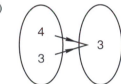   (b) 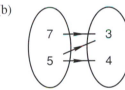   (c)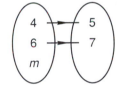

(d)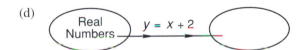

solution   Diagrams **(a)** and **(d)** depict functions because both diagrams show exactly one answer for each member of the domain. In (a) the domain is specified to be the numbers 4 and 3, and each member of the domain has one image. The image for 4 is 3, and the image for 3 is 3. Thus both images are the same, but this is acceptable. In (d) the domain is specified, and a way is given to find any image.

Diagrams (b) and (c) do not depict functions because in (b) 5 has two images and in (c) no image is designated for *m*. Thus, (c) is not even a relation.

example 94.2   Which of the following sets of ordered pairs of *x* and *y* are functions?

(a)   (4, 3), (7, 5), (−3, −2), (−6, −4)
(b)   (4, 3), (5, −2), (7, 3), (6, 3)
(c)   (5, −2), (4, −3), (7, −2), (5, 4)
(d)   (7, −2), (3, −2), (6, 5), (4, 5)

solution   **(a)**, **(b)**, and **(d)** are functions. **(c)** is not a function because 5 has two images.

94.C
functional notation

If we are given the two equations

$$y = x + 2 \qquad \text{and} \qquad y = x - 5$$

and are asked to find the value of *y* when *x* equals 2, we would want to know which equation to use. If we use functional notation, we can name the left-hand equation the *f* equation and the right-hand equation the g equation. If we do this, we will use the notations $f(x)$ and $g(x)$ instead of using y.

$$f(x) = x + 2 \qquad g(x) = x - 5$$

We read the left-hand equation as "f of x equals x plus 2" and the right-hand equation as "g of x equals x minus 5." Now if we are asked to find $g(2)$, we are asked to find the value of the g equation when x equals 2.

$$g(2) = (2) - 5 \qquad \text{replaced } x \text{ with 2}$$

$$g(2) = -3 \qquad \text{simplified}$$

Since the answer is -3 when x equals 2, we say that "g of 2 equals -3."

example 94.3 If $h(x) = 4x - 3$ and $p(x) = x^2 - 3x$, find $p(-3)$.

solution We are asked to find "p of -3," which is the value of the p equation when x equals -3.

$$p(-3) = (-3)^2 - 3(-3) \qquad \text{replaced } x \text{ with } -3$$

$$p(-3) = 9 + 9 \qquad \text{simplified}$$

$$p(-3) = 18 \qquad \text{simplified}$$

practice Which of the following sets of ordered pairs of x and y are functions?

a. $(5, 1), (4, 0), (-3, -2), (-4, 2)$ **b.** $(1, 1), (2, -1), (-5, 1), (3, 1)$

c. $(1, -1), (-1, -1), (6, -2), (4, -2)$ **d.** $(3, 2), (0, -2), (3, 0), (-2, -2)$

e. If $h(x) = 7x - 2$ and $p(x) = x^2 - 5x$, find $p(-5)$.

problem set 94

1. Use the relationship $PV = nRT$ to find the number of moles in a quantity of an ideal gas when the temperature is 473 K, the pressure is 2 newtons per square meter, and the volume is 10 liters ($R = 0.0821$). Begin by solving the abstract equation for n.

2. Analysis showed that the 40 gallons in the tank was only 20% disinfectant. A 44% disinfectant solution was needed. How many gallons of a 60% disinfectant solution should be added?

3. The *Suzie Q* could go 48 miles downstream in the same time that it took her to go 32 miles upstream. If her speed was 10 miles per hour in still water, what was the speed of the current?

4. The temperature of a quantity of an ideal gas was held constant at 430°C. The pressure was 700 newtons per square meter and the volume was 1400 ml. What was the volume when the pressure was increased to 2800 newtons per square meter? Begin by changing degrees Celsius to kelvins.

5. The number of purples varied inversely as the square of the number of reds. When 10 reds were present, the number of purples was 4. How many purples were present when the number of reds was only 5?

6. Which of the following sets of ordered pairs of x and y are functions?

    (a) $(3, 2), (4, 5), (-2, -1), (-5, -3)$

    (b) $(3, 2), (4, -1), (6, 2), (5, 2)$

    (c) $(6, -3), (5, -4), (7, -3), (6, 3)$

    (d) $(6, 2), (4, 2), (5, 4), (3, 4)$

7. If $h(x) = 3x - 1$ and $p(x) = x^2 - 2x$, find $p(-2)$.

8. Use the discriminant to determine the kinds of numbers that will satisfy the equation $-x^2 = 4x + 4$. Do not solve.

9. Graph: $\begin{cases} 2x + 3y > -6 \\ x - 3y \geq -6 \end{cases}$

10. Graph on a number line: $-x + 3 \not< -2$ or $-x + 3 < -5$; $D = \{\text{Integers}\}$

Solve:

11. $\begin{cases} x + 2y - z = 2 \\ 2x + y + z = 2 \\ 3x - y + 2z = 4 \end{cases}$ **12.** $\begin{cases} x^2 + y^2 = 3 \\ x - y = 2 \end{cases}$

Simplify:

13. $\dfrac{x^{a/3} y^2}{x^{3a/2}(y^2)^a}$ **14.** $\dfrac{p}{m + \dfrac{m}{p - \dfrac{1}{mp}}}$ **15.** $\dfrac{x}{a + \dfrac{b}{ab - \dfrac{1}{b}}}$

16. $\dfrac{2i + 7}{-3i}$ **17.** $\dfrac{-5i - 2}{-2i + 5}$

18. Solve: $\sqrt{k} + \sqrt{k - 21} = 7$ **19.** Add: $-20\underline{/-160°} + 20\underline{/200°}$

20. Write $-4R + 8U$ in polar form.

Simplify:

21. $\dfrac{2 - 5\sqrt{2}}{3\sqrt{2} - 2}$ **22.** $4\sqrt{\dfrac{5}{12}} + 3\sqrt{\dfrac{12}{5}} - 2\sqrt{240}$

23. Find c: $\dfrac{a + b}{x} = \left(\dfrac{1}{R_1} + \dfrac{c}{R_2}\right)$ **24.** Find R_1: $\dfrac{a + b}{x} = \dfrac{1}{R_1} + \dfrac{c}{R_2}$

25. Simplify: $3i^2 + 2i^5 - 2i + \sqrt{-9} - \sqrt{-2}\sqrt{2}$

26. Use similar triangles to find:
(a) M and N (b) C and D

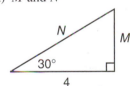

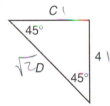

27. Solve $-2x^2 = -7 + x$ by using the quadratic formula.

28. Use unit multipliers to convert 20 liters per second to cubic inches per minute.

29. Estimate the position of the line suggested by the data points and write the equation that expresses the number of neutrons as a function of radiation: $N = mR + b$

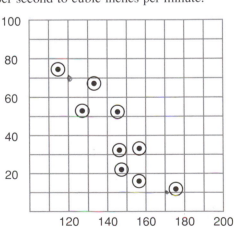

30. Add: $\dfrac{3}{a^4 x^3} + \dfrac{4}{x^2 - 9} - \dfrac{3x}{-3 + x}$

LESSON 95 *More nonlinear systems*

The nonlinear systems that we have investigated thus far have been like one of the following:

$$\text{(a)} \quad \begin{cases} BT_D + 5T_D = 25 \\ BT_D - 5T_D = 15 \end{cases} \qquad \text{(b)} \quad \begin{cases} x^2 + y^2 = 9 \\ 2x - y = 3 \end{cases}$$

On the left we have the equations of two hyperbolas, and on the right we have the equation of a circle and the equation of a straight line. In this lesson, we will look at two other types of nonlinear systems. The first will consist of the equation of a hyperbola and the equation of a straight line.

example 95.1 Solve the system: $\begin{cases} 6x - y = 5 & \text{(a)} \quad \text{(straight line)} \\ xy = 4 & \text{(b)} \quad \text{(hyperbola)} \end{cases}$

solution We cannot use elimination, for the terms in both equations are not alike. Thus, some form of substitution should work. We will solve equation (b) for x and substitute this expression for x in equation (a).

$$x = \frac{4}{y} \qquad \text{equation (b)}$$

$$6\left(\frac{4}{y}\right) - y = 5 \qquad \text{substituted}$$

$$\frac{24}{y} - y = 5 \qquad \text{multiplied}$$

Now, whenever an equation has denominators, we eliminate the denominators. Thus, we multiply every term by y and cancel the denominator.

$$y \cdot \frac{24}{y} - y \cdot y = 5y \qquad \text{multiplied every term by } y$$

$$24 - y^2 = 5y \qquad \text{simplified}$$

$$y^2 + 5y - 24 = 0 \qquad \text{rearranged}$$

$$(y - 3)(y + 8) = 0 \qquad \text{factored}$$

$$y = 3, -8 \qquad \text{zero factor theorem}$$

We finish by using both 3 and -8 in equation (b) to find the paired values of x.

$$\begin{array}{cc} \text{Using } 3 & \text{Using } -8 \\ x(3) = 4 & x(-8) = 4 \\ x = \dfrac{4}{3} & x = -\dfrac{1}{2} \end{array}$$

Thus, the ordered pairs of x and y that satisfy both equations are $\left(\frac{4}{3}, 3\right)$ and $\left(-\frac{1}{2}, -8\right)$. The graphs of the two equations are shown here, and we note that the line intersects the hyperbola at the coordinates we have found.

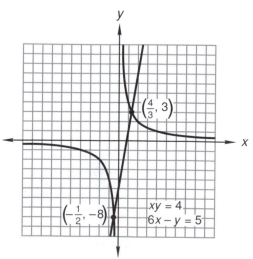

example 95.2   Solve the system: $\begin{cases} x^2 + y^2 = 9 & \text{(a)} \quad \text{(circle)} \\ 2x^2 - y^2 = -6 & \text{(b)} \quad \text{(hyperbola)} \end{cases}$

solution   This system can be solved by using either substitution or elimination. We must be careful to get all the answers because this circle and hyperbola intersect at four different points. We decide to use elimination. We can eliminate the y^2 terms if we add the equations just as they are. If we do this, we get

$$3x^2 = 3 \qquad \text{added}$$

$$x^2 = 1 \qquad \text{divided by 3}$$

Here we must be careful because this equation has both +1 and −1 as solutions.

$$x = \pm\sqrt{1} \quad \longrightarrow \quad x = 1, -1$$

Now we must use these values of x one at a time to solve for y. We will use equation (a) and begin by letting x equal +1.

$$(1)^2 + y^2 = 9 \qquad \text{substituted (1) for } x$$

$$y^2 = 8 \qquad \text{added } -1$$

$$y = \pm 2\sqrt{2} \qquad \text{solved}$$

Thus, there are two points of intersection when x equals 1. So two solutions of the system are

$$\textbf{(1, 2}\sqrt{\textbf{2}}\textbf{)} \qquad \text{and} \qquad \textbf{(1, } -\textbf{2}\sqrt{\textbf{2}}\textbf{)}$$

Next, we find the values of y that pair with a value of −1 for x. Again we use equation (a).

$$(-1)^2 + y^2 = 9 \qquad \text{substituted } (-1) \text{ for } x$$

$$y^2 = 8 \qquad \text{added } -1$$

$$y = \pm 2\sqrt{2} \qquad \text{solved}$$

Thus, our other two solutions to the system are

$$\textbf{(}-\textbf{1, 2}\sqrt{\textbf{2}}\textbf{)} \qquad \text{and} \qquad \textbf{(}-\textbf{1, } -\textbf{2}\sqrt{\textbf{2}}\textbf{)}$$

Here we show the graphs of the two curves and note that there are four points where the curves intersect.

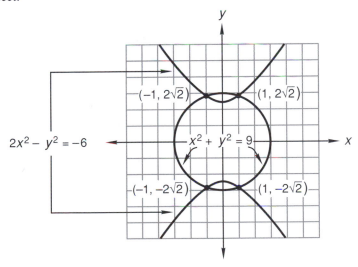

practice   Solve:

a. $\begin{cases} 3x - y = 4 \\ xy = 7 \end{cases}$

b. $\begin{cases} x^2 + y^2 = 10 \\ 3x^2 - y^2 = -2 \end{cases}$

problem set 95

1. Weatherman Bob walked to the campsite at 6 kilometers per hour. Then he caught a ride home on an old truck at 24 kilometers per hour. If he was gone for 10 hours, how far was it to the campsite?

2. The alcohol concentration had to be exactly 52%. How many milliliters of a 60% solution should be added to 200 ml of a 20% solution to get the proper concentration?

3. The pressure of a quantity of ideal gas was held constant at 764 newtons per square meter. The original temperature was 200°C, and the original volume was 400 ml. If the temperature was increased to 600°C, what was the final volume? Begin by solving for V_2. Then add 273 to convert degrees Celsius to kelvins.

4. Ronald could travel 360 miles in one-fourth the time it took Jimmy to travel 480 miles. This was because Ronald's speed exceeded that of Jimmy by 60 mph. What were the speeds and times of both?

5. Detia could row 60 miles downstream in 4 hours but required 8 hours to row 72 miles upstream. What was her speed in still water and what was the speed of the current?

Solve:

6. $\begin{cases} 4x - y = 3 \\ xy = 6 \end{cases}$

7. $\begin{cases} x^2 + y^2 = 16 \\ 2x^2 - y^2 = -1 \end{cases}$

8. $\begin{cases} x + 2y - z = 4 \\ 2x - y + z = -3 \\ x - y + z = -4 \end{cases}$

9. Which of the following depict functions?

(a)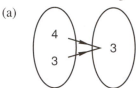

(b)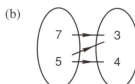

(c)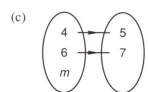

(d)
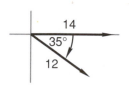

10. Use the discriminant to find the types of roots of $3x^2 - x + 5 = 0$. Do not solve.

11. Graph: $\begin{cases} x > -2 \\ 2x + 3y > -3 \end{cases}$

12. Graph on a number line: $-2 < -x - 2 < 0$; $D = \{$Integers$\}$

Simplify:

13. $\dfrac{(a^{x+2})^{1/2} x^{2a}}{a^x x^a}$

14. $\dfrac{x}{4 + \dfrac{4}{1 + \dfrac{x}{4}}}$

15. $\dfrac{5}{2 + \dfrac{1}{2 + \dfrac{2}{x}}}$

16. $\dfrac{3i - 5}{3 - 5i}$

17. $\dfrac{4 + i}{-i}$

18. Solve: $\sqrt{x^2 + 2x + 34} - x = 4$

19. The two forces are applied to the point as shown. Find the resultant force.

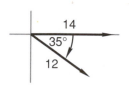

Simplify:

20. $\dfrac{2\sqrt{2} - 5}{2\sqrt{2} + 3}$

21. $\dfrac{3 - \sqrt{2}}{\sqrt{2} + 4}$

22. Find r: $\quad \dfrac{m(a+b)}{x} = \left(\dfrac{1}{p} + \dfrac{r}{q}\right)$ **23.** Find p: $\quad \dfrac{m(a+b)}{x} = \left(\dfrac{1}{p} + \dfrac{r}{q}\right)$

Simplify:

24. $4 - 3i^2 + \sqrt{-9} + \sqrt{-3}\sqrt{-3}$ **25.** $2\sqrt{\dfrac{7}{3}} - 3\sqrt{\dfrac{3}{7}} - 2\sqrt{84}$

26. Solve $-x = x^2 - 3x - 4$ by completing the square.

27. Use the formula $PV = nRT$ to find the volume of 1.32 moles of gas at a pressure of 5 atmospheres and a temperature of 600 K ($R = 0.0821$).

28. Find the equation of the line that passes through (4, 2) and is perpendicular to the line $3y - 2x = 5$.

29. Find x and y.

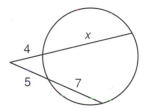   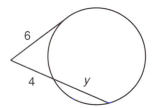

30. Use unit multpliers to convert 30 miles per hour to centimeters per second.

LESSON 96   *Joint and combined variation • More on irrational roots*

96.A
joint and combined variation

To review the concept of direct variation, we recall that if the number of peaches varies directly as the number of apples, either of the following equations may be used.

$$\text{(a)}\quad P = kA \qquad \text{(b)}\quad \frac{P_1}{P_2} = \frac{A_1}{A_2}$$

The form on the left is the direct variation form of the relationship. On the right we show the ratio form of the same relationship. Note that P_1 is on top on the left and that A_1 is on top on the right.

The word *inverse* means inverted or turned upside down, so if we are told that peaches vary inversely as apples, we remember that the relationships are the upside-down form of the direct variation equations.

$$\text{(c)}\quad P = \frac{k}{A} \qquad \text{(d)}\quad \frac{P_1}{P_2} = \frac{A_2}{A_1}$$

In equation (c) note that k remains on top and A goes below. On the right-hand side of equation (d), note that A_2 is above and A_1 is below. This is the inverse, or inverted form of equation (b).

Some statements of variation give the relationship between three or more variables. The words **varies jointly** imply a sort of double direct variation that has only one constant of proportionality. Thus, the statement that peaches vary jointly as apples and raisins implies the following relationships:

$$\text{(e)}\quad P = kAR \qquad \text{(f)}\quad \frac{P_1}{P_2} = \frac{A_1 R_1}{A_2 R_2}$$

If we are told that peaches vary inversely as apples and raisins, we must invert the variables on the right-hand side of both equations and get

$$\text{(g)} \quad P = \frac{k}{AR} \qquad\qquad \text{(h)} \quad \frac{P_1}{P_2} = \frac{A_2 R_2}{A_1 R_1}$$

Note that in (g) we have only one constant of proportionality k and that k is not inverted but remains on top.

In some relationships we have both direct and inverse variations in the same statement. For instance, the statement that girls vary inversely as boys and directly as teachers implies the equations

$$\text{(i)} \quad G = \frac{kT}{B} \qquad\qquad \text{(j)} \quad \frac{G_1}{G_2} = \frac{B_2 T_1}{B_1 T_2}$$

In equation (i), B for boys went below because boys varied inversely, and in equation (j), B_2 went above B_1 because boys varied inversely.

We note that either the variation form of the equation or the ratio form of the equation may be used. It is helpful to know how to use both forms because both approaches will be encountered in advanced courses in mathematics and science.

example 96.1   The number of girls varied inversely as the number of boys and directly as the number of teachers. When there were 50 girls, there were 20 teachers and 10 boys. How many boys were there when there were 10 girls and 100 teachers? Work the problem twice; first use the variation form and then use the ratio form.

solution   Since the boys varied inversely, B goes on the bottom.

$$G = \frac{kT}{B}$$

Now we solve for k.

$$50 = \frac{k(20)}{10} \quad\longrightarrow\quad k = 25$$

In the original equation, we replace k with 25 and get

$$G = \frac{25T}{B}$$

Next we use 10 for girls and 100 for teachers, and solve for boys.

$$10 = \frac{(25)(100)}{B} \quad\longrightarrow\quad B = \frac{2500}{10} \quad\longrightarrow\quad \mathbf{B = 250}$$

Now we will work the problem again, but this time we will use the ratio form of the equation. On the right-hand side, B_1 will go below because the variation is inverse for boys.

$$\frac{G_1}{G_2} = \frac{T_1 B_2}{T_2 B_1}$$

Now we replace G_1, T_1, and B_1 with 50, 20, and 10 and get

$$\frac{50}{G_2} = \frac{20 B_2}{T_2 (10)}$$

Now we use 10 for G_2 and 100 for T_2 and solve for B_2.

$$\frac{50}{10} = \frac{20 B_2}{100(10)} \quad\longrightarrow\quad 5 = \frac{B_2}{50} \quad\longrightarrow\quad \mathbf{B_2 = 250}$$

We could have made all the substitutions at one time, but we chose to do it in two steps because it is so easy to make a mistake when substituting for five variables.

example 96.2   Strawberries varied jointly as plums and tomatoes. If 500 strawberries went with 4 plums and 25 tomatoes, how many plums would go with 40 strawberries and 2 tomatoes? First work the problem using the variation form and then work it again using the equal ratio form.

solution   The words **varied jointly** implies the relationship

$$S = kPT$$

Next we replace S, P, and T with 500, 4, and 25 and solve for k.

$$500 = k(4)(25) \quad \longrightarrow \quad k = 5$$

Now we replace k in the original equation with 5.

$$S = 5PT$$

We finish by replacing S with 40 and T with 2 and solving for P.

$$40 = 5(P)(2) \quad \longrightarrow \quad P = \mathbf{4 \ plums}$$

Now we will work the problem again and use the ratio form. There is no inverse relationship, so S_1, P_1, and T_1 all go on top.

$$\frac{S_1}{S_2} = \frac{P_1 T_1}{P_2 T_2}$$

We first replace S_1, P_1, and T_1 with 500, 4, and 25; and we then replace S_2 and T_2 with 40 and 2.

$$\frac{500}{40} = \frac{4(25)}{P_2(2)}$$

Next we eliminate the denominators by multiplying both sides by $40P_2$.

$$40P_2 \cdot \frac{500}{40} = \frac{4(25)}{P_2(2)} \cdot 40P_2 \quad \longrightarrow \quad 500P_2 = 2000$$

We finish by dividing by 500.

$$\frac{500P_2}{500} = \frac{2000}{500} \quad \longrightarrow \quad P_2 = \mathbf{4}$$

Again we find that the answer is **4 plums.**

96.B
more
on irrational
roots

In Lesson 95 we found that the solution to the system

$$\begin{cases} 6x - y = 5 \\ xy = 4 \end{cases}$$

required the solution to the quadratic equation

$$y^2 + 5y - 24 = 0$$

This equation can be factored, and thus the solution to this system consists of ordered pairs of rational numbers (fractions). If a quadratic equation cannot be solved by factoring, we can always find the solutions by using the quadratic formula. **Trying the factor method to solve quadratic equations that result from real-life problems is usually a waste of time, and thus many people use the formula without even trying to factor.**

example 96.3   Solve: $\begin{cases} x - 2y = 3 & \text{(a)} \\ xy = 6 & \text{(b)} \end{cases}$

solution   We begin by solving equation (b) for x and substituting into equation (a).

$$x = \frac{6}{y} \qquad \text{solved for } x$$

$$\frac{6}{y} - 2y = 3 \qquad \text{substituted}$$

$$6 - 2y^2 = 3y \qquad \text{multiplied by } y$$

$$2y^2 + 3y - 6 = 0 \qquad \text{rearranged}$$

Now we use the quadratic formula to find the roots of this equation.

$$y = \frac{-b \pm \sqrt{b^2 - 4ac}}{2a} \qquad y = \frac{-3 \pm \sqrt{9 - 4(2)(-6)}}{4} \qquad y = -\frac{3}{4} \pm \frac{\sqrt{57}}{4}$$

Now we will use the linear equation $x - 2y = 3$ to find x.

$$x = 2y + 3 \qquad\qquad\qquad\qquad\qquad x = 2y + 3$$

$$x = 2\left(-\frac{3}{4} + \frac{\sqrt{57}}{4}\right) + 3 \quad \text{substituted} \qquad x = 2\left(-\frac{3}{4} - \frac{\sqrt{57}}{4}\right) + 3 \quad \text{substituted}$$

$$x = \frac{3}{2} + \frac{\sqrt{57}}{2} \qquad\qquad \text{simplified} \qquad x = \frac{3}{2} - \frac{\sqrt{57}}{2} \qquad\qquad \text{simplified}$$

Thus, the solutions are the following ordered pairs of x and y.

$$\left(\frac{3}{2} + \frac{\sqrt{57}}{2}, -\frac{3}{4} + \frac{\sqrt{57}}{4}\right) \qquad \text{and} \qquad \left(\frac{3}{2} - \frac{\sqrt{57}}{2}, -\frac{3}{4} - \frac{\sqrt{57}}{4}\right)$$

practice

a. Solve: $\begin{cases} x - 3y = 5 \\ xy = 3 \end{cases}$

b. The number of elk varied inversely as the number of deer and directly as the number of antelope. When there were 75 elk, there were 85 deer and 15 antelope. How many deer were there when there were 20 elk and 30 antelope? Work once using the variation form and again using the equal ratio form.

problem set 96

1. The number of girls varied inversely as the number of boys and directly as the number of teachers. When there were 65 girls, there were 15 teachers and 3 boys. How many boys were there when there were 5 girls and 100 teachers? Work the problem twice, once using the variation form and then using the equal ratio form.

2. Spanners varied jointly as ratchets and miters. If 100 spanners went with 4 ratchets and 5 miters, how many ratchets would go with 20 spanners and 2 miters? First work the problem using the variation form and then work it again using the equal ratio form.

3. The current in the Bolibee River flows at 6 kilometers per hour. The boat can go 40 kilometers upstream in twice the time it takes to go 80 kilometers downstream. How fast can the boat go in still water?

4. The 300-mile trip to Aunt Lucy's took 5 hours longer than the trip home. If the speed coming back was twice as great as the speed going, find both speeds and both times.

5. The only way to mix a 32% antiseptic solution was to mix a 20% solution and a 60% solution. How much of each should be used to get 500 ml of the 32% solution?

Solve:

6. $\begin{cases} x - 3y = 2 \\ xy = 4 \end{cases}$

7. $\begin{cases} x^2 + y^2 = 4 \\ 4x^2 - y^2 = -4 \end{cases}$

8. $\begin{cases} x + 3y - z = 2 \\ x + y + 2z = 6 \\ 2x + 2y - z = 2 \end{cases}$

9. Which of these sets of ordered pairs are functions?

   (a) $(4, -3), (5, -3), (-5, 2), (7, -3)$

   (b) $(6, -2), (-2, 6), (4, 6), (5, -3)$

   (c) $(4, 2), (6, 2), (5, -3), (4, 3)$

10. If $g(x) = x^2 - 4$; $D = \{\text{Integers}\}$, find $g(-2)$.

11. Graph: $\begin{cases} x - y < -2 \\ y \geq -2 \end{cases}$

12. Graph on a number line: $x + 4 \not> 2$ or $x - 4 > -1$; $D = \{\text{Reals}\}$

Simplify:

13. $\dfrac{(a^{x+4})^{1/2}\, b^{2x}}{a^{3/2}\, b^x}$

14. $\dfrac{x}{a + \dfrac{b}{a^2 + \dfrac{1}{ab}}}$

15. $2\sqrt{8\sqrt[3]{2}}$

16. $\dfrac{2i - 8}{4 - 6i}$

17. $\dfrac{i - i^2}{i}$

18. Solve: $\sqrt{s - 39} = 13 - \sqrt{s}$

19. The two forces act on the point as shown. Find the resultant force.

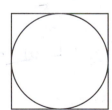

Simplify:

20. $\dfrac{3\sqrt{5} - 1}{1 - \sqrt{5}}$

21. $\dfrac{2 - \sqrt{7}}{1 + \sqrt{7}}$

22. Find p: $c = m\left(\dfrac{a + d}{p} - r\right)$

23. Find a: $c = m\left(\dfrac{a + d}{p} - r\right)$

24. Simplify: $2i^4 - i^2 - \sqrt{-16} - \sqrt{-4}\sqrt{-4}$

25. Use the relationship $PV = nRT$ to find the number of moles in a quantity of an ideal gas when the temperature is 673 K, the pressure is 5 atmospheres, and the volume is 20 liters ($R = 0.0821$).

26. Solve $4x^2 + 6 = -x$ by using the quadratic formula.

27. Solve by graphing and then find an exact solution by using either substitution or elimination:
$$\begin{cases} x + 2y = 6 \\ 2x - 5y = -10 \end{cases}$$

28. Use unit multipliers to convert 40 inches per hour to kilometers per minute.

29. Find C.

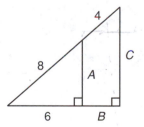

30. The area of the inscribed circle is 100π square meters. What is the area of the square?

LESSON 97 *Advanced substitution*

We have found that we can solve a system of two equations in two unknowns by using either the substitution method or the elimination method. We normally use elimination to solve systems such as

$$\begin{cases} 3x + 2y = -3 \\ 4x - 3y = 13 \end{cases}$$

in which every variable has a coefficient that is greater than 1. Substitution can be used to solve these systems if we remember to eliminate the denominators as the first step after we substitute.

example 97.1 Use substitution to solve: $\begin{cases} 3x + 2y = -3 & \text{(a)} \\ 4x - 3y = 13 & \text{(b)} \end{cases}$

solution We decide to solve equation (a) for y and substitute for y in equation (b). First, we solve for y.

$$3x + 2y = -3 \qquad\qquad \text{equation (a)}$$

$$2y = -3 - 3x \qquad\quad \text{added } -3x$$

$$y = \frac{-3 - 3x}{2} \qquad\quad \text{divided by 2}$$

Now we will substitute for y in equation (b).

$$4x - 3\left(\frac{-3 - 3x}{2}\right) = 13$$

Next we eliminate the denominator 2 by multiplying every term on both sides by 2.

$$4x(2) - 3\left(\frac{-3 - 3x}{2}\right)(2) = 13(2)$$

We cancel, multiply, and then solve.

$$8x + 9 + 9x = 26 \qquad \text{multiplied}$$

$$17x = 17 \qquad \text{simplified}$$

$$x = 1 \qquad \text{divided}$$

Now we use 1 for x in equation (a) and solve for y.

$$3(1) + 2y = -3 \qquad \text{substituted}$$

$$3 + 2y = -3 \qquad \text{multiplied}$$

$$2y = -6 \qquad \text{added } -3$$

$$y = -3 \qquad \text{divided}$$

Thus, the solution is the ordered pair **(1, −3)**.

example 97.2 Use substitution to solve: $\begin{cases} 5x - 3y = 9 & \text{(a)} \\ 2x - 4y = -2 & \text{(b)} \end{cases}$

solution We decide to solve equation (a) for x and substitute for x in equation (b). First, we solve for x.

$$5x - 3y = 9 \qquad\qquad \text{equation (a)}$$

$$5x = 9 + 3y \qquad\quad \text{added } 3y$$

$$x = \frac{9 + 3y}{5} \qquad\quad \text{divided by 5}$$

Now we will substitute for *x* in equation (b).

$$2\left(\frac{9 + 3y}{5}\right) - 4y = -2 \quad \text{substituted}$$

Next we will eliminate the denominator 5 by multiplying every term by 5.

$$(5)(2)\left(\frac{9 + 3y}{5}\right) - (5)(4y) = -2(5) \quad \text{multiplied by 5}$$

Now we cancel, expand, and solve.

$$18 + 6y - 20y = -10 \quad \text{canceled and expanded}$$
$$-14y = -28 \quad \text{added } -18$$
$$y = 2 \quad \text{divided}$$

Now we use 2 for *y* in equation (a) and solve for *x*.

$$5x - 3(2) = 9 \quad \text{substituted}$$
$$5x - 6 = 9 \quad \text{multiplied}$$
$$5x = 15 \quad \text{added } 6$$
$$x = 3 \quad \text{divided}$$

Thus, the solution is the ordered pair **(3, 2)**.

practice   Use substitution to solve: $\begin{cases} 3x - 5y = 11 \\ 2x - 4y = -6 \end{cases}$

problem set 97

1. More were talented than not. Twice the number of talented exceeded 3 times the number of untalented by 12. Also, 4 times the number of untalented was only 48 less than 3 times the number of talented. How many were talented and how many were untalented?

2. The strength of the solution had to be increased from 20% to 24%. How many milliliters of a 30% solution should be added to 300 ml of the 20% solution to get the desired result?

3. The volume of a quantity of an ideal gas was held constant at 7.4 liters. The original temperature was 227°C and the original pressure was 10 centimeters of mercury. What was the final temperature in kelvins if the pressure was increased to 30 centimeters of mercury? Remember to add 273 to convert degrees Celsius to kelvins.

4. Blues varied directly as greens and inversely as whites squared, and 4 blues and 2 whites went with 3 greens. How many greens were required for 2 blues and 4 whites? Work the problem once using the variation method and again using the equal ratio method.

5. Cheers varied jointly as the number of fans and the square of the jubilation factor. When there were 100 fans and the jubilation factor was 4, there were 1000 cheers. How many cheers were there when there were only 10 fans and the jubilation factor was 20? Work the problem two ways.

Use substitution to solve:

6. $\begin{cases} 2x + 3y = -7 \\ 3x - 3y = 12 \end{cases}$ 7. $\begin{cases} 4x - 2y = 8 \\ 3x - 2y = -4 \end{cases}$

Solve:

8. $\begin{cases} y - 2x = 3 \\ xy = 4 \end{cases}$ 9. $\begin{cases} x^2 + y^2 = 11 \\ 2x^2 - y^2 = -2 \end{cases}$ 10. $\begin{cases} 3x + y + z = 2 \\ 2x - y - z = 3 \\ x + 2y - z = 8 \end{cases}$

11. If $g(x) = x^2 - 2x + 2$; $D = \{\text{Reals}\}$, find $g(5)$.

12. Graph: $\begin{cases} x - 2y < 2 \\ y \geq 0 \end{cases}$

13. Graph on a number line: $4 \not> x + 3 < 7$; $D = \{\text{Integers}\}$

Simplify:

14. $\dfrac{(x^{a+2})^{1/2}\, x^{3a/2}\, y^{b}}{y^{-b/2}}$

15. $\dfrac{p}{mx - \dfrac{m}{x + \dfrac{1}{mx}}}$

16. $\dfrac{5i - i^2}{2i^2 + i^3}$

17. $\sqrt[6]{8\sqrt{2}}$

18. $\dfrac{5i - 2i^2}{-i}$

19. Solve: $\sqrt{z - 33} + \sqrt{z} = 11$

20. Find the resultant of the two forces shown.

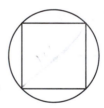

Simplify:

21. $\dfrac{3 - 2\sqrt{2}}{4 + 2\sqrt{2}}$

22. $\dfrac{3 - \sqrt{7}}{-\sqrt{7}}$

23. $\sqrt{\dfrac{9}{3}} + 2\sqrt{\dfrac{3}{9}} - 5\sqrt{27}$

24. $-4i^2 - \sqrt{-9} - \sqrt{2}\sqrt{-2} - 5i^5$

25. Find y: $\dfrac{a}{x + y} - c = \dfrac{1}{r^2}$

26. Find a: $\dfrac{p}{x} = my\left(\dfrac{1}{a} + \dfrac{1}{c}\right)$

27. A square is inscribed in a circle as shown. The area of the circle is 49π cm². What is the area of the square?

28. Begin with $ax^2 + bx + c = 0$ and derive the quadratic equation by completing the square.

29. Use unit multipliers to convert 400 milliliters per second to cubic inches per hour.

30. Solve: $\dfrac{x - 4}{2} - \dfrac{x - 6}{3} = 7$

LESSON 98 *Relationships of numbers*

We say that the real numbers constitute an ordered set because the real numbers can be arranged in definite order. Every real number has a definite relationship to every other real number. We use the number line to give us a visual representation of how real numbers can be ordered.

On this number line, we can see that 1 is greater than -3 because the graph of 1 lies to the right of the graph of -3. Also, we can see that the graph of 1 is four-sevenths of the distance from -3 to $+4$ on the number line. Furthermore, we can see that the name of a number designates its distance and direction from the origin on the number line. The graph of −3 is 3 units to the left of the origin while the graph of +4 is 4 units to the right of the origin. We will use these facts to solve problems about the order relationships of numbers.

example 98.1 Find the number that is $\frac{7}{10}$ of the way from 30 to 40.

solution First we locate the numbers on a number line.

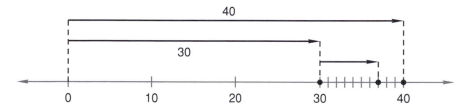

From the diagram we see that 40 is 40 units to the right of the origin and that 30 is 30 units to the right of the origin, so the distance between the numbers is $40 - 30 = 10$. Seven-tenths of 10 is 7, so the number in question lies 7 units to the right of 30 and its distance from the origin is $30 + 7 = 37$. We can write the solution to this problem in a more compact form by writing

$$N = 30 + \frac{7}{10}(40 - 30) \quad \longrightarrow \quad N = 37$$

example 98.2 Find the number that is $\frac{1}{5}$ of the way from $\frac{1}{8}$ to $\frac{9}{11}$.

solution We will refer to the solution to the last problem and see that the name of a number is its distance from the origin. We will use S to represent the smaller number and L to represent the larger number.

$$\overset{\displaystyle \overset{L-S}{\longrightarrow}}{\underset{\displaystyle \underset{S}{\bullet} \qquad\qquad\qquad \underset{L}{\bullet}}{}}$$

We see that the distance between two positive numbers is the larger number minus the smaller number. A fraction, F, of this distance is

$$F(L - S)$$

In this problem, the fraction is $\frac{1}{5}$ and the numbers are $\frac{1}{8}$ and $\frac{9}{11}$, so we have

$$\frac{1}{5}\left(\frac{9}{11} - \frac{1}{8}\right)$$

This is the length of the second arrow shown below. To this we must add the distance from the origin to the smaller number, so the number we are looking for is

$$\frac{1}{8} + \frac{1}{5}\left(\frac{9}{11} - \frac{1}{8}\right)$$

$$\overset{\displaystyle \overset{\bullet}{\longrightarrow} \; \longrightarrow}{\underset{\displaystyle \frac{1}{8}}{}} \qquad\qquad\qquad\qquad\qquad\qquad \underset{\displaystyle \frac{9}{11}}{\bullet}$$

We finish by simplifying this expression.

$$\frac{1}{8} + \frac{1}{5}\left(\frac{72}{88} - \frac{11}{88}\right) \qquad \text{common denominator}$$

$$\frac{1}{8} + \frac{1}{5}\left(\frac{61}{88}\right) \qquad\qquad \text{added}$$

$$\frac{1}{8} + \frac{61}{440} \qquad \text{multiplied}$$

$$\frac{55}{440} + \frac{61}{440} \qquad \text{common denominator}$$

$$\mathbf{\frac{29}{110}} \qquad \text{added and simplified}$$

example 98.3 Find the number that is $\frac{2}{3}$ of the way from $2\frac{1}{4}$ to $3\frac{5}{6}$.

solution The distance between the two numbers is

$$3\frac{5}{6} - 2\frac{1}{4}$$

and $\frac{2}{3}$ of this distance is

$$\frac{2}{3}\left(3\frac{5}{6} - 2\frac{1}{4}\right)$$

But this is not the distance from the origin, so to this we add $2\frac{1}{4}$. Thus, the number we want is

$$N = 2\frac{1}{4} + \frac{2}{3}\left(3\frac{5}{6} - 2\frac{1}{4}\right) \qquad \text{added } 2\frac{1}{4}$$

$$= \frac{9}{4} + \frac{2}{3}\left(\frac{23}{6} - \frac{9}{4}\right) \qquad \text{improper fractions}$$

$$= \frac{9}{4} + \frac{2}{3}\left(\frac{46}{12} - \frac{27}{12}\right) \qquad \text{common denominators}$$

$$= \frac{9}{4} + \frac{2}{3}\left(\frac{19}{12}\right) \qquad \text{added}$$

$$= \frac{9}{4} + \frac{19}{18} \qquad \text{multiplied}$$

$$= \frac{81}{36} + \frac{38}{36} = \mathbf{\frac{119}{36}} \qquad \text{added}$$

practice Find the number that is $\frac{1}{8}$ of the way from $3\frac{1}{2}$ to $5\frac{1}{6}$.

problem set 98

1. Johnny ran for a while at 12 miles per hour and then walked the rest of the way at 6 miles per hour. If he covered the 96 miles in 12 hours, how far did he run and how far did he walk?

2. The 15% alcohol solution had to be mixed by using a 10% alcohol solution and a 60% alcohol solution. How much of each should be used to get 600 ml of the 15% solution?

3. Charles and Maria found that the 1200-mile drive to the city took 4 times as long as the 360-mile drive to the mountains. If the speed driving to the mountains was 10 mph greater than the speed driving to the city, find both times and both rates.

4. Mako could row 28 miles downstream in 4 hours but required 8 hours to go 40 miles upstream. What was the speed of the current and how fast could he row in still water?

5. The number of rabbits varied directly as the number of squirrels and inversely as the number of raccoons. When there were 10 rabbits and 40 squirrels, there were only 2 raccoons. How many raccoons went with 5 rabbits and 20 squirrels? Work the problem two ways.

6. Find the number that is $\frac{1}{2}$ of the way from $\frac{1}{4}$ to $\frac{7}{12}$.

7. Find the number that is $\frac{2}{5}$ of the way from $1\frac{1}{5}$ to $2\frac{5}{6}$.

8. Use substitution to solve: $\begin{cases} 3x + 2y = 5 \\ 5x + 6y = 7 \end{cases}$

Solve:

9. $\begin{cases} y - 3x = 5 \\ xy = 6 \end{cases}$ **10.** $\begin{cases} x^2 + y^2 = 16 \\ 2x^2 - y^2 = -4 \end{cases}$

11. Which of the following sets of ordered pairs are functions?

    (a)   (5, 7), (7, 5), (−3, −2)

    (b)   (5, 7), (4, 7), (−3, 7)

    (c)   (−2, 5), (4, −2), (4, −2)

12. Graph: $\begin{cases} x + 2y > 4 \\ y \geq 1 \end{cases}$

13. Graph on a number line: $4 \not< -x + 2$ or $x + 3 < -1$; $D = \{\text{Integers}\}$

Simplify:

14. $\dfrac{(x^{2a})^{1/3} x^{2a}}{x^{a/2}}$ **15.** $\dfrac{xy}{x + \dfrac{xy}{x + \dfrac{1}{y}}}$ **16.** $\sqrt[7]{4\sqrt[3]{2}}$

17. $\dfrac{2i - 3i^2}{-i}$ **18.** $\dfrac{2i^4 - i^3}{-2i}$

19. Estimate the location of the line suggested by the data points and write the equation that expresses salt as a function of pepper: $S = mP + b$

20. Find the resultant of the forces shown.

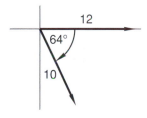

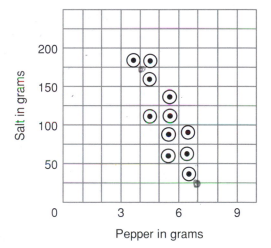

Simplify:

21. $\dfrac{5 - \sqrt{2}}{2 - 5\sqrt{2}}$ **22.** $\dfrac{3 - \sqrt{3}}{3 - 2\sqrt{3}}$

23. Find m: $\dfrac{a}{x + y} = z\left(\dfrac{1}{m} + \dfrac{1}{n}\right)$ **24.** Find y: $\dfrac{a}{x + y} = z\left(\dfrac{1}{m} + \dfrac{1}{n}\right)$

Simplify:

25. $-\sqrt{-9} - \sqrt{-2}\sqrt{-2} + 3i^2 - 2i^3 + 4$ **26.** $\sqrt{\dfrac{7}{3}} - 2\sqrt{\dfrac{3}{7}} + 5\sqrt{84}$

27. Solve $-5x^2 - x - 5 = 0$ by completing the square.

28. Use unit multipliers to convert 60 inches per second to centimeters per minute.

29. Simplify: $\dfrac{\dfrac{a}{xy} - \dfrac{x}{y^2}}{\dfrac{4}{x} - \dfrac{3}{xy^2}}$

30. Use a calculator to simplify. Estimate first.

(a) $\dfrac{47,162 \times 10^{-12}}{50,132 \times 10^5}$

(b) $\sqrt[3.4]{311}$

LESSON 99 *Absolute value inequalities • Negative numbers and absolute value*

99.A
absolute value inequalities

Absolute value inequalities are either conjunctions or disjunctions. The reason for this is evident if we draw a number line and indicate the absolute value of each number.

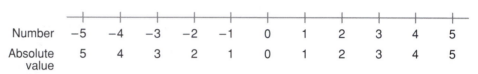

We note that except for zero, every absolute value is positive and that the numbers closest to zero have the smallest absolute values. For example, we see that every number between 3 and −3 has an absolute value that is less than 3. Further, we see that every number to the right of 3 and to the left of −3 has an absolute value that is greater than 3.

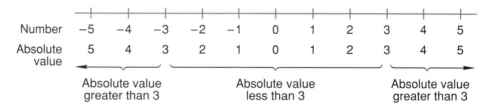

99.B
absolute value conjunctions

The graph above shows that all numbers between −3 and 3 have an absolute value that is less than 3. We can indicate these numbers in two ways. The first way is with a conjunction that does not contain absolute value.

$$x > -3 \quad \text{and} \quad x < 3$$

and the second way is to write the absolute value inequality.

$$|x| < 3$$

Thus, we see that a conjunction can be written that places the same restrictions on a variable as does an absolute value statement of *less than*.

example 99.1 Graph: $|x| \le 4$; $D = \{\text{Integers}\}$

solution This less-than absolute value inequality designates the same numbers as does the conjunction

$$x \ge -4 \quad \text{and} \quad x \le 4$$

The domain restricts the values of x to the integers, so our graph shows the integers that are greater than or equal to −4 **and** are also less than or equal to +4.

example 99.2   Graph: $-|x| + 3 > 0$; $D = \{$Integers$\}$

solution   First we add -3 to both sides. We **do not** reverse the inequality symbol when we add negative quantities to both sides.

$$\begin{array}{rcr} -|x| + 3 &>& 0 \\ -\ 3 && -3 \\ \hline -|x| &>& -3 \end{array}$$

Now we mentally multiply both sides by -1 and reverse the inequality symbol.

$$|x| < 3$$

Now we can replace this absolute value inequality with a conjunction that places the same restrictions on the variable.

$$x > -3 \quad \text{and} \quad x < 3$$

Thus, we are asked to indicate the integers that are less than $+3$ and that are greater than -3. The graph is

99.C
absolute value disjunctions

Since the numbers whose absolute values are greater than 3

$$|x| > 3$$

graph to the right of 3 or to the left of -3, we can make the same statement by writing the disjunction

$$x < -3 \quad \text{or} \quad x > 3$$

Thus, we see that a disjunction can be written that places the same restriction on the variable as does an absolute value statement of *greater than*.

example 99.3   Graph: $-|x| + 2 < -2$; $D = \{$Reals$\}$

solution   We begin by adding -2 to both sides of the inequality. We **do not** reverse the inequality symbol when we add a negative quantity.

$$\begin{array}{rcr} -|x| + 2 &<& -2 \\ -\ 2 && -2 \\ \hline -|x| &<& -4 \end{array}$$

Now we mentally multiply both sides by -1 and reverse the inequality symbol.

$$|x| > 4$$

This absolute value statement of greater than can be replaced with the disjunction

$$x > 4 \quad \text{or} \quad x < -4$$

The graph of this disjunction, using the real numbers as the domain, is

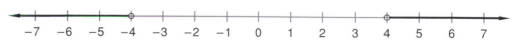

99.D
negative numbers and absolute value

The preceding absolute value inequalities stated that the absolute value of x was greater than or less than a given positive number. If we look at a number line on which the absolute values have been indicated,

Number	-5	-4	-3	-2	-1	0	1	2	3	4	5
Absolute value	5	4	3	2	1	0	1	2	3	4	5

we note that, except for zero, all the absolute values are positive numbers. **There are no absolute values that are less than zero.** Thus, a notation such as

$$|x| < -3$$

has no solution. **If the smallest absolute value is zero, then there is no absolute value less than zero and certainly no absolute value that is less than -3.** In the same way, the solution to the inequality

$$|x| > -3$$

is any number because any number (even zero) has an absolute value greater than -3.

example 99.4   Graph: (a) $|x| < -2$; $D = \{\text{Reals}\}$ (b) $|x| > -2$; $D = \{\text{Reals}\}$

solution   These can be thought of as trick questions because the answers are either all the numbers or none of the numbers. The answer to (a) is none of the numbers. We write this answer by using the symbol for the null set or the empty set.

(a) $x = \emptyset$ or $x = \{\ \}$

(b) Every real number has an absolute value greater than -2, so the graph shows every real number.

example 99.5   Graph: $-|x| - 5 > -3$; $D = \{\text{Integers}\}$

solution   We begin by adding $+5$ to both sides, and we get

$$-|x| > 2$$

Now we mentally multiply both sides by -1 and reverse the inequality symbol.

$$|x| < -2$$

There are no real numbers that satisfy this inequality, and thus the original inequality has no solution.

practice   Graph: $-|x| + 3 < -2$; $D = \{\text{Integers}\}$

problem set 99

1.   If the silver could be separated from the sulfur, the reaction would be a success. What is the percentage by weight of the silver (Ag) in the compound Ag_2S? (Ag, 108; S, 32)

2.   Some walked purposefully, and some merely maundered. Twice the number of the purposeful walkers exceeded 10 times the number of maunderers by 16. Also, 13 times the number of maunderers exceeded the number of purposeful walkers by only 8. How many walked purposefully and how many just maundered?

3.   The current in the river was 8 miles per hour. The boat could go 60 miles upstream in one-half the time it took to go 280 miles downstream. How fast could the boat go in still water?

4.   The work accomplished varied jointly as the number of people and their average productivity factor. If 100 people with an average productivity factor of 20 could produce 8000 units on one shift, how many people whose factor was only 2 would be required to produce 16,000 units on one shift? Work the problem two ways.

5.   The first container held a 5% iodine solution and the second container held a 10% iodine solution. How much of each should Nadine and Bob use to get 1200 ml of an 8% iodine solution?

Graph on a number line:

6. $-|x| + 5 > 0$; $D = \{\text{Integers}\}$ 7. $-|x| + 1 < -3$; $D = \{\text{Reals}\}$

8. Use similar triangles to find:

(a) m and n

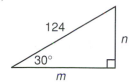

(b) p and q

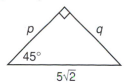

9. Find the number that is $\dfrac{2}{7}$ of the way from $\dfrac{1}{2}$ to $2\dfrac{1}{3}$.

10. Use substitution to solve: $\begin{cases} 3x + 3y = 9 \\ 4x - 6y = -8 \end{cases}$

Solve:

11. $\begin{cases} x - 3y = 2 \\ xy = 8 \end{cases}$

12. $\begin{cases} x^2 + y^2 = 8 \\ 2x^2 - y^2 = 7 \end{cases}$

13. $\begin{cases} 3x + 2y - z = 1 \\ x + y - z = -1 \\ 5x + 2y + 2z = 8 \end{cases}$

14. If $h(x) = x^2 - 4$; $D = \{$Negative integers$\}$, find $h(4)$.

Graph:

15. $\begin{cases} x - y < -2 \\ 3x + 5y \le -5 \end{cases}$

16. $x + 4 \not> 3$ or $x + 1 \not\le 2$; $D = \{$Reals$\}$

Simplify:

17. $(x^{2-a})^2 x^{a/4}$

18. $\dfrac{ab}{a + \dfrac{b}{1 + \dfrac{a}{b^2}}}$

19. $2\sqrt[5]{16\sqrt[3]{2}}$

20. $\dfrac{2 - 3i}{2i^3}$

21. $\dfrac{2i - i^3}{-i}$

22. Estimate the location of the line suggested by the data points and write the equation that expresses output as a function of input: $O = mI + b$

23. Find the resultant of the forces shown.

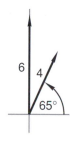

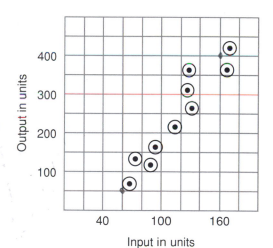

Simplify:

24. $\dfrac{3 - \sqrt{2}}{\sqrt{2} + 3}$

25. $\dfrac{4 - 2\sqrt{3}}{-\sqrt{3}}$

26. Find c: $\dfrac{m}{c} = p\left(\dfrac{1}{a} + \dfrac{1}{b}\right)$

Simplify:

27. $-\sqrt{-9} - 3i^3 + 2i - \sqrt{-4}\sqrt{4} + 3\sqrt{-4}$

28. $-\sqrt{\dfrac{3}{10}} + 4\sqrt{\dfrac{10}{3}} - \sqrt{120}$

29. Use the formula $PV = nRT$ to find the volume of 0.0163 mole of ideal gas at 10 atmospheres of pressure and a temperature of 870 K ($R = 0.0821$).

30. Use unit multipliers to convert 40 centimeters per second to miles per hour.

LESSON 100   *Graphs of parabolas*

We remember that a polynomial equation in one variable in which the highest power of the variable is 2 is called a quadratic equation. Thus the following equations are quadratic equations in x.

(a) $x^2 + 4x + 4 = 0$ 　　　 (b) $-x^2 + 4x - 3 = 0$ 　　　 (c) $x^2 - 3 = 0$

These equations are conditional equations. The values or value of x that makes one of these equations a true equation is called the solution of the equation and is said to satisfy the equation. We can find the solutions to these quadratic equations by factoring, completing the square, or using the quadratic formula. **These equations cannot be graphed on a coordinate plane because they contain only one unknown.**

If we set each of these expressions equal to y instead of zero, we will have written special kinds of quadratic equations that define quadratic functions.

(d) $y = x^2 + 4x + 4$ 　　　 (e) $y = -x^2 + 4x - 3$ 　　　 (f) $y = x^2 - 3$

These equations can be graphed because we can assign values to x and find the resulting values for y. **Any function can be graphed by using the laborious method of assigning many values to x and finding the resulting values for y.** To graph equations (d) and (e), we would make a table for each equation and choose values for x.

(d) $y = x^2 + 4x + 4$ 　　　　　　　　　　 (e) $y = -x^2 + 4x - 3$

x	0	2	3	-2	-3
y					

x	0	2	3	-2	-3
y					

Next, we use the equations to find the value of y that is paired with each value of x. Then we enter these values in the table.

(d) $y = x^2 + 4x + 4$ 　　　　　　　　　　 (e) $y = -x^2 + 4x - 3$

x	0	2	3	-2	-3
y	4	16	25	0	1

x	0	2	3	-2	-3
y	-3	1	0	-15	-24

We can immediately see the disadvantage of point-by-point graphing. In the table on the left, only the ordered pairs $(0, 4)$, $(-2, 0)$, and $(-3, 1)$ are usable because the other ordered pairs have y values too large to plot on a small graph. To get a good graph, we would have to use other values of x until we found more usable ordered pairs. We would also need to use other values of x to get more usable pairs for the table on the right. Finally, we would get enough usable ordered pairs to draw the following graphs. We call the graph of a quadratic equation a **parabola.**

(d)   Opens upward   (e)   Opens downward

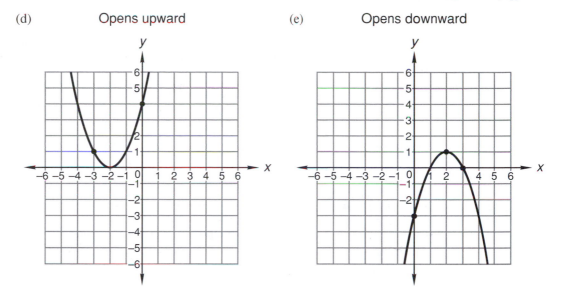

If we graph many, many quadratic equations, we will find that the following relationships between the function and the graph of the function are always true.

1. Every parabola is symmetric about some vertical line. This vertical line is called the **axis of symmetry.**

2. Every parabola opens upward or it opens downward. If the coefficient of x^2 is a positive number, the graph opens upward. If the coefficient of x^2 is a negative number, the graph opens downward. This is true regardless of the values of the other numbers in the equation.

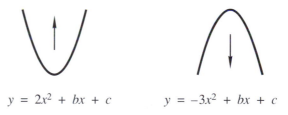

$$y = 2x^2 + bx + c \qquad\qquad y = -3x^2 + bx + c$$

3. The parabola crosses the y axis at the value of the constant term in the equation. This is because the value of x is zero on the y axis. If we let x equal zero, all terms that contain x equal zero and all that is left is the constant term.

$$y = -3x^2 + 14x - 5 \qquad\qquad \text{equation}$$
$$y = -3(0)^2 + 14(0) - 5 \qquad \text{let } x = 0$$
$$y = 0 + 0 - 5 \qquad\qquad\quad \text{simplified}$$
$$y = -5 \qquad\qquad\qquad\quad 0 + 0 = 0$$

   If we change the value of the constant term in the equation, the graph will have the same centerline and the same shape but will be shifted up or down. The only difference in each of the following equations is the constant term. Note that the change in the constant term shifts the curves up or down.

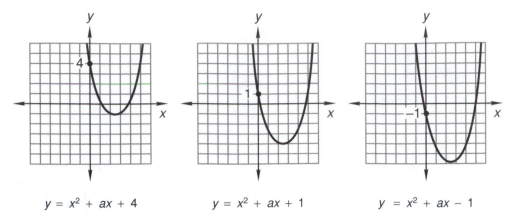

$$y = x^2 + ax + 4 \qquad y = x^2 + ax + 1 \qquad y = x^2 + ax - 1$$

Thus, we can look at a quadratic equation and determine the y intercept and determine whether the graph opens upward or downward. Let's look at two more equations.

(a) $y = -x^2 - 8x - 13$ (b) $y = x^2 - 8x + 13$

The graph of equation (a) will open downward because the coefficient of x^2 is -1 and the y intercept is $(0, -13)$. The graph of equation (b) will open upward because the coefficient of x^2 is $+1$ and the y intercept is $(0, 13)$.

If we complete the square on equations (a) and (b), we can change the forms of the equations so that by inspection we can determine the following.

1. The equation of the axis of symmetry
2. The y coordinates of the vertex

The new forms of the equations are shown here, along with an analysis of the new forms.

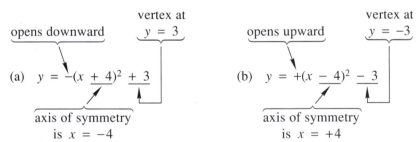

Note that when we have $(x + 4)^2$, the x value of the axis of symmetry is -4, not $+4$! Also, when we have $(x - 4)^2$, the x value of the axis of symmetry is $+4$, not -4!

The squared term is always positive, and for large values of x, this term approximates the value of y. When the squared term is preceded by a plus sign, y is positive for both large positive and negative values of x; and thus the curve opens upward. When the squared term is preceded by a minus sign, y is negative for both large positive and negative values of x; and the curve opens downward.

When the value of x is such that the squared term equals zero, the curve is at a maximum or minimum point; and the y coordinate of the vertex is the constant at the end of the expression.

We will show how to use completing the square as an aid to graphing in the next three examples.

example 100.1   Complete the square to graph $y = x^2 - 4x + 2$.

solution   We begin by noting the following.

(a) The graph will open upward (the coefficient of x^2 is positive).

(b) The y intercept is $(0, 2)$ (the constant term is $+2$).

Now we want to rearrange the equation into the form

$$y = (x + a)^2 + k$$

so we place parentheses around the x^2 term and the x term.

$$y = (x^2 - 4x \quad) + 2$$

Now, to make the expression inside the parentheses a perfect square, it is necessary to add the square of one-half the coefficient of x

$$\left(-4 \cdot \frac{1}{2}\right)^2 = 4$$

which is 4. **Thus, we add +4 inside the parentheses and −4 outside the parentheses. This addition of +4 and −4 to the same side of the equation is a net addition of zero.**

$$y = (x^2 - 4x + 4) + 2 - 4$$

Now the term in the parentheses is a perfect square and we write it as such.

$$y = (x - 2)^2 - 2$$

From this form we can determine the three things necessary to sketch the curve.

(a)   Opens upward

(b)   Axis of symmetry is $x = 2$ $\qquad y = +(x - 2)^2 - 2$

(c) y coordinate of vertex is -2

We use this information to draw the axis of symmetry and the vertex of the curve, as we show in the following figures. We also put a dot at $(0, 2)$, which is the y intercept. Now we get the coordinates of one more point on the graph and make a sketch. Let's take $x = 4$ in the original equation and solve for y.

$$y = (4)^2 - 4(4) + 2 \qquad \text{substituted}$$

$$y = 16 - 16 + 2 \qquad \text{multiplied}$$

$$y = 2 \qquad \text{simplified}$$

Thus, the point $(4, 2)$ lies on the curve. We remember that the curve is symmetric about the line $x = 2$ and complete the sketch.

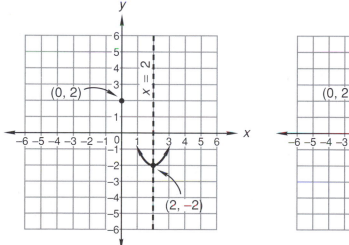
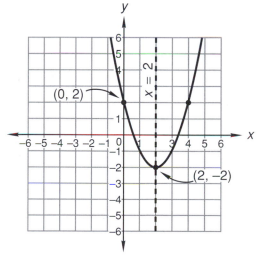

example 100.2   Complete the square to graph $y = -x^2 + 2x + 2$.

solution   We begin by noting that the parabola will open downward, and that the y intercept is $(0, 2)$. The rest of the procedure will be the same, except that, as the first step, we will multiply both sides by –1 so that the coefficient of x^2 will be +1. As the last step, we will again multiply both sides by –1.

$$-y = x^2 - 2x - 2 \qquad \text{multiplied by } -1$$

Now to change the right side into the form

$$(x + a)^2 + k$$

we place parentheses around $x^2 - 2x$

$$-y = (x^2 - 2x \qquad) - 2$$

and add +1 inside the parentheses and –1 outside the parentheses.

$$-y = (x^2 - 2x + 1) - 2 - 1$$

Now the expression in the parentheses is a perfect square.

$$-y = (x - 1)^2 - 3$$

As the last step, we multiply both sides by –1 so that y will be positive. **Note that we do not change the sign inside the parentheses!**

$$y = -(x - 1)^2 + 3$$

Now we can read the salient features of the graph.

(a) Opens downward

(b) Axis of symmetry is $x = 1$ $\qquad y = -(x - 1)^2 + 3$

(c) y coordinate of vertex is +3

We already know that the y intercept is (0, 2).

To find another point on the curve, we let $x = -2$ in the original equation and solve for y.

$$y = -(-2)^2 + 2(-2) + 2 \qquad \text{substituted}$$
$$y = -4 - 4 + 2 \qquad \text{multiplied}$$
$$y = -6 \qquad \text{simplified}$$

We complete the graph by using the point $(-2, -6)$, remembering that the curve is symmetric about the line $x = 1$.

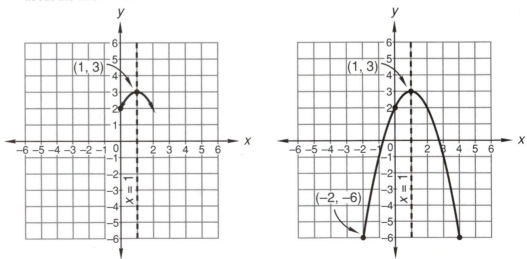

example 100.3 Complete the square to graph $y = -x^2 - 6x - 8$.

solution We begin by noting that the graph of the parabola will open downward, and that the y intercept is $(0, -8)$. Since the x^2 term is negative, we will begin and end by multiplying both sides of the equation by –1.

$$-y = (x^2 + 6x \qquad) + 8 \qquad \text{multiplied by } -1$$

$$-y = (x^2 + 6x + 9) + 8 - 9 \qquad \text{added } +9 \text{ and } -9$$

$$-y = (x + 3)^2 - 1 \qquad \text{simplified}$$

$$y = -(x + 3)^2 + 1 \qquad \text{multiplied by } -1$$

Now we can diagnose the salient features of the graph.

(a)   Opens downward ————————————┐

(b)   Axis of symmetry is $x = -3$ ╲ $\quad y = \overset{\downarrow}{-}(x + 3)^2 + 1$

(c) y coordinate of vertex is $+1$ ╲

On the left, we use these facts to begin the curve. To find another point on the curve, we replace x with -1 and find that y equals -3. Then we use the point $(-1, -3)$ and symmetry to complete the graph.

$$y = -(-1)^2 - 6(-1) - 8 \qquad \text{substituted}$$

$$y = -1 + 6 - 8 \qquad \text{multiplied}$$

$$y = -3 \qquad \text{simplified}$$

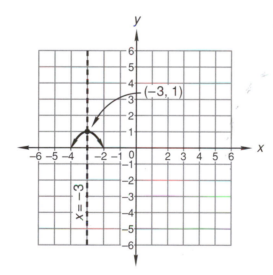
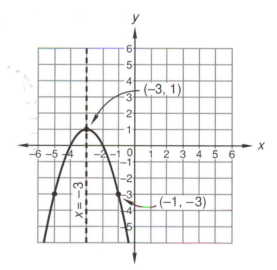

practice   Complete the square as an aid in graphing:

   a. $y = x^2 - 2x + 3$ **b.** $y = -x^2 - 4x - 3$

problem set 100

1.   Bickford traveled twice as fast as Shawn traveled. Thus, Bickford could travel the 320 miles to the reef in only 2 hours less than it took Shawn to travel the 240 miles to Jane's house. Find the rates and times of both boys.

2.   The solution was 68% alcohol, and it came up to the 600-ml mark on the beaker. How much alcohol should be evaporated so that the remainder would be only 20% alcohol?

3.   Some sparkled, and the rest coruscated. Ten times the number of sparklers exceeded 6 times the number that coruscated by 40. But 4 times the number that coruscated exceeded twice the number that sparkled by only 160. How many were in each category?

4.   Reds varied directly as yellows and inversely as greens squared; 100 reds and 40 yellows went with 10 greens. How many reds went with 20 yellows and only 5 greens?

5.   Find three consecutive integers such that the product of the first and the third exceeds the product of 8 and the second by 32.

Complete the square as an aid in graphing:

6. $y = x^2 - 6x + 1$

7. $y = -x^2 + 4x + 4$

8. Use similar triangles to solve for:

(a) m and n

(b) p and q

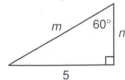

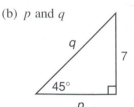

9. Graph on a number line: $-|x| + 5 \nleq 3$; $D = \{\text{Reals}\}$

10. Find the number that is $\frac{5}{11}$ of the way from $2\frac{1}{3}$ to $3\frac{1}{6}$.

11. Use substitution to solve: $\begin{cases} 2x - 2y = -1 \\ 4x + 3y = 5 \end{cases}$

Solve:

12. $\begin{cases} x - y = 5 \\ xy = 2 \end{cases}$

13. $\begin{cases} x^2 + y^2 = 5 \\ 2x^2 - y^2 = 4 \end{cases}$

14. $\begin{cases} 2x + 2y - z = 0 \\ x + y - 2z = -12 \\ 2x - y + z = 10 \end{cases}$

15. Which of the following sets of ordered pairs are functions?

(a) (4, 2), (5, 2), (2, 5), (2, 4)

(b) (−7, 2), (4, 2), (2, 4), (2, −7)

(c) (−3, 2), (3, −2), (−3, 4), (−2, 3)

Graph:

16. $\begin{cases} y \le -3 \\ 4x + y < -2 \end{cases}$

17. $x + 2 \nleq 5$ or $x + 3 < 3$; $D = \{\text{Integers}\}$

Simplify:

18. $\dfrac{(a^{x+2})^2 a^{b/2}}{(a^{2-b})^{1/2}}$

19. $\dfrac{x}{x^2 y - \dfrac{1}{1 + \dfrac{1}{xy}}}$

20. $3\sqrt{9\sqrt{3}}$

21. $\dfrac{2 - i^3}{-i}$

22. $\dfrac{3 - i}{i^2 - 3i}$

23. Determine the resultant of the force vectors shown.

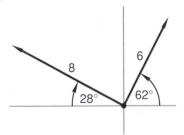

Simplify:

24. $\dfrac{5 - 2\sqrt{2}}{\sqrt{2}}$

25. $\dfrac{3 - 4\sqrt{5}}{\sqrt{5} - 1}$

26. $\sqrt{\dfrac{2}{13}} + 3\sqrt{\dfrac{13}{2}} - 2\sqrt{104}$

27. $-\sqrt{-7}\sqrt{-7} + 2\sqrt{-16} - 3i^5 + 2i^2$

28. Find y: $\dfrac{a}{k + c} = m\left(\dfrac{x}{y} + d\right)$

29. Use unit multipliers to convert 400 cubic feet per minute to cubic inches per second.

30. Use a calculator to simplify. Estimate first.

(a) $\dfrac{0.001852 \times 10^{-3}}{-47.3 \times 10^5}$

(b) $(4.15)^{-5.06}$

LESSON 101   *Percent markups*

The difference between the selling price of a piece of merchandise and the price paid for it is called **markup**. For example, if a dealer purchased an item from the factory for $40 and sold it for $50, we say that the markup was $10. Ten dollars is 25 percent of $40 but is only 20 percent of $50. Thus, the markup was 20 percent of the selling price and 25 percent of the purchase price. The store owner would say that the markup was only 20 percent, but the customer would say that the markup was 25 percent.

The selling price is the purchase price plus the markup.

$$\text{Selling price} = \text{purchase price} + \text{markup}$$

Any one of these components can be expressed as a percentage of any other component.

When we work these problems, we will use rate instead of percent. Rate is percent divided by 100 and is the decimal form of the relationship. Thus,

20 percent	equals a rate of	0.2
415 percent	equals a rate of	4.15
2 percent	equals a rate of	0.02

example 101.1   The selling price was $48. If the markup was 20 percent of the purchase price, what was the purchase price? What was the markup?

solution   First we write

$$\text{Selling price} = \text{purchase price} + \text{markup}$$

The selling price was 48 and the markup was 20 percent of the purchase price, or $0.2P_P$. We substitute and find

$$48 = P_P + 0.2P_P \qquad \text{substituted}$$
$$48 = 1.2P_P \qquad \text{simplified}$$
$$\mathbf{40} = \mathbf{P_P} \qquad \text{divided}$$

Thus the purchase price was **$40**, and the markup was **$8**.

example 101.2   The purchase price of the item was $1800. If the markup was 40 percent of the selling price, what was the selling price?

solution   First we write

$$\text{Selling price} = \text{purchase price} + \text{markup}$$

The markup was 40 percent of the selling price, or $0.4S_P$, and the purchase price was $1800. We make these substitutions and solve.

$$S_P = 1800 + 0.4S_P \qquad \text{substituted}$$
$$0.6S_P = 1800 \qquad \text{added } -0.4S_P$$
$$\mathbf{S_P = 3000} \qquad \text{divided}$$

Thus, the markup was $1200, which is 40 percent of $3000.

example 101.3   The sports car retailed for $10,368. What was the purchase price if the car had been marked up 8 percent of the purchase price?

solution   First we write

$$\text{Selling price} = \text{purchase price} + \text{markup}$$

The markup was 8 percent of the purchase price, so we get

$$10{,}368 = P_P + 0.08P_P \quad \text{substituted}$$

$$10{,}368 = 1.08P_P \quad \text{simplified}$$

$$\mathbf{9600 = P_P} \quad \text{divided}$$

Thus the car had been purchased for $9600, and the markup was $768.

practice The sedan retailed for $16,295. What was the purchase price if the car had been marked up 25 percent of the purchase price?

**problem set
101**

1. The selling price was $78. If the markup was 30 percent of the purchase price, what was the purchase price? What was the markup?

2. The purchase price of the item was $2100. If the markup was 60 percent of the selling price, what was the selling price?

3. The sports car retailed for $16,535. What was the purchase price if the car had been marked up 25 percent of the purchase price?

4. Schneider has 400 liters of a solution that is 60% antifreeze. How many liters of an 80% solution should he add to make a solution that is 72% antifreeze?

5. The boat could go 104 miles downstream in the same time it took to go 56 miles upstream. If the speed of the boat was 20 miles per hour in still water, what was the speed of the current?

Complete the square as an aid in graphing:

6. $y = -x^2 - 4x - 5$ 7. $y = x^2 + 2x + 2$

8. The volume of a right circular cone whose radius is 3 cm is $60\pi \text{ cm}^3$. What is the height of the cone?

9. Graph on a number line: $-|x| + 3 \not< 2; \ D = \{\text{Integers}\}$

10. Find the number that is $\dfrac{2}{5}$ of the way from $3\dfrac{1}{3}$ to $4\dfrac{5}{6}$.

11. Use substitution to solve: $\begin{cases} 6x + 5y = 8 \\ 4x + 2y = 3 \end{cases}$

Solve:

12. $\begin{cases} 3x - y = 4 \\ xy = 5 \end{cases}$ 13. $\begin{cases} x^2 + y^2 = 16 \\ 2x^2 - y^2 = 2 \end{cases}$ 14. $\begin{cases} 3x + y + z = 7 \\ x - 2y - z = 2 \\ -x + y - z = -5 \end{cases}$

15. If $p(x) = x^2 - 4; \ D = \{\text{Reals}\}$, find $p\left(\frac{1}{2}\right)$

Graph:

16. $\begin{cases} 3y - 2x > -3 \\ x \geq -2 \end{cases}$ 17. $-10 \not\geq x + 2 < -4; \ D = \{\text{Reals}\}$

Simplify:

18. $\dfrac{(a^{\,x+2})^{1/2}\, y}{(y^2)^a}$ 19. $\dfrac{m}{my - \dfrac{1}{1 - \dfrac{1}{my}}}$ 20. $7\sqrt{49\sqrt[3]{7}}$

21. $\dfrac{-2 - 3i}{3 + 2i}$ 22. $\dfrac{i}{2i - 3}$

23. Determine the resultant of the two vectors.

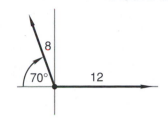

Simplify:

24. $\dfrac{2\sqrt{2} - \sqrt{3}}{\sqrt{6}}$ **25.** $\dfrac{3 - \sqrt{2}}{\sqrt{2} + 2}$

26. $2\sqrt{\dfrac{2}{5}} - 4\sqrt{\dfrac{5}{2}} + 2\sqrt{40}$ **27.** $-3i^2 - 2i^3 + 4 - i^5 + \sqrt{-9}$

28. Find R_1: $\dfrac{m}{x} - c = \dfrac{1}{R_1} + \dfrac{1}{R_2}$

29. Solve $-2x^2 = x + 4$ by using the quadratic formula.

30. Multiply: $\dfrac{2x^0 y^{-2} p^{-4}}{x^2 p^{-4}} \left(\dfrac{yp^4 m}{x^{-2}} - 3xy^0 \right)$

LESSON *102* *Sums of functions* • *Products of functions*

102.A
sums of functions

Recall that functional notation has several advantages. The first is that it allows us to identify the equations being considered. For instance, if we have the three equations

$$y = x + 3 \qquad y = x^2 - 6 \qquad y = 2x^2 + 5$$

we can name them from left to right as, say, equation h, equation ϕ, and equation g. If we use functional notation, we would use $h(x)$, $\phi(x)$, and $g(x)$ instead of y when we write the equations.

$$h(x) = x + 3 \qquad \phi(x) = x^2 - 6 \qquad g(x) = 2x^2 + 5$$

These notations in themselves do not completely define the function, for the domain must be specified or implied for every function. Thus, we will arbitrarily assign a domain to each equation.

$$h(x) = x + 3 \qquad \phi(x) = x^2 - 6 \qquad g(x) = 2x^2 + 5$$
$$D = \{\text{Reals}\} \qquad D = \{\text{Integers}\} \qquad D = \{\text{Negative integers}\}$$

If we add two of these equations, we get an equation for the sum. The sum of equation h and equation g is

$$\begin{aligned} h(x) &= x + 3 \\ g(x) &= 2x^2 + 5 \\ \hline h(x) + g(x) &= 2x^2 + x + 8 \end{aligned}$$

We see that $h(x) + g(x)$ means that we have added the h equation to the g equation. **Often we use the notation**

$$(h + g)(x)$$

to mean the same thing. The notation $(h + g)(x)$ means that we have added the h equation to the g equation. The domain for the new equation is all numbers that were common to both of the original domains.

example 102.1   Given $h(x) = x + 3$; $D = \{\text{Reals}\}$, and $\phi(x) = x^2 - 6$; $D = \{\text{Integers}\}$, find $(h + \phi)(2)$.

solution   We can find the answer two ways. First we will find $h(2)$ and $\phi(2)$ and add.

$$h(2) = 2 + 3 \qquad \phi(2) = (2)^2 - 6$$
$$h(2) = 5 \qquad \phi(2) = -2$$

Thus, since $(h + \phi)(2)$ means $h(2) + \phi(2)$, we have

$$(h + \phi)(2) = (5) + (-2) = \mathbf{3}$$

The second way is to add the equations to find the equation $(h + \phi)(x)$.

$$h(x) = x + 3$$
$$\underline{\phi(x) = x^2 - 6}$$
$$(h + \phi)(x) = x^2 + x - 3$$

Now we use 2 for x and find the value of the $h + \phi$ equation when x equals 2.

$$(h + \phi)(2) = (2)^2 + (2) - 3$$
$$= 4 + 2 - 3$$
$$\mathbf{(h + \phi)(2) = 3}$$

We get the same answer both ways.

example 102.2   Given $h(x) = x + 3$; $D = \{\text{Reals}\}$, and $g(x) = 2x^2 + 5$; $D = \{\text{Negative integers}\}$, find $(h + g)(5)$.

solution   We can find the equation $(h + g)(x)$ by adding the h equation and the g equation.

$$h(x) = x + 3$$
$$\underline{g(x) = 2x^2 + 5}$$
$$(h + g)(x) = 2x^2 + x + 8$$

However, we cannot use this equation to find $(h + g)(5)$ because 5 was not a member of the domain of $g(x)$ so 5 is not a member of the domain of $(h + g)(x)$. Thus, we say that the problem has no answer, or we can say that the answer is the null set $\varnothing$ or the empty set $\{\ \}$. Remember that null set and empty set mean the same thing.

102.B
products of functions

When we multiply two functions, the product is also a function. If we have the equations

$$h(x) = x + 3 \qquad \text{and} \qquad g(x) = x^2 - 6$$
$$D = \{\text{Reals}\} \qquad\qquad D = \{\text{Negative integers}\}$$

and we multiply the h equation by the g equation, we get the product $h(x)g(x)$.

$$h(x)g(x) = (x + 3)(x^2 - 6) \qquad \text{product of functions}$$
$$h(x)g(x) = x^3 - 6x + 3x^2 - 18 \qquad \text{multiplied}$$
$$h(x)g(x) = x^3 + 3x^2 - 6x - 18 \qquad \text{rearranged}$$

Instead of writing $h(x)g(x)$ to designate the product of the two functions, we find it convenient to write

$$hg(x)$$

instead. The notation hg means that the h equation has been multiplied by the g equation in the same way that

$$(h + g)(x)$$

means that the h equation and the g equation have been added.

example 102.3   Find $hg(-4)$ if $h(x) = x + 3$; $D = \{\text{Reals}\}$, and $g(x) = x^2 - 6$; $D = \{\text{Negative integers}\}$.

solution   As in Example 102.1, we can find the answer two ways. The first is to find $h(-4)$ and $g(-4)$ and then multiply these answers.

$$h(-4) = -4 + 3 \qquad g(-4) = (-4)^2 - 6$$
$$h(-4) = -1 \qquad g(-4) = 10$$

so

$$hg(-4) = (-1)(10)$$
$$\mathbf{hg(-4) = -10}$$

The second way is to find $hg(x)$ and then find $hg(-4)$.

$$hg(x) = (x + 3)(x^2 - 6)$$
$$hg(x) = x^3 + 3x^2 - 6x - 18$$

Now to find $hg(-4)$, we use -4 for x in the equation.

$$hg(-4) = (-4)^3 + 3(-4)^2 - 6(-4) - 18$$
$$hg(-4) = -64 + 48 + 24 - 18$$
$$\mathbf{hg(-4) = -10}$$

example 102.4   Find $fg(-4)$ if $f(x) = x + 3$; $D = \{\text{Reals}\}$, and $g(x) = x - 5$; $D = \{\text{Positive integers}\}$.

solution   We begin by multiplying the equations to find $fg(x)$.

$$f(x)g(x) = (x + 3)(x - 5)$$
$$fg(x) = x^2 - 2x - 15$$

We cannot use this function to find $fg(-4)$ because -4 is not a member of the domain of $g(x)$, so it is not a member of the domain of $fg(x)$. Thus, we may say that this problem has no answer, or we may say that the answer is either

$$\varnothing \qquad \text{or} \qquad \{\,\}$$

practice   **a.** Find $(h + g)(5)$ if $h(x) = x + 1$; $D = \{\text{Reals}\}$, and $g(x) = x^2 - 1$; $D = \{\text{Integers}\}$.

b. Find $hg(-2)$ if $h(x) = x + 2$; $D = \{\text{Reals}\}$, and $g(x) = x^2 - 7$; $D = \{\text{Negative integers}\}$.

c. Find $fg(x)$ if $f(x) = x + 6$; $D = \{\text{Reals}\}$, and $g(x) = x - 4$; $D = \{\text{Positive integers}\}$.

problem set 102

1. Sarah had a 40-mile head start and was driving north at 46 miles per hour when James and Renee began their pursuit at 50 miles per hour. How much farther did Sarah go before James and Renee caught up?

2. The speed of the boat in still water was 10 miles per hour. The boat could go 78 miles down the Lazy River in the same time it took to go 42 miles up the Lazy River. How fast did the current flow in the Lazy River?

3. Horses varied directly as goats and inversely as pigs squared. When the barnyard contained 5 horses, there were 4 pigs and only 2 goats. How many goats went with 6 pigs and 10 horses?

4. A 70 percent markup of the purchase price made the selling price of the item $1666. What did the store owner pay for the item?

5. The selling price was $1680. This low price was possible because Audry and Sam only marked the item up 40 percent of the purchase price. What did they pay for the item?

6. The pressure of a quantity of an ideal gas was held constant at 453 newtons per square meter. The original temperature was 503°C, and the original volume was 450 ml. If the temperature was increased to 743°C, what was the final volume? Remember to add 273 to convert degrees Celsius to kelvins.

7. Find $hg(-5)$ if $h(x) = x + 1$; $D = \{\text{Reals}\}$, and $g(x) = x^2 - 6$; $D = \{\text{Negative integers}\}$.

8. Find $fg(-3)$ if $f(x) = x + 4$; $D = \{\text{Reals}\}$, and $g(x) = x - 1$; $D = \{\text{Positive integers}\}$.

Complete the square as an aid in graphing:

9. $y = x^2 + 4x + 2$

10. $y = -x^2 - 4x - 2$

11. Graph on a number line: $-|x| - 2 \leq -5$; $D = \{\text{Integers}\}$

12. Use substitution to solve: $\begin{cases} 2x + 3y = -3 \\ 4x - 2y = 18 \end{cases}$

Solve:

13. $\begin{cases} 2x - y = 6 \\ xy = 4 \end{cases}$

14. $\begin{cases} x^2 + y^2 = 12 \\ 3x^2 - y^2 = 4 \end{cases}$

15. $\begin{cases} x + y + z = 8 \\ x + y - z = 0 \\ 2x - y + z = 3 \end{cases}$

Graph:

16. $\begin{cases} 3x - 5y < 10 \\ y \geq -2 \end{cases}$

17. $x + 2 < 0$ or $x + 3 \nleq 3$; $D = \{\text{Integers}\}$

18. Find the number that is $\frac{1}{8}$ of the way from $\frac{1}{5}$ to $2\frac{1}{3}$.

Simplify:

19. $\dfrac{x^{a/2}\, y^{2a}}{x^{3a}\, y^{-2a/3}}$

20. $\dfrac{kx}{x - \dfrac{kx}{k - \dfrac{1}{x}}}$

21. $\sqrt[3]{x^5 y}\,\sqrt[4]{xy^2}$

22. $\dfrac{3i + 4}{-i^2 - i^5}$

23. $\dfrac{4i - 3}{i^3 - 2i^2}$

24. Write $4R + 10U$ in polar form.

25. Find c: $\dfrac{a}{x} + \dfrac{b}{m + c} = ax$

Simplify:

26. $\dfrac{2 - 2\sqrt{2}}{3\sqrt{2} - 2}$

27. $3i^5 - \sqrt{-2}\,\sqrt{-2} + \sqrt{3}\,\sqrt{-3} - \sqrt{-9}$

28. $4\sqrt{\dfrac{5}{12}} + 3\sqrt{\dfrac{12}{5}} - 3\sqrt{60}$

29. Begin with $ax^2 + bx + c = 0$ and derive the quadratic formula by completing the square.

30. If the line AB intersects the line CD at point E, which of the following pairs of angles need not be equal? Begin by drawing a diagram of the problem.

   (a) $\angle AEB$ and $\angle CED$
   (b) $\angle AEC$ and $\angle BED$
   (c) $\angle AED$ and $\angle CEA$
   (d) $\angle BEC$ and $\angle DEA$
   (e) $\angle DEC$ and $\angle BEA$

LESSON 103   *Advanced polynomial division*

We can divide polynomials that have more than one variable by using the same method that we use when only one variable is present.

example 103.1   Divide $x^3 + y^3$ by $x + y$.

solution   We use the format for long division.

$$x + y \overline{\smash{)}x^3 + y^3}$$

x^3 divided by x is x^2, so we record an x^2 above

$$\begin{array}{r} x^2 \\ x + y \overline{\smash{)}x^3 + y^3} \end{array}$$

and multiply x^2 by $x + y$ and record.

$$\begin{array}{r} x^2 \\ x + y \overline{\smash{)}x^3 + y^3} \\ \underline{x^3 + x^2 y } \end{array}$$

Now we mentally change the signs and add.

$$\begin{array}{r} x^2 \\ x + y \overline{\smash{)}x^3 + y^3} \\ \underline{x^3 + x^2 y } \\ - x^2 y \end{array}$$

Now $-x^2 y$ divided by x equals $-xy$, so we record $-xy$ above and then multiply and add.

$$\begin{array}{r} x^2 - xy \\ x + y \overline{\smash{)}x^3 + y^3} \\ \underline{x^3 + x^2 y } \\ - x^2 y \\ \underline{- x^2 y - xy^2 } \\ xy^2 \end{array}$$

Finally, xy^2 divided by x equals y^2. We record a y^2 above and multiply to finish.

$$\begin{array}{r} \mathbf{x^2 - xy + y^2} \\ x + y \overline{\smash{)}x^3 + y^3} \\ \underline{x^3 + x^2 y } \\ - x^2 y \\ \underline{- x^2 y - xy^2 } \\ xy^2 + y^3 \\ \underline{xy^2 + y^3} \end{array}$$

example 103.2   Divide $x^3 - y^3$ by $x - y$.

solution   The procedure is the same, and the answer is the same, except that the sign of the middle term is different.

$$\begin{array}{r} \mathbf{x^2 + xy + y^2} \\ x - y \overline{\smash{)}x^3 - y^3} \\ \underline{x^3 - x^2 y } \\ x^2 y \\ \underline{x^2 y - xy^2 } \\ xy^2 - y^3 \\ \underline{xy^2 - y^3} \end{array}$$

practice   Use long division to divide:

    a. $8x^3 + 64y^3$ by $2x + 4y$ **b.** $8x^3 - 64y^3$ by $2x - 4y$

problem set
103

1. A 60 percent markup of the purchase price was necessary to pay the rent, utilities, and the workers and still make a small profit. If an item sold for $1424, what did the storekeeper pay for it?

2. Sister Baby's boat could attain a speed of 18 miles per hour on a lake. If the boat took the same time to go 132 miles down the river as it took to go 84 miles up the river, how fast was the current in the river?

3. Donna took twice as long to drive 720 miles as Maple took to drive 200 miles. Find the rates and times of both if Donna's speed exceeded that of Maple by 40 miles per hour.

4. The initial pressure and temperature of a quantity of an ideal gas was 400 millimeters of mercury and 300 K. If the volume was held constant, what would the final temperature be in kelvins if the pressure was increased to 600 millimeters of mercury?

5. David and Le Van found three consecutive multiples of 11 such that 4 times the sum of the first and third was 66 less than 10 times the second. What were the numbers?

6. Use long division to divide $27x^3 + 8y^3$ by $3x + 2y$.

7. Find x and y. Then find the perimeter of the triangle.

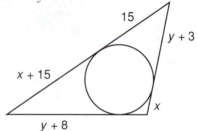

8. Find $ab(2)$ if $a(x) = x - 5$; $D = \{\text{Reals}\}$, and $b(x) = x^2 + 4$; $D = \{\text{Negative integers}\}$.

Complete the square as an aid in graphing:

9. $y = x^2 + 4x + 6$ **10.** $y = -x^2 + 4x - 6$

11. Graph on a number line: $x + 3 \geq 5$; $D = \{\text{Reals}\}$

12. Find the number that is $\dfrac{2}{3}$ of the way from $\dfrac{1}{4}$ to $2\dfrac{1}{2}$.

13. Use substitution to solve: $\begin{cases} 4x + 3y = 17 \\ 2x - 3y = -5 \end{cases}$

Solve:

14. $\begin{cases} x^2 + y^2 = 6 \\ x - y = 2 \end{cases}$ **15.** $\begin{cases} x^2 + y^2 = 10 \\ 2x^2 - 2y^2 = 5 \end{cases}$ **16.** $\begin{cases} x + 2y + z = -1 \\ 3x - y + z = 6 \\ 2x - 3y - z = 8 \end{cases}$

Graph:

17. $\begin{cases} x - 4y \leq -4 \\ x < 3 \end{cases}$ **18.** $-3 \leq x - 3 \not> 4$; $D = \{\text{Integers}\}$

Simplify:

19. $\dfrac{(x^{2a-2})^b}{x^{b/2}}$ **20.** $\dfrac{m}{m^2 + \dfrac{m}{m^2 + \dfrac{1}{m}}}$ **21.** $\sqrt[5]{x^2 y^3} \sqrt[4]{xy}$

22. $\dfrac{2i^2 + i^3}{i^3 + 2}$ **23.** $\dfrac{2i - 5}{5i^2 - 2i}$ **24.** $\dfrac{3 + 2\sqrt{5}}{5 - \sqrt{20}}$

25. The two vectors act on the point as shown. Find the resultant vector.

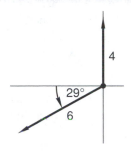

26. Find x: $a\left(\dfrac{b}{c} - \dfrac{1}{x}\right) = \dfrac{m}{p}$

27. Solve: $\sqrt{z} + \sqrt{z + 33} = 11$

Simplify:

28. $3\sqrt{\dfrac{4}{3}} - 2\sqrt{\dfrac{3}{4}} + 5\sqrt{48}$

29. $\sqrt{-16} - \sqrt{-2}\sqrt{2}\sqrt{-3}\sqrt{-3} - i^5$

30. In this diagram, $AB = AC$, angle $A = 40°$, and $\overline{BD}$ is perpendicular to $\overline{AC}$ at D. How many degrees are there in angle DBC?

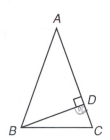

LESSON 104 *Complex numbers, rational numbers, and decimal numerals*

104.A
complex numbers

We review complex numbers by remembering that this number

$$\frac{4}{7} + \sqrt{2}\,i$$

is a complex number written in standard form. The real part is $\frac{4}{7}$ and is written first. The imaginary part is $\sqrt{2}\,i$ and is written after the real part. We often use the letters a and b to designate the standard form of a complex number by writing

$$a + bi$$

and say that a and b can be any real numbers. Since zero is a real number, either a or b can be zero. If b is 0, then only a is left, and thus,

$$4 \qquad \frac{3}{4} \qquad -5\sqrt{2} \qquad -\frac{19}{3}$$

are all complex numbers whose imaginary parts are zero. If a is zero, then only the imaginary part b remains. Thus, the imaginary numbers

$$-\sqrt{2}\,i \qquad \frac{4\sqrt{2}}{3}i \qquad -3i \qquad i$$

are all complex numbers whose real parts equal zero.

104.B
subsets of the set of real numbers

The set of real numbers has an infinite number of members, and these can be used to form an infinite number of subsets. Normally, however, we restrict our attention to five major subsets of the set of real numbers. The first three are

The counting (natural) numbers	$\{1, 2, 3, \ldots\}$
The whole numbers	$\{0, 1, 2, 3, \ldots\}$
The integers	$\{\ldots, -3, -2, -1, 0, 1, 2, 3, \ldots\}$

These three sets account for the numbers that are designated when the number line is drawn because we usually designate the location of the integers below a number line.

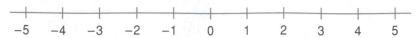

All integers can be written as fractions of other integers. For example, -4, 0, and 13 can be written as fractions, as shown here.

$$-4 = \frac{12}{-3} \qquad 0 = \frac{0}{2} \qquad 13 = \frac{-39}{-3}$$

We say that a number that can be written as a fraction of integers is a *rational number* because *ratio* is a another name for **fraction**.

The rest of the set of real numbers is made up of all the positive numbers of arithmetic and their negative counterparts. Some of these numbers can be written as fractions of integers and thus are rational numbers. The rest cannot be written as fractions of integers and are irrational numbers.[†]

Irrational numbers cannot be represented exactly with decimal numerals that contain a finite number of digits. The square root of 2 is an irrational number and thus can only be approximated with a decimal numeral. A pocket calculator gives an approximation of the square root of 2 as

$$\sqrt{2} \approx 1.4142136$$

The complete representation of this number would require a numeral with an infinite number of digits, and the digits would occur in a nonrepeating pattern.

1. **If the digits in a decimal numeral terminate, the number is a rational number.**

2. **If the digits in a nonterminating decimal numeral repeat in a pattern, the number is a rational number.**

example 104.1 Show that 0.00314 is a rational number by writing it as a fraction of integers.

solution Terminating decimal numerals can be written as fractions by multiplying above and below by a judiciously chosen power of 10. If we move the decimal point five places to the right in this example, we get 314. Thus, we will multiply above and below by 10^5.

$$0.00314 \times \frac{10^5}{10^5} = \frac{314}{100{,}000}$$

We could reduce this fraction to lowest terms, but this reduction is not required because any fraction of integers satisfies the requirement.

example 104.2 Show that 0.00000623 is a rational number by writing it as a fraction of integers.

solution We will multiply above and below by 10^8.

$$0.00000623 \times \frac{10^8}{10^8} = \frac{623}{100{,}000{,}000}$$

104.C

repeating digits

We indicate that digits in a decimal fraction repeat by drawing a bar over the repeating digits. Thus, in the numerals $0.01\overline{623}$ and $1.0031\overline{543}$, the digits under the bars repeat in an endless pattern, as follows:

(a) $0.01\overline{623} = 0.01623232323232323 \cdots$

(b) $1.0031\overline{543} = 1.0031543543543543543543 \cdots$

[†]At this level, it is helpful to think of half the real numbers as being rational numbers and half as being irrational numbers although this is not true because an infinite set cannot be divided into halves!

Each of these numerals represents a rational number and any rational number can be written as a quotient of integers. To write (a) as a quotient of integers, we must get rid of the repeating digits. **We can eliminate these repeating digits by subtracting (a) from the product of (a) and 100. This product has the same repeating digits that (a) has,**

$$100(a) \qquad 100N = 1.623 \ 23 \ 23 \ 23 \ 23 \ 23 \cdots$$
$$(a) \qquad \underline{\quad N = 0.016 \ 23 \ 23 \ 23 \ 23 \cdots}$$
$$99N = 1.607 \qquad \text{(repeating digits elimimated)}$$

The equation with $100N$ is the same as the equation with N except that each side has been multiplied by 100. We multiplied by 100 because there were two repeating digits. Three repeating digits would require a multiplier of 1000, four repeating digits would require a multiplier of 10,000, etc. We will investigate this procedure in the next three examples.

example 104.3 Show that $0.016\overline{23}$ is a rational number by writing it as a fraction of integers.

solution We will use three steps. The first is to write the number and indicate the repeating digits. **Then we record another decimal point above the decimal point in the number.**

$$N = 0.016|23|23|23 \cdots$$

Now we mentally move the digits up and shift them two places to the left.

$$100N = 1.623 \ | \ 23 \ | \ 23 \ | \ 23 \cdots$$
$$N = 0.016 \ | \ 23 \ | \ 23 \ | \ 23 \cdots$$

Moving the digits two places to the left while holding the decimal point fixed is the same as multiplying by 100, so we record 100N on the left side. Now we can subtract the lower equation from the upper equation.

$$100N = 1.623 \ 23 \ 23 \cdots$$
$$\underline{\quad N = 0.016 \ 23 \ 23 \cdots}$$
$$99N = 1.607$$

We have eliminated the repeating digits. Now we divide both sides of the equation by 99 to find N.

$$N = \frac{1.607}{99}$$

Finally, we can get a fraction of integers if we multiply above and below by 1000.

$$N = \frac{1607}{99,000}$$

example 104.4 Show that $1.0031\overline{543}$ is a rational number by writing it as a quotient of integers.

solution We begin by recording the number in expanded form and writing a decimal point above the decimal point in the number.

$$N = 1.0031|543|543|543 \cdots$$

Now we mentally move the digits up and shift them three places to the left. Since this is the same as multiplying by 1000, we multiply N by 1000 and then subtract the lower equation from the upper equation.

$$1000N = 1003.1543 \ 543 \ 543 \ 543 \cdots$$
$$\underline{\quad N = \quad \ 1.0031 \ 543 \ 543 \ 543 \cdots}$$
$$999N = 1002.1512$$

Next, we divide both sides by 999.

$$N = \frac{1002.1512}{999}$$

We can make this a fraction of integers if we multiply above and below by 10,000. We do this, and the fraction of integers is

$$N = \frac{10,021,512}{9,990,000}$$

example 104.5 Show that $13.01\overline{2}$ is a rational number by writing it as a fraction of integers.

solution We record the number in expanded form, indicate the repeating digits, and write the new decimal point.

$$N = 13.01|2|2|2|2|2 \cdots$$

This time only one digit repeats, so when we mentally move the digits up, we shift them one place to the left. Shifting the digits one place is equivalent to multiplying by 10, so we also multiply N by 10 and then subtract.

$$
\begin{array}{r}
10N = 130.12\ 2\ 2\ 2\ 2\ 2 \cdots \\
N = \ \ 13.01\ 2\ 2\ 2\ 2\ 2 \cdots \\
\hline
9N = 117.11
\end{array}
$$

We finish by dividing by 9 and then mentally multiplying above and below by 100 to get

$$N = \frac{117.11}{9} \quad \longrightarrow \quad N = \frac{11,711}{900}$$

practice Show that each number is a rational number by writing it as a fraction of integers:

    a. 0.00000513 **b.** $0.015\overline{24}$

problem set 104

1. Twice the number of ducks was only 6 less than 24 times the number of geese. Also, 10 times the number of geese was only 5 less than the number of ducks. How many ducks and geese were there?

2. Up was very far, so Van and Samuel began early and traveled at 240 miles per hour. Back happened to be the same distance, but the speed back was 360 miles per hour. If the time back was 4 hours less than the time up, how far was up?

3. Pamela and Gail were puzzled. They needed 400 ml of a solution that was 11% salt. They had two solutions to work with. One solution was 10% salt and the other was 20% salt. How much of each solution should they use?

4. The riverboat *Emily-Eleanor* could go 168 miles downstream in 12 hours, but it took her 9 hours to go 54 miles upstream. What was her speed in still water and what was the speed of the current in the river?

5. At the bazaar Deeb paid the Burk $2400 for the rug. If the Burk had only paid $400 for the rug, what was the markup as a percentage of cost, and what was the markup as a percentage of the selling price?

Show that each number is a rational number by writing it as a fraction of integers:

    6. 0.00000512 **7.** $0.014\overline{32}$

8. Find the area of the equilateral triangle whose sides are $4\sqrt{3}$ centimeters long.

9. Use similar triangles to find:

    (a) m and n (b) x and y

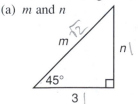
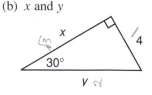

10. Divide $m^3 - p^3$ by $m - p$.

11. Which of the following sets of ordered pairs are functions?

    (a) $(7, 11), (11, 7), (4, -4), (-4, 4)$

    (b) $(5, -2), (-2, 5), (4, -2), (-2, 4)$

    (c) $(7, -3), (-3, 7), (5, 7), (3, 7)$

12. Complete the square as an aid in graphing: $y = -x^2 + 4x - 2$

13. Graph on a number line: $-|x| + 2 > -2$; $D = \{\text{Integers}\}$

14. Find the number that is $\frac{3}{7}$ of the way from $\frac{1}{8}$ to $3\frac{1}{4}$.

Solve:

15. $\begin{cases} \frac{1}{3}x - \frac{2}{5}y = -5 \\ 0.005x - 0.04y = -0.755 \end{cases}$

16. $\begin{cases} 4x - y = 2 \\ xy = 3 \end{cases}$

17. $\begin{cases} x - y - 2z = -14 \\ 2x + y - z = 2 \\ -x + y - z = -4 \end{cases}$

Graph:

18. $\begin{cases} x - 3y \leq -6 \\ x \geq -2 \end{cases}$

19. $0 < x + 2 \not> 4$; $D = \{\text{Reals}\}$

Simplify:

20. $\dfrac{(x^{-2})^{a+2} y^{-3a}}{y^{-a/2}}$

21. $\dfrac{k}{k^2 x - \dfrac{1}{x - \dfrac{1}{k^2}}}$

22. $\sqrt[5]{4 \sqrt[3]{2}}$

23. $\dfrac{6i - i^2}{-i^3 + 3}$

24. $\dfrac{2 + 3\sqrt{2}}{4 - \sqrt{18}}$

25. $-i^5 + \sqrt{-4}\sqrt{4} - 3\sqrt{-9} + 2i^4$

26. $3\sqrt{\dfrac{5}{2}} + 2\sqrt{\dfrac{2}{5}} - 4\sqrt{40}$

27. Add: $-30/\underline{135°} + 20/\underline{-20°}$

28. Find c: $m\left(\dfrac{1}{x} + \dfrac{1}{y}\right) = \dfrac{1}{c} + \dfrac{a}{b}$

29. Solve $-4x^2 = x - 5$ by completing the square.

30. Use unit multipliers to convert 4000 milliliters per second to cubic feet per minute.

LESSON 105   *Advanced factoring*

When two or more quantities are multiplied to form a product, we say that each of the quantities is a **factor** of the product. Thus, since

$$x(x + 3)(x + 2) = x^3 + 5x^2 + 6x$$

we can say that x and $x + 3$ and $x + 2$ are factors of $x^3 + 5x^2 + 6x$. This is the reason that we use the word *factoring* to describe the procedure of writing a sum as the product of the quantities that can be multiplied to form the sum. Thus far, our factoring of trinomials has been restricted to those trinomials whose lead coefficient is 1. Now we will investigate the procedure used to factor trinomials in which the lead coefficient is a number other than 1. This type of trinomial is the product of two binomials, at least one of which has a lead coefficient that is not 1. Let's look at the pattern that evolves when binomials of this type are multiplied.

$$\begin{array}{r} 3x + 2 \\ 2x - 3 \\ \hline 6x^2 + 4x \\ -9x - 6 \\ \hline 6x^2 - 5x - 6 \end{array}$$

We note that the first term of the trinomial is the product of the first terms of the binomials and that the last term of the trinomial is the product of the last terms of the binomials. However, the coefficient of the middle term of the trinomial is not the sum of the last two terms of the binomials. **The middle term is the sum of the product of the first term of the first binomial and the last term of the second binomial and the product of the last term of the first binomial and the first term of the second binomial.** It is easier to see this if we write the original indicated multiplication in horizontal form and note that the middle term is the sum of the products of the means and the extremes.[†]

$$\overset{\text{extremes}}{\underset{\text{means}}{(3x + 2)(2x - 3)}} = 6x^2 - 5x - 6$$

$$\begin{aligned} \text{Product of means} &= 4x \\ \text{Product of extremes} &= -9x \\ \hline \text{Sum} &= -5x \end{aligned}$$

We can devise a method of factoring trinomials whose lead coefficient is not 1 by observing the pattern that occurs when we multiply $ax + b$ by $cx + d$.

$$\begin{array}{r} ax + b \\ cx + d \\ \hline acx^2 + bcx \\ + adx + bd \\ \hline acx^2 + adx + bcx + bd \end{array}$$

We note that the binomials have a total of four constants, a, b, c, and d. Two of these constants form the coefficient of the x^2 term. The other two form the last term of the product. If we multiply the coefficient of the x^2 term by the last term, we get

$$abcd$$

Factoring a trinomial is the task of finding the value of each of these constants.

We also note that the constants in the two middle terms are products of pairs of these constants. Thus, we can factor a trinomial by

1. Multiplying the coefficient of the first term by the last term.
2. Finding two factors of this product whose sum is the coefficient of the middle term.
3. Factoring the resultant expression.

example 105.1 Factor $6x^2 + x - 2$.

solution First we multiply the 6 by −2.

$$(6)(-2) = -12$$

[†]Mathematicians sometimes use the word *mean* to mean *middle* and the word *extreme* to mean *end*. Thus the mean terms in the multiplication shown are the middle terms, and the extreme terms are the end terms.

Now we search for two factors of -12 whose sum is $+1$.

FACTORS OF -12	SUM OF THE FACTORS
$(12)(-1)$	11
$(6)(-2)$	4
$(4)(-3)$	1

Since 4 plus -3 equals $+1$, we have our middle terms.

$$6x^2 + x - 2 \qquad \text{original trinomial}$$

$$6x^2 + 4x - 3x - 2 \qquad \text{substituted } 4x - 3x \text{ for } x$$

Now we factor this expression and get

$$2x(3x + 2) - 1(3x + 2)$$

Finally, we factor out $3x + 2$ and get

$$\mathbf{(2x - 1)(3x + 2)}$$

example 105.2   Factor $2x^2 - 7x - 15$.

solution   First we multiply the coefficient of x^2 by the constant term.

$$(2)(-15) = -30$$

Now we find two factors of -30 whose sum is -7. Since the sum is a negative number, we will let the factor with the greatest absolute value be negative.

FACTORS OF -30	SUM OF THE FACTORS
$(-30)(1)$	-29
$(-15)(2)$	-13
$(-10)(3)$	-7

We can stop here.

We can stop because $-10 + 3$ equals -7. Now, in the trinomial $2x^2 - 7x - 15$, we replace $-7x$ with $-10x + 3x$.

$$2x^2 - 7x - 15 \qquad \text{original trinomial}$$

$$2x^2 - 10x + 3x - 15 \qquad \text{substituted}$$

Next we factor and get

$$2x(x - 5) + 3(x - 5)$$

Now we factor one last time and get

$$\mathbf{(2x + 3)(x - 5)}$$

Thus, we see that $2x^2 - 7x - 15$ can be factored over the integers as $(2x + 3)(x - 5)$.

example 105.3   Factor to find the solutions of $-5x - 6 = -6x^2$.

solution   We begin by writing the equation in standard form as

$$6x^2 - 5x - 6 = 0$$

The product of the lead coefficient and the constant term is

$$(6)(-6) = -36$$

Now we look for a pair of factors of -36 whose sum is -5.

FACTORS OF -36	SUM OF THE FACTORS
$(-36)(1)$	-35
$(-18)(2)$	-16
$(-9)(4)$	-5

We can stop here.

Now in the trinomial, we replace $-5x$ with $-9x + 4x$.

$$6x^2 - 9x + 4x - 6 = 0$$

Next we factor once

$$3x(2x - 3) \neq 2(2x - 3) = 0$$

and factor again and get

$$(3x + 2)(2x - 3) = 0$$

We complete the solution by using the zero factor theorem.

If $3x + 2 = 0$ and if $2x - 3 = 0$

$3x = -2$ $2x = 3$

$x = -\dfrac{2}{3}$ $x = \dfrac{3}{2}$

practice   Solve by factoring:

       a. $-5x - 12 + 2x^2 = 0$ **b.** $-7x - 6 = -3x^2$

problem set   **1.**   The weight of the chlorine (Cl) in a quantity of the compound $C_3H_3Cl_5$ was 1050 grams.
105   What was the total weight of the compound? (C, 12; H, 1; Cl, 35)

      2.   Weasel headed for Table Rock Lake in a 10-mile-per-hour trot. But soon short wind forced her to slow to a 6-mile-per-hour walk. If she covered 64 miles in 8 hours, how far did she walk and how far did she trot?

      3.   In the flask was 140 ml of a 20% alcohol solution. How much pure alcohol should be added to get a final solution that is 44% alcohol?

      4.   The number of potatoes varied jointly as the number of mules and the number of farmers squared. If 5 mules and 5 farmers went with 750 potatoes, how many potatoes went with 10 mules and 10 farmers?

      5.   The markup was $800, and the cost was $2400. What was the markup as a percentage of selling price, and what was the markup as a percentage of the cost?

      6.   Find $(g + h)(2)$ if $g(x) = x^2 + 1$; $D = \{\text{Integers}\}$ and $h(x) = x - 5$; $D = \{\text{Reals}\}$.

      7.   Find $hg(x)$ for the function defined in Problem 6.

Show that each number is a rational number by writing it as a fraction of integers:

      8. $0.0001\overline{234}$ **9.** $0.01\overline{651}$

     10.   Divide $m^3 + p^3$ by $m + p$.   **11.**   Divide $x^3 + y^3$ by $x + y$.

     12.   Complete the square as an aid in graphing: $y = x^2 + 6x + 8$

     13.   Graph on a number line: $-|x| + 2 > 1$; $D = \{\text{Integers}\}$

     14.   Find the number that is $\dfrac{3}{5}$ of the way from $2\dfrac{1}{4}$ to $4\dfrac{1}{2}$.

Solve:

     15. $\begin{cases} \dfrac{2}{7}x - \dfrac{1}{4}y = -6 \\ 0.07x + 0.14y = 6.58 \end{cases}$ **16.** $\begin{cases} 5x - y = 2 \\ xy = 6 \end{cases}$

     17. $\begin{cases} x^2 + y^2 = 8 \\ x - y = 2 \end{cases}$ **18.** $\begin{cases} 2x - y - 2z = 2 \\ x + y - z = 7 \\ 2x + y - z = 0 \end{cases}$

     19.   Use substitution to solve: $\begin{cases} 3x + 5y = 4 \\ 10x - 15y = -50 \end{cases}$

20. Graph: $\begin{cases} x - 3y \geq 6 \\ x \geq -3 \end{cases}$

Solve by factoring. Rearrange if necessary.

21. $-5x^2 - 2x + 3x^3 = 0$

22. $-10x - 4 + 6x^2 = 0$

23. $8x + 4 + 3x^2 = 0$

24. $24x^2 + 9x^3 + 12x = 0$

25. $2p^2 - 3p - 5 = 0$

26. $8 + 18x + 4x^2 = 0$

Simplify:

27. $\dfrac{x^a \left(x^{a/2+4}\right)^2 y^b}{y^{b/3} x^{a/6}}$

28. $\dfrac{2 - 3i^3}{i + 2i^2 + 3i^3}$

29. $\dfrac{4 + 2\sqrt{5}}{5 - 3\sqrt{5}}$

30. The radii of the circles are 1 unit, 1 unit, and 2 units, as shown. The base of the triangle is 3 units long. The area of the triangle equals the sum of the areas of the three circles. What is the altitude of the triangle?

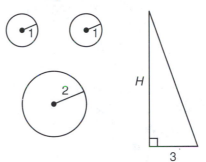

LESSON *106* *More on systems of three equations*

Some systems of three equations in three unknowns do not have all three variables in each of the equations. These equations can also be solved by using the substitution method or the elimination method.

example 106.1   Solve: $\begin{cases} 2x + 3y = -4 & \text{(a)} \\ x - 2z = -3 & \text{(b)} \\ 2y - z = -6 & \text{(c)} \end{cases}$

solution   One variable is missing in each equation. We can see this better if we write the equations in expanded form.

$$\text{(a)} \quad 2x + 3y \qquad = -4$$

$$\text{(b)} \quad x \qquad - 2z = -3$$

$$\text{(c)} \qquad 2y - z = -6$$

The first step is to add any two of the equations so that one variable is eliminated. We could add (a) and (b) to eliminate x; or add (b) and (c) to eliminate z; or add (a) and (c) to eliminate y. We choose to eliminate x, so we add equation (a) to the product of equation (b) and (-2).

$$
\begin{array}{lrr}
\text{(a)} & 2x + 3y & = -4 \\
(-2)\text{(b)} & -2x \qquad + 4z = & 6 \\
\hline
\text{(d)} & 3y + 4z = & 2
\end{array}
$$

The resulting equation, (d), has y and z as variables. So does equation (c). We use equations (d) and (c) to eliminate z by adding equation (d) to the product of (4) and equation (c).

$$
\begin{array}{rl}
(4)(c) & 8y - 4z = -24 \\
(d) & \underline{3y + 4z = 2} \\
& 11y = -22 \\
& y = -2
\end{array}
$$

Now we replace y with -2 in equation (a) and solve for x. Then we replace y with -2 in equation (c) and solve for z.

$$
\begin{array}{ll}
\text{(a)} \quad 2x + 3(-2) = -4 & \text{(c)} \quad 2(-2) - z = -6 \\
\qquad\quad 2x - 6 = -4 & \qquad\quad -4 - z = -6 \\
\qquad\qquad\quad 2x = 2 & \qquad\qquad\quad -z = -2 \\
\qquad\qquad\quad\ x = 1 & \qquad\qquad\qquad\ z = 2
\end{array}
$$

Thus, our solution is the ordered triple **(1, −2, 2)**.

example 106.2   Solve: $\begin{cases} 3y - 2z = -12 & \text{(a)} \\ 2x - 3z = -5 & \text{(b)} \\ x - 2y = 6 & \text{(c)} \end{cases}$

solution   Although it is not necessary, we will begin by writing the equations in expanded form.

$$
\begin{array}{lll}
\text{(a)} & & 3y - 2z = -12 \\
\text{(b)} & 2x & - 3z = -5 \\
\text{(c)} & x - 2y & = 6
\end{array}
$$

This time we decide to eliminate y, so we will add the product of 2 and equation (a) to the product of 3 and equation (c).

$$
\begin{array}{rl}
(2)(a) & 6y - 4z = -24 \\
(3)(c) & \underline{3x - 6y = 18} \\
(d) & 3x - 4z = -6
\end{array}
$$

Now equation (b) also is an equation in x and z, so we decide to add the product of -3 and equation (b) to the product of 2 and equation (d).

$$
\begin{array}{rl}
(-3)(b) & -6x + 9z = 15 \\
(2)(d) & \underline{6x - 8z = -12} \\
& z = 3
\end{array}
$$

Now we will replace z with 3 in equation (a) and (b) and solve for x and y.

$$
\begin{array}{ll}
\text{(a)} \quad 3y - 2(3) = -12 & \text{(b)} \quad 2x - 3(3) = -5 \\
\qquad\quad 3y - 6 = -12 & \qquad\quad 2x - 9 = -5 \\
\qquad\qquad\ 3y = -6 & \qquad\qquad\ 2x = 4 \\
\qquad\qquad\ y = -2 & \qquad\qquad\ x = 2
\end{array}
$$

So the solution to this system is the ordered triple **(2, −2, 3)**.

It was not necessary to use elimination as the first step. For example, we could have solved equation (c) for x.

$$ x = 6 + 2y $$

and substituted $6 + 2y$ for x in equation (b).

$$
\begin{array}{ll}
\text{(b)} & 2(6 + 2y) - 3z = -5 \\
& 12 + 4y - 3z = -5
\end{array}
$$

(e) $4y - 3z = -17$

Now equation (e) could be used with equation (a) to solve for y and z.

practice   Solve: $\begin{cases} 3x + 2y = 6 \\ 2x - z = 9 \\ 3y - z = 10 \end{cases}$

problem set 106

1. Jerry started a used parts establishment. He marked up the items 30 percent of the purchase price. If one item sold for $715, what did Jerry pay for it?

2. The small plane could go 6 times as fast as the small car. Thus, the plane could go 1200 miles in only 1 hour less than it took the car to go 250 miles. Find the rate and the time of the small plane and the rate and the time of the small car.

3. The near-stagnant river flowed at only 2 miles per hour. Harold's boat could go 56 miles down the river in one-half the time it took to go 80 miles up the river. What was the speed of his boat in still water?

4. The temperature of a quantity of an ideal gas was held constant at 500°C. The initial pressure and volume were 700 torr and 500 ml. What would the final pressure be if the volume were increased to 1000 ml?

Solve:

5. $\begin{cases} 3x - 3y = 9 \\ 4x + z = 5 \\ 4y + 2z = -10 \end{cases}$ 6. $\begin{cases} 2x - 2y - z = 9 \\ 3x + 3y - z = 6 \\ x + y + z = -2 \end{cases}$

7. Find the area of this isosceles triangle. Units are in centimeters.

8. Use similar triangles to find:
    (a) m and n (b) p and q

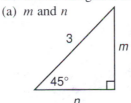   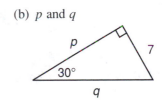

Show that each number is a rational number by writing it as a quotient of integers:

9. $0.0007\overline{013}$ 10. $4.10\overline{26}$

11. Complete the square as an aid in graphing: $y = -x^2 - 2x - 3$

12. Graph on a number line: $-|x| - 4 \not> 0$; $D = \{\text{Integers}\}$

13. Find the number that is $\frac{1}{10}$ of the way from $3\frac{1}{3}$ to $6\frac{1}{2}$.

Solve:

14. $\begin{cases} \frac{3}{5}x - \frac{1}{4}y = 5 \\ 0.012x + 0.07y = 2.20 \end{cases}$ 15. $\begin{cases} 5x - y = 3 \\ xy = 4 \end{cases}$

16. $\begin{cases} x^2 + y^2 = 7 \\ 2x - y = 2 \end{cases}$

17. Use substitution to solve: $\begin{cases} 5x - 3y = 27 \\ 2x - 5y = 26 \end{cases}$

18. Graph: $\begin{cases} 3x - 4y \geq 8 \\ y > -2 \end{cases}$

19. Simplify: $\dfrac{3 - 2i^2 - i}{3i^3 + 3i + 2}$

20. Solve $3x^2 - x - 7 = 0$ by completing the square.

21. Find the resultant of the vectors shown.

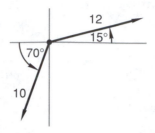

22. Does this set of ordered pairs designate a function?

$(4, 6), (-3, -2), (8, 5), (4, 6), (3, -2), (8, 5), (11, -3)$

23. Find $\psi\theta\left(\frac{1}{2}\right)$ if $\psi(x) = x + 2$; $D = \{\text{Reals}\}$, and $\theta(x) = x^2$; $D = \{\text{Integers}\}$.

24. Divide $m^3 - p^3$ by $m - p$.

Solve by factoring. Rearrange as necessary. Always look for a common factor.

25. $3x^2 + 7x + 2 = 0$

26. $3x^2 + x - 2 = 0$

27. $2z^2 + 13z + 15 = 0$

28. $33p^2 + 45p + 6p^3 = 0$

29. $3p^2 - 13p - 10 = 0$

30. $-11a + 15 = -2a^2$

LESSON 107 · Numbers, numerals, and value · Number word problems

107.A
numbers, numerals, and value

We remember that a numeral is a single symbol or a meaningful arrangement of symbols that we use to represent a particular number. We say that the value of each of the following numerals is three because each numeral represents the number 3.

$$3 \qquad 1 + 1 + 1 \qquad \frac{81}{27} \qquad 2^3 - 5 \qquad 2^2 - 1$$

Thus, we see that **number** and **value** mean the same thing. Also, we see that it would be redundant to speak of the value of a number because that is the same thing as saying the number of a number. But we can speak of the value of a numeral because this is the number represented by the numeral. Since paying excessive attention to the difference between a number and a numeral is often counterproductive, we sometimes use the word number when we should use the word numeral. See if you can find where this mistake is made in this lesson. The mistake is not serious.

107.B
number word problems

The value of a digit in a decimal numeral depends on the position of the digit with respect to the decimal point. The digit 6 in the numeral

$$496.0$$

has a value of 6 times 1, or 6, because it is in the units' place, which is the first place to the left of the decimal point. The digit 9 has a value of 9 times 10, or 90, because it is in the tens' place, which is two places to the left of the decimal point. The digit 4 has a value of 400 because it is in the hundreds' place, which is three places to the left of the decimal point. We can use the fact that the value of a digit depends on its position to solve some rather interesting word problems.

To solve these problems, we will use U to represent the units' digit; T to represent the tens' digit, and H to represent the hundreds' digit. Also, we will say that

The value of the units' digit is $1U$
The value of the tens' digit is $10T$
The value of the hundreds' digit is $100H$

example 107.1   The sum of the digits in a two-digit counting number is 11. If the digits are reversed, the new number is 27 greater than the original number. What was the original number?

solution   The original number is written with the tens' digit followed by the units' digit, as

$$TU$$

and the sum of T and U is 11. This gives us

$$\text{(a)} \quad U + T = 11$$

The value of the original number is

$$\text{(b)} \quad 10T + U$$

The value of the number with the digits reversed is

$$\text{(c)} \quad 10U + T$$

Since (c) is 27 greater than (b), we add 27 to (b) when we write the equation

$$10T + U + 27 = 10U + T$$

Now from (a) we substitute $11 - T$ for U and solve.

$10T + (11 - T) + 27 = 10(11 - T) + T$	substituted
$10T + 11 - T + 27 = 110 - 10T + T$	removed parentheses
$9T + 38 = 110 - 9T$	added like terms
$18T = 72$	simplified
$T = 4$	divided

and since $T + U = 11$, then $U = 7$, and thus the original number was **47.**

example 107.2   The sum of the digits of a two-digit counting number was 9. When the digits were reversed, the new number was 45 less than the original number. What was the original number?

solution   The original number is written as TU, and the sum of the digits is 9.

$$\text{(a)} \quad T + U = 9$$

The original number had a value of

$$\text{(b)} \quad 10T + U$$

and when the digits were reversed, the value was

$$\text{(c)} \quad 10U + T$$

Now (c) is 45 less than (b), so we add 45 to (c) to write our equation.

$$10U + T + 45 = 10T + U$$

Then we substitute $9 - U$ for T and get

$$10U + (9 - U) + 45 = 10(9 - U) + U$$

Now we solve the equation.

$$9U + 54 = 90 - 10U + U$$
$$9U + 54 = 90 - 9U$$
$$18U = 36$$
$$U = 2$$

Thus, since $T + U = 9$, T equals 7, and the original number was **72.**

practice   The sum of the digits of a two-digit counting number was 7. When the digits were reversed, the new number was 27 less than the original number. What was the number?

problem set
107

1. Lynn's mixture was 30% formaldehyde, and Lucy's mixture was 60% formaldehyde. How much of each should they use to get a 400 ml of a mixture that is 36% formaldehyde?

2. The beaker was filled with methylbromide, CH_3Br. What percent by weight of this compound was bromine (Br)? (C, 12; H, 1; Br, 80)

3. The sum of the digits in a two-digit counting number was 15. If the digits were reversed, the new number was 9 greater than the original number. What was the original number?

4. The sum of the digits of a two-digit counting number was 13. When the digits were reversed, the new number was 9 less than the original number. What was the original number?

5. A 70% markup on the selling price brought the selling price to $1400. What did the shopkeeper pay for the item and what was the markup?

Solve:

6. $\begin{cases} 2x - y = -6 \\ 3y + 2z = 12 \\ x - 3z = -11 \end{cases}$

7. $\begin{cases} 5x - y - z = 2 \\ x - 5y + z = -2 \\ -x + y - z = -2 \end{cases}$

8. Show that $0.001\overline{213}$ is a rational number by writing it as a fraction of integers.

9. Divide $x^3 + y^3$ by $x + y$.

10. Complete the square as an aid in graphing: $y = -x^2 + 2x + 1$

11. Graph on a number line: $x - 5 \not< -4$; $D = \{$Reals$\}$

12. Find the number that is $\frac{3}{4}$ of the way from 2 to $6\frac{2}{3}$.

Solve:

13. $\begin{cases} \frac{2}{3}x - \frac{1}{3}y = 5 \\ -0.006x - 0.04y = -0.432 \end{cases}$

14. $\begin{cases} 4y - x = 2 \\ xy = 5 \end{cases}$

15. $\begin{cases} x^2 + y^2 = 4 \\ x - 2y = 1 \end{cases}$

16. Use substitution to solve: $\begin{cases} 5x - 3y = 32 \\ 2x - 2y = 16 \end{cases}$

17. Graph: $\begin{cases} 3x - 8y > -x \\ x \leq y \end{cases}$

Simplify:

18. $\dfrac{2i^3 - i + 2}{3 + 4i}$

19. $\sqrt[3]{x^5y^2}\,\sqrt[4]{xy^3}$

20. Solve: $\sqrt{s - 48} = 8 - \sqrt{s}$ **21.** Write $-4R + 8U$ in polar form.

22. Find the equation of the line that passes through $(-2, 3)$ and is perpendicular to $5x - 3y = 4$.

23. Find r: $mx = \dfrac{1}{m}\left(\dfrac{1}{r} + \dfrac{1}{p}\right)$

24. Use unit multipliers to convert 800 liters per minute to cubic feet per second.

25. The rectangle and the right triangle have equal areas. What is the length of FG? Dimensions are in meters.

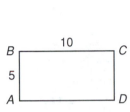 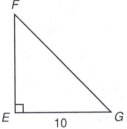

26. The radius of a circle is $2\sqrt{\pi}$ centimeters long. What is the length of a side of a square whose area equals the area of the circle?

Solve by factoring. Rearrange as necessary. Always look for a common factor.

27. $2x^2 - 30x + 4x^3 = 0$ **28.** $4 + 12x + 8x^2 = 0$

29. Write $74.\overline{213}$ as a quotient of integers.

30. Find $(h + g)(-3)$ if $h(x) = x + 2$; $D = \{\text{Reals}\}$, and $g(x) = x^3 + 2$; $D = \{\text{Integers}\}$.

LESSON 108   *Sum and difference of two cubes*

Expressions that are the sum of two squares, such as

$$x^2 + y^2 \quad \text{and} \quad 9x^2y^2 + 4p^2$$

cannot be factored, but we can factor expressions that are the difference of two squares, such as

$$x^2 - y^2 \quad \text{and} \quad 9x^2y^2 - 4p^2$$

To factor these expressions, we must recognize that each one is the difference of two squares. **There is no procedure to follow.** We just write down the factored forms by inspection.

$$x^2 - y^2 = (x + y)(x - y) \quad \text{and} \quad 9x^2y^2 - 4p^2 = (3xy + 2p)(3xy - 2p)$$

If we had not recognized the forms, we could not have factored the expressions. Two other forms that require recognition for factoring are the sum and difference of two cubes. Both of these can be factored.

$$a^3y^3 - p^3 \quad \text{and} \quad a^3y^3 + p^3$$

Unfortunately, the factored forms of these expressions are sometimes difficult to remember. However, they can be easily derived when needed by using simple expressions and polynomial division. To find the factored forms, we remember that $a^3 + b^3$ is evenly divisible by $a + b$ and that $a^3 - b^3$ is evenly divisible by $a - b$. We will do both of these divisions here.

$$
\begin{array}{r}
a^2 - ab + b^2 \\
a + b\overline{)a^3 \qquad\qquad + b^3} \\
\underline{a^3 + a^2b} \\
- a^2b \\
\underline{- a^2b - ab^2} \\
ab^2 + b^3 \\
\underline{ab^2 + b^3}
\end{array}
\qquad
\begin{array}{r}
a^2 + ab + b^2 \\
a - b\overline{)a^3 \qquad\qquad - b^3} \\
\underline{a^3 - a^2b} \\
a^2b \\
\underline{a^2b - ab^2} \\
ab^2 - b^3 \\
\underline{ab^2 - b^3}
\end{array}
$$

Thus, we see that we can factor as follows:

$$(1) \quad a^3 + b^3 = (a + b)(a^2 - ab + b^2)$$

$$(2) \quad a^3 - b^3 = (a - b)(a^2 + ab + b^2)$$

To extend these forms to more complicated expressions, some people find that it is helpful to think F_T for *first thing* and S_T for *second thing* instead of a and b. If we do this in equations (1) and (2) above, we get

$$(1') \quad F_T^3 + S_T^3 = \left(F_T + S_T\right)\left(F_T^2 - F_T S_T + S_T^2\right)$$

$$(2') \quad F_T^3 - S_T^3 = \left(F_T - S_T\right)\left(F_T^2 + F_T S_T + S_T^2\right)$$

example 108.1 Factor: $x^3y^3 - p^3$

solution We recognize that this expression can be written as the difference of two cubes:

$$(xy)^3 - (p)^3$$

The first thing that is cubed is xy, and the second thing that is cubed is p. If we use form (2′) above,

$$(2') \quad F_T^3 - S_T^3 = \left(F_T - S_T\right)\left(F_T^2 + F_T S_T + S_T^2\right)$$

and replace F_T with xy and S_T with p, we can write the given expression in factored form.

$$x^3y^3 - p^3 = (xy - p)(x^2y^2 + xyp + p^2)$$

example 108.2 Factor: $8m^3y^6 + x^3$

solution We recognize this as the sum of two cubes

$$(2my^2)^3 + (x)^3$$

and note that the first thing that is cubed is $2my^2$ and that the second thing that is cubed is x. Thus if we use form (1′),

$$(1') \quad F_T^3 + S_T^3 = \left(F_T + S_T\right)\left(F_T^2 - F_T S_T + S_T^2\right)$$

and replace F_T with $2my^2$ and S_T with x, we get

$$8m^3y^6 + x^3 = (2my^2 + x)(4m^2y^4 - 2my^2x + x^2)$$

example 108.3 Factor: $a^{12} + b^{12}$

solution We can write the expression as the sum of two cubes as

$$(a^4)^3 + (b^4)^3$$

We see that the first thing that is cubed is a^4, and the second thing that is cubed is b^4. Thus if we use the form (1′),

$$(1') \quad F_T^3 + S_T^3 = \left(F_T + S_T\right)\left(F_T^2 - F_T S_T + S_T^2\right)$$

and use a^4 for F_T and b^4 for S_T, we get

$$a^{12} + b^{12} = (a^4 + b^4)(a^8 - a^4b^4 + b^8)$$

practice Factor: $64p^6a^9 - x^3y^{12}$

problem set
108

1. The break-even markup was 40 percent of the selling price. If the total income for a day on which the store broke even was $6400, what did the store owner pay for the items sold?

2. Ross and Thais had a two-digit counting number. They saw that the sum of the digits was 6, and if the digits were reversed, the new number was 18 less than the original number. What was the original number?

3. Jeff and Cindy noted that the sum of the digits of a two-digit counting number was 15. If the digits were reversed, they found that the new number was 27 less than the original number. What was the original number?

4. Yellows varied directly as greens squared and inversely as blues. When there were 100 yellows, there were 5 blues but only 1 green. How many blues went with 10 yellows and 10 greens? Solve two ways.

5. The current in the river was only 3 miles per hour, so the fast boat could go 92 miles downstream in the same time it took to go 68 miles upstream. How long did each trip take, and what was the speed of the boat in still water?

Factor:

6. $27a^6p^{12} + y^3$

7. $8x^{12}z^6 - m^3y^9$

8. $m^6y^9 - z^6$

9. If $\sqrt{13^2 - 12^2} = \sqrt[n]{125}$, what number does n represent?

10. Show that $4.12\overline{3}$ is a rational number by writing it as a fraction of integers.

11. Complete the square as an aid in graphing: $y = x^2 - 2x - 1$

12. Graph on a number line: $-|x| - 3 < -5$; $D = \{\text{Reals}\}$

Solve:

13. $\begin{cases} \dfrac{2}{5}x - \dfrac{1}{4}y = 2 \\ -0.008x - 0.2y = -1.68 \end{cases}$

14. $\begin{cases} x^2 + y^2 = 5 \\ y - 2x = 2 \end{cases}$

15. $\begin{cases} 4x + 2y = 8 \\ 3x - 3z = -9 \\ -3x + z = 1 \end{cases}$

16. $\begin{cases} x - y - 3z = -2 \\ 3x + y + z = 12 \\ 2x - y + z = 5 \end{cases}$

17. Graph: $\begin{cases} -x - 3y \geq -9 \\ y < 2x \end{cases}$

Simplify:

18. $\dfrac{-i^3 - \sqrt{-2}\,\sqrt{-2}}{i^2 - 2i}$

19. $\dfrac{4 + 3\sqrt{5}}{2 + \sqrt{5}}$

20. $\sqrt[6]{9\sqrt[3]{3}}$

21. $5\sqrt{\dfrac{3}{5}} + 2\sqrt{\dfrac{5}{3}} - \sqrt{60}$

22. Convert $4R - 14U$ to polar form.

23. Use unit multipliers to convert 40 feet per second to miles per hour.

24. Find the distance between $(-3, 7)$ and $(5, 3)$.

25. Draw the line suggested by the data points and write the equation that expresses work as a function of energy: $W = mE + b$

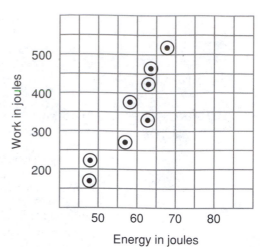

26. Simplify: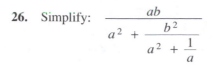

27. Begin with $ax^2 + bx + c = 0$ and derive the quadratic formula by completing the square.

Solve by factoring. Rearrange as necessary. Always look for a common factor.

28. $9x + 2x^2 + 9 = 0$ 29. $6b + 20b^2 + 6b^3 = 0$

30. Find $(h + p)(-3)$ if $h(x) = x^2$; $D = \{\text{Reals}\}$, and $p(x) = x^3$; $D = \{\text{Negative integers}\}$.

LESSON 109   *More on fractional exponents*

In Lesson 38 we noted that the power rule for exponents cannot be used when the base is a sum. This is very important so we will review the reason here. Note that when we write

$$x^2$$

we indicate that x is to be multiplied by itself, so

$$x^2 = x \cdot x$$

In the same way, if we write

$$(2x^2y^3z^{-4})^2$$

we indicate that $2x^2y^3z^{-4}$ is to be multiplied by itself, and we get

$$(2x^2y^3z^{-4})^2 = (2x^2y^3z^{-4})(2x^2y^3z^{-4}) = 4x^4y^6z^{-8}$$

The power rule permits this expansion directly if we multiply the exponents to get

$$(2x^2y^3z^{-4})^2 = 4x^4y^6z^{-8}$$

The power rule cannot be used when we have a sum. Thus, $(x^2 + y^2)^2$ does not equal $x^4 + y^4$.

$$(x^2 + y^2)^2 \neq x^4 + y^4$$

To expand $(x^2 + y^2)^2$, we must multiply $x^2 + y^2$ by itself as follows.

$$
\begin{array}{r}
x^2 + y^2 \\
x^2 + y^2 \\
\hline
x^4 + x^2y^2 \\
x^2y^2 + y^4 \\
\hline
x^4 + 2x^2y^2 + y^4
\end{array}
$$

The power rule can be used only when the expression raised to a power is a product of factors.

example 109.1   Expand: $(2x^{1/2}y^{1/4}z)^3$

solution   We apply the power rule and get

$$8x^{3/2}y^{3/4}z^3$$

example 109.2 Expand: $(x^{1/2} + y^{1/2})^2$

solution The power rule cannot be used for a sum. We must multiply $x^{1/2} + y^{1/2}$ by $x^{1/2} + y^{1/2}$.

$$
\begin{array}{r}
x^{1/2} + y^{1/2} \\
x^{1/2} + y^{1/2} \\
\hline
x + \quad x^{1/2}y^{1/2} \\
x^{1/2}y^{1/2} + y \\
\hline
x + 2x^{1/2}y^{1/2} + y
\end{array}
$$

example 109.3 Expand: $(x^{1/2} + y^{-1/2})^2$

solution We must multiply $x^{1/2} + y^{-1/2}$ by $x^{1/2} + y^{-1/2}$.

$$
\begin{array}{r}
x^{1/2} + y^{-1/2} \\
x^{1/2} + y^{-1/2} \\
\hline
x + \quad x^{1/2}y^{-1/2} \\
x^{1/2}y^{-1/2} + y^{-1} \\
\hline
x + 2x^{1/2}y^{-1/2} + y^{-1}
\end{array}
$$

practice Expand:

    a. $(m^{1/2} + a^{1/2})^2$ **b.** $(z^{1/2} + p^{1/3})^2$

problem set 109

1. A two-digit counting number has a value that is 8 times the sum of its digits. If 6 times the units' digit is 5 more than the tens' digit, what is the number?

2. The sum of the digits of a two-digit counting number was 11. If the digits were reversed, the new number would be 27 less than the original number. What was the original number?

3. Lucretius could cover the 63 miles to Pompeii in 11 more hours than it took Cassius to drive the 60 miles to Rome. Find the rates and times of both if Lucretius traveled at half the speed of Cassius.

4. Doctor Steve held the volume of a quantity of an ideal gas constant at 400 ml. The original pressure and temperature were 800 torr and 400 K. What would be the final pressure if the temperature were raised to 1200 K?

5. Tom and Zollie had 600 ml of a 46% alcohol solution that had to be reduced to a 40% solution by extracting pure alcohol. How much pure alcohol should be extracted?

Expand:

    6. $(3x^{1/4}y^{1/2}m)^3$ **7.** $(x^{1/4} + y^{1/4})^2$ **8.** $(x^{1/4} + y^{-1/4})^2$

Factor:

    9. $x^3y^6 - 27m^3$ **10.** $64x^9y^6 + p^{12}z^3$

11. Show that $1.02\overline{342}$ is a rational number by writing it as a fraction of integers.

12. Complete the square as an aid in graphing: $y = -x^2 + 4x - 1$

13. Graph on a number line: $-2 < x + 2 > -4; \; D = \{\text{Reals}\}$

Solve:

    14. $\begin{cases} \dfrac{2}{5}x - \dfrac{1}{3}y = 1 \\ 0.3x - 0.05y = 2.55 \end{cases}$ **15.** $\begin{cases} x - 2y = 5 \\ xy = 3 \end{cases}$

    16. $\begin{cases} x + y - z = 7 \\ 4x + y + z = 4 \\ 3x + y - z = 9 \end{cases}$ **17.** $\begin{cases} 2x + 3y = 15 \\ x - 2z = -3 \\ 3y - z = 6 \end{cases}$

18. Graph: $\begin{cases} 2x - 5y \geq 15 \\ x \leq -y \end{cases}$

Simplify:

19. $\dfrac{2i^2 - \sqrt{-9} + 2}{3 - \sqrt{-2}\sqrt{2}}$

20. $\dfrac{3 + 2\sqrt{2}}{5\sqrt{2} - 2}$

21. $\sqrt[3]{27}\sqrt[3]{3}$

22. $\sqrt{x^5 y}\,\sqrt{x^2 y}$

23. $\dfrac{(x^{a-2})^{1/3}(y^b)^{1/2}}{x^{2a}\,y^{-b}}$

24. $\sqrt{\dfrac{2}{3}} + 4\sqrt{\dfrac{3}{2}} - 6\sqrt{24}$

25. $\dfrac{ka^2}{ka - \dfrac{a^2}{k^2 - \dfrac{1}{a}}}$

26. Find the resultant of the force vectors shown.

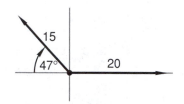

27. Use unit multipliers to convert 1000 liters per minute to milliliters per second.

Solve by factoring. Rearrange as necessary. Always look for a common factor.

28. $-3p + 6p^2 - 30 = 0$

29. $-8x - 14x^2 + 4x^3 = 0$

30. $3x^2 - 7x - 6 = 0$

LESSON 110 *Quadratic inequalities (greater than)*

If we wish to say that x is a positive number, we can write

$$x \text{ is a positive number}$$

or we can use the greater than/less than symbol and write

$$x > 0$$

This is the symbolic way to designate a positive number because all the numbers that are greater than zero are positive numbers.

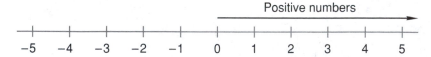

If the product of two quantities is greater than zero,

$$(\quad)(\quad) > 0$$

then the product is a positive number. There are two ways that a product of two factors can be positive. Either both factors are positive or both factors are negative.

$$(\text{Pos})(\text{Pos}) > 0 \qquad \text{and} \qquad (\text{Neg})(\text{Neg}) > 0$$

We will use this fact to graph the solution to the following inequality.

example 110.1   Graph the solution on a number line: $(x + 2)(x - 3) > 0$; $D = \{\text{Reals}\}$

solution   The product of the factors is greater than zero. This means that the product is a positive number. For a product of two factors to be positive, both factors must be positive *or* both factors must be negative. There are two possibilities.

(Pos)	**and**	(Pos)	**or**	(Neg)	**and**	(Neg)
$x + 2$		$x - 3$		$x + 2$		$x - 3$

Since each of these factors is positive, each must be greater than zero.

$$\begin{array}{rcl} x + 2 > & 0 \\ -2 & -2 \\ \hline x > & -2 \end{array} \quad \text{and} \quad \begin{array}{rcl} x - 3 > & 0 \\ +3 & +3 \\ \hline x > & 3 \end{array}$$

Since each of these factors is negative, each must be less than zero.

$$\begin{array}{rcl} x + 2 < & 0 \\ -2 & -2 \\ \hline x < & -2 \end{array} \quad \text{and} \quad \begin{array}{rcl} x - 3 < & 0 \\ +3 & +3 \\ \hline x < & 3 \end{array}$$

The numbers that satisfy the conjunction on the left are the numbers that are greater than -2 and that are also greater than 3. **Of course, all numbers that are greater than 3 are also greater than -2, so the numbers that are greater than 3 satisfy both conditions on the left.**

The numbers that satisfy both conditions of the conjunction on the right are the numbers that are less than -2 and are also less than 3. **Of course, all numbers that are less than -2 are also less than 3, so the numbers that are less than -2 satisfy both conditions.**

Thus, our solution is

$$x > 3 \quad \text{or} \quad x < -2$$

Now we graph all the real numbers that satisfy either one of these conditions.

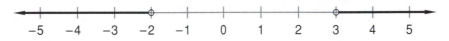

example 110.2   Graph the solution on a number line: $x^2 - 2x \geq 3$; $D = \{\text{Integers}\}$

solution   We begin the solution of this quadratic inequality by writing it in standard form.

$$x^2 - 2x - 3 \geq 0$$

Next we factor and get

$$(x - 3)(x + 1) \geq 0$$

This notation uses symbols to tell us that the product is equal to or is greater than zero. This is another way of saying that the product is zero or is a positive number. If the product is a positive number, both factors must be positive or both factors must be negative.

(Pos)	**and**	(Pos)	**or**	(Neg)	**and**	(Neg)

$$\begin{array}{rcl} x - 3 \geq & 0 \\ +3 & +3 \\ \hline x \geq & 3 \end{array} \quad \text{and} \quad \begin{array}{rcl} x + 1 \geq & 0 \\ -1 & -1 \\ \hline x \geq & -1 \end{array} \quad \text{or}$$

$$\begin{array}{rcl} x - 3 \leq & 0 \\ +3 & +3 \\ \hline x \leq & 3 \end{array} \quad \text{and} \quad \begin{array}{rcl} x + 1 \leq & 0 \\ -1 & -1 \\ \hline x \leq & -1 \end{array}$$

Now, both of the inequalities on the left must be true, **or** both the inequalities on the right must be true. On the left, if $x \geq 3$ and $x \geq -1$, then

$$x \geq 3$$

On the right, if $x \leq 3$ and $x \leq -1$, then

$$x \leq -1$$

Thus, the solution is the graph of the disjunction

$$x \geq 3 \quad \text{or} \quad x \leq -1$$

When we graph, we remember that the domain is the set of integers.

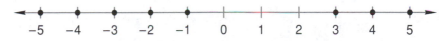

practice   Graph the solution on a number line:

    a. $(x + 1)(x - 2) > 0$; $D = \{\text{Reals}\}$ **b.** $x^2 - 3x \geq 4$; $D = \{\text{Integers}\}$

problem set 110

1.   RJ and Cheng noted that the sum of the digits of their two-digit counting number was 9. If the digits were reversed, they found that the new number was 27 less than the original number. What was the original number?

2.   Suzanne had a two-digit counting number, and the sum of the digits was 7. If she reversed the digits, she found that the new number was 45 greater than the original number. What was the original number?

3.   The weight of the oxygen in a quantity of the compound KCr_2O_7 was 336 grams. What was the total weight of the compound? (K, 39; Cr, 52; O, 16)

4.   The *Delta Queen* could go 120 miles downstream in 8 hours, but it took her 9 hours to go 63 miles upstream. What was her speed in still water, and what was the speed of the current in the river?

5.   Steve and Susu found that the cost of the machine was $1400, and that it was marked up $700 over what it had cost. What was the markup as a percentage of the selling price, and what was the markup as a percentage of the cost?

Graph the solution on a number line:

6. $(x + 2)(x - 4) > 0$; $D = \{\text{Reals}\}$ **7.** $x^2 - 4x \geq 5$; $D = \{\text{Integers}\}$

Expand:

8. $(-2x^3y^2z^3)^2$ **9.** $(x^{1/2} - y^{1/4})^2$

10. $(x^{1/2} + y^{1/2})(x^{1/2} - y^{1/2})$

Factor:

11. $8x^9 - y^6p^3$ **12.** $27x^{12}y^9 + p^6m^{15}$

13.   Show that $0.01\overline{362}$ is a rational number by writing it as a fraction of integers.

14.   Complete the square as an aid in graphing: $y = x^2 - 4x + 3$

15.   Graph on a number line: $3 \not\leq x + 2$ or $x + 5 \not< 8$; $D = \{\text{Reals}\}$

Solve:

16. $\begin{cases} \dfrac{2}{7}x - \dfrac{1}{5}y = 2 \\ 0.03x + 0.07y = 1.12 \end{cases}$ **17.** $\begin{cases} x + y + z = 1 \\ 4x - 2y - z = 6 \\ 3x - y + z = -1 \end{cases}$

18. $\begin{cases} x^2 + y^2 = 6 \\ x - y = 1 \end{cases}$ **19.** $\begin{cases} x - z = 3 \\ x + 2y = 5 \\ y + z = 0 \end{cases}$

20.   Graph: $\begin{cases} 2x \geq 6 \\ x + y < 3 \end{cases}$

Simplify:

21. $\dfrac{2i^2 - 2i + 2}{\sqrt{-9} - \sqrt{-3}\sqrt{-3}}$ **22.** $\dfrac{3 + 2\sqrt{20}}{1 - \sqrt{5}}$ **23.** $\dfrac{(a^{2-b})^2 x^2}{x^{b/2} a^{b/2}}$

24. $\sqrt[4]{8\sqrt{2}}$ **25.** $\sqrt{\dfrac{2}{3}} - 5\sqrt{\dfrac{3}{2}} + 3\sqrt{24}$

26.   Solve: $\sqrt{k} = 6 + \sqrt{k - 48}$ **27.**   Write $4R - 15U$ in polar form.

Solve by factoring:

28. $11s + 3s^2 + 10 = 0$ **29.** $-2x + 8x^3 + 6x^2 = 0$

30.   Find $(h + g)(-2)$ if $h(x) = x$; $D = \{\text{Reals}\}$, and $g(x) = x^2$; $D = \{\text{Positive integers}\}$.

LESSON 111   *Three statements of equality*

Many word problems contain three statements of equality. We can solve these problems by writing an equation for each statement of equality. If we have to use only three unknowns and if a solution to the system of equations exists, it can be found by using substitution, elimination, or both.

example 111.1   There were 26 nickels, dimes, and quarters in all, and their value was $2.25. How many coins of each type were there if there were 10 times as many nickels as quarters?

solution   We have three statements of equality, and each one can be written as an equation.

Number of nickels + number of dimes + number of quarters = 26

$$\text{(a)} \quad N_N + N_D + N_Q = 26$$

Value of nickels + value of dimes + value of quarters = 225 pennies

$$\text{(b)} \quad 5N_N + 10N_D + 25N_Q = 225$$

Ten times the number of quarters equals the number of nickels

$$\text{(c)} \quad 10N_Q = N_N$$

We begin by substituting using equation (c) $10N_Q$ for N_N in equations (a) and (b).

$$\text{(a)} \quad (10N_Q) + N_D + N_Q = 26 \qquad \longrightarrow \qquad N_D + 11N_Q = 26 \qquad \text{(a}'\text{)}$$

$$\text{(b)} \quad 5(10N_Q) + 10N_D + 25N_Q = 225 \qquad \longrightarrow \qquad 10N_D + 75N_Q = 225 \qquad \text{(b}'\text{)}$$

To solve, we will multiply equation (a′) by –10 and add the equations.

$$
\begin{array}{rl}
-10 \ (a') & -10N_D - 110N_Q = -260 \\
(b') & \underline{10N_D + 75N_Q = 225} \\
& -35N_Q = -35 \\
& N_Q = 1
\end{array}
$$

Now we can state that there are 10 nickels because $N_N = 10N_Q$, and this means 15 dimes because there are 26 coins in all. So

$$N_D = 15 \qquad N_N = 10 \qquad N_Q = 1$$

example 111.2   The total number of blues, greens, and yellows in the pot was 7. The blues weighed 1 pound each, the greens weighed 4 pounds each, and the yellows weighed 5 pounds each. The total weight was 25 pounds. If there was 1 more yellow than green, how many of each color were there?

solution   Again we get three statements of equality that we can write as equations.

Number of blues + number of greens + number of yellows = 7

$$\text{(a)} \quad N_B + N_G + N_Y = 7$$

Weight of blues + weight of greens + weight of yellows = 25

$$\text{(b)} \quad N_B + 4N_G + 5N_Y = 25$$

There was 1 more yellow than green

$$\text{(c)} \quad N_G + 1 = N_Y$$

To begin, we will replace N_Y in (a) and (b) with $N_G + 1$ and then simplify.

$$\text{(a)} \quad N_B + N_G + (N_G + 1) = 7 \qquad \longrightarrow \qquad N_B + 2N_G = 6 \qquad \text{(a}'\text{)}$$

$$\text{(b)} \quad N_B + 4N_G + 5(N_G + 1) = 25 \qquad \longrightarrow \qquad N_B + 9N_G = 20 \qquad \text{(b}'\text{)}$$

Now we multiply (a′) by –1 and add the result to (b′).

$$-1 \text{ (a′)} \quad -N_B - 2N_G = -6$$
$$\text{(b′)} \quad \underline{N_B + 9N_G = 20}$$
$$7N_G = 14$$
$$N_G = 2$$

Now since $N_Y = N_G + 1$, there were 3 yellows.

$$N_Y = 3$$

There must have been 2 blues because the total was 7.

$$N_B = 2 \qquad N_Y = 3 \qquad N_G = 2$$

practice There were 35 nickels, dimes, and quarters in all, and their value was $5.00. How many coins of each type were there if there were 2 times as many quarters as nickels?

problem set 111

1. There were 28 nickels, dimes, and quarters in all, and their value was $2.50. How many coins of each type were there if there were 5 times as many nickels as quarters?

2. The total number of blues, greens, and yellows in the pot was 10. The blues weighed 1 pound, the greens weighed 4 pounds, and the yellows weighed 5 pounds. The total weight was 39 pounds. If there were 2 more yellows than greens, how many of each color were there?

3. A two-digit counting number has a value that equals 4 times the sum of its digits. If the units' digit is 1 greater than the tens' digit, what is the number?

4. Find three consecutive integers such that the product of the first and the third is 35 greater than the product of the second and 5.

5. The number of students varied directly as the number of teachers and as the number of administrators squared. One thousand students were present when there were 5 teachers and 2 administrators. How many students were there when there were 8 teachers and only 1 administrator? Solve two ways.

Graph the solution on a number line:

6. $(x + 4)(x - 2) > 0$; $D = \{\text{Integers}\}$ 7. $x^2 > -6 + 5x$; $D = \{\text{Integers}\}$

Expand:

8. $(x^{1/2} + y^{1/4})^2$ 9. $(x^{1/2} - y^{-1/2})^2$ 10. $(x^{1/2}y^{-1/2})^2$

Factor:

11. $x^3 - m^6y^6$ 12. $8x^6y^3 - 27m^3p^{12}$

13. Show that $1.02\overline{13}$ is a rational number by writing it as a quotient of integers.

14. Complete the square as an aid in graphing: $y = -x^2 - 4x - 1$

Graph on a number line:

15. $-|x| - 3 \not< -7$; $D = \{\text{Reals}\}$ 16. $-2 \not> x + 5 < 4$; $D = \{\text{Integers}\}$

Solve:

17. $\begin{cases} \dfrac{3}{5}x - \dfrac{2}{5}y = -10 \\ 0.003x + 0.2y = 1.97 \end{cases}$ 18. $\begin{cases} x + 2y = 10 \\ x - 3z = -16 \\ y + 2z = 16 \end{cases}$ 19. $\begin{cases} x^2 + y^2 = 4 \\ x - y = 1 \end{cases}$

20. Solve $-2x^2 + 3x + 5 = 0$ by completing the square.

21. Use unit multipliers to convert 40 inches per second to meters per hour.

Simplify:

22. $\dfrac{2i^3 - \sqrt{-3}\,\sqrt{-3}}{4 - 3i^2}$ 23. $\dfrac{2\sqrt{3} + 2}{3 - \sqrt{3}}$ 24. $\dfrac{a^{x/2}(y^{2-x})^{1/2}}{a^{3x}\,y^{-2x}}$

25. $\sqrt{xy}\,\sqrt{x^2 y}$

26. $\sqrt{\dfrac{2}{7}} - 3\sqrt{\dfrac{7}{2}} + 2\sqrt{126}$

27. Graph: $\begin{cases} -y < 3 \\ 3x + y \le 3 \end{cases}$

28. Find the resultant of the two force vectors shown.

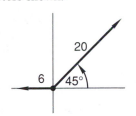

Solve by factoring:

29. $2x^2 = x + 10$

30. $-15x = 7x^2 - 2x^3$

LESSON 112 *Quadratic inequalities (less than)*

In Lesson 110 we discovered that the solution to a greater-than quadratic inequality is a disjunction. We found that the solution to the inequality

$$(x + 2)(x - 3) > 0$$

is the disjunction

$$x < -2 \quad \textbf{or} \quad x > 3$$

If we reverse the inequality symbol in the original inequality, we get

$$(x + 2)(x - 3) < 0$$

This product is a negative number because all numbers that are less than zero are negative numbers. We will find that the solution to a less-than inequality such as this one will be a conjunction.

example 112.1 Solve the inequality and graph the solution on a number line: $(x + 2)(x - 3) < 0$; $D = \{\text{Reals}\}$

solution For a product to be less than zero, the product must be negative, for all real numbers that are less than zero are negative numbers. **Thus the first factor must be negative and the second factor positive, or the first factor must be positive and the second factor negative.**

(Neg)	**and**	(Pos)	**or**	(Pos)	**and**	(Neg)
$x + 2$		$x - 3$		$x + 2$		$x - 3$

For $x + 2$ to be negative, $x + 2$ must be less than zero; and for $x - 3$ to be positive, $x - 3$ must be greater than zero. Thus,

For $x + 2$ to be positive, $x + 2$ must be greater than zero; and for $x - 3$ to be negative, $x - 3$ must be less than zero. Thus,

$$\begin{array}{rl} x + 2 < 0 \quad \textbf{and} \quad & x - 3 > 0 \\ \underline{-2 \quad -2} \quad & \underline{+3 \quad +3} \\ x \quad\quad < -2 \quad \textbf{and} \quad & x \quad\quad > 3 \end{array}$$

$$\begin{array}{rl} x + 2 > 0 \quad \textbf{and} \quad & x - 3 < 0 \\ \underline{-2 \quad -2} \quad & \underline{+3 \quad +3} \\ x \quad\quad > -2 \quad \textbf{and} \quad & x \quad\quad < 3 \end{array}$$

The numbers that satisfy the conjunction on the left are the real numbers that are less than -2 and that are also greater than 3. **There are no numbers that are less than -2 and that are also greater than 3. Thus, there are no numbers that satisfy the left-hand side,** and the total solution must come from the right-hand side. The numbers that satisfy the conjunction on the right are the real numbers greater than -2 that are also less than 3.

$$-2 < x < 3$$

```
    +    +    ○----+----+----+----+----●    +
   -4   -3   -2   -1    0    1    2    3    4
```

example 112.2   Solve the inequality and graph the solution on a number line:

$$x^2 + 2x - 8 < 0; \quad D = \{\text{Integers}\}$$

solution   We begin by factoring to get

$$(x + 4)(x - 2) < 0$$

The product of these two factors is less than zero. **This means that the product is a negative number.** Thus, the first factor must be negative **and** the second factor must be positive, **or** the first factor must be positive **and** the second factor must be negative.

(NEG)	**and**	(POS)	**or**	(POS)	**and**	(NEG)
$x + 4$		$x - 2$		$x + 4$		$x - 2$

For $x + 4$ to be negative, $x + 4$ must be less than zero; and for $x - 2$ to be positive, $x - 2$ must be greater than zero. Thus,	For $x + 4$ to be positive, $x + 4$ must be greater than zero; and for $x - 2$ to be negative, $x - 2$ must be less than zero. Thus,

$$
\begin{array}{llll}
x + 4 < 0 & \text{and} & x - 2 > 0 \\
\underline{-4 \quad -4} & & \underline{+2 \quad +2} \\
x \quad\; < -4 & \text{and} & x \quad\; > 2 \quad \text{or}
\end{array}
\qquad
\begin{array}{lll}
x + 4 > 0 & \text{and} & x - 2 < 0 \\
\underline{-4 \quad -4} & & \underline{+2 \quad +2} \\
x \quad\; > -4 & \text{and} & x \quad\; < 2
\end{array}
$$

There are no integers that satisfy the conjunction on the left because there are no integers that are less than -4 that are also greater than 2. Thus, the total solution must come from the conjunction on the right and consists of the integers that are greater than -4 and that are also less than 2.

$$-4 < x < 2$$

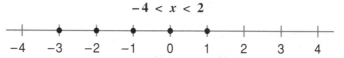

practice   Solve each inequality and graph the solution on a number line:

    a. $(x + 4)(x - 1) < 0; \quad D = \{\text{Reals}\}$ **b.** $x^2 - 3x - 10 < 0; \quad D = \{\text{Integers}\}$

problem set 112

1. There were 20 nickels, dimes, and quarters whose value was $3.25. If there were twice as many quarters as dimes, how many coins of each kind were there?

2. The sum of the digits of a two-digit counting number is 7. If the digits are reversed, the new number is 9 less than the original number. What is the original number?

3. The bookstore went out of business because the markup they used was only 20 percent of the selling price of the books. If they sold one shipment for a total price of $1800, what did they pay for these books?

4. The Ochlochnee River meandered at 3 miles per hour. A fast boat could go 230 miles down the river in twice the time it took to go 85 miles up the river. What was the speed of the boat in still water, and what were the times?

5. Fabian took 3 times as long to fly 1800 miles as it took Hamilcar to fly 1200 miles. Find the rates and times of both if Hamilcar's speed was 200 miles per hour greater than that of Fabian.

Graph the solution on a number line:

    6. $(x + 3)(x - 4) < 0; \quad D = \{\text{Reals}\}$ **7.** $x^2 - 5x - 6 < 0; \quad D = \{\text{Integers}\}$

    8. $(x + 2)(x - 3) > 0; \quad D = \{\text{Reals}\}$

    9. Multiply: $(x^{1/2} + y^{1/2})(x^{1/2} - y^{-1/4})$ **10.** Factor: $p^6 x^6 - k^3$

11. Show that $4.01\overline{43}$ is a rational number by writing it as a fraction of integers.

12. Complete the square as an aid in graphing: $y = x^2 + 2x + 3$

Graph on a number line:

13. $-|x| + 3 \leq 0;\ D = \{\text{Reals}\}$

14. $x - 2 \not> 0$ or $x + 4 > 8;\ D = \{\text{Integers}\}$

15. Which of the following sets of ordered pairs are functions?

    (a)   (4, 2), (2, 4), (5, 7), (7, 5)

    (b)   (4, −2), (−2, 6), (5, 7), (7, −5)

    (c)   (−4, 2), (5, 3), (−4, 7), (3, 5)

16. Find the number that is $\frac{1}{3}$ of the way from $2\frac{1}{8}$ to 5.

17. Solve $-5x^2 - x = 4$ by using the quadratic formula.

18. Find the equation of the line through $(5, -7)$ that is perpendicular to $x + 3y - 4 = 0$.

Solve:

19. $\begin{cases} \frac{1}{2}x + \frac{1}{4}y = 7 \\ 0.7x - 0.02y = 5.36 \end{cases}$

20. $\begin{cases} x - 2y = 10 \\ 3x - z = 11 \\ 2y - 3z = -9 \end{cases}$

21. $\begin{cases} 2x - y = 7 \\ xy = 4 \end{cases}$

22. $\begin{cases} x + y + z = 1 \\ 2x - y + z = -5 \\ 3x + y + z = 5 \end{cases}$

Simplify:

23. $\dfrac{-3i^2 - 2i^3}{\sqrt{-3}\,\sqrt{-3} - \sqrt{-9}}$

24. $\dfrac{\left(a^{x-2}\right)^{1/2} y^2}{y^{x/2}\, a^{x/2}}$

25. $\sqrt[3]{x^5 y}\,\sqrt{x^5 y^2}$

26. $\sqrt{\dfrac{2}{7}} + \sqrt{\dfrac{7}{2}} - 3\sqrt{56}$

27. $\dfrac{3\sqrt{2} - 2}{7\sqrt{2} - 3}$

Solve by factoring:

28. $3x^3 + 5x^2 = -2x$

29. $2 = 2x^2 - 3x$

30. $3x^2 + 8x + 4 = 0$

LESSON 113   *Logarithms • Antilogarithms*

113.A

logarithms   We begin by reviewing the rules for exponents.

    1. If two numbers are written as powers of the same base, we can multiply the numbers by adding the exponents.

$$10^7 \cdot 10^2 = 10^{7+2} = 10^9$$

Also $4^7 \cdot 4^2 = 4^{7+2} = 4^9$

    2. If two numbers are written as powers of the same base, we can divide the numbers by subtracting the exponent in the denominator from the exponent in the numerator.

$$\frac{10^7}{10^2} = 10^{7-2} = 10^5$$

$$\frac{4^7}{4^2} = 4^{7-2} = 4^5$$

3. A number written as a power is raised to a power by multiplying the powers.

$$(10^7)^2 = 10^{7 \cdot 2} = 10^{14}$$

$$(4^7)^2 = 4^{7 \cdot 2} = 4^{14}$$

These are the three rules for exponents. These are also the three rules for logarithms, for *logarithm* **is simply another name for exponent. The three rules illustrated above are the only rules for logarithms. That's all there is to logarithms.**

What makes logarithms difficult for some people is the use of the word *logarithm*. We can remember that logarithm is another name for exponent, but it is difficult to remember that an antilogarithm is the name for the number.

$$\text{base} \longrightarrow 10^2 \longleftarrow \text{logarithm} = 100 \longleftarrow \text{antilogarithm}$$

To make things worse, we often write this equality in another form. We note that the base is 10. The logarithm is 2 and the antilogarithm is 100. Thus, we can say

The logarithm of 100 to the base 10 is 2.

We can write this in an abbreviated form as

$$\log_{10} 100 = 2$$

The base of a logarithm cannot be 1 because 1 to any power is 1.

$$1^2 = 1 \qquad 1^{100.5} = 1 \qquad 1^{2\sqrt{\pi}} = 1$$

The base of a logarithm cannot be 0, because zero to a positive power is 0. Zero to the zero power is undefined; zero to a negative power is also undefined.

$$0^2 = 0 \qquad 0^0 = ? \qquad 0^{-4} = \frac{1}{0^4} = ?$$

The base of a logarithm can be any positive number except 1. Bases other than 10 and e are rarely encountered in algebra, and we will concentrate on the bases 10 and e. If a number is entered on a calculator and the log key is depressed, the calculator will give us the proper exponent (log) such that 10 to that power will equal the number. To find the base 10 logarithm of 105, we enter 105 and depress the log key.

Enter	Display
105	105
log	2.021189299

This means that

$$105 \approx 10^{2.021189299}$$

Since the exponent is a logarithm, we can also write

$$\log_{10} 105 \approx 2.021189299$$

When the abbreviation *log* is used and no base is indicated, the base is understood to be 10. Therefore, we can also write the above as

$$\log 105 \approx 2.021189299$$

The number e, known as the **natural logarithm** and abbreviated **ln**, is an irrational number that is very important in calculus and other advanced math courses. The decimal approximation of e to 10 decimal places is

$$e \approx 2.7182818285$$

If a number is entered on a calculator and the ln key is depressed, the calculator will give us the proper exponent such that e to that power equals the number. To find the base e logarithm of 105, we enter 105 and press the ln key.

ENTER	DISPLAY
105	105
ln	4.65396035

This means that

$$105 \quad e^{4.65396035}$$

Since the exponent is a logarithm, we can also write

$$\log_e 105 = 4.65396035$$

Instead of writing $\log_e$, we usually write the abbreviation *ln*, which tells us the base is *e*, and write

$$\ln 105 = 4.65396035$$

113.B
antilogarithms

An **antilogarithm** is the number we get if we raise the base to the logarithm, or power. If we are asked to find the base 10 antilog of 4.316, we remember that the number is the antilog, so 4.316 must be the exponent. Thus we want to evaluate

$$10^{4.316}$$

Since we are looking for the antilog, we use the inverse log sequence.

ENTER	DISPLAY
4.316	4.316
inv log	**20,701.41**

This means that $10^{4.316} = 20,701.41$. We could also have found the inverse logarithm of 4.316 by using the y^x key to raise 10 to the 4.316 power.

To find the base *e* antilogarithm of 4.316, we note that we have been given the exponent and have been asked to raise *e* to that power. If your calculator has an e^x key, you can use that. Otherwise you have to use the inv ln sequence, which requires only one step.

ENTER	DISPLAY
4.316	4.316
inv ln	**74.88847449**

example 113.1   Use your calculator to get a value for *e*.

solution   We note that *e* raised to the 1 power equals *e*.

$$e^1 = e$$

This means that 1 is the logarithm of *e* to the base *e*. Thus *e* is the inverse logarithm of 1.

ENTER	DISPLAY
1	1
inv ln	**2.718281828**

Logarithms often give students trouble, not because logarithms are difficult, but because students forget that a logarithm is an exponent, and that the rules of exponents apply to logarithms. In the past, logarithms were used to compute numerical answers. Today we can do the numerical computations on a calculator. Nonetheless, we will begin our work with logarithms by using them to find numerical answers. **These problems are introduced not to teach a new method of computation but to emphasize that a logarithm is just an exponent.**

example 113.2   Use base 10 logarithms to simplify: $\dfrac{(47,832)(59,463)}{0.000817}$

solution First we use the [log] key on the calculator to find the proper exponent to write each number as a power of 10. For convenience, we will round each exponent to two or three decimal places. By using the [log] key, we find that

47,832 can be written as $10^{4.68}$

59,463 can be written as $10^{4.77}$

0.000817 can be written as $10^{-3.09}$

Now we have

$$\frac{(10^{4.68})(10^{4.77})}{10^{-3.09}} = 10^{4.68+4.77+3.09} = 10^{12.54}$$

We could use the [y^x] key to evaluate 10 to this power, but we will use the [inv] [log] keys to find the antilogarithm.

ENTER	DISPLAY
12.54	12.54
[inv] [log]	3.47×10^{12}

example 113.3 Use base e logarithms to simplify: $\dfrac{(47,832)(59,463)}{0.000817}$

solution We use the [ln] key on the calculator to write each number as e raised to a power. We get

$$\frac{(e^{10.78})(e^{10.99})}{e^{-7.11}} = e^{10.78+10.99+7.11} = e^{28.88}$$

We will find the antilogarithm of 28.88 by using the [inv] [ln] keys.

ENTER	DISPLAY
28.88	28.88
[inv] [ln]	3.49×10^{12}

This example is the same as Example 113.2 except that we used base e this time. Our answers are different because of errors that accumulate when we round logarithms.

example 113.4 If $-0.0325 = \log_e y$, find y.

solution In a logarithm problem there is a base and two other numbers. The base in this problem is e. The problem states that -0.0325 is a logarithm. Thus, y must be the antilog of -0.0325.

ENTER	DISPLAY
-0.0325	-0.0325
[inv] [ln]	0.968

This tells us that

$$e^{-0.0325} \approx 0.968$$

example 113.5 If $y = 4302$, find: (a) $\log y$ (b) $\ln y$

solution

ENTER	DISPLAY	ENTER	DISPLAY
4302	4302	4302	4302
[log]	3.63	[ln]	8.37

This tells us that

$$10^{3.63} \approx 4302 \quad \text{and} \quad e^{8.37} \approx 4302$$

practice Find x. Remember that if we know the base there are only two other numbers. One number is the logarithm, and the other number is the antilogarithm. Thus we will always use the log or ln key or the inv log or inv ln keys.

    a. $x = \ln 0.0052$ **b.** $e^x = 51.4$

    c. $\ln x = -4.16$ **d.** $\log x = -4.16$

    e. Use the ln key to express each number as a power. Then use the rules of exponents to find the answer.

$$\frac{(0.000612)(576)}{0.0512 \times 10^{-14}}$$

problem set 113

1. There were 19 nickels, dimes, and quarters in the pot. James noted that their value was $2. How many coins of each type were there if there were twice as many nickels as dimes?

2. Reds varied directly as blues and inversely as mauves squared. When there were 10 reds, there were 2 mauves and 4 blues. How many blues were there when there were 20 mauves and 3 reds?

3. It was necessary to mix 200 ml of a solution in which the key ingredient made up exactly 63 percent of the total. One container held a solution that contained 70% key ingredient, and the other container held a solution that was only 60% key ingredient. How much of each solution should be used?

4. Carlene made the trip in only 8 hours, whereas Dan took 12 hours to make the same trip. This was because Dan dawdled and drove 20 miles per hour slower than Carlene drove. How fast did each drive and how long was the trip?

5. The pressure of a quantity of an ideal gas was held constant at 1400 torr. The initial volume and temperature were 1000 ml and 1700 K. If the volume were increased to 2000 ml, what would the final temperature be in kelvins?

Find x. Remember that in a logarithm problem we find either a logarithm or an antilogarithm.

6. (a) $x = \ln 0.0093$ (b) $e^x = 62.5$

7. (a) $\ln x = 5163$ (b) $\log x = 2.136$

8. In the figure shown, $AE = 6, AD = 4$, and $AB = 24$. What is the length of the segment AC? Dimensions are in meters.

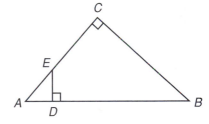

Graph the solution on a number line:

9. $(x + 2)(x - 3) < 0$; $D = \{\text{Reals}\}$ **10.** $x^2 - x - 6 \geq 0$; $D = \{\text{Integers}\}$

11. $x^2 - 8 \leq 2x$; $D = \{\text{Integers}\}$ **12.** $|x| - 1 \not\leq 0$; $D = \{\text{Integers}\}$

13. $7 \not> x - 2 < 10$; $D = \{\text{Reals}\}$

14. Multiply: $(x^{1/3} + y^{2/3})(x^{2/3} + y^{1/3})$ **15.** Factor: $8p^6k^{15} - x^3m^6$

16. Show that $0.003\overline{16}$ is a rational number by writing it as a fraction of integers.

17. Complete the square as an aid in graphing: $y = -x^2 + 4x - 1$

18. Find the number that is $\dfrac{2}{11}$ of the way from 2 to $4\dfrac{1}{6}$.

19. Solve $3x^2 - x - 7 = 0$ by completing the square.

Solve:

20. $\begin{cases} 1\frac{1}{5}x + \frac{2}{3}y = 30 \\ -0.18x - 0.02y = -3.78 \end{cases}$

21. $\begin{cases} x^2 + y^2 = 4 \\ 3x - y = 2 \end{cases}$

22. $\begin{cases} x - 4y = -15 \\ 3x + z = 20 \\ 2y - z = 5 \end{cases}$

23. $\begin{cases} x - 2y - z = -8 \\ 3x - y - 2z = -5 \\ x + y + z = 9 \end{cases}$

Simplify:

24. $\dfrac{2i^3 - i}{-\sqrt{-3}\,\sqrt{-3} + 3}$

25. $\dfrac{\sqrt{2} - 5}{2\sqrt{2} - 4}$

26. $\sqrt[6]{xy^3}\,\sqrt[3]{xy^2}$

27. $\sqrt{\dfrac{4}{3}} + 2\sqrt{\dfrac{3}{4}} - 3\sqrt{48}$

Solve by factoring:

28. $2x^3 = -x^2 + 3x$ **29.** $3x^2 - 2 = x$ **30.** $-7x^2 - 2x = 3x^3$

LESSON 114 *Nonlinear inequalities*

Remember that a straight line divides the set of all points in the plane into three disjoint subsets. These are the set of points that lie on the line and the sets of points that lie on either side of the line. In the figure we show the graph of the equation $y = -\frac{1}{2}x - 1$. All points whose coordinates satisfy this equation lie on this line. The rest of the points lie on one side of the line or the other side of the line and will satisfy one of the following inequalities.

(a) $y > -\dfrac{1}{2}x - 1$

(b) $y < -\dfrac{1}{2}x - 1$

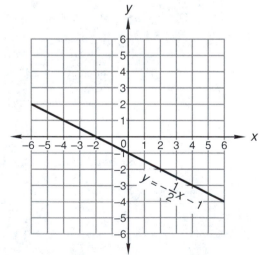

If the coordinates of a point satisfy one of these inequalities, the coordinates of all the points on the same side of the line will satisfy the same inequality. Inequality (a) is read "y is greater than negative one-half x minus 1" and since y is "greater than" as we move up, we surmise that this inequality designates the points that lie above the line. We can always use a test point to check our surmise.

The same thoughts apply to curved lines, as will be demonstrated in the following examples.

example 114.1   The line and the parabola graphed in the figure divide the coordinate system into four distinct regions, which have been labeled $A, B, C,$ and D. Which region contains the coordinates of the points that satisfy the given system of inequalities?

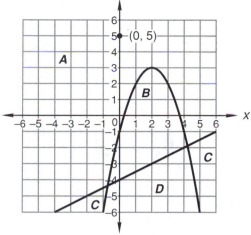

(a) $\quad \begin{cases} y \geq \dfrac{1}{2}x - 4 & \text{(line)} \\[2mm] y \geq -x^2 + 4x - 1 & \text{(parabola)} \end{cases}$

(b)

solution   Inequality (a) is read "y is greater than or equal to $\frac{1}{2}x - 4$." Since y is "greater than" as we move up, we suspect that this inequality designates the points on or above the line. Inequality (b) reads "y is greater than or equal to $-x^2 + 4x - 1$." Thus we suspect that this inequality designates the points on or above the parabola. The region above both the line and the parabola is region A. We will use the point $(0, 5)$ as a test point.

Line: $\qquad y \geq \dfrac{1}{2}x - 4 \quad \longrightarrow \quad 5 \geq \dfrac{1}{2}(0) - 4 \quad \longrightarrow \quad 5 \geq -4 \qquad$ True

Parabola: $\quad y \geq -x^2 + 4x - 1 \quad \longrightarrow \quad 5 \geq -(0)^2 + 4(0) - 1$

$\qquad\qquad\qquad\qquad\qquad\qquad \longrightarrow \quad 5 \geq -1 \qquad\qquad\qquad$ True

Thus the solution is **region A, including the bordering points that lie on the line or on the parabola.** The coordinates of any point in area A or on the border will satisfy both of the inequalities given.

example 114.2   In the figure we show the graphs of the given line and parabola. Which region contains the coordinates of the points that will satisfy both inequalities?

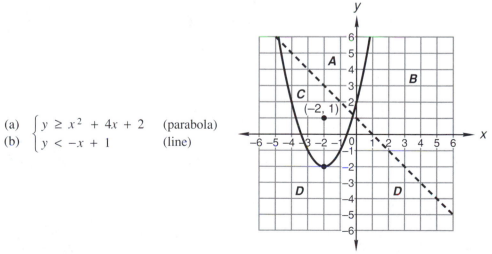

(a) $\begin{cases} y \geq x^2 + 4x + 2 & \text{(parabola)} \\ y < -x + 1 & \text{(line)} \end{cases}$

(b)

solution   The quadratic inequality reads "y is greater than or equal to," which indicates the points above or on the parabola; and the linear inequality reads "y is less than," which indicates the points below the line. This indicates region C. We will use the point $(-2, 1)$ as a test point.

Parabola: $y \geq x^2 + 4x + 2$ → $1 \geq (-2)^2 + 4(-2) + 2$

→ $1 \geq 4 - 8 + 2$ → $1 \geq -2$ True

Line: $y < -x + 1$ → $1 < -(-2) + 1$ → $1 < 2 + 1$

→ $1 < 3$ True

Thus the coordinates of all points that lie below the line and on or above the parabola satisfy this system of inequalities, and the answer is **region C.**

example 114.3   Which region of the graph satisfies this system of nonlinear inequalities?

(a) $\begin{cases} x^2 + y^2 \leq 16 \quad \text{(circle)} \\ y \geq x \quad\quad\quad\ \text{(line)} \end{cases}$
(b)

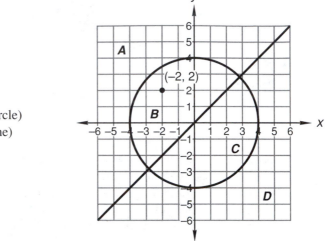

solution   The inequality $x^2 + y^2 \leq 16$ designates the points that lie on or inside the circle (try a few points to check this out), and the linear inequality designates the points on or above the line. We will use the point $(-2, 2)$ as a test point.

Circle: $x^2 + y^2 \leq 16$ → $(-2)^2 + (2)^2 \leq 16$

→ $4 + 4 \leq 16$ True

Line: $y \geq x$ → $2 \geq -2$ True

This verifies our surmise, and thus the region designated is **region B, including the points on the boundary of this area.**

practice   Which region of the graph satisfies this system of nonlinear inequalities?

$\begin{cases} x^2 + y^2 > 9 \quad \text{(circle)} \\ y \geq x \quad\quad\quad\ \text{(line)} \end{cases}$

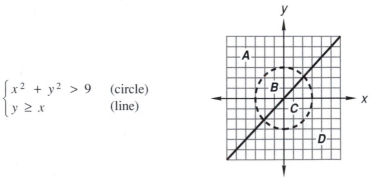

**problem set
114**

1.   Find three consecutive even integers such that the product of the first and the third exceeds the product of the second and 8 by 16.

2.   The druggist began with 240 ml of a 20% antiseptic solution. How much pure antiseptic should be added to get a solution that is 52% antiseptic?

3. The long trip was 4800 miles, and for this trip Ken used fast transportation, which took 1 hour more than the slower conveyance took to cover 2000 miles. Find the rates and times of both methods of transportation if the faster moved at twice the speed of the slower.

4. The current in the river flowed at 4 miles per hour. The steamboat could go 34 miles downstream in one-third the time it took to go 54 miles upstream. What was the speed of the boat in still water?

5. The sum of the digits of a two-digit counting number was 8. If the digits were reversed, the new number would be 54 greater than the original number. What was the original number?

For each figure, designate the region in which the coordinates of the points satisfy the given system of inequalities.

6. $\begin{cases} y \geq \frac{1}{4}x - 2 \\ y \geq -x^2 + 3x - 2 \end{cases}$
   7. $\begin{cases} y \geq x^2 + 6x + 3 \\ y < -x + 2 \end{cases}$
   8. $\begin{cases} x^2 + y^2 \leq 36 \\ y \geq x \end{cases}$

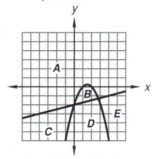
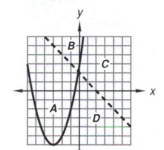
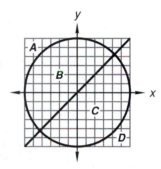

Find x. Remember that in a logarithm problem we find either a logarithm or an antilogarithm.

9. (a) $x = \ln 0.0085$ (b) $e^x = 87.4$

10. (a) $\ln x = 9185$ (b) $\log x = 3.188$

11. The area of the square is 9 cm². What is the length of the segment AB? Dimensions are in centimeters.

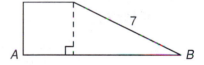

12. Use a calculator to write 0.000715×10^{-4} as an exponential expression whose base is 10. Then use logarithm notation to make the same statement.

13. The markup was 20 percent of the selling price. If the selling price was $140, what did the store owner have to pay for the item?

Graph the solution on a number line:

14. $x^2 \geq -3x + 4$; $D = \{\text{Integers}\}$ 15. $x^2 - 4 \leq 3x$; $D = \{\text{Integers}\}$

16. $-|x| - 3 > -7$; $D = \{\text{Integers}\}$ 17. $6 \not> x - 4 < 8$; $D = \{\text{Integers}\}$

18. Show that $0.001\overline{056}$ is a rational number by writing it as a quotient of integers.

19. Complete the square as an aid in graphing: $y = x^2 - 4x + 7$

20. Find the number that is $\frac{3}{8}$ of the way from $4\frac{1}{2}$ to $6\frac{1}{4}$.

Solve:

21. $\begin{cases} 2\frac{1}{3}x + \frac{1}{5}y = 10 \\ 0.03x - 0.03y = -0.36 \end{cases}$
    22. $\begin{cases} 3x - z = 8 \\ 2x - 2y = -4 \\ 2y + 3z = 2 \end{cases}$

23. $\begin{cases} x + y = 6 \\ xy = -1 \end{cases}$

24. $\begin{cases} x - y + z = 3 \\ 2x - y + 2z = 9 \\ -x + y + z = 1 \end{cases}$

25. Estimate the location of the line suggested by the data points and write the equation that expresses output as a function of input: $O = mI + b$

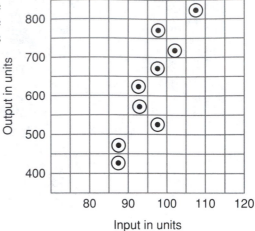

Simplify:

26. $\dfrac{i^3 - i^2}{i^5 + 2}$

27. $\sqrt{\dfrac{5}{8}} + 2\sqrt{\dfrac{8}{5}} - 3\sqrt{40}$

28. Solve: $\sqrt{x - 15} + \sqrt{x} = 5$

Solve by factoring:

29. $5x^3 + 2x = -7x^2$

30. Expand: $(x^{1/2} - y^{-1/2})^2$

LESSON 115 · *Exponential equations · Exponential functions · Compound interest*

115.A
exponential equations

We know that 10^2 equals 100. Thus, if we write

$$10^2 = 10^x$$

the left-hand side of the equation equals 100. For the right-hand side of the equation to equal 100, x must equal 2.

$$10^2 = 10^2$$

We will generalize this observation by saying that **if two powers of the same base are equal, the exponents must be equal. If we write**

$$10^{1.23} = 10^{x+2}$$

the equals sign tells us that the expressions are equal. The bases are equal, so the exponents must be equal.

$$1.23 = x + 2$$

We will solve this equation in the following example.

example 115.1   Solve: $17 = 10^{x+2}$

solution   We can solve this exponential equation by writing 17 as a power of 10. If we use the $\boxed{\text{log}}$ key on the calculator, we can find the proper exponent. The calculator tells us that $\log_{10} 17$ is approximately 1.23, so we write

$$10^{1.23} = 10^{x+2}$$

Since these expressions are equal and the bases are equal, the exponents must be equal.

$$1.23 = x + 2 \qquad \text{equated exponents}$$
$$-0.77 = x \qquad \text{solved}$$

115.B
exponential
functions

Exponential functions are important because they help us understand problems such as the growth of bacteria in biology, the voltage on a capacitor in engineering, radioactive decay in physics, and the growth of money in finance. The equations on the left define two exponential functions.

$$A_t = 200e^{0.4t}$$

$$A_t = 200e^{0.2t}$$

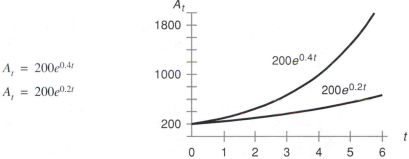

In these equations the symbol A_t, read "A of t," represents the amount of money, bacteria, electrons, etc., present at some time t. The graphs of the equations above show that when time equaled zero, the amount present was 200, and that the amount increased **exponentially** with time. These equations describe an exponential increase because the coefficients of t are 0.2 and 0.4, which are positive numbers. **The graph of $200e^{0.4t}$ goes up faster than the graph of $200e^{0.2t}$ because 0.4 is greater than 0.2.** When t equals zero, the amount present is called "A of zero" and equals 200 for both of these functions because e^0 equals 1 and 1 times 200 equals 200.

$$A_0 = 200e^{0.4(0)} \qquad\qquad A_0 = 200e^{0.2(0)}$$
$$= 200(1) \qquad\qquad\qquad = 200(1)$$
$$= 200 \qquad\qquad\qquad\quad = 200$$

If the exponential constant is a negative number, the graph of the curve goes down as t is increased.

$$A_t = 200e^{-0.4t}$$

$$A_t = 200e^{-0.2t}$$

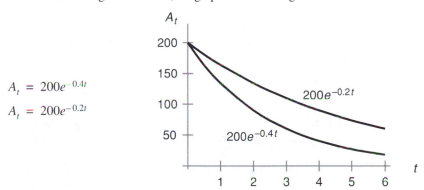

These equations describe exponential decreases because the amount present decreases as time increases. **The graph of $200e^{-0.4t}$ goes down faster than the graph of $200e^{-0.2t}$ because the absolute value of -0.4 is greater than that of -0.2.**

When we write the general form of an exponential function, we use A_0 (A of zero) to represent the amount present when t equals zero and we use k to represent the coefficient of time in the exponent. If k is greater than zero (a positive number), the curve goes up. If k is less than zero (a negative number), the curve goes down.

$$A_t = A_0 e^{kt}$$

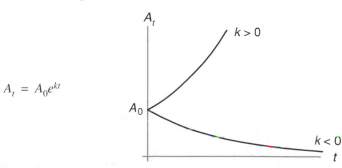

example 115.2   In the beginning, there were 400 bacteria in the dish. The number of bacteria increased exponentially. Three days later there were 2000 bacteria in the dish. How many bacteria would there be in 8 days?

solution   The general exponential equation is

$$A_t = A_0 e^{kt}$$

These problems are two-step problems. The first step is to find the values of A_0 and k. Then we will use the finished equation to find A_8, the amount present when t equals 8. Since A_0 is the initial amount, we can replace A_0 with 400.

$$A_t = 400 e^{kt}$$

Next we will find k. We substitute 2000 for A_t and 3 for t because 2000 were present when t equaled 3. **After we find k, we will not use the numbers 2000 and 3 again in this problem.**

$$2000 = 400 e^{k(3)} \qquad \text{substituted}$$

$$5 = e^{k(3)} \qquad \text{divided by 400}$$

Now we use the $\boxed{\text{ln}}$ key to find the exponent and write 5 as a power of e.

$$e^{1.61} = e^{k(3)}$$

The quantities are equal and the bases are the same, so the exponents must be equal.

$$1.61 = 3k \qquad \text{equated exponents}$$

$$0.537 = k \qquad \text{solved for } k$$

If we replace k with 0.537, we have

$$A_t = 400 e^{0.537t}$$

Now we have an equation that can be evaluated for any value of t. The problem asked for A_8, which is A_t when t equals 8.

$$A_8 = 400 e^{0.537(8)} \qquad \text{substituted}$$

$$A_8 = 400 e^{4.296} \qquad \text{multiplied}$$

$$A_8 = 400(73.4) \qquad \text{used } \boxed{\text{inv}}\ \boxed{\text{ln}}\ \text{keys}$$

$$A_8 \approx \textbf{29,362 bacteria} \qquad \text{multiplied}$$

example 115.3   In the beginning there were 4.25 ounces of radioactive material. It decayed exponentially. Ten years later 3.92 ounces remained. How much would remain after 30 years?

solution   We see that A_0 equals 4.25, so we can substitute 4.25 for A_0.

$$A_t = A_0 e^{kt} \quad \longrightarrow \quad A_t = 4.25 e^{kt}$$

In order for us to find k, the problem must always give us A_t for some t. **This problem tells us that A_{10} equals 3.92, so we can find k.** After we find k, we will not use these numbers again. We substitute 10 for t and 3.92 for A_t.

$$3.92 = 4.25 e^{k(10)} \qquad \text{substituted}$$

$$0.9223 = e^{10k} \qquad \text{divided}$$

$$e^{-0.08} = e^{10k} \qquad \text{used } \boxed{\text{ln}}\ \text{key}$$

Since the expressions are equal and the bases are the same, the exponents must be equal. This lets us solve for k.

$$-0.08 = 10k \qquad \text{equated exponents}$$

$$-0.008 = k \qquad \text{solved}$$

Now we know both A_0 and k, so we can write

$$A_t = 4.25 e^{-0.008t}$$

Finally, to find A_{30}, we replace t with 30 and evaluate $e^{-0.008(30)}$ by using the $\boxed{\text{inv}}$ $\boxed{\text{ln}}$ keys.

$$A_{30} = 4.25e^{-0.008(30)} \qquad \text{substituted}$$

$$A_{30} = 4.25(0.7866) \qquad \text{used } \boxed{\text{inv}} \boxed{\text{ln}} \text{ keys}$$

$$A_{30} = \textbf{3.34 oz} \qquad \text{multiplied}$$

115.C
compound interest

James puts $100 in a bank account at 8 percent interest compounded yearly. At the end of the first year, his account had accrued $8 interest.

$$\$100 \times 0.08 = \$8$$

Thus the total amount in his account was $108. This is 1.08 times the amount he deposited.

$$\text{Total amount}_1 = \$100(1.08)$$

The next year he was paid 8 percent interest on $108, so the total amount at the end of the second year was

$$\text{Total amount}_2 = [\$100(1.08)](1.08) = \$100(1.08)^2$$

At the end of the third year, he would have

$$\text{Total amount}_3 = \$100(1.08)^3$$

and at the end of n years, he would have

$$\text{Total amount}_n = 100(1.08)^n$$

Following this reasoning, we can see that the total amount in a bank account at the end of n years can be written as

$$\text{Total amount}_n = P(1 + r)^n$$

where P is the **principal,** or the amount deposited, r is the rate of interest, and n is the number of years.

In this problem the interest was compounded yearly. Advanced mathematics books will develop the equation for the amount present if the interest is compounded continuously. This equation is the **exponential equation**

$$A_t = Pe^{rt}$$

This function describes an exponential increase where A_0 is equal to P and k is equal to the rate, which is a positive number. The development of this equation for continuously compounded interest is beyond the scope of this book.

example 115.4   Roger deposited $5000 in an interest-bearing account that paid $7\frac{1}{2}$ percent interest compounded continuously. How much money did he have in 10 years?

solution   The equation is

$$A_t = Pe^{rt}$$

The value of P is $5000 and r is 0.075.

$$A_t = 5000e^{0.075t}$$

To find A_{10}, we let t equal 10.

$$A_{10} = 5000e^{0.075(10)}$$

We can use the $\boxed{\text{inv}}$ $\boxed{\text{ln}}$ keys to evaluate.

$$A_{10} = 5000e^{0.75} \qquad \text{multiplied}$$

$$A_{10} = 5000(2.12) \qquad \text{used } \boxed{\text{inv}} \boxed{\text{ln}} \text{ keys}$$

$$A_{10} = \textbf{\$10,600}$$

practice

a. Solve: $27 = 10^{x+5}$

b. Mary deposited $980 at 7 percent interest compounded continuously. How much money did she have at the end of 9 years? ($A_t = Pe^{rt}$)

c. The number of rabbits increased exponentially. At first there were 400. Three years later there were 1600. How many rabbits would there be at the end of 10 years?

problem set 115

1. A chemist had a container of the compound $KMnO_4$ that weighed 790 grams. What was the weight of the oxygen in the container? (K, 39; Mn, 55; O, 16)

2. Marsha rode for a while at 20 miles per hour with Bertha and completed the 280-mile trip by riding with Sherri at 45 miles per hour. If the total trip took 9 hours, how far did she ride with each friend?

3. Hathaway flew in the fast airplane, traveling 1200 miles in only 1 more hour than it took Beauregard to travel 480 miles. If Hathaway's speed was twice that of Beauregard, find the rates and times of both.

4. There were 14 nickels, dimes, and quarters whose total value equaled $1.05. How many coins of each kind were there if there were 3 times as many dimes as quarters?

5. A two-digit counting number has a value that is 4 times the sum of its digits. If 4 times the units' digit is 14 greater than the tens' digit, what is the number?

Find x. Remember that in a logarithm problem the task is usually to find either a logarithm or an antilogarithm.

6. $x = \ln 0.0076$

7. $e^x = 92.6$

8. $34 = 10^{x+3}$

9. $\log x = 3.412$

10. Use the $\boxed{\ln}$ key to express each number as a power. Then use the rules of exponents to find the answer.
$$\frac{(0.000374)(485)}{0.0618 \times 10^{-16}}$$

11. LaFarge deposited $100 at 9 percent interest compounded continuously. How much money did she have at the end of 9 years? ($A_0 = 100$, $r = 0.09$)

12. The population increased exponentially. At first there were 16,000. Three years later there were 60,000. How many inhabitants would there be at the end of 10 years?

13. In this figure, $ABCD$ is a square with sides of length s. The points $E, F, G,$ and H are midpoints of $\overline{AB}, \overline{BC}, \overline{CD},$ and $\overline{AD}$, respectively. This tells us that the length of the legs of each of the four right triangles is $s/2$. Find the length of one hypotenuse. What is the area of quadrilateral $HEFG$?

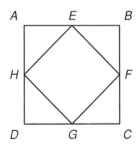

14. The coordinates of the points in which of the areas designated in the figure satisfy the given system of inequalities?
$$\begin{cases} x^2 + y^2 \le 16 & \text{(circle)} \\ y \le x^2 + 4x + 2 & \text{(parabola)} \end{cases}$$

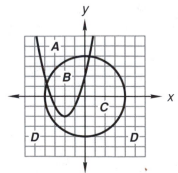

Graph the solution on a number line:

15. $(x + 5)(x - 2) > 0;$ $D = \{$Reals$\}$ **16.** $x^2 < -7x - 10;$ $D = \{$Reals$\}$

17. $|x| + 2 \not> 4;$ $D = \{$Integers$\}$ **18.** $6 \not\geq x - 5 \not\geq 10;$ $D = \{$Integers$\}$

19. Expand: $(x^{1/4} - y^{-1/4})^2$ **20.** Factor: $x^3 m^{15} - 8p^3 y^6$

21. Show that $1.0\overline{47}$ is a rational number by writing it as a fraction of integers.

22. Complete the square as an aid in graphing: $y = x^2 + 2x + 1$

23. Find the number $\dfrac{1}{10}$ of the way from $5\dfrac{2}{3}$ to $6\dfrac{1}{2}$.

Solve:

24. $\begin{cases} 1\dfrac{2}{3}x - 2\dfrac{1}{4}y = -7 \\ -0.2x + 0.05y = -1.8 \end{cases}$ **25.** $\begin{cases} 4x - z = 12 \\ x - 3y = -10 \\ 3y + z = 8 \end{cases}$

26. $\begin{cases} x + y - z = 4 \\ 2x + y + z = 10 \\ 3x + y + z = 14 \end{cases}$

Simplify:

27. $3i^2 - 2i^5 - \sqrt{-2}\,\sqrt{-2} + (i + 2)(i - 3)$ **28.** $\dfrac{4\sqrt{2} + 1}{1 - 3\sqrt{2}}$

29. Solve $-3x^2 = -x - 6$ by completing the square.

30. Solve $3x^2 + 2x = 8$ by factoring.

LESSON 116 *Fundamental counting principle and permutations •*
Probability • Independent events

116.A
fundamental counting principle and permutations

The letters A, B, and C can be arranged six different ways. If we put A first, we can get

| A | B | C |

and

| A | C | B |

If we put the B first, we can get

| B | A | C |

and

| B | C | A |

If we put the C first, we can get

| C | A | B |

and

| C | B | A |

All the arrangements use the same letters, but the order of the letters is different in each arrangement. We call each different arrangement of the members of a set in a definite order a **permutation** of the members of the set.

We see that any one of the three letters can be put in the first box.

Now either of the two remaining letters can be put in the second box.

3	2	

And then the last letter goes in the third box.

3	2	1

Thus, any one of the 3 letters can be used first, and any one of the 2 letters that remain can be used second for a total of 3 times 2, or 6, ways. This example illustrates the **fundamental counting principle.**

> FUNDAMENTAL COUNTING PRINCIPLE
>
> If one choice can be made in A ways and another choice can be made in B ways, then the number of possible choices, in order, is A times B different ways.

If repetition is permitted, the number of possible permutations is even greater. For example, if the letters A, B, and C can be used more than once, then any of the 3 letters can be used in the first, second, or third position for a total of

3	3	3

$3 \times 3 \times 3$, or 27, possible ways. If we list the ways, we get

$$\begin{array}{ccccccccc}
\text{AAA} & \text{AAB} & \text{AAC} & \text{ABA} & \text{ABB} & \text{ABC} & \text{ACA} & \text{ACB} & \text{ACC} \\
\text{BBB} & \text{BBA} & \text{BBC} & \text{BAB} & \text{BAA} & \text{BAC} & \text{BCB} & \text{BCA} & \text{BCC} \\
\text{CCC} & \text{CCB} & \text{CCA} & \text{CBC} & \text{CBB} & \text{CBA} & \text{CAC} & \text{CAB} & \text{CAA}
\end{array}$$

The fundamental counting principle can be extended to any number of choices in order. If the first choice can be made in 2 ways, the second choice in 3 ways, the third choice in 5 ways, and the fourth choice in 7 ways, then there are

$$2 \cdot 3 \cdot 5 \cdot 7 = \textbf{210 ways}$$

that the choices can be made in order.

example 116.1   How many different ways can the numbers 3, 5, 7, and 8 be arranged in order if no repetition is permitted?

solution   Any one of the 4 numbers can be put in the first box.

4			

Then any one of the remaining 3 numbers can be in the next box,

4	3		

then 2 in the next, and 1 in the last.

4	3	2	1

By the fundamental counting principle, there are

$$4 \cdot 3 \cdot 2 \cdot 1 = \textbf{24 ways}$$

that the numbers can be arranged in order. Each of these 24 ways is a permutation.

example 116.2   How many 4-letter signs can be made from the letters in the word EQUAL if repetition is permitted?

solution   Any one of the 5 letters can be used in any of the positions.

| 5 | 5 | 5 | 5 | → $5 \cdot 5 \cdot 5 \cdot 5 = \mathbf{625}$

So we have a total of 625 possible arrangements.

example 116.3   A multiple-choice quiz has 8 questions, and there are 4 possible answers to each question. How many permutations of the answers are possible?

solution   There are 8 questions, and 4 answers are possible to each question.

| 4 | 4 | 4 | 4 | 4 | 4 | 4 | 4 | → $4 \cdot 4 \cdot 4 \cdot 4 \cdot 4 \cdot 4 \cdot 4 \cdot 4 = \mathbf{65,536}$

Thus, there are 65,536 possible sets of answers to a multiple-choice test that has only 8 questions!

example 116.4   How many 3-letter signs can be made from the letters in the word NUMERAL if no repetition is permitted?

solution   This problem is a little different, as only 3 of 7 letters will be used in each arrangement. So we have only 3 positions, and any one of the 7 letters can be used in the first position.

| 7 | | |

Now, any one of the 6 that are left can be used in the next position,

| 7 | 6 | |

and any of the 5 that remain can be used in the last box.

| 7 | 6 | 5 |

$7 \cdot 6 \cdot 5 = \mathbf{210}$

So 210 three-letter permutations of the 7 letters are possible.

example 116.5   Jamie had 5 places to put items in her display. She had 38 different items. How many arrangements were possible?

solution   We begin by drawing a diagram of the 5 places.

| | | | | |

There are 38 items that can go in the first space. This leaves 37 for the second space, 36 for the third space, etc.

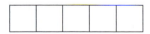

Thus the number of permutations of 38 things taken 5 at a time is

$$38 \cdot 37 \cdot 36 \cdot 35 \cdot 34 = \mathbf{60,233,040}$$

116.B
probability

The study of probability is based on the study of outcomes that have an **equal chance of occurrence**. We call each of the equally probable outcomes an **event**. We call the **set of equally probable outcomes** the **sample space**.

If we flip a fair coin, there are two equally probable outcomes, so our sample space looks like this:

H	T

We define the probability of a particular result as the number of outcomes that satisfy the requirement divided by the total number of outcomes. The probability of getting a head on one flip of a coin is one-half.

$$P(H) = \frac{\text{number of outcomes that are H}}{\text{total number of outcomes in the sample space}} = \frac{1}{2}$$

If we roll a single dice, the sample space is

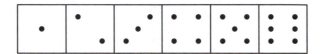

If we define the event as getting a 5, the probability of getting a 5 is

$$P(5) = \frac{\text{number of outcomes that are 5}}{\text{total number of outcomes in the sample space}} = \frac{1}{6}$$

The probability of getting a number greater than 3 is

$$P(>3) = \frac{\text{number of outcomes that are greater than 3}}{\text{total number of outcomes in the sample space}} = \frac{3}{6} = \frac{1}{2}$$

For a single roll of a dice, the denominator of the probability fraction is 6, and the numerator is some number between 0 and 6 inclusive. The list of probabilities is

$$\frac{0}{6}, \quad \frac{1}{6}, \quad \frac{2}{6}, \quad \frac{3}{6}, \quad \frac{4}{6}, \quad \frac{5}{6}, \quad \frac{6}{6}$$

Thus we see that a probability of 4.6 or $-3\frac{1}{2}$ is not possible because the probability of any single event or combination of events is a number between 0 and 1 inclusive.

example 116.6 Two dice are rolled. What is the probability of getting (a) a 7, and (b) a number greater than 8?

solution First we draw a diagram of our sample space. The outcomes are the sums of the values on the individual dice, and there are 36 outcomes in our sample space.

Outcome of second die

	1	2	3	4	5	6
1	2	3	4	5	6	7
2	3	4	5	6	7	8
3	4	5	6	7	8	9
4	5	6	7	8	9	10
5	6	7	8	9	10	11
6	7	8	9	10	11	12

Outcome of first die

(a)   The event is rolling a 7, and we see that 6 of these outcomes are 7, so

$$P(7) = \frac{\text{number of outcomes that equal 7}}{\text{total number of outcomes in the sample space}} = \frac{6}{36}$$

Thus, we find that the probability of rolling a 7 is $\frac{1}{6}$.

(b)   The event is rolling a number greater than 8, and we see that 10 of these outcomes are greater than 8, so

$$P(>8) = \frac{\text{number of outcomes that are greater than 8}}{\text{total number of outcomes in the sample space}} = \frac{10}{36}$$

Thus, the probability of rolling a number greater than 8 is $\frac{5}{18}$.

116.C
independent events

We say that events that do not affect one another are **independent events.** If Denise flips a dime and Paul flips a penny, the outcome of Denise's flip does not affect the outcome of Paul's flip. Thus, we say that these events are independent events. **The probability of independent events occurring in a designated order is the product of the probabilities of the individual events.**

A tree diagram can always be used to demonstrate the probability of independent events occurring in a designated order. This diagram shows the possible outcomes if a coin is tossed 3 times.

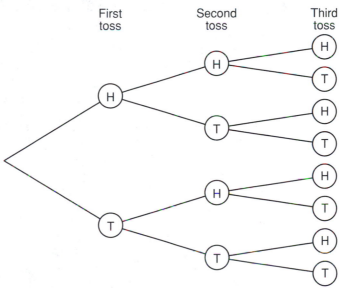

The first toss can be heads or tails and the second and third tosses can be heads or tails. Thus, there are 8 possible outcomes, and the probability of each of these outcomes is $\frac{1}{8}$.

example 116.7   A fair coin is tossed 3 times. What is the probability that it comes up heads every time?

solution   Coin tosses are independent events because the result of one toss has no effect on the result of the next toss. Since the probability of independent events occurring in a designated order is the product of the individual probabilities, we have

$$P(3 \text{ heads}) = \frac{1}{2} \cdot \frac{1}{2} \cdot \frac{1}{2} = \mathbf{\frac{1}{8}}$$

example 116.8   A fair coin is tossed 4 times and it comes up heads each time. What is the probability it will come up heads on the next toss?

solution   The results of past coin tosses do not affect the outcome of future coin tosses. Thus, the probability of getting a head on the next toss is $\frac{1}{2}$.

practice   a.   There are 3 chairs in a row. There are 7 students. How many ways can these students sit in the 3 chairs?

b.   Roger rolls a pair of dice twice. What is the probability that in two rolls he will get a total of 7 on the first roll and a total greater than 9 on the second roll?

problem set   1.   Find three consecutive multiples of 6 such that 6 times the sum of the first and the third
116   is 84 less than 10 times the second.

2.   The pressure of a quantity of an ideal gas was held constant at 600 torr. The initial volume and temperature were 400 ml and 800 K. What was the final temperature in kelvins when the volume was decreased to 200 ml?

3.   The *Natchez Belle* could go 65 miles downstream in 5 hours, but it took her 8 hours to go 56 miles upstream. What was her speed in still water and what was the speed of the current in the river?

4.   The number of tomatoes varied jointly as the rain and as the fertilizer squared. If 1000 tomatoes resulted from 2 inches of rain and 1 ton of fertilizer, how many tomatoes would result from 2 tons of fertilizer and 1 inch of rain? Solve two ways.

5.   A two-digit counting number has a value that is 7 greater than twice the sum of its digits. If the units' digit is 3 greater than 3 times the tens' digit, what is the number?

6.   Find x. Remember that in a logarithm problem we find either a logarithm or an antilogarithm.
(a) $x = \ln 0.0071$ (b) $\ln x = -5.14$ (c) $83 = 10^{x+4}$

7.   Use the [ln] key to express each number as a power. Then use the rules of exponents to find the answer.
$$\frac{(0.000612)(576)}{\left(0.0512 \times 10^{-14}\right)}$$

8.   A logarithmic function is used in chemistry to determine the relative acidity of a liquid. The relative acidity, called pH, is written with a small "p" and a large "H." This symbol combines the letter "H" for hydrogen and the letter "p" from the Latin word *potentia* which means "power" or "strength." We use pH as a measure of the relative concentration of hydrogen ions in a liquid. The equation that defines pH is

$$\text{pH} = -\log \text{H}^+$$

The pH of liquids ranges from about 0 to +14, and the pH of water, considered to be neutral, is 7. Since investigation of the theory of pH is a topic for chemistry, we will limit our discussion and concentrate on the mathematical problem of finding H^+ if given pH and of finding pH if given H^+. If we are given H^+, we can use the [log] key and change the sign to get pH. If we are given pH, we change its sign and use the [inv] [log] keys to find H^+.
(a)   If $\text{H}^+ = 0.00204$ mole/liter, find pH.   (b)   If $\text{pH} = 3.2$, find H^+.

9.   Zuniga deposited $1400 at 11 percent interest compounded continuously. How much money did he have at the end of 7 years? $(A_t = Pe^{rt})$

10.   The number of paramecia increased exponentially. At first there were 400,000. Three days later there were 1,200,000. How many paramecia would there be at the end of 8 days?

11.   There are 5 chairs in a row. There are 9 students. How many ways can these students sit in the 5 chairs?

12.   How many different 7-digit telephone numbers (for example, 923-5678) can be written if the digit 0 cannot appear among the first 3 digits? Remember that digits can be repeated.

13. In $\triangle ABC$ the measure of $\angle 1$ and the measure of $\angle 2$ each equals $35°$. What is the sum of the measure of $\angle 3$ and the measure of $\angle 4$?

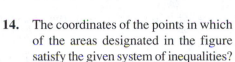

14. The coordinates of the points in which of the areas designated in the figure satisfy the given system of inequalities?

$$\begin{cases} y \geq \frac{2}{3}x + 2 & \text{(line)} \\ y \geq x^2 + 2x & \text{(parabola)} \end{cases}$$

Graph the solution on a number line:

15. $x^2 - 3 \geq -2x$; $D = \{\text{Integers}\}$

16. $x^2 + 2x < 3$; $D = \{\text{Integers}\}$

17. $-|x| + 2 \not\leq -1$; $D = \{\text{Integers}\}$

18. Show that $1.0\overline{476}$ is a rational number by writing it as a fraction of integers.

19. Complete the square as an aid in graphing: $y = -x^2 - 4x - 6$

20. Expand: $(x^{1/2} - y^{3/4})^2$ 21. Factor: $27m^9p^3 - x^{12}y^3$

22. Find the number that is $\frac{2}{9}$ of the way from $\frac{1}{4}$ to $3\frac{1}{2}$.

Solve:

23. $\begin{cases} \frac{3}{5}x - \frac{1}{7}y = 6 \\ -0.21x + 0.02y = -2.73 \end{cases}$ 24. $\begin{cases} x - 2z = 7 \\ y + 2z = -9 \\ -x + 2y = -7 \end{cases}$

25. $\begin{cases} 2x - y + 2z = -9 \\ 2x + 2y + z = -15 \\ x - 2y + z = 0 \end{cases}$

Simplify:

26. $\dfrac{2i^3 - 3}{1 - \sqrt{-4}\sqrt{4}}$ 27. $\dfrac{3 - \sqrt{24}}{2 - \sqrt{6}}$ 28. $\dfrac{(a^{x/2 - 2})^2 m^x}{(m^2 a^2)^{x/4}}$

29. Solve: $\sqrt{p} = 9 - \sqrt{p - 45}$

30. Solve $3x^3 = x^2 + 2x$ by factoring.

LESSON 117   *Letter symbols for sets • Set-builder notation*

117.A
letter symbols for sets

The most commonly encountered sets of numbers are the sets of natural or counting numbers, whole numbers, integers, rational numbers, irrational numbers, real numbers, and complex numbers. While most mathematicians have agreed on the definition of and composition of these sets of numbers, they have not been able to agree totally on a capital letter to assign to each set. The capital letters shown below are those that seem to be used most frequently.

$$N = \{1, 2, 3, \ldots\} \quad \text{Natural or counting numbers}$$

The set of natural numbers is a subset of the set of complex numbers, has an infinite number of members, and is a subset of all the sets listed above except the set of irrational numbers.

$$W = \{0, 1, 2, 3, \ldots\} \qquad \text{Whole numbers}$$

We note that the set of whole numbers contains the set of natural numbers and also contains the number zero.

$$J = \{\ldots, -3, -2, -1, 0, 1, 2, 3, \ldots\} \quad \text{Integers}$$

The set of integers contains the set of whole numbers and also contains the negative of each member of the set of natural numbers.

$$Q = \{\text{Rational numbers}\}$$

The set of rational numbers contains the set of integers and also contains all other numbers that can be written as a fraction (quotient) of integers. Of course, any integer can be written as a fraction of integers. For example, -4 can be written as $\frac{40}{-10}$ or $\frac{240}{-60}$. Both represent the number -4. The rational numbers can also be described as real numbers that can be represented by a terminating or repeating decimal numeral.

$$P = \{\text{Irrational numbers}\}$$

The set of irrational numbers consists of all numbers that cannot be written as fractions of integers. Numbers such as $\sqrt{2}, \pi, \sqrt[3]{4}, \sqrt[5]{3}$ are examples of the infinite number of numbers that cannot be written as fractions of integers and are therefore irrational numbers. The irrational numbers can also be described as real numbers that cannot be represented by repeating or terminating decimal numerals. The decimal representation of these numbers consists of nonrepeating decimal numerals of infinite length.

$$R = \{\text{Real numbers}\}$$

The set of real numbers consists of all members of the set of rational numbers and all members of the set of irrational numbers. A real number can be thought of as any number that does not have i as a factor. Any real number can be paired with a unique point on the number line, and any point on the number line can be paired with a unique real number. **To use simpler, nonrigorous language, we may say that any number on the number line is a real number.**

$$C = \{\text{Complex numbers}\}$$

The set of complex numbers consists of all numbers of the form $a + bi$, where a and b are real numbers. Any real number is also a complex number. For instance, the number 3 may be thought of as being in the form $a + bi$ where the value of b is zero; thus $3 + 0i$ equals 3.

117.B

set-builder notation

Use of **set-builder notation** allows us to describe a set completely and exactly. Its compactness appeals to mathematicians, and it is used extensively in more advanced courses in mathematics. The first component is the leading half of the set of braces, or the *open brace*. This is read as *the set whose members are*. The second component is a statement about the variable and is followed by the third component, a vertical line that is read *such that*. This is followed by one or more restrictions on the variable, and the last component is the terminal half of the set of braces, or the *close brace*.

$$A = \{\, x \in J \mid x + 2 > 4 \,\}$$

This would be read as: "A is the set whose members x are integers, such that x plus 2 is greater than 4." Previously in this book we would have described this set by writing

$$x + 2 > 4; \quad D = \{\text{Integers}\}$$

The graph of the solution set is

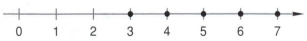

Set-builder notation can also be used to designate ordered pairs and to state the restrictions on the ordered pairs.

example 117.1 Graph the solution: $\{(x, y) \in R \mid y > x + 2 \text{ and } y < -x\}$

solution This is read as: "The set whose members are ordered pairs of x and y, where x and y are real numbers such that $y > x + 2$ and $y < -x$." Previously, this problem would have been stated as follows:

Graph the solution set to this system of linear inequalities:

$$\begin{cases} y > x + 2 \\ y < -x \end{cases} \quad D = \{\text{Reals}\}$$

Thus, we see that set-builder notation is just another way to describe a particular set, and the reader should not be confused by the notation. The graph of the solution set of this problem is shown here.

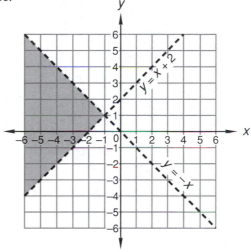

practice Graph the solution: $\{x \in J \mid x + 3 > 5\}$

problem set 117

1. There were 16 nickels, dimes, and quarters whose total value was $1.50. How many coins of each kind were there if there were 3 times as many dimes as quarters?

2. The dealer paid $4000 for the car and sold it for $5000. What was the percent markup of the purchase price, and what was the percent markup of the selling price?

3. Gabriel could drive the 200 miles to the seashore in one-half the time it took Martha to drive the 650 miles to the mountains. Find the speeds and times of both if Martha's speed was 25 miles per hour greater than Gabriel's speed.

4. The two containers were side by side on the shelf. The first one was 30% medicine, and the second one was 60% medicine. How much of each should be used to mix 600 ml of a solution that is 39% medicine?

5. Kenneth and David rebelled at paying $4320 because this was a markup of 80 percent of the selling price. What was the markup, and what did the store pay for the item?

6. Rosabelle has 5 different flags and 8 flagpoles in a row. How many different ways can she fly her flags if she flies all 5 flags every time?

7. Roger rolls a pair of dice twice. What is the probability he will get both a total of 5 on the first roll and a total greater than 5 on the second roll?

Find x:

8. $x = 3 \ln 0.0041$

9. $4e^x = 24$

10. The pH of a solution is given by the equation pH $= -\log$ H$^+$, where H$^+$ is the concentration of hydrogen ions in moles per liter. Find H$^+$ in moles per liter when the pH of the liquid is

(a)   5.34   (b)   0.00263

11. Lorien deposited \$1260 at 8 percent interest compounded continuously. How much money did she have at the end of 11 years? ($A_t = Pe^{rt}$)

12. The number of rabbits increased exponentially. At first there were 85 rabbits. Twelve years later there were 1700 rabbits. How many rabbits would there be at the end of 130 years?

13. The coordinates of the points in which of the areas designated in the figure satisfy the given system of inequalities?

$$\begin{cases} x^2 + y^2 \geq 16 & \text{(circle)} \\ y \leq -x^2 + 2 & \text{(parabola)} \end{cases}$$

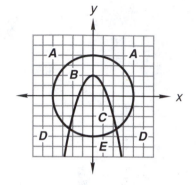

14. The figure shown is the base of a container 10 meters high. Dimensions are in meters. Find the volume in cubic feet.

15. Find C.

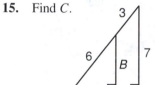

Graph the solution:

16. $A = \{x \in J \mid x + 1 > 1\}$ **17.** $B = \{x \in R \mid x^2 \geq -4x - 3\}$

18. $C = \{x \in R \mid x^2 + 4x < -3\}$ **19.** $\{(x, y) \in R \mid y > x + 3 \text{ and } y < -x\}$

20. Show that $2.04\overline{25}$ is a rational number by writing it as a fraction of integers.

21. Complete the square as an aid in graphing: $y = -x^2 + 4x - 2$

22. Expand: $(x^{1/2} - y^{-1/2})^2$ **23.** Factor: $m^3 - 8p^6k^9$

24. Find the number that is $\dfrac{4}{9}$ of the way from $\dfrac{3}{5}$ to $1\dfrac{2}{5}$.

Solve:

25. $\begin{cases} \dfrac{5}{8}x - \dfrac{2}{3}y = 4 \\ -0.15x + 0.2y = -0.6 \end{cases}$ **26.** $\begin{cases} 3x + y = 2 \\ 2x - z = 0 \\ 2y + z = -4 \end{cases}$

Simplify:

27. $\dfrac{\sqrt{-2}\,\sqrt{-2} - 3i^3}{i + 2}$ **28.** $\dfrac{2\sqrt{2} - 4}{2\sqrt{4} - \sqrt{2}}$ **29.** $\sqrt[3]{x^5}\,\sqrt[4]{x^3 y}$

30. Solve $6 = 2x^2 - 11x$ by factoring.

LESSON *118* *Logarithmic equations*

We have been working with the logarithms of numbers. When we use logarithmic equations, we encounter expressions such as

$$\log (x^2 + 4x) \qquad \log (x - 2) \qquad \log 3$$

In these expressions, $x^2 + 4x$, $x - 2$, and 3 are called the **arguments** of the logarithmic functions. Finding the solution of a logarithmic equation is easy if we can remember the following bold-face statement and remember how to use the rules for exponents (logarithms) backwards.

> **If the bases are the same and the logarithms are equal, the numbers are equal.**

$$7^2 = 7^2$$

Here the bases are both 7 and the exponents are equal, so the number on the left, which is 49, equals the number on the right, which is also 49. If we write

$$\log_7 x = \log_7 y$$

the bases are both 7, and the equals sign tells us that the logarithms are equal so the arguments must be equal. This means that x equals y.

If we write

$$\log_5 (x + 4) = \log_5 7$$

the bases are the same, and the equals sign tells us that the logarithms are equal so the arguments must be equal. Thus

$$x + 4 = 7 \qquad \text{equal argument}$$

$$x = 3 \qquad \text{solved}$$

We remember that we can multiply numbers by adding their logarithms.

$$10^2 \times 10^3 = 10^{2+3}$$

To use this rule backwards, we note that if the bases are the same and the logarithms are added, the arguments are multiplied. To illustrate we note that

$$\log_{10} 5 + \log_{10} 7 \text{ has the same value as } \log_{10} 5 \cdot 7$$

$$\log_8 (x + 2) + \log_8 3 \text{ has the same value as } \log_8 (x + 2)(3)$$

We remember that we can divide numbers by subtracting the lower logarithm from the top logarithm.

$$\frac{10^5}{10^2} = 10^{5-2}$$

To use this rule backwards, we note if the bases are the same and the logarithms are subtracted, the arguments are divided. To illustrate we note that

$$\log_3 14 - \log_3 9 \text{ has the same value as } \log_3 \frac{14}{9}$$

We remember that we can raise a number to a power by multiplying the logarithm by the power.

$$(10^5)^3 = 10^{15}$$

To use this rule backwards, we note that if the logarithms are multiplied by a constant or a variable, the argument is raised to that power. To illustrate, we note the following.

Since

$$10^2 = 100$$

the logarithm of 100 to the base 10 is 2. If we multiply by 4 the logarithm 2, we have raised the number 100 to the fourth power.

$$10^{2(4)} \qquad \text{means} \qquad (10^2)^4 \qquad \text{or} \qquad (100)^4$$

In the language of logarithms we can write

$$4 \log_{10} 100 \qquad \text{means} \qquad \log_{10} 100^4 \quad \text{and}$$

$$x \log_{10} 100 \qquad \text{means} \qquad \log_{10} 100^x$$

This is a little difficult to assimilate when first encountered and a rote rule is helpful.

The coefficient of a logarithm can be turned into an exponent.

$$④ \log_3 5 \quad \longrightarrow \quad \log_3 5^4$$

An exponent can be turned into a coefficient of a logarithm.

$$\log_3 5^④ \quad \longrightarrow \quad 4 \log_3 5$$

Thus the expression $3 \log_{10} x$ can be rewritten

$$3 \log_{10} x \quad \longrightarrow \quad \log_{10} x^3$$

and the expression $\log_5 4^x$ can be rewritten

$$\log_5 4^x \quad \longrightarrow \quad x \log_5 4$$

example 118.1   Solve: $\log_3 (x + 7) + \log_3 2 = \log_3 20$

solution   We begin by noting that we have never used base 3 for computation with logarithms. Authors of mathematics books and standardized tests like to vary their algebra of logarithm problems by using unusual bases. The unusual base tells us this is probably a problem about the algebra of logarithms. We note that the logs of $x + 7$ and 2 are added so $x + 7$ and 2 are multiplied.

$$\log_3 (x + 7)(2) = \log_3 20$$

The bases are the same. The equation says that the logarithms are equal. Thus the arguments must be equal.

$$(x + 7)(2) = 20 \qquad \text{equal arguments}$$

$$2x + 14 = 20 \qquad \text{multiplied}$$

$$2x = 6 \qquad \text{added } -14$$

$$x = 3 \qquad \text{divided and solved}$$

example 118.2   Solve: $\log_{14} (x + 3) - \log_{14} (x - 3) = \log_{14} 7$

solution   On the left side of the equation, the bases are equal and the logarithms are subtracted; thus, the arguments are divided.

$$\log_{14} \frac{x + 3}{x - 3} = \log_{14} 7$$

The bases are equal. The logarithms are equal. Thus the arguments must be equal.

$$\frac{x + 3}{x - 3} = 7 \qquad \text{equal arguments}$$

$$x + 3 = 7x - 21 \qquad \text{multiplied}$$

$$24 = 6x \qquad \text{simplified}$$

$$x = 4 \qquad \text{solved}$$

example 118.3   Solve: $3 \log_b x = \log_b 64$

solution   In the expression on the left, we note that $\log_b x$ has a coefficient of 3. We use the power rule of logarithms to rewrite this expression by turning the coefficient into an exponent.

$$\log_b x^3 = \log_b 64$$

The bases are equal. The equation says the logarithms are equal. Thus the arguments are equal.

$$x^3 = 64 \qquad \text{equal arguments}$$

$$x = 4 \qquad \text{solved}$$

practice Solve:

a. $\log_5 (x + 8) + \log_5 4 = \log_5 80$

b. $\log_7 (x + 7) - \log_7 (x - 2) = \log_7 10$ c. $2 \log_3 x = \log_3 16$

problem set 118

1. Seven-sixteenths of the assembled throng squatted in place. If 6399 did not squat in place, how many did squat in place?

2. The jar was half full of the compound C_2H_4O. The total weight of the compound was 396 grams. What was the weight of the hydrogen (H) in the jar? (C, 12; H, 1; O, 16)

3. The class had a collection of nickels and quarters that totaled 18 coins and had a value of $2.70. How many were nickels, and how many were quarters?

4. The current in the Ogeechee River flowed at 5 miles per hour. The boat could go 160 miles downstream in twice the time it took to go 40 miles upstream. What was the speed of the boat in still water, and what were the times?

5. A two-digit counting number has a value that is 1 greater than 8 times the sum of its digits. If 3 times the tens' digit is 11 greater than the units' digit, what is the number?

6. Daphne deposited $15,000 in an account that paid $9\frac{1}{2}$ percent interest compounded continuously. How much money did she have in 10 years? $(A_t = Pe^{rt})$

7. There are 6 chairs in a row. There are 11 students. How many ways can these students sit in the 6 chairs?

8. Melanie rolls a pair of dice twice. What is the probability she will get a total of 4 on the first roll and a total greater than 6 on the second roll?

9. Find x: $5 \ln x = 0.072$

10. Use the In key to express each number as a power. Then use the rules of logarithms to find the answer.
$$\frac{(23,354 \times 10^{-5})(45,633 \times 10^4)}{35,139}$$

11. In the beginning there were 4. When time equaled 5, there were 20. How many would there be when time equaled 40?

We remember that the pH of a solution is a measure of the relative acidity of the solution. The equation for pH is

$$\text{pH} = -\log \text{H}^+$$

where pH is a number and H^+ is the concentration of hydrogen ions in moles per liter.

12. Find the pH if $\text{H}^+ = 3.14 \times 10^{-3}$ mole per liter.

13. Find H^+ if pH is 5.042.

14. We remember that a rhombus is a parallelogram with four equal sides. *ABCD* is a rhombus. Find $m\angle BDA$.

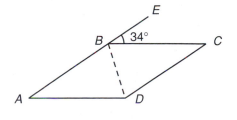

Solve for x:

15. $\log_7 (x + 5) + \log_7 3 = \log_7 60$

16. $4 \log_6 x = \log_6 64$

17. Show that $0.00\overline{163}$ is a rational number by writing it as a fraction of integers.

18. Complete the square as an aid in graphing: $y = -x^2 - 6x - 10$

Graph the solutions:

19. $\left\{ x \in J \mid |x| - 3 > -7 \right\}$

20. $\left\{ (x, y) \in R \mid x + y \geq -3 \text{ and } x - 2y < -4 \right\}$

21. $\left\{ x \in J \mid x^2 \geq 2x + 3 \right\}$ 22. $\left\{ x \in J \mid x^2 - 3 < 2x \right\}$

23. Solve by completing the square: $6x + x^2 = -10$

24. Expand: $(x^{1/4} - y^{1/4})^2$ 25. Factor: $x^6y^3 - 27p^6m^9$

26. Find the number that is $\dfrac{3}{8}$ of the way from $4\dfrac{1}{5}$ to $6\dfrac{1}{10}$.

27. Solve: $\begin{cases} 3x - y - z = 9 \\ 2x + y - z = 12 \\ 2x - y + z = 0 \end{cases}$

Simplify:

28. $\dfrac{-2i^3 + 2}{i - i^2}$ 29. $\sqrt[3]{27\sqrt{3}}$ 30. $\dfrac{\sqrt{2} - 5}{3 - 2\sqrt{2}}$

LESSON 119 *Absolute value inequalities*

From Lesson 99 we recall that the absolute value notation with a variable usually implies two answers.[†] For instance,

$$|\text{Something}| = 4 \qquad D = \{\text{Reals}\}$$

has two values that satisfy, for both $+4$ and -4 have an absolute value of 4.

An absolute value statement of **less than** tells us that the value of the variable lies between a positive number and a negative number. For instance,

$$|\text{Something}| < 4 \qquad D = \{\text{Reals}\}$$

tells us that the value of something is between 4 and -4 and is described by the conjunction

$$\text{Something} > -4 \quad \textbf{and} \quad \text{Something} < 4$$

Thus, we see that an absolute value statement of *less than* can be replaced with two statements that do not contain absolute value but place the same restrictions on the variable.

example 119.1   Graph: $\left\{ x \in R \mid |x - 2| < 4 \right\}$

solution   We know from the discussion above if the absolute value of something is less than 4, then something is greater than -4 **and** something is less than 4.

$$\text{Something} > -4 \quad \textbf{and} \quad \text{Something} < 4$$

[†]The absolute value of zero is zero so $|x| = 0$ has only one answer.

In this problem, something is $x - 2$. So we replace something with $x - 2$ and solve.

$$
\begin{array}{lll}
x - 2 > -4 & \text{and} & x - 2 < 4 \\
\underline{+ 2 \quad +2} & & \underline{+ 2 \quad +2} \\
x \quad\;\; > -2 & \text{and} & x \quad\;\; < 6
\end{array}
$$

Thus, all real numbers greater than -2 and less than 6 will satisfy the given inequality.

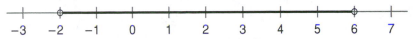

Sometimes it helps understanding to check one of the solutions. From the graph, we see that $+3$ satisfies the condition. If we replace x in the original problem with $+3$, we get

$$|(3) - 2| < 4$$

$$|1| < 4$$

$$1 < 4 \qquad \text{True}$$

example 119.2   Graph: $\{x \in J \mid |x + 2| \le 3\}$

solution   This problem in the old notation would have been written as follows.

$$\text{Graph} \quad |x + 2| \le 3; \; D = \{\text{Integers}\}$$

Here our "something" is less than or equal to 3, so the absolute value inequality can be replaced with the conjunction

$$\text{Something} \ge -3 \quad \text{and} \quad \text{Something} \le 3$$

Since our something is $x + 2$, we get

$$
\begin{array}{lll}
x + 2 \ge -3 & \text{and} & x + 2 \le 3 \\
\underline{- 2 \quad -2} & & \underline{- 2 \quad -2} \\
x \quad\;\; \ge -5 & \text{and} & x \quad\;\; \le 1
\end{array}
$$

Thus, our solution consists of the integers that are greater than or equal to -5 and that are also less than or equal to 1.

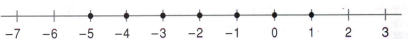

  When the absolute value is greater than a given positive number, then a *disjunction* is implied. The disjunction consists of two statements of greater than, neither of which use the absolute value notation. For instance, if

$$|\text{Something}| > 3 \qquad D = \{\text{Reals}\}$$

then something is less than -3 *or* is greater than 3.

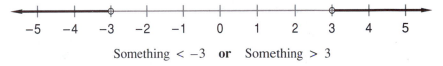

$$\text{Something} < -3 \quad \text{or} \quad \text{Something} > 3$$

example 119.3   Graph: $|x - 2| > 3; \; D = \{\text{Reals}\}$

solution   This is the same statement as the statement above except here something is $|x - 2|$. If we replace something with $x - 2$, we get

$$
\begin{array}{lll}
x - 2 < -3 & \text{or} & x - 2 > 3 \\
\underline{+ 2 \quad +2} & & \underline{+ 2 \quad +2} \\
x \quad\;\; < -1 & \text{or} & x \quad\;\; > 5
\end{array}
$$

The graph of the solution is

All of the numbers shown used for x satisfy the original inequality

$$|x - 2| > 3$$

example 119.4 Graph: $\{x \in J \mid |x + 2| > 4\}$

solution If the absolute value of something is greater than 4, then something is less than -4 **or** is greater than 4.

$$\text{Something} < -4 \quad \textbf{or} \quad \text{Something} > 4$$

Since our something is $x + 2$, we get

$$
\begin{array}{ccccc}
x + 2 < -4 & \textbf{or} & x + 2 > & 4 \\
\underline{-2 \quad -2} & & \underline{-2 \quad -2} \\
x \quad < -6 & \textbf{or} & x \quad > & 2
\end{array}
$$

Thus, the graph indicates all integers that are less than -6 **or** are greater than 2.

practice Graph the solution on the number line:

   a. $\{x \in R \mid |x - 3| < 5\}$ **b.** $\{x \in J \mid |x + 3| \le 2\}$

problem set 119

1. The pressure of a quantity of an ideal gas was held constant at 475 torr. The initial volume and temperature were 500 ml and 700 K. If the temperature were raised to 2100 K, what would the final volume be?

2. Gomez could drive the 250 miles to the mountains in one-half the time it took de la Tore to drive the 400 miles to the seashore. Gomez drove 10 miles per hour faster than de la Tore. Find the rates and times of both.

3. Pinks varied inversely as blacks squared and directly as whites. Two pinks and 10 blacks went with 4 whites. How many pinks went with 1 black and 20 whites? Work the problem once using the equal ratio format and once using the variation format.

4. There were 24 nickels, dimes, and quarters whose total value equaled $4.25. How many coins of each kind were there if the number of nickels equaled the number of dimes?

5. The sum of the digits of a two-digit counting number was 5. If the digits were reversed, the new number would be 27 less than the original number. What was the original number?

6. The number of amoebae increased exponentially. At first there were 40. Three hours later there were 640. How many amoebae would there be at the end of 20 hours?

Graph the solution on a number line:

   7. $\{x \in J \mid |x + 1| \le 3\}$ **8.** $\{x \in R \mid |x - 3| > 5\}$

Solve:

   9. $\ln(x + 3) + \ln 4 = \ln 40$ **10.** $3 \log_{15} x = \log_{15} 27$

   11. Find the number that is $\dfrac{4}{5}$ of the way from $6\dfrac{1}{10}$ to $12\dfrac{3}{20}$.

   12. Expand: $(x^{1/2} - y^{1/2})^2$ **13.** Factor: $x^3 y^9 - 64 p^{12} m^9$

   14. Find x. Remember that in a logarithm problem we find either a logarithm or an antilogarithm.

     (a) $x = \ln 0.0043$ (b) $\ln x = -4.13$

15. Use the ⬛ **In** key to express each number as a power. Then use the rules of exponents to find the answer.

$$\frac{(0.000416)(431)}{0.0432 \times 10^{-17}}$$

16. If AC is a straight line, what is the measure of angle x?

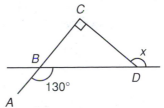

17. In which of the areas designated in the figure do the coordinates satisfy the given system of inequalities?

$$\begin{cases} y \geq -\dfrac{1}{4}x - 2\dfrac{1}{2} & \text{(line)} \\ y \leq -x^2 + 2 & \text{(parabola)} \end{cases}$$

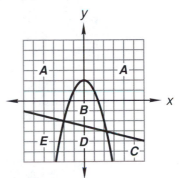

18. Find the pH (pH $= -\log$ H$^+$ or $10^{-\text{pH}} = $ H$^+$) of the solution when the concentration of hydrogen ions (H$^+$) is 9.52×10^{-12} mole per liter.

19. Find the concentration of hydrogen ions (H$^+$) in moles per liter when the pH of the liquid is 2.23.

20. Use unit multipliers to convert 5000 liters per minute to cubic inches per second.

21. A circle is centered at the origin. The coordinates of one endpoint of a diameter are $(3 + \sqrt{2}, -4)$. What are the coordinates of the other endpoint of the diameter? (A sketch is helpful.)

22. $OE = 3$ cm. $\overline{OA}$ is the bisector of $\angle DOE$. Use similar triangles to find the area of the shaded portion of the circle.

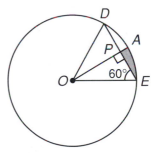

23. Solve: $\begin{cases} 2x - y - z = 8 \\ 3x + y - z = 9 \\ 6x - y + z = 0 \end{cases}$

24. Show that $0.0\overline{13}$ is a rational number by writing it as a fraction of integers.

25. Complete the square as an aid in graphing: $y = -x^2 - 2x + 1$

26. Begin with $ax^2 + bx + c = 0$ and derive the quadratic formula by completing the square.

27. Find the number that is $\dfrac{1}{5}$ of the way from $3\dfrac{1}{3}$ to $5\dfrac{2}{5}$.

Simplify:

28. $\dfrac{2i^3 - 2}{3 - \sqrt{-2}\sqrt{2}}$

29. $\dfrac{(a^2)^{b/2} x^b}{(ax)^{b/2}}$

30. Solve $3x^3 - 3x^2 = 6x$ by factoring.

LESSON 120 *Age word problems*

Most age word problems discuss the present ages of two or more people and their ages at some given time in the past and/or in the future. **The key to these problems is the proper choice of unknowns. Subscripted variables are very helpful.**

example 120.1 A man is 13 times as old as his son. In 10 years he will be 3 times as old as his son will be. How old are they now?

solution We will use these variables.

	Age Now	Age 10 Years from Now
Man	M_N	$M_N + 10$
Son	S_N	$S_N + 10$

The first statement is that the man now is 13 times as old as his son is now. This gives the equation

$$\text{(a)} \quad M_N = 13S_N$$

Now we use the same variables to say that in 10 years he will be 3 times as old as his son will be then.

$$\text{(b)} \quad M_N + 10 = 3(S_N + 10)$$

Substitute in equation (b) the value for M_N from equation (a) and solve.

$$
\begin{aligned}
13S_N + 10 &= 3(S_N + 10) & &\text{substituted} \\
13S_N + 10 &= 3S_N + 30 & &\text{multiplied} \\
10S_N &= 20 & &\text{simplified} \\
\mathbf{S_N = 2,} \quad &\textbf{so} \quad \mathbf{M_N = 26} & &\text{solved}
\end{aligned}
$$

example 120.2 Five years ago Brenda was $\frac{4}{5}$ as old as Layton. Ten years from now she will be $\frac{7}{8}$ as old as Layton. How old is each now?

solution We will use these variables.

	Now	5 Years Ago	10 Years from Now
Brenda	B_N	$B_N - 5$	$B_N + 10$
Layton	L_N	$L_N - 5$	$L_N + 10$

Brenda's age 5 years ago equaled $\frac{4}{5}$ Layton's age 5 years ago.

$$\text{(a)} \quad B_N - 5 = \frac{4}{5}(L_N - 5)$$

Note that Layton's age then is multiplied by $\frac{4}{5}$, but not Brenda's.

Brenda's age 10 years from now equals $\frac{7}{8}$ of Layton's age 10 years from now.

$$\text{(b)} \quad B_N + 10 = \frac{7}{8}(L_N + 10)$$

To clear fractions, multiply every term in equation (a) by 5 and every term in equation (b) by 8 to get equations (a′) and (b′):

$$\text{(a′)} \quad (5)(B_N - 5) = (5)\left(\frac{4}{5}\right)(L_N - 5)$$

$$(b')\quad (8)(B_N + 10) = (8)\left(\frac{7}{8}\right)(L_N + 10)$$

Simplify and get equations (a″) and (b″)

$$(a'')\quad 5B_N - 4L_N = 5 \qquad (b'')\quad 8B_N - 7L_N = -10$$

which can be solved by using elimination.

$$
\begin{array}{lllll}
(a'') & 5B_N - 4L_N = 5 & \rightarrow & (7) & \rightarrow & 35B_N - 28L_N = 35 \\
(b'') & 8B_N - 7L_N = -10 & \rightarrow & (-4) & \rightarrow & -32B_N + 28L_N = 40 \\
\hline
& & & & & 3B_N \qquad\qquad = 75
\end{array}
$$

$$B_N = 25$$

Now we will replace B_N with 25 in equation (a″) and solve for L_N.

$$(a'')\quad 5(25) - 4L_N = 5$$
$$125 - 4L_N = 5$$
$$-4L_N = -120$$
$$L_N = 30$$

example 120.3   Thirty years ago Barbie was 1 year older than twice Mary's age then. Twenty years ago Mary was $\frac{4}{5}$ as old as Barbie was then. How old is each girl now?

solution   We will use the variables B_N for Barbie now and M_N for Mary now.

Now	30 Years Ago	20 Years Ago
B_N	$B_N - 30$	$B_N - 20$
M_N	$M_N - 30$	$M_N - 20$

The first sentence gives us the equation

$$(a)\quad B_N - 30 - 1 = 2(M_N - 30)$$

which simplifies to

$$B_N - 2M_N = -29$$

The second sentence gives us the equation

$$(b)\quad M_N - 20 = \frac{4}{5}(B_N - 20)$$

which simplifies to

$$4B_N - 5M_N = -20$$

Next we will use elimination to solve the two equations for M_N.

$$
\begin{array}{lllll}
B_N - 2M_N = -29 & \rightarrow & (-4) & \rightarrow & -4B_N + 8M_N = 116 \\
4B_N - 5M_N = -20 & \rightarrow & (1) & \rightarrow & 4B_N - 5M_N = -20 \\
\hline
& & & & 3M_N \qquad = 96
\end{array}
$$

$$M_N = 32$$

Now we replace M_N in equation (a) with 32 to find B_N.

$$(B_N - 30) - 1 = 2(32 - 30) \quad\rightarrow\quad B_N = 35$$

practice   Five years ago Ben was $\frac{2}{3}$ as old as Kris. Ten years from now he will be $\frac{5}{6}$ as old as Kris. How old are they now?

problem set 120

1. Three hundred liters of a 76% antifreeze solution had to be reduced to a 20% solution. How many liters of pure antifreeze should be extracted?

2. The sum of the digits of a two-digit counting number is 11. If the digits are reversed, the new number is 5 greater than 3 times the original number. What was the original number?

3. A man is 18 times as old as his son. In 15 years, he will be 3 times as old as his son will be then. How old are they now?

4. The perimeter of an equilateral triangle is 24 meters. What is the area of the triangle?

5. Twenty years ago Lucie was 2 years older than twice Myrna's age then. Six years ago Myrna was $\frac{3}{4}$ as old as Lucie was then. How old is each girl now?

Graph the solution on a number line:

6. $\{x \in J \mid |x - 3| < 2\}$

7. $\{x \in R \mid |x + 3| \le 2\}$

8. Mary deposited $2000 at 6 percent annual interest compounded continuously. How much money did she have at the end of 8 years? $(A_t = Pe^{rt})$

9. How many different auto tags can be made by 3 letters followed by 3 numbers if no repetition is permitted? (Both the letter O and the number 0 are allowed.)

Use the ⃞In key to express each number as a power. Then use the rules of exponents to find the answer.

10. $\dfrac{0.0123 \times 10^{-5}}{375,000}$

11. $(0.0123 \times 10^5)^{1.5}$

Find x:

12. $4x = 3 \ln 0.0037$

13. $3 \ln x = -4.13$

14. Find the pH (pH $= -\log H^+$ or $10^{-pH} = H^+$) of a solution when the concentration of hydrogen ions $H^+ = 0.062$ mole per liter.

15. Find the concentration of hydrogen ions (H^+) in moles per liter when the pH of the liquid is 3.13.

16. Use unit multipliers to convert 40 cubic feet per minute to cubic inches per second.

17. Show that $0.02\overline{163}$ is a rational number by writing it as a fraction of integers.

18. A circle is centered at the origin. The coordinates of one endpoint of a diameter is $(3 - \sqrt{2}, -4 + \sqrt{2})$. What are the coordinates of the other endpoint of the diameter?

19. Complete the square as an aid in graphing: $y = -x^2 + 2x - 3$

20. Solve $-5x^2 = 2x - 1$ by completing the square.

21. Solve: $\begin{cases} \dfrac{3}{2}x + y = 13 \\ 0.2x - 0.02y = 1.12 \end{cases}$

22. Write $4R - 6U$ in polar form.

23. Add: $4\underline{/20°} - 6\underline{/230°}$

24. Solve $5x^3 + 9x^2 - 2x = 0$ by factoring.

Solve for x:

25. $\log_4 (x - 3) - \log_4 7 = \log_4 31$

26. $\ln (x + 5) + \ln 5 = \ln 65$

27. Expand: $(x^{3/2} - y^{3/2})^2$

28. Factor: $m^6 p^{12} - 8y^3 z^{15}$

29. In the figure shown, the length of the diameter BC is 12. The length of $\overline{AC}$ is 9. What is the length of $\overline{OB}$? Find the length of the hypotenuse AB. Use similar triangles to find the length of $\overline{OP}$.

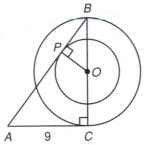

30. In the figure shown, $\overline{AC}$ is the bisector of angle BAD. Angle B intercepts an arc whose measure is 180°. What is the measure of angle B? What is the measure of angle BAD? What is the measure of half of angle BAO? Triangle ACO is isosceles. What is the value of x?

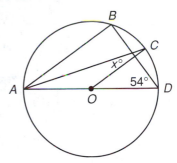

LESSON 121 *Rational inequalities*

We have learned to graph the solution to quadratic inequalities such as

$$x^2 - x - 6 > 0 \quad \text{and} \quad x^2 - x - 6 < 0$$

by factoring first.

(a) $(x - 3)(x + 2) > 0$ and (b) $(x - 3)(x + 2) < 0$

In inequality (a), we note that the product is greater than zero and thus is a positive number. For this to be true, both factors must represent positive numbers *or* both factors must represent negative numbers.

(Pos)	and	(Pos)	**or**	(Neg)	and	(Neg)
$(x - 3)$		$(x + 2)$		$(x - 3)$		$(x + 2)$

In inequality (b), we note that the product is less than zero and thus is a negative number. For this to be true, one factor must represent a positive number, while the other factor must represent a negative number.

(Pos)	and	(Neg)	**or**	(Neg)	and	(Pos)
$(x - 3)$		$(x + 2)$		$(x - 3)$		$(x + 2)$

We will use the same method to solve quadratic inequalities that are written in fractional form, such as

$$1 \le \frac{-1}{x - 4}$$

As the first step, we must eliminate the denominator. **If we multiply both sides by $x - 4$ to do this,**

$$(x - 4)1 \le \frac{-1}{x - 4}(x - 4) \qquad \text{incorrect}$$

we have a problem. We must reverse the inequality symbol if $x - 4$ represents a negative quantity, but we don't know if $x - 4$ is negative or not because the value of $x - 4$ is determined by the replacement value of x. We can resolve our dilemma if instead of multiplying by $x - 4$, we multiply by $(x - 4)^2$ because the square of any nonzero quantity always represents a positive number. We will demonstrate this in the next example.

example 121.1   Graph: $1 \le \dfrac{-1}{x - 4}$; $D = \{\text{Reals}\}$

solution   **We begin by noting that $x = 4$ cannot be a solution because division by zero is not defined.** To eliminate the denominator, we will multiply both sides by $(x - 4)^2$ and cancel the denominator.

$$(x - 4)^2(1) \le \dfrac{-1}{x - 4}(x - 4)(x - 4)$$

For clarity, on the right-hand side we used $(x - 4)(x - 4)$ to represent $(x - 4)^2$. Now we simplify and get

$$x^2 - 8x + 16 \le -x + 4$$

Then we add $x - 4$ to both sides to get

$$x^2 - 7x + 12 \le 0$$

which we factor as

$$(x - 4)(x - 3) \le 0$$

This product is less than or equal to zero. If the product is less than zero, it must be a negative number because all real numbers that are less than zero are negative. Thus, if the product is negative, the first factor is positive **and** the second factor is negative; **or** the first factor is negative **and** the second factor is positive. So we get

(Pos)		(Neg)	**or**	(Neg)		(Pos)
$x - 4$	**and**	$x - 3$		$x - 4$	**and**	$x - 3$
$x - 4 \ge 0$	**and**	$x - 3 \le 0$		$x - 4 \le 0$	**and**	$x - 3 \ge 0$
$+ 4 \quad +4$		$+ 3 \quad +3$		$+ 4 \quad +4$		$+ 3 \quad +3$
$x \quad\ge\ 4$	**and**	$x \quad\le\ 3$	**or** x	$x \quad\le\ 4$	**and**	$x \quad\ge\ 3$

Impossible. There is no real number that is greater than 4 and also less than 3.

Thus, the total solution comes from the conjunction stated on this side.

$$x \le 4 \quad \textbf{and} \quad x \ge 3$$

Note that the graph excludes 4, an answer that we rejected at the outset.

example 121.2   Graph: $\dfrac{m - 2}{m + 2} \le 2$; $D = \{\text{Reals}\}$

solution   **As the first step, we note that -2 cannot be a solution, for this would make the denominator equal to zero.** Next we multiply both sides by $(m + 2)^2$ and do not reverse the inequality symbol because $(m + 2)^2$ is always a positive quantity.

$$(m + 2)(m + 2)\left(\dfrac{m - 2}{m + 2}\right) \le 2(m + 2)(m + 2)$$

$$m^2 - 4 \le 2m^2 + 8m + 8$$

Now we add $-m^2 + 4$ to both sides and get

$$m^2 + 8m + 12 \ge 0$$

which factors as

$$(m + 2)(m + 6) \ge 0$$

Now, for the product to be positive, both factors must be positive **or** both factors must be negative.

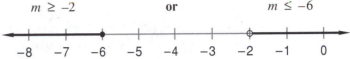

	(Pos)		(Pos)	or	(Neg)		(Neg)
	$m + 2$	**and**	$m + 6$		$m + 2$	**and**	$m + 6$

$$
\begin{array}{lll}
m + 2 \geq 0 & \text{and} & m + 6 \geq 0 \\
\underline{-2 \quad -2} & & \underline{-6 \quad -6} \\
m \qquad \geq -2 & \text{and} & m \qquad \geq -6
\end{array}
\quad \text{or} \quad
\begin{array}{lll}
m + 2 \leq 0 & \text{and} & m + 6 \leq 0 \\
\underline{-2 \quad -2} & & \underline{-6 \quad -6} \\
m \qquad \leq -2 & \text{and} & m \qquad \leq -6
\end{array}
$$

Any number greater than –2 is certainly greater than –6, so the values of *m* that satisfy this conjunction are the values of *m* such that

$$m \geq -2$$

or

Any number less than –6 is certainly less than –2, so the values of *m* that satisfy this conjunction are the values of *m* such that

$$m \leq -6$$

In the graph we have an open circle at –2 because at the beginning of the problem we noted that this number could not be a part of the solution.

practice Graph the solution on a number line: $1 < \dfrac{-2}{x - 3}$; $D = \{\text{Reals}\}$

problem set 121

1. Eighty liters of a 40% antifreeze solution had to be strengthened so that it contained 52% antifreeze. How many liters of pure antifreeze should be added?

2. The volume of a quantity of an ideal gas was held constant at 500 ml. The initial pressure and temperature were 10 atmospheres and 600 K. What would the final temperature be in kelvins if the pressure were increased to 20 atmospheres?

3. The *Bayou Belle* could go 84 miles downstream in 6 hours, but it took her 7 hours to go 42 miles upstream. What was her speed in still water, and what was the speed of the current in the river?

4. Garfunkel was twice as old as his dog, Spot. Ten years later, he found that 4 times his age exceeded 3 times his dog's age by only 15 years. How old were both in the beginning?

5. Rover Boy was 5 years older than Yolanda. In 10 years he found that 4 times his age exceeded twice Yolanda's age by only 50. How old were Rover Boy and Yolanda in the beginning?

6. Mary had 4 tables in a row. She had 9 dolls. How many ways could she arrange the dolls if she put 1 doll on each table?

7. Raoul rolls a pair of dice twice. What is the probability he will get a total of 4 on the first roll and a total greater than 5 on the second roll?

8. J. P. deposited $1,000,000 at 13 percent interest compounded continuously. How much money did he have at the end of 7 years? $(A_t = Pe^{rt})$

9. In the beginning there were 40. Their number increased exponentially. Five years later there were 140. How many would there be in 5 more years (a total of 10 years)?

Graph the solutions on a number line:

10. $\{x \in R \mid |x + 2| \leq 3\}$

11. $\dfrac{m - 3}{m + 3} \leq 2$; $D = \{\text{Reals}\}$

Solve for *x*:

12. $\ln (x + 2) + \ln 6 = \ln 36$

13. $3 \log_{11} x = \log_{11} 27$

14. $x = 6 \ln 0.003$

15. $6 \ln x = -2.78$

16. Use the $\boxed{\ln}$ key to help express each number as a power. Then use the rules of exponents to find the answer.

$$\frac{1.115 \times 10^5}{0.03 \times 10^{-2}}$$

17. In the figure shown, $\overline{QS}$ and $\overline{SR}$ are angle bisectors. How many degrees are there in $\angle PQR + \angle PRQ$? How many degrees are there in half of $\angle PQR + \angle PRQ$? This is the sum of the base angles in $\triangle QRS$. What is the measure of $\angle QSR$?

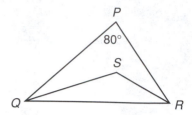

18. Find the pH ($pH = -\log H^+$ or $10^{-pH} = H^+$) of the solution when the concentration of hydrogen ions (H^+) = 0.053 mole per liter.

19. Find the concentration of hydrogen ions (H^+) in moles per liter when the pH of the liquid is 7.24×10^{-5}.

20. In the figure shown, $m\angle ABD = 62°$ and $m\angle BDC = 28°$. Find P. Find x. Find y, if $y = \frac{1}{2}(P - m\,\widehat{BD})$. Find $m\widehat{BD}$. Find $m\angle A$. Find $m\angle BDA$.

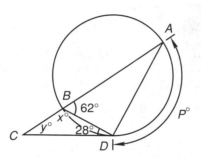

21. If the measure of the vertex angle of an isosceles triangle is 60°, the base angles are 60° angles and the triangle is equilateral, as shown. This is a regular hexagon inscribed in a circle whose radius is 8 cm. Find the area of the triangle. Find the area of the hexagon.

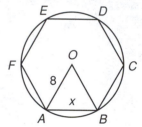

22. In which of the areas designated in the figure do the coordinates satisfy the given system of inequalities?

$$y > -\frac{2}{3}x + 2 \quad \text{(line)}$$
$$y < -x^2 + 4x \quad \text{(parabola)}$$

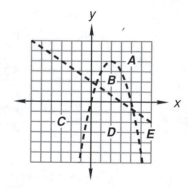

23. Complete the square as an aid in graphing: $y = -x^2 + 4x - 7$

24. Find the resultant vector of the two vectors shown.

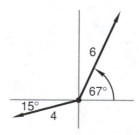

Solve:

25. $\begin{cases} 3x - y + z = 1 \\ x - y - z = 1 \\ x - 2y - z = -2 \end{cases}$

26. $\begin{cases} \frac{1}{2}x + \frac{1}{3}y = 5 \\ 0.4x - 0.2y = -0.2 \end{cases}$

27. $\begin{cases} 5x + y = 7 \\ 2x - z = -1 \\ y + z = 5 \end{cases}$

28. Simplify: $\dfrac{4 - \sqrt{5}}{\sqrt{5} + 2}$

29. Solve $3x^3 = 4x^2 + 7x$ by factoring.

30. Use a calculator to simplify. Estimate first.

    (a) $\dfrac{0.00763 \times 10^{-22}}{4,396,209 \times 10^{-35}}$

(b) $(3.91)^{-3.91}$

LESSON 122   *Laws of logarithms • Intersection of sets • Union of sets • Venn diagrams*

122.A
laws of logarithms

We remember that a logarithm is just an exponent. We know that 100 equals 10^2, so the logarithm of 100 to the base 10 is 2. We know that 1000 equals 10^3, so the logarithm of 1000 to the base 10 is 3.

Because $(10^2)(10^3)$ equals 10^{2+3}

we say $\log_{10}(100)(1000) = \log_{10} 100 + \log_{10} 1000$

If we let b represent the base, let M represent 100, and let N represent 1000, we can write a general statement of the rule for the logarithm of a product.

If $b^a = M$ and $b^c = N$, we say

$$\log_b (MN) = \log_b M + \log_b N$$

because $(b^a)(b^c) = b^{a+c}$.

    To use numbers to illustrate the rule for the logarithm of a quotient, we can write:

Because $\dfrac{10^5}{10^3} = 10^{5-3}$

we say $\log_{10} \dfrac{100,000}{1000} = \log_{10} 100,000 - \log_{10} 1000$

We can use letters to write a general rule for the logarithm of a quotient.

If $b^a = M$ and $b^c = N$, we say

$$\log_b \dfrac{M}{N} = \log_b M - \log_b N$$

because $\dfrac{b^a}{b^c} = b^{a-c}$.

To use numbers to illustrate the rule for the logarithm of a power we can write:

Because $$(10^2)^4 = 10^{2 \cdot 4}$$

we say $$\log_{10}(100)^4 = 4\log_{10}100$$

If we use letters to write a general rule for the logarithm of a power and if $b^a = M$, we say

$$\log_b M^x = x \log_b M$$

because $(b^a)^x = b^{ax}$.

The statements of the rules for logarithms that use letters instead of numbers are abstract and can be confusing. We can always use numbers that are powers of 10 (as we did above) to straighten us out if we get confused. These rules are not difficult to understand if we use numbers and remember that a logarithm is just an exponent.

122.B
intersection of sets

If we have the two sets

$$A = \{1, 2, 3, 6, 7\} \qquad B = \{1, 3, 8, 9\}$$

we see that the numbers 1 and 3 are members of both sets. We say the set $\{1, 3\}$ is the **intersection** of sets A and B. If we use the symbol $\cap$ to represent the word **intersection**, we can write

$$A \cap B = \{1, 3\}$$

This is read as the intersection of sets A and B and often as

$$A \text{ intersection } B$$

We see that the *intersection* of two sets is the set whose members are members of both of the given sets.

example 122.1   Given $P = \{1, 2, 3, 4, 7, 9, 13\}$ and $K = \{2, 5, 7, 8, 10, 13, 15\}$, find $P \cap K$.

solution   We are asked to find P intersection K, which is the set whose members are members of set P and are also members of set K.

$$P \cap K = \{2, 7, 13\}$$

122.C
union of sets

If we look at sets A and B

$$A = \{1, 2, 3, 7\} \qquad B = \{1, 3, 8, 9\}$$

and list all the members of both sets, we would write

$$1, 2, 3, 7, 1, 3, 8, 9$$

If we list these numbers using set notation, we would write

$$\{1, 2, 3, 7, 8, 9\}$$

for we only write a number once when we use set notation. **This set is called the *union* of sets A and B and consists of all the members of set A and all the members of set B, none listed more than once.** We use the symbol $\cup$ to represent the word **union**.

$$A \cup B = \{1, 2, 3, 7, 8, 9\}$$

example 122.2   Given $P = \{1, 2, 3, 4, 7, 9, 13\}$ and $K = \{2, 5, 7, 8, 13, 15\}$, find $P \cup K$.

solution   The union of the sets consists of all the members of both sets, none listed more than once. Thus,

$$P \cup K = \{1, 2, 3, 4, 5, 7, 8, 9, 13, 15\}$$

122.D
Venn diagrams

Diagrams can be used to enhance our understanding of intersection and union. In the diagram, we have designated set *A* by drawing a circle around the members of this set. We have also circled the members of set *B*. We can see that the numbers 15 and 17 are members of both set *A* and set *B*, so these numbers are the intersection of sets *A* and *B*.

$$A \cap B = \{15, 17\}$$

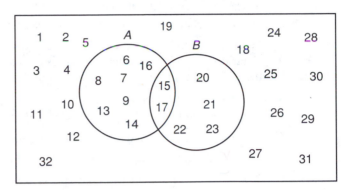

The union of sets *A* and *B* consists of all the members of both sets. Thus, we can write

$$A \cup B = \{6, 7, 8, 9, 13, 14, 15, 16, 17, 20, 21, 22, 23\}$$

example 122.3

The circles contain the members of sets *A*, *B*, and *C*, as indicated. Designate the areas that represent:

(a) *A* ∩ *B*   (b) *B* ∪ *C*   (c) *B* ∩ *C*

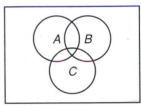

solution

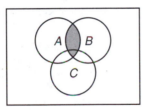   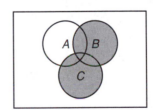   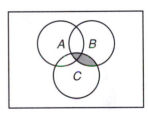

(a) *A* ∩ *B*
The points that are members of both *A* and *B*

(b) *B* ∪ *C*
The points that are members of either *B* or *C* or both

(c) *B* ∩ *C*
The points that are members of both *B* and *C*

practice

Designate the areas that represent:

  a. *M* ∩ *P*

  b. *M* ∩ *X*

  c. *X* ∪ *M*

  d. *X* ∪ *P*

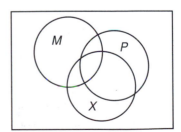

problem set 122

1. Citronella dropped the last coin into her bank and smiled. She had kept a tally and knew that her bank contained $4.55 in nickels and dimes and that there were 25 more nickels than dimes. How many coins of each type did Citronella have?

2. Monkeys varied jointly as apes and edible vines squared. If 400 monkeys went with 2 apes and 2 pounds of edible vines, how many apes were there when there were 1600 monkeys and $\frac{1}{2}$ pound of edible vines? Work the problem once using the ratio method and once using the variation method.

3. A two-digit counting number has a value that is 8 more than twice the sum of its digits. If 4 times the units' digit is 30 greater than the tens' digit, what is the number?

4. Yehudi is 4 years older than his brother Mohab. In 10 years, twice his age will exceed his brother's age by 24. How old are the boys now?

5. Petunia was proud because she was twice as old as Daisy. Ten years later she was chagrined to realize that twice her age exceeded Daisy's age by 25 years. How old were the girls at the outset?

6. The euglenae increased exponentially. At first there were 1000. Three hours later there were 1,500,000. How many euglenae would there be at the end of 8 hours?

7. There were 4 chairs in a row and 7 students. How many ways could the students occupy the 4 chairs if 4 students sat down every time?

8. A multiple-choice quiz has 6 questions, and there are 3 possible answers to each question. How many permutations of the answers are possible?

9. Use M and N to represent positive numbers and write:

    (a) the product rule for logarithms.
    (b) the quotient rule for logarithms.
    (c) the power rule for logarithms.

Graph the solution on a number line:

10. $2 \leq \dfrac{-2}{x - 2}$; $D = \{\text{Reals}\}$

11. $\dfrac{m + 3}{m - 3} \leq 1$; $D = \{\text{Integers}\}$

12. $\{x \in R \mid |x - 2| < 1\}$

13. $\{x \in J \mid |x + 3| \leq 4\}$

Use logarithms as required to perform the following operations. Begin by writing each number as an exponential expression whose base is 10. Show your work.

14. $\dfrac{516 \times 10^7}{0.00713 \times 10^{-5}}$

15. $(321,000)^{2/7}$

16. $(4.6 \times 10^{14})(3.02 \times 10^{-20})$

Find the pH ($pH = -\log H^+$ or $10^{-pH} = H^+$) of the solution when the concentration of hydrogen ions (H^+) in moles per liter is:

17. 3.26×10^{-9}

18. 7.04×10^{-5}

19. 0.0016

Find the concentration of hydrogen ions (H^+) in moles per liter when the pH of the liquid is:

20. 4.02

21. 8.23

22. 10.13

23. Find the area of the shaded portion of the figure in square inches. Dimensions are in yards.

24. Use unit multipliers to convert 1000 inches per second to kilometers per hour.

25. Show that $0.001\overline{68}$ is a rational number by writing it as a fraction of integers.

26. Solve $-5x^2 - x = 4$ by completing the square.

Simplify:

27. $\dfrac{\sqrt{-3}\sqrt{-3} - i^3}{2 - \sqrt{-2}\sqrt{2}}$

28. $\dfrac{4\sqrt{2} - 5}{1 - \sqrt{2}}$

29. $\sqrt[4]{xy^5}\,\sqrt[6]{x^3 y}$

30. Designate the areas that represent:

(a) $P \cap Z$

(b) $P \cap S$

(c) $Z \cup P$

(d) $Z \cup S$

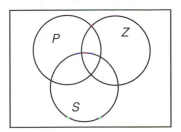

LESSON 123 *Locus • Basic construction*

123.A

locus   The English word *locus* is spelled just like the Latin word *locus*, which means place. We use the word locus to mean all the places and only the places (points) that satisfy a particular condition. Examples are the best ways to explain how we use the word locus.

The locus of points that are 3 cm from a given point is a circle whose radius is 3 cm.

The locus of points less than 5 cm from point *A* is the interior of a circle whose center is *A* and whose radius is 5 cm.

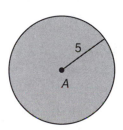

example 123.1   What is the locus of points that are 3 cm from a given line *x*?

solution   The locus is the two lines that are parallel to line *x* and are 3 cm from line *x*.

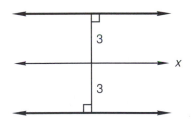

example 123.2   A circle has a radius of 1 foot. What is the locus of all points that are 10 inches from the given circle?

solution   The locus is **two** circles as shown. The smaller circle has a radius of 2 inches and the larger circle has a radius of 22 inches.

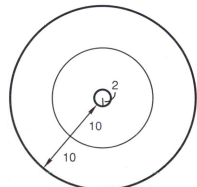

123.B
basic construction

Constructions using a compass and a straightedge help us understand some of the fundamental concepts of geometry.

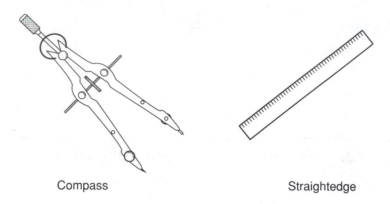

Compass   Straightedge

The straightedge of the Greeks did not have markings as do our rulers, so the Greek straightedge could only be used to draw a line segment between or through two points or to extend a given line segment. Sometimes we will use the ruler as a straightedge, and at other times we will use a ruler to measure distances. The modern compass shown above has capabilities that the Greek compasses did not have. We can use a modern compass to copy a given length, to draw a full circle, or to draw an arc of a circle.

The most useful constructions are copying line segments, copying angles, bisecting line segments, bisecting angles, and erecting perpendiculars. These basic constructions can be used in combination to perform more involved constructions, as we shall see. We will learn to do the constructions first. Then we will practice their use.

copying a line segment

To copy a line segment, we use a compass. On the left is segment *AB*. To copy this segment, we use a straightedge to draw ray *XY*. Then we place one end of the compass on point *A* and adjust the compass so that the other end is at point *B*.

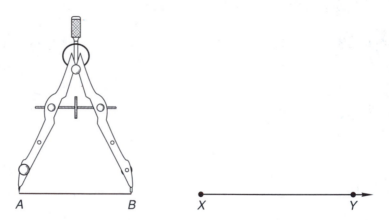

A   *B*   *X*   *Y*

Then we place one end of the compass at *X* and draw an arc that intersects $\overrightarrow{XY}$ at a point we call *P*.

X   *P*   *Y*

The distance *XP* is equal to the distance *AB*, so the segments *AB* and *XP* have equal lengths. Segments that have equal lengths are congruent, so we may write

$$\overline{AB} \cong \overline{XP}$$

copying an angle To copy the angle *BAC* on the left, we first draw ray *XY* on the right. Then we use the compass to draw equal arcs with center at *A* and *X*.

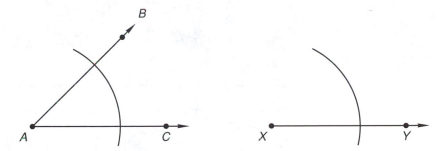

In the figure on the left below, the arc intersects the sides of the angle at points we call *M* and *N*. In the figure on the right, the arc intersects $\overrightarrow{XY}$ at *Z*.

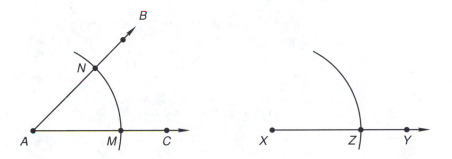

Next we adjust the compass to length *MN* and draw from *Z* an arc whose radius is *MN*.

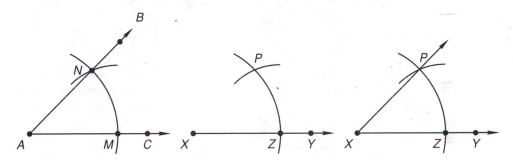

As the last step, we use the straightedge to draw ray *XP*. Angle *PXY* has the same measure as angle *BAC*, so the angles are congruent.

bisecting an angle To bisect angle *BAC* on the left below, we first draw an arc whose center is *A* as we show in the center figure. This arc intercepts the sides of the angle at points we call *X* and *Y*.

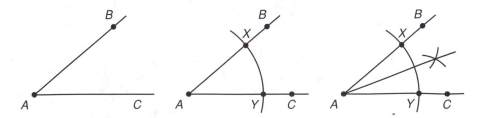

Then we draw arcs of equal radii whose centers are *X* and *Y*. The intersection of these arcs lies on the ray that is the bisector of the angle.

constructing
perpendiculars

To construct the perpendicular bisector of segment *BK*, we draw equal intersecting arcs from both *B* and *K* as shown.

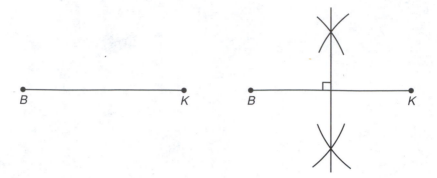

The line that connects the points of intersection of the arcs is the perpendicular bisector of segment *BK*.

To erect a perpendicular from point *P* on line *AC* below, we draw equal arcs on either side of point *P* that intersect $\overleftrightarrow{AC}$ at points we label *M* and *N*, as we see on the right.

Then we widen the compass and construct the perpendicular bisector of $\overline{MN}$.

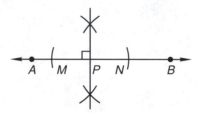

Now to construct a perpendicular from line *MN* on the left below that passes through point *P*, we draw two equal arcs whose center is *P* that intersect the line at *R* and *S*, as shown on the right.

Then we construct the perpendicular bisector of *RS*. This line will pass through · oint *P*.

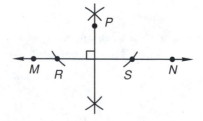

example 123.3

Given segments *BC*, *AX*, and *MN*, construct a triangle whose sides have these lengths.

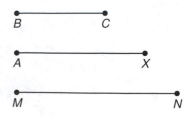

solution **There is only one triangle whose sides have these lengths.** First we draw a ray. Then draw on this ray an arc whose length is either *MN*, *AX*, or *BC*. We decide to use length *MN*, so we label the origin of the ray as *M*. The other end is *N*. *M* and *N* will be two vertices of the triangle.

From one end of $\overline{MN}$ we draw an arc whose length is *AX*. From the other end of $\overline{MN}$ we draw an arc whose length is *BC*. The intersection of these arcs is the other vertex of the triangle.

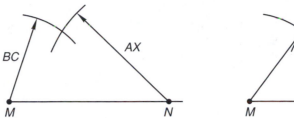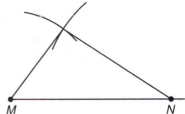

practice **a.** What is the locus of all points that are 3 feet from a given line?

b. What is the locus of all points that are 3 cm from a circle whose radius is 8 cm?

problem set 123

1. Draw an angle and copy the angle.

2. Draw an angle and construct its bisector.

3. Draw a line segment and construct the perpendicular bisector of the segment.

4. Construct a perpendicular to a line from a designated point on the line.

5. Construct a perpendicular to a line passing through a designated point that is not on the line.

6. Construct a triangle whose sides have these lengths.

7. Find three consecutive odd integers such that the product of the first and third is 13 less than the product of the second and 10.

8. Mulberry was exasperated because his plane took twice as long to cover 800 miles as it took the other plane to cover 650 miles. If the other plane was only 50 miles per hour faster, find the rates and times of both.

9. Tommy and Sarah huffed off in a hurry at 16 miles per hour. After a time they broke down and had to walk home at 10 miles per hour. If they were gone for 13 hours, how far did they get before they broke down?

10. A two-digit counting number has a value that is 13 greater than 3 times the sum of the digits. If the units' digit is 1 greater than the tens' digit, what is the number?

11. A man is 6 times as old as his son. In 5 years he will be 2 years older than 3 times his son's age then. How old are both now?

12. Given that $P = \{3, 5, 7, 9, 11, 13\}$ and $K = \{2, 5, 7, 8, 10, 13, 15\}$, find $P \cap K$.

13. Given that $A = \{1, 3, 8, 10\}$ and $B = \{1, 5, 7, 10\}$, find $A \cup B$.

14. Given *P* and *K* as defined in Problem 12, find $P \cup K$.

Graph the solution on a number line:

15. $1 \le \dfrac{-1}{x - 1}$; $D = \{$Reals$\}$

16. $\dfrac{m + 1}{m - 1} \le 1$; $D = \{$Reals$\}$

17. $\{x \in R \mid |x + 3| \leq 3\}$

18. Find the pH (pH $= -\log$ H$^+$ or $10^{-\text{pH}} =$ H$^+$) of the solution when the concentration of hydrogen ions H$^+$ $= 1.42 \times 10^{-11}$ mole per liter.

19. Find the concentration of hydrogen ions (H$^+$) in moles per liter when the pH of the liquid is 3.97.

20. On day zero there were 240 bacteria in the dish. Their number increased exponentially. After 10 days there were 480 bacteria in the dish. How many bacteria would there be after 18 days?

21. Designate the areas in which the coordinates of the points satisfy the given system of inequalities.

$$y \leq \frac{1}{3}x$$
$$x^2 + y^2 < 9$$

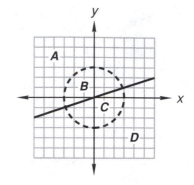

Solve for x:

22. $\log_7 (x + 2) + \log_7 3 = \log_7 15$ 23. $\ln (x - 9) + \ln 2 = \ln 45$

24. Expand: $(x^{4/5} - y^{4/5})^2$ 25. Factor: $a^3 m^{27} - p^6 y^{36}$

26. Show that $0.0012\overline{352}$ is a rational number by writing it as a fraction of integers.

27. Complete the square as an aid in graphing: $y = -x^2 + 4x - 1$

28. Solve $3x^2 = -5x - 2$ by using the quadratic formula.

29. Simplify: $-3i^2 - 2\sqrt{-9} + \sqrt{-4} - \sqrt{-2}\sqrt{2} + 2i^5$

30. Use similar triangles to find:
     (a) m and n (b) c and d

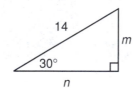

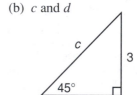

LESSON 124   *Conditions of congruence • Proofs of congruence • Isosceles triangles*

124.A
conditions of congruence

A triangle has three sides and three angles. Two triangles are congruent if any one of four conditions stated here are met.

   1. **SSS. If the lengths of the sides in one triangle are equal to the lengths of the sides in a second triangle, the triangles are congruent. We call this condition** *side, side, side* **(SSS).**

2. **AAAS. If the angles in one triangle have the same measures as the angles in a second triangle, the triangles are similar. If a side in one of a pair of similar triangles has the same length as the corresponding side in the other triangle, the scale factor is 1 and the triangles are congruent. We call this condition** *angle, angle, angle, side* **(AAAS).**

3. **SAS. If two sides and the included angle in one triangle have the same measures as two sides and the included angle in a second triangle, the triangles are congruent. We call this condition** *side, angle, side* **(SAS).**

4. **HL. If the lengths of the hypotenuse and a leg in one right triangle equal the lengths of the hypotenuse and a leg in a second right triangle, the right triangles are congruent. We call this condition** *hypotenuse, leg* **(HL).**

The proof of these conditions is a topic for a more advanced course. The SSS condition results from the fact that only one triangle can be formed from three sides of designated lengths. The AAAS reasoning is explained in 2 above. We can illustrate the fact that SAS determines congruence by considering this pair of congruent triangles.

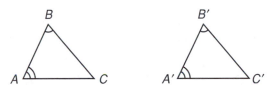

If we erase the bottom sides and the bottom angles, we get

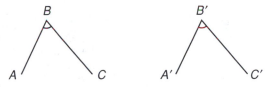

We see that there is only one way to draw the missing segments *AC* and *A'C'* to reform the triangles. If we draw these segments, we see that the triangles will be the same triangles, *ABC* and *A'B'C'*, with which we began. This illustrates the fact that if we fix the lengths of two sides and the measure of the included angle, we have completely defined the triangle.

The reason for the HL (hypotenuse-leg) condition is a little harder to see because the fixed angle is the right angle at the intersection of the two legs.

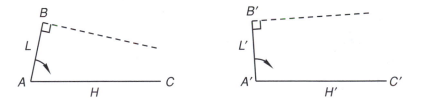

In these figures *L* and *L'* are given legs and *H* and *H'* are the given hypotenuses. If we rotate the legs *L* and *L'* about the points labeled *A* and *A'* until the dotted lines intersect the endpoints of *H* and *H'*, we can see that the triangles formed will be congruent.

The four conditions for triangle congruence are pictured here.

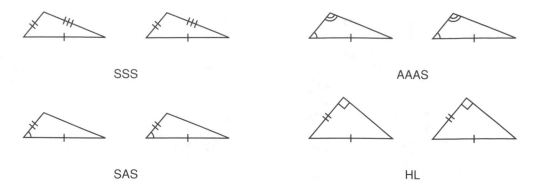

SSS   AAAS

SAS   HL

These are also the conditions for triangle similarity. If the lengths in SSS, SAS, and HL are not equal but are proportional, the triangles are similar. As a statement of similarity, the "side" in AAAS is redundant because AAA is sufficient to designate similarity.

We remember that when we write a statement of similarity or a statement of congruence we must list corresponding vertices in the same order. Here we show two congruent triangles.

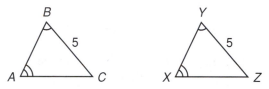

Triangle *ABC* is congruent to triangle *XYZ* by AAAS. The triangles are similar because the angles have equal measures. The scale factor is 1 because the sides opposite the equal angles *A* and *X* have the same length. We can write

$$\triangle ABC \cong \triangle XYZ \qquad \text{correct}$$

When we write a statement of congruence, we designate corresponding angles by the order in which the vertices are listed. Vertex *A* and vertex *X* are corresponding vertices, and they are given corresponding positions in the statement of congruence. This statement also indicates that *B* and *Y* are corresponding vertices and that vertex *C* corresponds to vertex *Z*. The following statement of congruence for the triangle above is not correct

$$\triangle ACB \cong \triangle XYZ \qquad \text{not correct}$$

because it indicates that corresponding vertices are at *C* and *Y* and *B* and *Z*, which is not so.

124.B

proofs of congruence

Many proofs require that we show that two segments have equal lengths or that two angles have equal measures. This is often accomplished by showing that two triangles are congruent, and thus the components have equal measures because corresponding parts of congruent triangles are congruent. If we can outline a proof first, the formal proof is easy. To outline a proof, we first sketch the figure and use tick marks to record the given information on the figure. Then we write the statement of congruence, being careful to list the vertices in corresponding order. Then write AAAS, SSS, SAS, or HL to show why the triangles are congruent.

example 124.1   Given: $\overline{AD} \cong \overline{DB}$
$\overline{DC} \perp \overline{AB}$

Outline a proof that shows: $\overline{AC} \cong \overline{BC}$

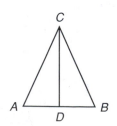

solution   The solution consists of a figure with tick marks and a simple statement.

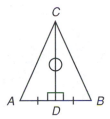

$\triangle ADC \cong \triangle BDC$ SAS
$\overline{AC} \cong \overline{BC}$ CPCTC

example 124.2   Given: $\angle E \cong \angle H$
$\overline{EF} \cong \overline{GH}$

Outline a proof that shows $\overline{DF} \cong \overline{DG}$.

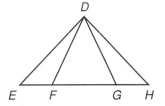

solution   We begin by copying the figure above using tick marks to indicate equal measures.

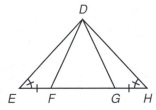

If we can show that $\triangle EFD$ is congruent to $\triangle HGD$, then $\overline{ED} \cong \overline{HD}$ by CPCTC. Because angle *E* has the same measure as angle *H*, triangle *EDH* is isosceles and side *DE* has the same length as side *DH*. We use tick marks to show this.

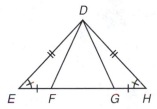

$\triangle FED \cong \triangle GHD$ SAS
$\overline{DF} \cong \overline{DG}$ CPCTC

124.C
isosceles triangles

There are several basic proofs about isosceles triangles. **Each proof requires that we show that an isosceles triangle can be separated into two congruent triangles.**

example 124.3   Outline a proof that shows the base angles of an isosceles triangle are equal.

solution   *Iso-* means "equal" and *skelos* means "legs." Thus an isosceles triangle is defined to be a triangle that has two sides whose lengths are equal. On the left we show triangle *ABC* and use tick marks to indicate that sides *AB* and *BC* have equal lengths.

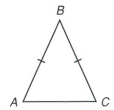

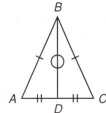

$\triangle ADB \cong \triangle CDB$ SSS
$\angle BAD \cong \angle BCD$ CPCTC

In the right-hand figure we have drawn a line between the vertex *B* and the point *D*, which is the midpoint of $\overline{AC}$. Thus $\overline{AD}$ has the same length as $\overline{DC}$. The segment *BD* is a side of both small triangles. The triangles are congruent by SSS. **When we write the statement of congruence,**

we are careful to list the vertices of equal angles in the same order. We see that the angle at *A* has the same measure as angle *C* because corresponding parts of congruent triangles are congruent.

example 124.4 Outline a proof that shows that if two angles of a triangle are equal, the triangle is an isosceles triangle.

solution We draw triangle *ABC* on the left and use tick marks to indicate that angle *A* has the same measure as angle *C*.

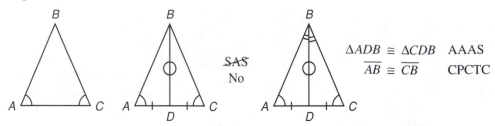

We know we are going to need two triangles for the proof. Thus, in the center we show median $\overline{BD}$, so we can show the triangles congruent by SAS. But this doesn't work because in SAS the angle must be between the sides. So we try again. In the right-hand figure we form the two triangles by bisecting angle *B*. Two angles in $\triangle ADB$ have the same measures as two angles in $\triangle CDB$, so the third angles are equal and the triangles are similar by AAA. Side *DB* is a side in both triangles and is opposite equal angles. Thus the scale factor is 1 and the triangles are congruent by AAAS. Thus $\overline{AB}$ and $\overline{CB}$ have equal lengths by CPCTC. This tells us the triangle is isosceles.

example 124.5 Prove that the bisector of the vertex angle of an isosceles triangle is perpendicular to the base.

solution This one is easy. On the left we show isosceles triangle *ABC*. On the right we bisect angle *B*.

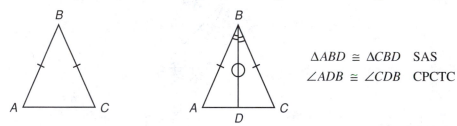

Since angles *ADB* and *CDB* have equal measures and they form a straight line, each of the angles is a right angle and thus $\overline{BD}$ is perpendicular to $\overline{AC}$.

practice **a.** Use letters to list the four conditions of congruence.

b. Given: $\angle E \cong \angle H$; $\overline{EF} \cong \overline{GH}$
Outline a proof that shows: $\triangle EDF \cong \triangle HDG$

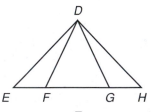

problem set 124 **1.** Given: $\overline{AX} \cong \overline{BY}$
$\angle 3 \cong \angle 4$
$\overline{AW} \cong \overline{BZ}$

Outline a proof that shows: $\angle W \cong \angle Z$

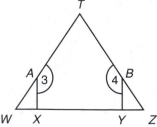

2. Given: $\overline{AB} \cong \overline{BC}$
             $\angle ABD \cong \angle CBD$

    Outline a proof that shows: $\overline{AD} \cong \overline{DC}$

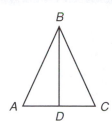

3. In this problem we will remember that if two segments have equal lengths, the halves of the segments have equal lengths.

    Given: $\overline{AG} \cong \overline{EF}$
             $\overline{CA} \cong \overline{CE}$

    B is the midpoint of $\overline{AC}$.

    D is the midpoint of $\overline{CE}$.

    Outline a proof that shows: $\overline{BG} \cong \overline{DF}$

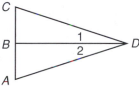

4. The following proof requires that we show two triangles are congruent and then state that two angles whose measures are equal and whose sum is a straight angle are both right angles.

    Given: $\overline{CD} \cong \overline{AD}$

    B is midpoint of $\overline{CA}$.

    Outline a proof that shows: $\overline{BD} \perp \overline{CA}$

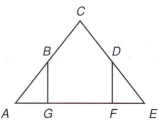

5. Given: *ADCB* is a rectangle.

    E is midpoint of $\overline{AD}$.

    Outline a proof that shows: $\overline{BE} \cong \overline{CE}$

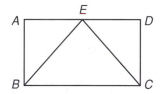

6. Draw a 5-centimeter line segment with a ruler and construct the perpendicular bisector of the line segment.

7. Use a protractor to draw an 80° angle. Then use a straightedge and a compass to construct the bisector of the angle.

8. Use a protractor to draw a 50° angle. Then use a straightedge and a compass to copy the angle.

Graph the solution on a number line:

9. $\dfrac{m + 3}{m - 3} \geq 1; \; D = \{\text{Reals}\}$ 10. $2 \leq \dfrac{-2}{x - 1}; \; D = \{\text{Integers}\}$

11. $\{x \in R \mid |x + 2| < 3\}$

Solve for *x*:

12. $38 = 10^{x+3}$ 13. $\ln (x + 2) + \ln 3 = \ln 39$

14. $4 \log_8 x = \log_8 48$

15. Use *M* and *N* to represent positive numbers and write:
    (a) the product rule for logarithms.
    (b) the quotient rule for logarithms.
    (c) the power rule for logarithms.

16. A man is 11 times as old as his son. In 12 years he will be 3 times as old as his son will be. How old are both now?

17. How many 4-letter signs can be made from the letters in the word CIRCLE if repetition of letters is permitted?

18. How many 3-digit combinations can be made with the digits 1, 2, 3, 4, 5, and 6 if repetition of digits is permitted?

19. Find the number that is $\frac{2}{7}$ of the way from $\frac{3}{8}$ to $2\frac{1}{5}$.

20. Complete the square as an aid to graphing: $y = -x^2 + 2x - 4$

LESSON 125 *Distance defined • Equidistance • Circle proofs*

125.A
distance defined

We remember the basic definition of distance.

1. The distance between two points is the length of the segment that connects the points.
2. The distance from a point to a line is measured along the segment through the point that is perpendicular to the line.

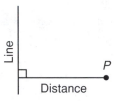

125.B
equidistant from two points

Any two points that are equidistant from the ends of a segment lie on the perpendicular bisector of the segment. We will prove this in two ways: once informally and once formally.

On the left, we show segment *BD* and two points, *A* and *C*, that are equidistant from *B* and *D*. We use these points to draw the figure on the right.

In the right-hand figure, we complete the drawing and use two tick marks to indicate equal distances from *A* and single tick marks to indicate equal distances from *C*. The centerline *AC* is a side of both the upper and lower big triangles $\triangle ABC$ and $\triangle ADC$. Thus these triangles are congruent by SSS. This means that the two angles at *A* are equal, CPCTC. If we use tick marks to mark these angles as equal angles and look only at the left-hand part of the figure, we get

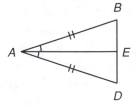

We see that the upper and lower triangles are congruent by SAS because $\overline{AE}$ is a side of both triangles. Thus $\overline{BE} \cong \overline{ED}$ and the two angles at *E* are equal, CPCTC. Thus the two angles at

E are right angles because they are equal and form a straight angle. This means that $\overline{AE}$ is perpendicular to $\overline{BD}$ and bisects $\overline{BD}$.   QED

Now we will do a two-column proof that laboriously and rigorously follows this same reasoning process.

1.	$\overline{AB} \cong \overline{AD}$	1.	Given
2.	$\overline{BC} \cong \overline{CD}$	2.	Given
3.	$\overline{AC} \cong \overline{AC}$	3.	Reflexive property
4.	$\triangle ABC \cong \triangle ADC$	4.	SSS (1, 2, 3)
5.	$\angle BAC \cong \angle DAC$	5.	CPCTC
6.	$\overline{AE} \cong \overline{AE}$	6.	Reflexive property
7.	$\triangle ABE \cong \triangle ADE$	7.	SAS (1, 5, 6)
8.	$\overline{BE} \cong \overline{ED}$	8.	CPCTC
9.	$\overline{AC}$ bisects $\overline{BD}$	9.	Forms two equal segments
10.	$\angle AEB \cong \angle AED$	10.	CPCTC
11.	$\overline{AE} \perp \overline{BD}$	11.	Two equal adjacent angles whose sum is 180° are right angles
12.	$\overline{AC}$ is $\perp$ bisector of $\overline{BD}$	12.	From steps 9 and 11

From this we see that a two-column proof is just a rigorous presentation of a proof that we have already completed in a nonrigorous fashion.

125.C
equidistant from sides of an angle

Any point that is equidistant from the two sides of an angle lies on the ray that is the angle bisector.

On the left, we show point X that is equidistant from the sides of the angle BAC. On the right, we draw perpendicular segments whose lengths equal these distances and also draw ray AX.

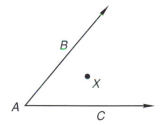 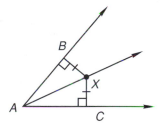

The segment AX is the hypotenuse of both right triangles and is congruent to itself. Thus the two triangles are congruent by HL, and angle BAX is congruent to angle CAX, CPCTC. Since these two angles have equal measures and are adjacent angles, ray AX is the bisector of angle BAC and X lies on the angle bisector. This proof can be repeated for any point that is equidistant from the sides of an angle.

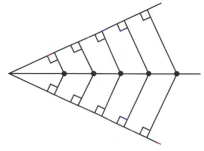

Thus we see that the locus of points that are equidistant from the sides of an angle is the ray that bisects the angle.

125.D

circle proofs Proofs involving the relationships between circles, chords, and tangents use the fact that all radii of a circle have equal lengths. Thus, if two circles have radii whose lengths are equal, the circles are congruent circles. These statements are definitions. We remember that all definitions are reversible. So we can say that if two circles are congruent, their radii have equal lengths.

All radii are equal   Congruent circles

Another useful fact is that **a tangent to a circle is perpendicular to a radius of the circle at the point of tangency.**

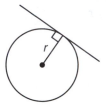

It would seem that this simple fact could be proved, but mathematicians have not found a proof that they find acceptable so this fact is postulated and accepted without proof.

POSTULATE

A tangent to a circle is perpendicular to the radius at the point of tangency.

Many theorems about segments and circles can be proved by proving that two triangles are congruent. We present several of these proofs here.

The lengths of tangent segments that intersect at a point outside the circle are equal. In the left-hand figure, we show two tangents to circle O that intersect at point P. In the right-hand figure, we draw radii to the points of tangency and draw segment OP.

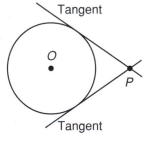
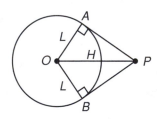

The radii are the legs of the two right triangles. Segment OP is the hypotenuse of both right triangles. Thus the triangles are congruent by HL, and $\overline{AP} \cong \overline{BP}$ by CPCTC.

If a radius is perpendicular to a chord, then it bisects the chord. The figure on the left shows a radius that is perpendicular to a chord. In the right-hand figure, we draw radii to A and B.

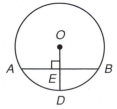
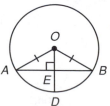

Since $\overline{OE}$ is a leg of both right triangles, the triangles are congruent by HL. Thus the two angles at E are congruent and $\overline{AE} \cong \overline{EB}$, by CPCTC. Because the angles are equal and sum to 180°, they are right angles. Thus $\overline{OD}$ and $\overline{OE}$ are perpendicular bisectors of $\overline{AB}$.

If a radius bisects a chord that is not a diameter, it is perpendicular to that chord. On the left, we show the basic figure. In the right-hand figure, we have drawn two radii to E and F.

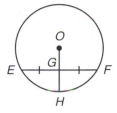 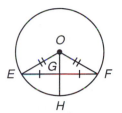

Segment OG is congruent to itself. Thus these triangles are congruent by SSS. It follows that the angles at G are congruent, by CPCTC, and because they sum to 180°, the angles are right angles.

The perpendicular bisector of a chord passes through the center of the circle. Again we draw two radii in the right-hand figure.

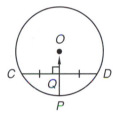 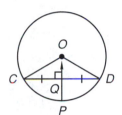

This time we do not have to prove that two triangles are congruent. All we have to do is remember that we have already proved that two points that are equidistant from the sides of a segment lie on the perpendicular bisector of the segment. Since O is equidistant from C and D, it lies on the perpendicular bisector of $\overline{CD}$.

practice   **a.** The proof that tangent segments from a point outside a circle to a circle have equal lengths was outlined in this lesson. Do a two-column proof of this theorem.

b. The proof that if a radius of a circle bisects a chord that is not a diameter it is perpendicular to that chord was outlined in this lesson. Do a two-column proof of this theorem.

c. Given: Circle Q
$\overline{PR} \perp \overline{ST}$

Prove that: $\angle S \cong \angle T$

Do a two-column proof. Begin by drawing radii SQ and TQ. First prove that the small triangles are congruent right triangles by HL, and then prove that the big triangles are congruent by HL.

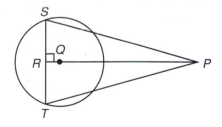

problem set   **1.** In this problem it will be necessary to draw radii $\overline{QS}$ and $\overline{QT}$ and prove that the small triangles formed are congruent. Then we can say that $\overline{RS}$ is congruent to $\overline{RT}$ and prove **125**   that the big triangles are congruent. Do a formal proof and list all the steps.

Given: Circle Q
$\overline{PR} \perp \overline{ST}$

Outline a proof that shows:

$\angle S \cong \angle T$

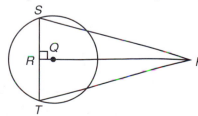

2. Given: $\overline{BE} \cong \overline{BD}$
              $\overline{BE} \perp \overline{AE}$
              $\angle BDC = 90°$

    Outline a proof that shows:
              $\angle AED \cong \angle CDE$

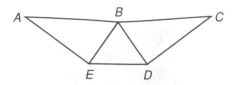

3. Given: $\overline{BE}$ bisects $\angle ABC$
              $\overline{BE}$ bisects $\angle CEA$

    Outline a proof that shows:
              $\overline{CE} \cong \overline{AE}$

4. Given: Circle O
              M is midpoint of $\overline{AB}$

    Outline a proof that shows:
              $\overline{OM} \perp \overline{AB}$

    Begin by drawing radii AO and OB.

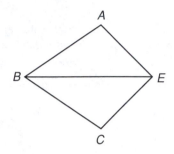

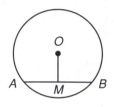

5. Given: Circle O
              Tangents PA and PB

    Outline a proof that shows: $\overline{AP} \cong \overline{PB}$

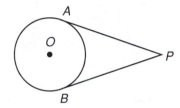

6. Construct a triangle whose sides are 4 cm, 3 cm, and 2 cm.

7. Use a ruler to draw a line segment 4 centimeters long. Construct a perpendicular to the line at a point 1 centimeter from the left endpoint.

8. Draw a line segment and a point outside the line. Construct a perpendicular to the line which passes through the point.

9. Use a protractor to draw a 36° angle. Then use a straightedge and a compass to construct the bisector of the angle.

10. Stephen deposited $1150 at 8 percent interest compounded continuously. How much money did he have at the end of 11 years?

11. The number of synaptic reactions increased exponentially. At first there were 1000. Three minutes later there were 1600. How many synaptic reactions would there be at the end of 10 minutes?

Solve for x:

12. $5x = 3 \ln 0.0035$

13. $3 \ln x = 0 - 5.13$

Graph the solution on a number line:

14. $\{x \in J \mid |x - 5| < 3\}$

15. $\{x \in R \mid x^2 - 6x > -9\}$

16. Twenty years ago, Melvina was 4 years older than twice Beula's age then. Six years ago, Beula was $\frac{3}{5}$ as old as Melvina was then. How old is each girl now?

17. A true-false quiz has 12 questions. How many permutations of the answers are possible?

18. How many different ways can the letters a, b, c, and d be arranged if no repetition is allowed?

19. Use the ⎡ln⎤ key to express each number as a power. Then use the rules of exponents to find the answer.

$$\frac{(0.000523)(916)}{0.0769 \times 10^{-11}}$$

20. Show that $0.03\overline{154}$ is a rational number by writing it as a fraction of integers.

LESSON 126   *Rectangles • Squares • Isosceles trapezoids • Chords and arcs*

We remember that a **quadrilateral** is a **polygon** that has four sides. A **parallelogram** is a quadrilateral that has five properties which we discussed in Lessons 37 and 39: (1) The opposite sides of a parallelogram are parallel. (2) The pairs of opposite sides have equal lengths (are congruent). (3) The pairs of opposite angles have equal measures (are congruent).

(4) The sum of any two consecutive angles is 180°. (5) The diagonals bisect each other.

Remember that a **rhombus** is a parallelogram whose sides have equal lengths. Thus every rhombus has the five properties of parallelograms just listed. In Lesson 39, we used AAAS to prove that (6) the diagonals of a rhombus bisect the angles of the rhombus and that (7) the diagonals are perpendicular bisectors of each other.

126.A
rectangles

A **rectangle** is a parallelogram (five properties). A rectangle has two additional properties. The first is contained in the definition of a rectangle. **A rectangle is a parallelogram in which all four angles are right angles.** Authors of many geometry books say that a rectangle is a parallelogram that has **at least one right angle** and require the students to prove that the other three angles are also right angles. This proof requires only that the student remember one of the properties possessed by all parallelograms.

The figure shown at the bottom of page 495 is a parallelogram with a right angle at *A*. The angles marked 2 must also be right angles because the sum of the measures of any two consecutive angles in a parallelogram is 180°. If these angles are right angles, then the angle marked 3 must also be a right angle for the same reason.

A rectangle has one other property that other parallelograms do not have. (6) **The diagonals of a rectangle have equal lengths.** To prove that the diagonals of a rectangle have equal lengths, we draw diagonals *AC* and *DB* in rectangle *ABCD*.

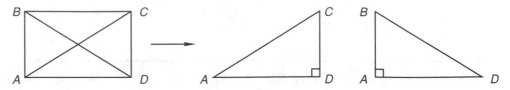

Now we break out two triangles from rectangle *ABCD* as shown above. Both triangles are right triangles because all the angles of a rectangle are right angles. Also, $DC \cong AB$ because opposite sides of a rectangle are congruent. Now, *AD* is a side of both triangles. By SAS, $\triangle ACD \cong \triangle DBA$. By CPCTC, we have $AC \cong DB$.

126.B

squares

A **square** is a parallelogram (five properties), a rhombus (two properties), and a rectangle (two properties), for a total of nine properties. **A square has no additional properties.**

We do note that the diagonals of a square bisect the four right angles at the corners so that the eight angles formed each measure 45°. Also, the diagonals are perpendicular bisectors of each other.

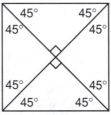

Thus the square and its diagonals form four isosceles right triangles.

126.C

isosceles trapezoids

Remember that a trapezoid is a quadrilateral that **has exactly two parallel sides [property (1)].** The parallel sides are called the bases of the trapezoid. We remember that we can find the area of a trapezoid by drawing a diagonal and forming two triangles that have the same altitude. The area of the trapezoid is the sum of the areas of the two triangles.

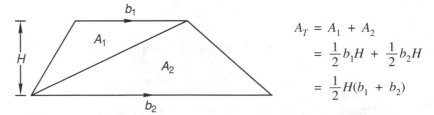

$$A_T = A_1 + A_2$$
$$= \frac{1}{2}b_1H + \frac{1}{2}b_2H$$
$$= \frac{1}{2}H(b_1 + b_2)$$

The nonparallel sides of a trapezoid are called the **legs.** The Greek prefix *iso-* means equal, and the Greek word for leg is *skelos.* We put these together to form the word *isosceles*, which means equal legs. Thus the second property of an **isosceles trapezoid is that the legs of an isosceles trapezoid have equal lengths.**

The angles at the ends of the shortest base of an isosceles are called **upper base angles and have equal measures [property (3)].** The angles at the ends of the longest base are called **lower base angles and have equal measures [property (4)].** Also, **the sum of two consecutive angles in a trapezoid is 180°** [property (5)].

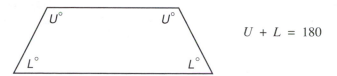

$$U + L = 180$$

We can prove the three angle properties by drawing perpendiculars and noting that the two right triangles formed are congruent by HL.

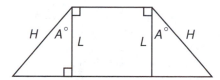

The lower base angles are equal by CPCTC. The upper base angles are equal because each of their measures is the sum of a right angle and one of the pair of the congruent angles marked *A*. Since the sum of the angles of a quadrilateral is 360°, and the upper base angles and lower base angles have equal measures, the sum of the measures of an upper base angle and a lower base angle is 180°, as we show on the right.

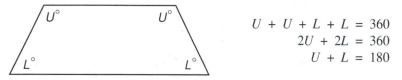

$$U + U + L + L = 360$$
$$2U + 2L = 360$$
$$U + L = 180$$

To prove that the diagonals of an isosceles trapezoid have equal lengths, we use two representations of the same isosceles trapezoid. The shaded triangles are congruent by SAS. The diagonals are corresponding sides because they are the sides opposite an upper base angle. They have equal lengths by CPCTC.

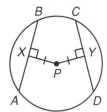   

126.D
chords and arcs

In Lesson 125, we noted that if a radius of a circle is perpendicular to a chord, the radius bisects the chord, and that if a radius bisects a chord, it is perpendicular to the chord. Now we will consider two more theorems about chords and four theorems about the arcs of a circle.

If two chords are equidistant from the center of a circle, the lengths of the chords are equal. On the left, we show two chords that are equidistant from the center *P* as measured along the perpendicular segments shown.

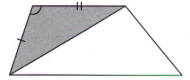   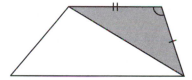

On the right, we draw four radii to form four right triangles that are congruent by HL. From this we can reason that the lengths of the chords are equal.

If two chords of a circle have equal lengths, they are equidistant from the center of the circle. On the left, we show the perpendiculars that bisect the two equal chords.

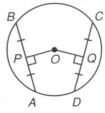   

On the right, we draw four radii that are the hypotenuses of four right triangles and are congruent by HL. Thus sides *PO* and *QO* have equal lengths, by CPCTC.

Remember that we have defined the degree measure of an arc to be the same as the measure of the central angle. Every definition is reversible, so the degree measure of a central angle is the same as the degree measure of the arc it intercepts. We note that two arcs can have the same degree measure but different lengths because they are in circles whose radii have different lengths.

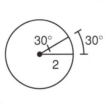   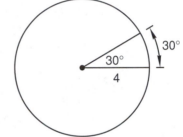

Congruent arcs are arcs of the same degree measure in the same circle or in circles of equal radii. We can use similar triangles to prove theorems about chords and arcs.

If two central angles of a circle (or of congruent circles) have equal measures, then the chords opposite the central angle have equal lengths. This figure shows two equal central angles and their chords. We note that the radii are equal.

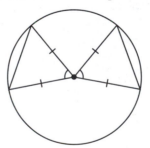

The two triangles are congruent by SAS, so the chords have equal lengths.

If two arcs of a circle (or of congruent circles) have equal lengths, then their corresponding chords also have equal lengths. If the two arcs have equal lengths, then the central angles have equal measures. Thus their corresponding chords have equal lengths by the proof above.

If two chords have equal lengths, the measures of the central angles are equal. In the following figure we show two chords of equal length connected to the center of the circle by equal radii.

The two triangles are congruent by SSS, and the central angles have equal measures by CPCTC.

If two chords of a circle (or of congruent circles) have equal lengths, then their corresponding arcs also have equal lengths. If the two chords of a circle have equal lengths, their central angles have equal measures. Thus their corresponding arcs have equal lengths by the proof above.

problem set 126

1. Given: $\overline{AB} \cong \overline{AD}$
   $\overline{BC} \cong \overline{CD}$

   Outline a proof that shows:
   $\overline{BE} \cong \overline{ED}$

   For the proof, remember that if the base angles of a triangle are congruent the sides opposite are congruent. Also, if two points are each equidistant from the endpoints of a segment, they lie on the perpendicular bisector of the segment.

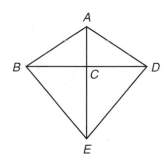

2. Given: Circle O
   $\angle ZWY \cong \angle ZXY$

   Outline a proof that shows:
   $\overline{OY} \perp \overline{WX}$

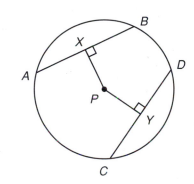

3. Given: Circle P
   $\overline{PX} \cong \overline{PY}$

   Outline a proof that shows:
   $\overline{AB} \cong \overline{CD}$

4. Given: $\angle 1 \cong \angle 2 \cong \angle 3 \cong \angle 4$
   $\overline{BE} \cong \overline{BF}$

   Outline a proof that shows:
   $\triangle BED \cong \triangle BFD$
   $\triangle BAE \cong \triangle BCF$

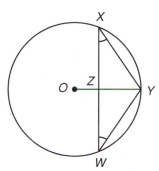

5. Given: $\overline{AB} \cong \overline{AF}$
   $\overline{BD} \cong \overline{DF}$
   $\angle 1 \cong \angle 2$
   $\overline{ED} \cong \overline{CD}$

   Outline a proof that shows:
   $\angle 3 \cong \angle 4$
   $\overline{EF} \cong \overline{BC}$

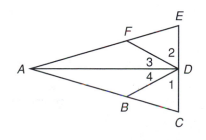

6. Construct a triangle whose sides are 3 cm, 4 cm, and 5 cm.

7. Use a ruler to draw a line segment 5 centimeters long. Construct a perpendicular to the line at a point 2 centimeters from the left endpoint.

8. Draw a line segment and a point outside the line. Construct a perpendicular to the line which passes through the point.

9. Use a protractor to draw a 110° angle. Then use a straightedge and a compass to construct the bisector of the angle.

10. How many 5-letter signs can be made from the word RAPSCALLION if no repetition of letters is permitted?

11. Each question on a survey has three possible answers: Yes, no, and maybe. The survey has 8 questions. How many possible permutations of responses to the survey are possible?

12. Use M and N to represent positive numbers and write:
    (a) the product rule for logarithms.
    (b) the quotient rule for logarithms.
    (c) the power rule for logarithms.

Solve for x:

13. $4x = 3 \ln 0.0069$

14. $4 \ln x = -4.98$

15. Find the pH (pH $= -\log H^+$ or $10^{-pH} = H^+$) of a solution when the concentration of hydrogen ions $H^+ = 0.0053$ mole per liter.

16. Find the concentration of hydrogen ions (H^+) in moles per liter when the pH of the liquid is 6.19.

Graph the solution on a number line:

17. $\dfrac{p + 2}{p - 2} \leq 2$; $D = \{\text{Reals}\}$

18. $\{x \in R \mid x^2 - 5x \leq -4\}$

19. The sum of the digits of a two-digit counting number is 14. If the digits are reversed, the new number is 23 less than 2 times the original number. What was the original number?

20. Complete the square as an aid to graphing: $y = x^2 - 6x + 3$

LESSON 127   *Lines and planes in space*

In Lesson A, we noted that we begin our study of geometry with terms that we are not able to define exactly. We call these terms **primitive terms** or **undefined terms**. We noted that the words *point*, *curve*, *line*, and *plane* are primitive terms. The complete meaning of primitive terms is found in the totality of the axoims that use these terms.

We think of a plane as a flat surface that has no thickness and that extends without bound in all directions. Geometric figures that can be drawn on flat surface are called **planar** figures, and the study of these figures is called **plane geometry**. The study of three-dimensional geometric figures is called **solid geometry**.

Two points determine a line, and if a third point lies on the same line, all three points are said to be **collinear**. Points A, B, and C shown here are on the same line and are thus collinear.

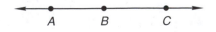

A line can be parallel to a plane, as we show in the following figure. The line and the plane have no points in common.

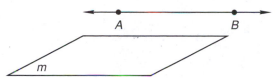

If two planes do not have any points in common, we say that the planes are **parallel planes**. In the figure on the left, planes m and n do not intersect, so these planes are parallel planes. If two planes intersect, their intersection is exactly one line, as we see in the figure on the right.

The angle made by two intersecting planes is called a **dihedral angle**. Thus angle MAD is a dihedral angle. We note that the line through A, B, and C is in plane m and also in plane n. Thus we see that three collinear points do not determine a plane. **We postulate that three points that are not collinear do determine a plane.** Since two of these points determine a line, we can also say that a plane is determined by a line and a point that is not on the line.

In the figure below, we see that the same three points determine two intersecting lines, so two intersecting lines also determine a plane.

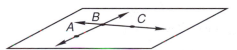

Two parallel lines also determine a plane. The definition of parallel lines requires that the lines be in the same plane. If lines AB and CD in the following figure are parallel, the lines must be in the same plane.

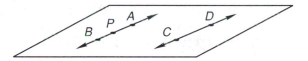

Since point P and points C and D determine a plane, there is only one plane that contains points P, C, and D; thus all five points lie in the same plane.

We also postulate that **if a line intersects a plane that does not contain the line, the line intersects the plane at only one point.** Line ST in the figure below intersects plane m only at point P.

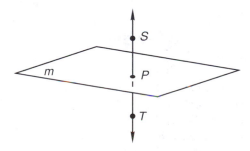

Parallel lines are defined to be lines in the same plane that do not intersect. *Skew lines* **are lines in space that are not parallel and do not intersect.**

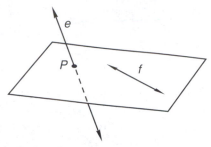

In this figure, lines *f* and *e* are not parallel and do not intersect, so these lines are skew lines. Another way to define skew lines is to say they are lines that are not in the same plane.

If a line is perpendicular to each of two intersecting lines at their point of intersection, the line is perpendicular to all lines in the plane that pass through the point of intersection. In the figure on the left, $\overleftrightarrow{ST}$ is perpendicular to $\overleftrightarrow{RT}$ but is not necessarily perpendicular to plane *m*.

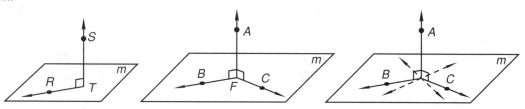

In the center figure, $\overleftrightarrow{AF}$ is perpendicular to two lines in the plane at their point of intersection and is thus perpendicular to all lines in the plane that pass through that point.

example 127.1   Given: $\overline{PF} \perp m$

$\overline{GF} \cong \overline{FH}$

Outline a proof that shows: $\overline{GP} \cong \overline{HP}$

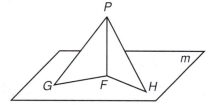

solution   We will outline the proof and assign the two-column proof as a practice problem. Since $\overline{PF}$ is perpendicular to plane *m*, it is perpendicular to $\overline{GF}$ and to $\overline{FH}$, so angles *GFP* and *HFP* are right angles. We are given that the bases of the triangle $\overline{GF}$ and $\overline{FH}$ have equal lengths. We see that $\overline{PF}$ is a side of both triangles. Thus

$$\triangle GFP \cong \triangle HFP \qquad \text{SAS}$$

and the hypotenuses $\overline{GP}$ and $\overline{HP}$ have equal lengths, by CPCTC.

example 127.2   Given: *B, C, D,* and *E* are in plane *k*

$\overline{AB} \perp k$

$\overline{BE}$ is a $\perp$ bisector of $\overline{CD}$

Outline a proof that shows:

$\triangle ADC$ is isosceles

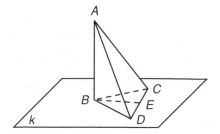

solution   First we show that the small triangles *BDE* and *BCE* in the base are congruent by SAS because $\overline{DE}$ and $\overline{EC}$ have equal lengths, the angles at *E* are both right angles, and $\overline{BE}$ is a side of both triangles. Thus $\overline{BC}$ and $\overline{BD}$ are congruent by CPCTC. Now we can show that the big triangles *ABD* and *ABC* are also congruent by SAS because the angles at *B* are right angles and $\overline{BA}$ is a side of both triangles. By CPCTC, $\overline{AD} \cong \overline{AC}$, and $\triangle ADC$ is isosceles.

practice   **a.**   Given: $\overline{PF} \perp m$
$\overline{PG} \cong \overline{PH}$

Outline a proof that shows:
$\angle G \cong \angle H$

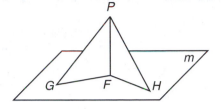

b.   Given: $B, C, D,$ and E lie in plane k
$\overline{AB} \perp k$
$\overline{BE}$ is a $\perp$ bisector of $\overline{CD}$

Outline a proof that shows:
$\triangle ADC$ is isosceles

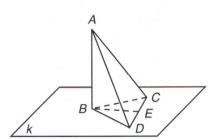

problem set   **1.**   Given: $\overline{PB} \perp m$
127 $\triangle APB \cong \triangle CPB$

Outline a proof that shows:
$\overline{AB} \cong \overline{CB}$

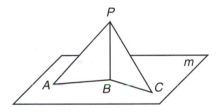

2.   Given: $\overline{PA} \perp s$
P is equidistant from B and C

Draw $\overline{BP}$ and $\overline{CP}$, and outline a proof
that shows A is equidistant from B
and C.

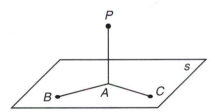

3.   Given: $\overline{CY} \cong \overline{AY}$
$\overline{YZ} \parallel \overline{CA}$

Outline a proof that shows:
$\overline{YZ}$ bisects $\angle AYB$

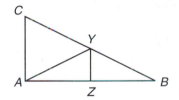

4.   Remember that if two points are each equidistant from the endpoints of a segment the
points lie on the perpendicular bisector of the segment. All points on the perpendicular
bisector are equidistant from the endpoints.

Given: $\overline{EX} \cong \overline{EY}$
$\overline{XP} \cong \overline{PY}$

Outline a two-step proof that shows:
$\overline{EZ} \perp \overline{XY}$

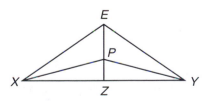

5.   Given: $\overline{AB} \cong \overline{AC}$
$\overline{BD}$ bisects $\angle ABE$
$\overline{CD}$ bisects $\angle ACE$

Outline a two-step proof that shows:
$\overline{AE}$ bisects $\overline{BC}$

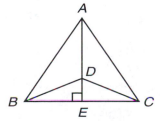

6. Construct a triangle whose sides are 4 cm, 3 cm, and 6 cm.

7. Use a ruler to draw a line segment 7 centimeters long. Construct a perpendicular to the line at a point 4 centimeters from the left endpoint.

8. Use a protractor to draw a 120° angle. Then use a straightedge and a compass to construct the bisector of the angle.

9. Draw a line segment and a point outside the line. Construct a perpendicular to the line which passes through the point.

10. Mary deposited $1460 at 9 percent interest compounded continuously. How much money did she have at the end of 9 years?

11. The incidents increased exponentially. At first there were 40 incidents. Three months later there were 560 incidents. How many incidents would there be at the end of 9 months?

12. Find the pH (pH = $-\log H^+$ or $10^{-pH} = H^+$) of a solution when the concentration of hydrogen ions $H^+ = 0.081$ mole per liter.

13. Find the concentration of hydrogen ions (H^+) in moles per liter when the pH of the liquid is 5.11.

14. Solve: $91 = 10^{x+2}$

15. Use the $\boxed{\text{ln}}$ key to express each number as a power. Then use the rules of exponents to find the answer.

$$\frac{0.01259 \times 10^{-7}}{519,000}$$

Graph the solution on a number line:

16. (a) $\{x \in R \mid |x + 1| \le 3\}$ (b) $\frac{p - 3}{p + 3} \ge 2$; $D = \{\text{Reals}\}$

17. Complete the square as an aid to graphing: $y = -x^2 - 2x + 4$

18. Find the number that is $\frac{3}{5}$ of the way from $\frac{5}{7}$ to $3\frac{1}{4}$.

19. Four chairs are placed against the wall. How many ways can the 7 boys sit in the four chairs if only 4 of them sit at one time?

20. Use similar triangles to help find the area of the triangle. Subtract this area from the area of the circle to find the area of the shaded portion. Dimensions are in centimeters.

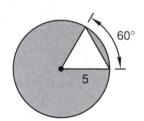

LESSON 128   *Circumscribed and inscribed • Inscribed triangles • Inscribed circles • Proof of the Pythagorean theorem • Inscribed angles*

128.A

circumscribed and inscribed

The Latin prefix for "around" is *circum-*, which comes from the Latin word *circus*, which means "circle." The Latin word that means "to write" is *scribere*. We put these words together to form the English word *circumscribe*, which means "to draw around." In the figure on the left,

we say that the circle is *circumscribed* about the triangle and that the triangle is *inscribed* in the circle.

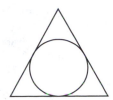

In the figure on the right, we say the triangle is circumscribed about the circle or the circle is inscribed in the triangle.

128.B
inscribed triangles

The Latin prefix *com-* means "together" and the Latin word *fluere* means "to flow." We put these together to form the English word *confluent*, which means "flow together." Thus we say that the Allegheny and Monongahela rivers in Pennsylvania are confluent at Pittsburgh. The word *concurrent* has a similar construction because the Latin word for "run" is *currere*. Thus the word *concurrent* means "running together."

Concurrent lines are lines that intersect at a single point. **The perpendicular bisectors of the sides of a triangle are concurrent at a point that is equidistant from all three vertices, as we show in the figure on the left.** We can use this point of intersection as the center of a circle that circumscribes the triangle, as we show on the right.

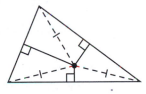
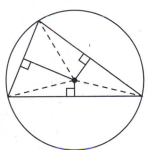

Thus to circumscribe a circle about a triangle we construct the perpendicular bisectors of the sides of the triangle to find the center of the circle.

example 128.1 Outline a proof showing that the perpendicular bisectors of the sides of a triangle intersect at a point that is equidistant from the vertices of the triangle.

solution In the figure on the left, the lines m, n, and p are perpendicular bisectors of the sides of the triangle as shown. Lines n and p intersect at T.

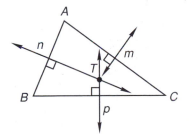
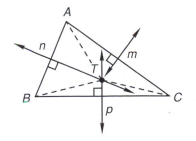

In the figure on the right, we note that the dotted paths from T to A and from T to B have equal lengths because all points on a perpendicular bisector are equidistant from the endpoints. The dotted path from T to C has the same length as path BT because T is on the perpendicular bisector of $\overline{BC}$. Because $\overline{TA} \cong \overline{TB}$ and $\overline{TB} \cong \overline{TC}$, $\overline{TC}$ has the same length as $\overline{TA}$. Thus T must lie on m because all points equidistant from A and C lie on m, the perpendicular bisector. Thus T is equidistant from A, B, and C.

128.C

inscribed circles

The bisectors of the angles of a triangle are concurrent at a point that is equidistant from the sides of the triangle. A circle with the proper radius whose center is at the point of concurrency will be tangent to the sides of the triangle, as we show on the right. This circle is inscribed in the triangle.

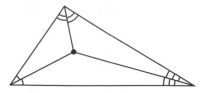 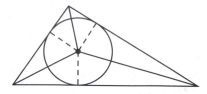

To inscribe a circle in a triangle, we bisect the angles of the triangle and use the point of intersection as the center of the circle.

example 128.2 Outline a proof showing that the angle bisectors of a triangle meet at a point that is equidistant from the sides of the triangle.

solution In the figure the angle bisectors of angles A, B, and C are drawn.

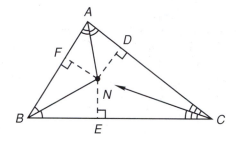

The bisectors of angles A and B meet at N. Dotted segment FN is congruent to dotted segment DN because N is on the bisector of angle A and thus is equidistant from both sides of angle A. Dotted segment FN is also congruent to dotted segment EN because N is on the bisector of angle B and thus is equidistant from both sides of angle B. Because both $\overline{DN}$ and $\overline{EN}$ have the same length as $\overline{FN}$, the length of $\overline{DN}$ equals the length of $\overline{EN}$. Thus N lies on the bisector of angle C because all points equidistant from the sides of angle C lie on its bisector.

128.D

proof of the Pythagorean theorem

We can use the three similar right triangles in this figure to do an easy proof of the Pythagorean theorem.

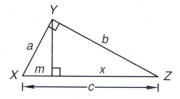

Each triangle has a right angle. Each triangle has an angle whose measure equals that of angle X. Each triangle has an angle whose measure equals that of angle Z. We can see this better if we draw the triangles separately.

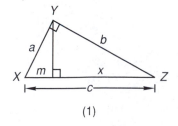

(1)

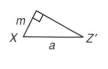

(2)

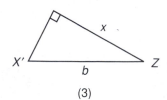

(3)

In triangle (2), one angle is a right angle and another angle is angle X. Thus the third angle must have the same measure as angle Z, and we label it Z'. In triangle (3), one angle is a right angle and another angle is angle Z. Thus the third angle must have the same measure as angle X, and we label it X'.

Now on the left below, we equate the ratios of the hypotenuse to the short side in triangle (1) and triangle (2). On the right, we equate the ratios of the hypotenuse to the long side in triangle (1) and triangle (3).

$$\frac{c}{a} = \frac{a}{m} \qquad\qquad \frac{c}{b} = \frac{b}{x}$$

Next, we cross-multiply and get

$$cm = a^2 \qquad\qquad cx = b^2$$

If we add these equations, we get $a^2 + b^2$ on the right-hand side.

$$cm + cx = a^2 + b^2$$

Now, on the left-hand side, we factor out c.

$$c(m + x) = a^2 + b^2$$

Then we replace $(m + x)$ with c. Thus the left-hand side becomes c^2, and we have our proof.

$$c(c) = a^2 + b^2 \qquad \text{substituted}$$
$$c^2 = a^2 + b^2 \qquad \text{simplified}$$

128.E
inscribed angles

We will prove that the measure of an angle inscribed in a circle equals half the measure of the intercepted arc. To do so, we will use variables instead of numbered values as we did in Lesson 11. The proof is in three parts.

Case I. The first part is the case where one side of the inscribed angle is a diameter.

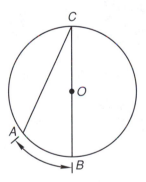

The first step is to draw radius AO, as shown below. We want to show that the measure of angle 4 is half the measure of arc AB.

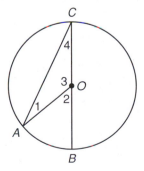

Now arc AB is defined to have the same measures as its central angle, which is angle 2.

$$m\widehat{AB} = m\angle 2 \qquad \text{definition}$$

Angle 2 is an exterior angle of triangle *OCA*. The measure of angle 2 equals the sum of the measures of the remote interior angles 1 and 4.

$$m\angle 1 + m\angle 4 = m\angle 2 \qquad \text{remote interior angles}$$

Angles 1 and 4 have equal measures because they are the base angles of an isosceles triangle. So the measure of angle 4 equals the measure of angle 1. If we substitute, we get

$$m\angle 4 + m\angle 4 = m\angle 2 \qquad \text{substituted}$$

This means that half the measure of angle 2 equals the measure of angle 4.

$$m\angle 4 = \frac{1}{2}m\angle 2$$

From the first step we know that the measure of arc *AB* equals the measure of angle 2. So we substitute to complete our proof.

$$m\angle 4 = \frac{1}{2}m\widehat{AB} \qquad \text{QED}$$

Case II. The second case is when the sides of the inscribed angles lie on either side of the center of the circle.

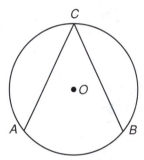

We want to prove that the measure of angle *C* is half the measure of arc *AB*. To do this, we draw a diameter from *C* through the center of the circle.

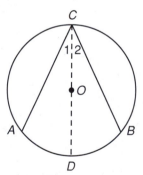

From Case I, we know that the measure of angle 1 equals half the measure of arc *AD* and that the measure of angle 2 equals half the measure of arc *DB*.

$$m\angle 1 = \frac{1}{2}m\widehat{AD} \qquad \text{from Case I}$$

$$m\angle 2 = \frac{1}{2}m\widehat{DB} \qquad \text{from Case I}$$

$$m\angle 1 + m\angle 2 = \frac{1}{2}m\widehat{AD} + \frac{1}{2}m\widehat{DB} \qquad \text{equals added to equals}$$

The sum of angles 1 and 2 equals angle *C*, and the sum of arcs *AD* and *DB* equals arc *AB*. So we substitute and get

$$m\angle C = \frac{1}{2}m\widehat{AB} \qquad \text{QED}$$

Case III. The third case is when both sides of the angle are on the same side of the center of the circle.

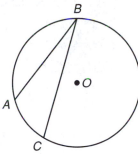

Again our first step is to draw a diameter from *B*.

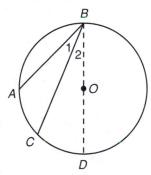

This time the measures of angles 1 and 2 equal half the measure of arc *AD*, and the measure of angle 2 equals half the measure of arc *CD*.

$$m\angle 1 + m\angle 2 = \frac{1}{2}m\overset{\frown}{AD} \qquad \text{Case I}$$

$$-\ m\angle 2 = -\frac{1}{2}m\overset{\frown}{CD} \qquad \text{subtracted}$$

$$m\angle 1 = \frac{1}{2}m(\overset{\frown}{AD} - \overset{\frown}{CD}) \qquad \text{substituted}$$

$\overset{\frown}{AD}$ minus $\overset{\frown}{CD}$ equals $\overset{\frown}{AC}$, so we have

$$m\angle 1 = \frac{1}{2}m\overset{\frown}{AC} \qquad \text{QED}$$

practice   **a.** Use a straightedge to draw a triangle. Use a compass to bisect all three angles. Use the point of intersection of the bisectors as the center of a circle that is inscribed in the triangle.

b. Use a straightedge to draw a triangle. Use a compass to draw the perpendicular bisectors of the three sides. Use the point of intersection of these bisectors as the center of a circle that circumscribes the triangle.

problem set   **1.** Given: $\overline{AB} \cong \overline{BC}$
128 $\overline{BD} \perp \overline{CA}$

Outline a proof that shows:

$$\angle ADB \cong \angle CDB$$

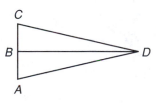

2. Do a two-column proof for Problem 1.

3. Given: *ABCD* is a rectangle.

        A is the midpoint of $\overline{XO}$

        B is the midpoint of $\overline{YO}$

        $\overline{XD} \cong \overline{CY}$

    Outline a proof that shows:

        $\triangle OXY$ is isosceles

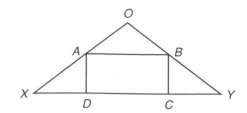

4. Do a two-column proof for Problem 3.

5. Given: $\angle KNL \cong \angle MNL$

        $\overline{KN} \cong \overline{MN}$

    Ouline a proof that shows:

        $\triangle KNL \cong \triangle MNL$

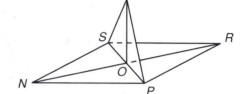

6. Given: *NPRS* is a parallelogram.

        Diagonals $\overline{SP}$ and $\overline{NR}$
        intersect at *O*.

        $\overline{TO} \perp$ plane of parallelogram
        NPRS.

    Outline a proof that shows:

        $\triangle STP$ is isosceles.

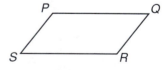

7. Do a two-column proof for Problem 6.

8. Draw a circle. Outline a proof that shows that chords equidistant from the center of the circle have the same length.

9. Draw a circle. Outline a proof that shows that tangents from a point outside the circle to the circle have equal lengths.

10. Draw a rectangle *ABCD*. Outline a proof that shows that angle *ADB* has the same measure as angle *BCA*.

11. *PQRS* is a parallelogram.

    $PQ = 2x + 6$

    $QR = 8$

    $RS = x + 8$

    Find the perimeter of *PQRS*.

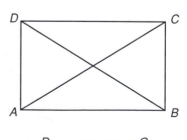

12. *ABCD* is a rectangle.

    $DB = 0.2x$

    $AC = x - 12$

    Find the length of $\overline{DB}$.

13. *DEFG* is a square.

    $m\angle D = (x + 60)°$

    $DE = x + 1$

    Find the perimeter of *DEFG*.

14. The volume of a right circular cylinder is 250π cm³. What is the radius of the cylinder if the height of the cylinder is 10 meters?

15. The area of a rectangle is $4\pi^2$ cm². What is the radius of a circle that has the same area?

16. Use similar triangles to find A and B.

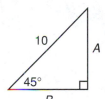

17. Use similar triangles to find C and D.

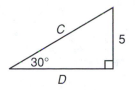

18. Find the area of this triangle. Dimensions are in inches.

19. Find x.

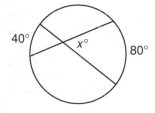

20. Find y.

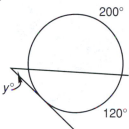

21. Find x.

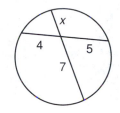

22. Find y.

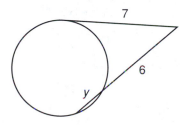

23. Find x and y.

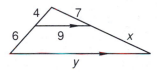

24. Find M and N.

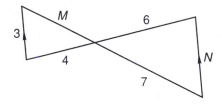

25. The radius of the circle is 6 cm. Find the length of arc ABC.

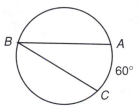

LESSON *129* *Stem and leaf plots • Measures of central tendency • The normal curve • Standard deviation*

129.A
stem and leaf plots

A stem and leaf plot is a method of arranging data so that it is easy to interpret. Suppose the weights of the 50 girls enrolled in a first-year gym class to the nearest pound are:

132	135	114	153	135	122	88	100	119	121
138	120	106	131	132	103	118	152	148	125
105	115	127	103	100	114	139	89	100	92
117	114	125	137	110	92	132	132	107	146
137	129	98	128	112	99	134	108	101	89

To arrange these numbers so that we can make sense of them, we note that the data could be grouped by tens. Some weights are in the 80s, some are in the 90s, etc. These groups provide our **stems**.

STEM

8
9
10
11
12
13
14
15

The first number in our list is 132, so we put a 2 in the 13 row. The next number is 138, so we put an 8 in the 13 row. The next two numbers are 105, which requires a 5 in the 10 row, and 117, which requires a 7 in the 11 row. We call the second part of each number a **leaf**.

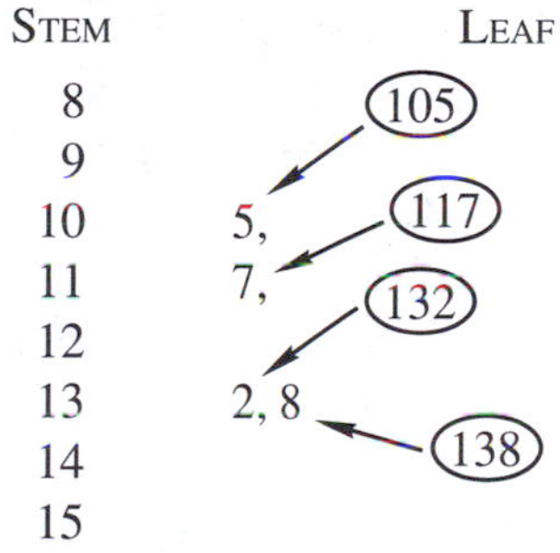

STEM	LEAF
8	(105)
9	
10	5, (117)
11	7, (132)
12	
13	2, 8 (138)
14	
15	

Note that we write a comma after each leaf. If we use the same procedure to record all the numbers, we get a completed stem and leaf plot.

STEM	LEAF
8	8, 9, 9
9	2, 9, 8, 2
10	5, 3, 6, 0, 8, 3, 0, 7, 1, 0
11	7, 4, 5, 4, 8, 4, 9, 0, 2
12	2, 0, 9, 7, 5, 8, 1, 5
13	2, 8, 7, 5, 9, 2, 4, 2, 1, 7, 5, 2
14	8, 6
15	3, 2

The leaves are not in order, as we note in the second row, the numbers recorded are 92, 98, 99, and 92. We could rearrange the leaves so that they are in order, but usually this is not necessary.

129.B
measures of central tendency

When we have data as in the problem above, we often want to know how much the numbers are spread out. We use the words *range*, *median*, *mode*, and *mean* to help us make these distinctions. If a group of numbers is arranged in order from the least to the greatest, we say that the **range** of the group is the difference between the first number and the last number. If there are an **odd number** of numbers, we say that the middle number is the **median** of the group. The range of the following group of numbers is 118 and the median is 85.

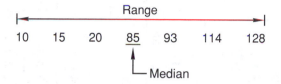

The word **median** is easy to remember because the grass strip in the middle of a divided highway is the median of the highway. If there is an **even number** of numbers in the group, the median is the number halfway between the two middle numbers. The median of the following group of eight numbers is 87.5.

$$10 \quad 15 \quad 20 \quad 85 \quad 90 \quad 93 \quad 114 \quad 128$$

$$\text{Median} = \frac{90 + 85}{2} = 87.5$$

The word *mode* is a French word that means "fashion." The number that appears the most often in a listing of numbers is the **mode** of the group of numbers. The average of a group of numbers is called the **mean** of the group of numbers. Thus, mean is another name for average.

example 129.1   Find the mean, median, mode, and range of the following group of numbers.

$$3, 8, 7, 4, 9, 10, 12, 9$$

solution   First we arrange the numbers in order.

$$3, 4, 7, 8, 9, 9, 10, 12$$

The mean of the group is the average, which is 7.75.

$$\text{Mean} = \frac{3 + 4 + 7 + 8 + 9 + 9 + 10 + 12}{8} = \frac{62}{8} = \textbf{7.75}$$

There are an even number of numbers, so the median is the number halfway between the middle two numbers.

$$\text{Median} = \frac{8 + 9}{2} = \textbf{8.5}$$

There are more 9s than any other, so **9** is the mode. The range is the difference between the least number and the greatest number, so the range is 9.

$$\text{Range} = 12 - 3 = \textbf{9}$$

129.C

the normal curve

The data from our stem and leaf plots of the girls' weights shows that three girls had weights of between 80 and 89 pounds; four had weights of between 90 and 99 pounds; ten had weights of between 100 and 109 pounds; etc. If we turn the plot sideways, we get

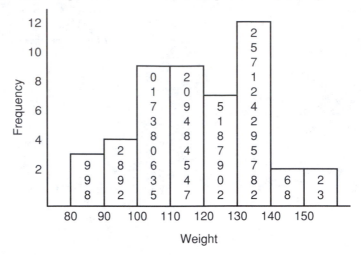

This graph is called a **frequency distribution graph** because it shows the number of girls in each 10-pound weight segment. If we were to find the weights of thousands of girls in gym classes all over the country, we could expect that the graph of their weights would look like the bell-shaped curve shown here.

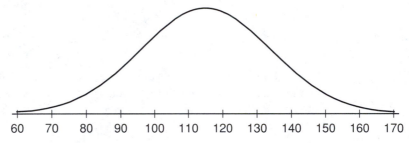

This graph is called a **normal curve** because this distribution is what you would normally expect to get if you weighed enough gym students. For a normal curve, the mean, median, and mode are all the same number. In a statistics course, much time is devoted to exploring frequency distributions that are best approximated by a normal curve.

129.D

standard deviation

We can compute a number called the **standard deviation** for a group of numbers that gives us a feel for how much the numbers are spread out. Let's consider the numbers 5, 7, 10, and 14, whose mean is 9.

$$\text{Mean} = \frac{5 + 7 + 10 + 14}{4} = 9$$

If we graph the numbers and measure the distance from each number to the mean, we get the following picture.

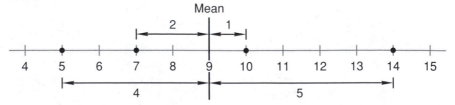

The distance from the graph of 5 to the mean is 4 units.
The distance from the graph of 7 to the mean is 2 units.
The distance from the graph of 10 to the mean is 1 unit.
The distance from the graph of 14 to the mean is 5 units.

The average of the square of the four distances is computed as follows.

$$\frac{1}{4}(4^2 + 2^2 + 1^2 + 5^2)$$

We call the square root of this average the **standard deviation** of the group of numbers.

$$\text{Standard deviation} = \sqrt{\frac{1}{4}(4^2 + 2^2 + 1^2 + 5^2)} \approx 3.39$$

Let's compare this with the standard deviation of 1, 2, 14, and 19. These numbers have the same mean but they are a little more spread out.

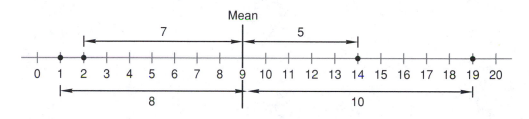

$$\text{Standard deviation} = \sqrt{\frac{1}{4}(8^2 + 7^2 + 5^2 + 10^2)} \approx 7.71$$

So both 5, 7, 10, 14 and 1, 2, 14, 19 have a mean of 9, but the first group has a standard deviation of 3.39 and the second group is spread out much more, with a standard deviation of 7.71. The standard deviation of a group of measurements that are normally distributed is useful because we can show that 68.2 percent of the measurements will be within one standard deviation of the mean, and 95.4 percent of the measurements will be within two standard deviations of the mean.

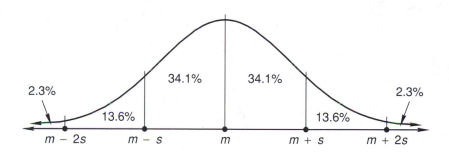

practice a. The following are the test grades of 50 students on a final examination. Make a stem and leaf plot of these data. What is the range of the scores?

70	72	74	74	48	63	52	98	90	99
68	98	92	86	70	88	77	74	62	86
87	77	68	80	96	89	60	49	75	75
81	74	67	47	84	73	79	61	65	83
97	63	77	63	74	88	89	83	87	66

b. Find the mean, median, mode, and range of these numbers.

$$4, 5, 2, 3, 10, 8, 15, 5$$

c. Find the standard deviation of these numbers.

$$5, 7, 9, 14, 8, 5$$

**problem set
129**

1. Find the area of this triangle. Dimensions are in centimeters.

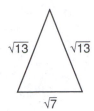

2. Find the perimeter of this triangle. Dimensions are in centimeters.

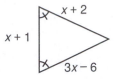

3. Find *x*.

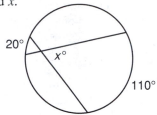

4. Find *x*.

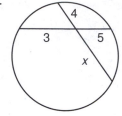

5. Find *z*.

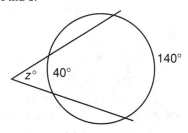

6. Find *x* and *y*.

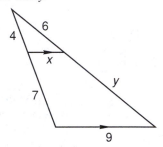

7. Find *P* and *Q*.

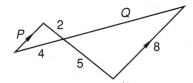

8. The diameter of the circle is 10 cm. Find the area of the 40° sector.

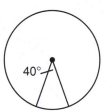

9. Use similar triangles to find *x* and *y*.

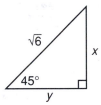

10. Use similar triangles to find *P* and *Q*.

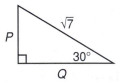

11. The area of the trapezoid equals the area of the circle. What is the radius of the circle?

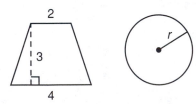

12. Use a compass and a straightedge to construct a triangle whose sides equal the lengths of these segments.

＿＿＿＿＿   ＿＿＿＿＿＿   ＿＿＿＿＿＿＿

13. Use a protractor to draw a 54° angle. Then use a compass and straightedge to bisect the angle.

14. Draw a segment that has the same length as this segment.

   _____

   Then use a compass and a straightedge to construct the perpendicular bisector of this segment.

15. Outline a proof that shows that the bisector of the vertex angle of an isosceles triangle is the perpendicular bisector of the base of the triangle.

16. The lateral surface area of a right circular cylinder whose radius is 3 centimeters is $6\pi^2$ square centimeters. What is the height of the cylinder?

17. The area of a rectangle is $25\pi^2$ cm². What is the diameter of a circle that has the same area as the area of the rectangle?

18. *PQRS* is a parallelogram.
    $PQ = 2x + 4$
    $QR = 7$
    $RS = 6x - 2$
    Find the perimeter of *PQRS*.

    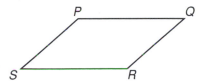

19. *ABCD* is a square.
    $m\angle D = (2x + 40)°$
    $AB = x - 15$
    Find the perimeter of *ABCD*.

    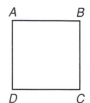

20. Frank deposited $1400 at 9 percent interest compounded continuously. How much money did he have at the end of 5 years?

21. The number of bugs increased exponentially. At first, there were 50 bugs. Three weeks later, there were 400 bugs. How many bugs would there be after a total of 12 weeks?

22. Solve: $75 = 10^{x+4}$

23. Use the ⬚In⬚ key to write each number as a power of *e*. Then use the rules of exponents to simplify. Then use the ⬚inv⬚ ⬚In⬚ keys to get the final answer.

$$\frac{0.00287 \times 10^{-8}}{620,000}$$

24. Graph: $\{x \in R \mid x + 3 \geq 2\}$

25. Complete the square as an aid in graphing: $y = x^2 - 2x - 2$

Glossary

abscissa The x coordinate of a point in a Cartesian (rectangular) coordinate system.

absolute value In reference to a number, the positive number that describes the distance on a number line of the graph of the number from the origin. The absolute value of zero is zero.

acute angle An angle whose degree measure is between $0°$ and $90°$.

acute triangle A triangle in which all the angles are acute.

additive inverse For any nonzero real number, the opposite of the number. The sum of any number and its additive inverse is zero.

additive property of inequality A property of real numbers such that, for any real numbers a, b, and c, if $a > b$, then $a + c > b + c$ and also $c + a > c + b$.

adjacent angles Two angles which have the same vertex and share a common side between them.

algebraic expression Any algebraic formula obtained by combining constants and/or variables using the arithmetic operators $+$, $-$, $\times$, or $\div$.

algebraic proof Use of definitions, axioms, and deductive reasoning to prove algebraic assertions.

altitude In reference to a triangle, the perpendicular distance from the base of the triangle or extension of the base to the opposite vertex. Any one of the three sides can be designated as the base.

angle The basic figure formed by two rays that have a common endpoint; also the measure of the rotation of a ray about its endpoint from an initial position to a final position.

associative property A property of real numbers that notes that, for any real numbers a, b, and c, $(a + b) + c = a + (b + c)$ and $(a \cdot b) \cdot c = a \cdot (b \cdot c)$.

axiom A statement that is accepted as true without proof, i.e., an assertion that is not proved. Also called *postulate*.

binomial A polynomial of two terms.

Cartesian coordinate system   A standard method of locating points in the plane that uses pairs of numbers denoting distances along two fixed intersecting number lines, called the *axes*. The axes are perpendicular to each other and intersect at the origin of both axes. The system is named for the French mathematician René Descartes. Also called a rectangular coordinate system.

centimeter   Metric unit of measurement: 1 centimeter = 10 millimeters; 100 centimeters = 1 meter.

chord   In reference to a circle, a straight-line segment whose endpoints are on the circle.

circle   A planar geometric figure in which every point on the figure is the same distance from a point called the *center* of the circle.

circumference   The distance around a circle. Also called the *perimeter* of the circle.

coefficient   Any factor or any product of factors in a product.

combined variation   A relation between one variable and a combination of other variables. For example, $F = k\dfrac{m_1 m_2}{r^2}$.

commutative property   A property of real numbers that notes that, for any real numbers a and b, $a + b = b + a$ and $a \cdot b = b \cdot a$.

complementary angles   Two angles whose sum is 90°.

complex number   A number of the form $a + bi$, where a and b are real numbers and i is the *imaginary unit* satisfying the equation $i^2 = -1$. The letter a represents the *real part* of the complex number and the letter b represents the *imaginary part* of the complex number.

concave polygon   A polygon that has at least one interior angle greater than 180°.

conditional equation   An equation whose truth or falsity depends on the numbers used to replace the variables in the equation.

congruent polygons   Two polygons in which the measures of the angles in one polygon are equal to the measures of the angles in the other polygon and the sides opposite equal angles have equal lengths.

congruent triangles   Two triangles in which the corresponding sides have equal lengths and the corresponding angles have equal measures.

conjunction   A statement of two conditions which must both be true in order for the statement to be true.

consecutive angles   Two angles in a polygon that share a common side.

consecutive sides   Two sides in a polygon that share a common vertex. Also called *adjacent sides*.

consecutive vertices   The endpoints of one side of a polygon.

constant   A quantity whose value does not change.

convex polygon   A polygon in which all interior angles have a measure less than or equal to 180°.

coordinate   A number that is associated with a point on a graph.

curve   The path traced by a moving point.

decagon   A polygon with 10 sides.

decimal fraction   A series of fractions; i.e., it is a number of tenths plus a number of hundredths plus a number of thousandths, etc. For example, 0.625 is a way of writing $0 + \dfrac{6}{10} + \dfrac{2}{100} + \dfrac{5}{1000}$, i.e., $(6 \times 10^{-1}) + (2 \times 10^{-2}) + (5 \times 10^{-3})$.

decimal number   A number designated by a linear arrangement of one or more of the 10 digits and that uses a decimal point to define the place value of the digits.

decimal system   The system of numeration that uses decimal numbers.

deductive reasoning   The process of reasoning logically from clearly stated premises to a conclusion.

degree   A unit of measure for angles. A right angle is a 90° angle and a straight angle is a 180° angle.

degree of a polynomial   The degree of the highest-degree term in the polynomial, calculated as follows: The degree of a term in a polynomial is the sum of the exponents in the term. For example, the terms x^5, y^2x^3, and xy^2mp are all fifth-degree terms.

denominate number   A combination of a number and a word that designates the units. For example, 5 feet represents a denominate number with units in feet.

denominator   The number under the fraction bar in a fraction, i.e., the divisor in a fraction.

dependent variable   When considering a function, the variable whose value depends on the value assigned to another variable, called the *independent variable*. For example, in the function $y = 2x + 3$, y is regarded as the dependent variable.

diagonal of a polygon   A line segment that connects any two nonconsecutive vertices.

diameter of a circle   The length of a chord of a circle that passes through the center of the circle.

direct variation   A relationship between two variables such that their ratio is constant. For example, the equation $y = kx$ defines a direct variation between x and y where k is the *constant of proportionality*.

disjunction   A statement of two conditions of which only one condition must be true in order for the statement to be true.

distributive property   A property of real numbers that notes that, for any real numbers a, b, and c, $a \cdot (b + c) = a \cdot b + a \cdot c$ and $(b + c) \cdot a = b \cdot a + c \cdot a$.

dividend   The number a in the expression $a \div b$.

division   The inverse operation of multiplication. If one number is divided by another number, the result is called the *quotient*.

divisor   The number b in the expression $a \div b$.

dodecagon   A polygon with 12 sides.

domain   The set of numbers which are permissible replacement values for the independent variable of a function.

element of a set   Any one of the individual objects belonging to a set.

empty set   The set that has no members, denoted by the symbol Ø. Also called the *null set*.

equality   The property of two things being equal and symbolized by the sign =.

equiangular polygon   A polygon in which all angles have the same measure. For example, in an *equiangular triangle* each angle has a measure of 60°.

equilateral polygon   A polygon that has all its sides of equal length.

equilateral triangle   A triangle that has three sides whose lengths are equal.

equivalent equations   Equations that have the same solution set.

even integer   Any member of the set { . . . , $-4, -2, 0, 2, 4, . . .$ }.

exponent   The number n in an expression of the form x^n, which indicates that the base x is to be used as a factor n times.

exponential function   A function of the form $y = kb^x$, where k and b are constants and $b \neq 1$ or $b \neq 0$.

exterior angle   An angle at a vertex of a polygon formed outside a polygon between one side and another side that has been extended. Also, two lines cut by a transversal form four exterior angles, those lying outside the two lines.

factor   One of two or more expressions that are multiplied to form a product. For example, 2 and 3 are factors of 6; $x - 2$ and $x + 2$ are factors of $x^2 - 4$.

function   A mapping between two sets that associates with each element of the first set a unique (one and only one) element of the second set. The first set is called the *domain* of the function. For each element x of the domain, the corresponding element y of the second set is called the *image* of x under the function. The set of all images of the elements of the domain is called the *range* of the function.

geometric solid   A three-dimensional geometric figure. For example, a sphere, a cube, and a prism are all geometric solids.

greatest common factor   Of two or more terms, the product of all prime algebraic factors common to every term, each to the highest power that it occurs in all of the terms.

heptagon   A polygon with seven sides.

hexagon   A polygon with six sides.

hypotenuse   The side opposite the right angle in a right triangle.

image   In a function, the element of the range that is paired with a particular element of the domain.

inconsistent system of equations Two or more simultaneous equations which have no common solution. The graphs of inconsistent equations are parallel lines.

independent variable When considering a function, the variable whose value can be chosen. For example, in the function $y = 2x + 3$, x is regarded as the independent variable.

integer Any member of the set { . . . , $-4, -3, -2, -1, 0, 1, 2, 3, 4, \ldots$ }.

intercept In a rectangular coordinate system the x intercept is the point at which the graph crosses the x axis and the y intercept is the point at which the graph crosses the y axis.

inverse operation An operation which "undoes" another operation. For example, addition and subtraction are inverse operations. Also, multiplication and division are inverse operations.

inverse variation A relationship between two variables such that their product is constant. For example, the equation $xy = k$ or $y = \dfrac{k}{x}$ defines an inverse variation between x and y where k is the *constant of proportionality.*

irrational number Any number that cannot be written as a quotient of integers. For example, the numbers π, e, $\sqrt{2}$, and $\sqrt[4]{17}$ are irrational numbers.

isosceles triangle A triangle that has at least two sides of equal length.

lead coefficient Of a polynomial, the coefficient of the term with the greatest exponent. For example, in the quadratic equation $5x^2 - 3x + 2$ the lead coefficient is 5.

least common multiple The smallest whole number that can be divided evenly by each of a group of specified whole numbers. For example 6, 12, and 27 have a least common multiple of 108.

like terms Terms whose literal components represent the same number regardless of the numbers used to replace the variables.

line segment A part of a line that consists of two endpoints and all points between the endpoints.

linear equation A first-degree polynomial equation in one or more variables.

median (statistics) The middle number of a group of numbers that are arranged in order from the least to the greatest. If there is an odd number of numbers in the group, the median is the middle number. If there is an even number of numbers in the group, the median is the average of the two middle numbers.

meter Metric unit of measurement; 100 centimeters = 1 meter.

millimeter Metric unit of measurement; 10 millimeters = 1 centimeter.

monomial A polynomial of one term.

multiplicative inverse For any nonzero real number, the reciprocal of the number. The product of any nonzero number and its multiplicative inverse is 1.

natural numbers The set of numbers that we use to count objects or things; also called the *positive integers*, i.e., any member of the set {1, 2, 3, . . .}.

nonagon A polygon with nine sides.

numerator   The number above the fraction bar in a fraction.

numerical coefficient   A coefficient that is a number.

numerical expression   A meaningful arrangement of digits and symbols that designate specific operations. Every numerical expression represents a particular number, and we say that this number is the *value* of the expression.

obtuse angle   An angle whose measure is between 90° and 180°.

obtuse triangle   A triangle which contains an obtuse angle.

parallel lines   Two lines in the same plane that do not intersect.

parallelogram   A quadrilateral that has two pairs of parallel sides.

pentagon   A polygon with five sides.

perimeter   The distance around the outside of a closed, planar geometric figure.

perpendicular bisector   A line that is perpendicular to a given line segment at the midpoint of the segment.

perpendicular lines   Two lines which intersect at right angles.

pi (π)   The ratio of the length of the circumference of a circle to the length of the diameter of that circle; $\pi \approx 3.14$.

polar coordinates   A method of locating a point in the plane in which the position of the point is determined by the length of the line segment from the origin to the point and the angle that the line segment makes with the positive x axis. Positive angles are measured counterclockwise from the positive x axis and negative angles are measured clockwise from the positive x axis.

polygon   Any simple, closed, flat geometric figure whose sides are straight lines.

polynomial   An algebraic expression with one or more variables having only terms with real number coefficients and whole number powers of the variables.

postulate   A statement that is accepted as true without proof, i.e., an assertion that is not proved. Also called *axiom*.

power rule for exponents   A rule for exponents: If $m, n,$ and x are real numbers and $x \neq 0$, then $(x^m)^n = x^{mn}$.

product of square roots rule   A rule for evaluating products of radical expressions: If m and n are nonnegative real numbers, then $\sqrt{m}\sqrt{n} = \sqrt{mn}$ and $\sqrt{mn} = \sqrt{m}\sqrt{n}$.

product rule for exponents   A rule for exponents: If $m, n,$ and x are real numbers and $x \neq 0$, then $x^m \cdot x^n = x^{m+n}$.

proportion   An equation or other statement which indicates that two ratios are equal.

Pythagorean theorem   In any right triangle, the square of the length of the hypotenuse equals the sum of the squares of the lengths of the other two sides.

quadratic equation   A polynomial equation in which the highest power of the variable is 2.

quadrilateral A polygon with four sides.

quotient The answer obtained when one number is divided by another number.

radical An expression for taking the root of a quantity indicated by the symbol $\sqrt{}$ called the radical sign. The number under the radical sign is called the radicand, and the little number that designates the root is called the index. If the index is not written, it is understood to be 2. For example, the radical $\sqrt{5}$ has index 2 and denotes the square root of 5. The radical $\sqrt[3]{5}$ has index 3 and denotes the cube root of 5. The radical $\sqrt[n]{5}$ has index n and denotes the nth root of 5.

radius of a circle The distance from the center of the circle to any point on the circle.

range The set of all images of the elements of the domain of a function.

rational number Any number that can be written as a quotient of integers (division by zero excluded).

ray An extension of a line segment in one direction. A ray is sometimes called a *half line*.

real numbers The set of numbers that includes all members of the set of rational numbers and all members of the set of irrational numbers.

reciprocal For any nonzero real number, the number in inverted form. The product of any nonzero number and its reciprocal is 1. The reciprocal of a number is often called the *multiplicative inverse* of the number. For example, the reciprocal of 3 is $\frac{1}{3}$ and the reciprocal of $\frac{3}{4}$ is $\frac{4}{3}$.

rectangle A parallelogram with four right angles.

reflex angle An angle whose measure is greater than a straight angle but less than two straight angles.

regular polygon A polygon whose interior angles have equal measures and whose sides have equal lengths.

relation A pairing that matches each element of the domain with one or more images in the range.

rhombus An equilateral parallelogram.

right angle An angle whose measure is 90°.

right geometric solid A geometric solid whose sides are perpendicular to the base. Also called *right solid*.

right triangle A triangle that has one right angle.

scalene triangle A triangle that has no sides of equal length.

scientific notation A method of writing a number as a product of a decimal number and a power of 10.

secant to a circle A line that intersects a circle at two points.

sector of a circle The area of the circle bounded by two radii and an arc of the circle.

set   A collection of objects. The individual objects that make up a set are called its *elements*.

similar triangles   Two triangles that have the same angles.

slope of a line   The ratio of the change in the y coordinate to the change in the x coordinate as we move from any point on the line to any other point on the line.

solution   Replacement values of the variable that make an equation a true equation. Also called *roots* of the equation. For example, in the equation $x + 5 = 9$ we say that the number 4 is a solution or root of the equation, and we also say that the number 4 satisfies the equation.

sphere   The figure defined by the set of all points in three-dimensional space that are equidistant from the point called the center of the sphere.

square   A rhombus with four right angles.

straight angle   An angle whose measure is 180°.

substitution axiom   If two expressions a and b are of equal value, $a = b$, then a may replace b or b may replace a in another expression without changing the value of the expression. Also a may replace b or b may replace a in any statement without changing the truth or falsity of the statement. Also a may replace b or b may replace a in any equation or inequality without changing the solution set of the equation or inequality.

supplementary angles   Two angles whose sum is 180°.

syllogism   A three-step deductive reasoning process consisting of a major premise, a minor premise, and a conclusion.

tangent to a circle   A line that intersects (touches) a circle at only one point.

theorem   An assertion that can be proved.

transitive axiom   For any real numbers a, b, and c:
$$\text{If } a > b \text{ and } b > c, \text{ then } a > c.$$
$$\text{If } a < b \text{ and } b < c, \text{ then } a < c.$$
$$\text{If } a = b \text{ and } b = c, \text{ then } a = c.$$

transversal   A line that cuts or intersects one or more other lines in the same plane.

trapezoid   A quadrilateral that has exactly two parallel sides.

triangle   A polygon with three sides.

trichotomy axiom   For any two real numbers a and b, exactly one of the following is true: $a < b$, $a = b$, or $a > b$.

trinomial   A polynomial of three terms.

undecagon   A polygon with 11 sides.

unit conversion   The process of changing a denominate number to an equivalent denominate number that has different units.

unit multiplier   A fraction that has units and has a value of 1. Unit multipliers are used to change the units of a number.

variable A letter used to represent a number.

vector A quantity that has both a magnitude and a direction.

vertex In reference to an angle, the point where the two rays of the angle intersect.

vertical angles Two nonadjacent angles formed by two intersecting lines. Vertical angles are equal angles.

whole number Any member of the set $\{0, 1, 2, 3, \ldots\}$.

zero factor theorem If p and q are any real numbers and if $p \cdot q = 0$, then either $p = 0$ or $q = 0$, or both p and q equal 0.

Answers

problem set A

1. 115   3. $x = 91$; $y = 89$; $p = 91$ 5. $140°$ 7. 0   9. 0   11. -10

13. 5   15. 4   17. -16 19. -66 21. 0   23. 35   25. 11   27. -10

29. 192

problem set B

1. 13.76 m^2 3. 136.96 cm^2 5. 8.72 m^2 7. 18.28 m^2; 146.24 m^3 9. 62.8 cm^2

11. $x = 35$; $y = 110$; $z = 110$ 13. 20   15. $120°$ 17. -4 19. 23

21. 26   23. -16 25. -11 27. 6   29. 0

practice   a. $m\angle C = 35°$; $m\angle B = 110°$ c. $A = 50$; $B = 65$; $C = 50$

problem set 1

1. $x = 45$; $y = 90$ 3. $A = 70$; $B = 110$; $C = 55$ 5. 17.49 cm^2 7. 60.56 ft

9. $x = 30$; $y = 30$; $p = 150$ 11. $73°$ 13. $16r^2$ 15. -100 17. -35

19. 87   21. 46   23. -87 25. -69 27. $-\dfrac{6}{13}$ 29. -15

practice   a. $-\dfrac{1}{16}$ c. $x^{-1}y^4$

problem set 2

1. $\dfrac{9}{2}$ 3. 178.99 m^3 5. $A = 120$; $B = 30$; $C = 40$ 7. 10 cm 9. $x^6 y^{-12}$

11. $x^{-1}y^7$ 13. $\dfrac{m^5 y^3}{x^2}$ 15. mn^4 17. $\dfrac{c^3}{b^6}$ 19. $\dfrac{k^{-2}}{L^{-3}}$ 21. $\dfrac{z^{-12}x^{-3}}{y^{-1}}$

23. $-\dfrac{1}{9}$ 25. -5 27. -38 29. -27

practice   a. 21   c. $-\dfrac{3x}{a^3 m} + \dfrac{a^3 m}{x}$

problem set 3

1. $x = \dfrac{15}{2}$; $y = 33$; $z = 9$ 3. $A = 40$; $B = 100$ 5. 54 cm 7. -23

9. 895   11. 2677   13. $-\dfrac{2p^2 x^4}{m^5} + 5p^4 m^5 x^4$ 15. $x^5 y^4 + 4xy$ 17. $x^8 y^{-8} p^6$

19. $x^{-7} m^{-7} p^2$ 21. x^3 23. $x^{-2} y^{12}$ 25. -1 27. -1 29. 100

practice   a. $\dfrac{4a^{-3} b^2}{x^2} - \dfrac{2a^{-6}}{cb^{-4}}$

problem set 4

1. $\pi r^2 \text{ cm}^2$; $4\pi r^2 \text{ cm}^2$ 3. $x = y = 40$; $P = 140$; $Q = R = 20$ 5. 2.09 cm

7. $\dfrac{1}{20}$ 9. -3 11. $-\dfrac{1145}{168}$ 13. $y^{-6} p^{-3} - 3y^{-3} p$ 15. $\dfrac{y^2 x^4}{4}$

17. $\dfrac{x^{-8}y^{-3}}{4}$ **19.** $6x^2y^2$ **21.** $7ay^2x^{-1} + 2xya^{-1}$ **23.** -18 **25.** -89

27. 48   **29.** 4

practice   **a.** -25

problem set 5

1. -14 **3.** -6 **5.** 40,000   **7.** $y = 52$; $x = 76$

9. 8 in.; 200.96 in.^{2}; 1004.8 in.^{3} **11.** $-\dfrac{13}{6}$ **13.** -6 **15.** 6   **17.** $2 - \dfrac{6a^2}{c}$

19. $4x^{-1}$ **21.** $\dfrac{m^2}{8p^7}$ **23.** $-8a^2x^3 + 2a^2x$ **25.** $\dfrac{7}{16}$ **27.** 10   **29.** -35

practice   **a.** 28,507

problem set 6

1. 30,000   **3.** -3 **5.** 5, 7, 9   **7.** $\dfrac{\sqrt{46}}{2}$ cm   **9.** $x = \dfrac{55}{6}$; $A = 25$; $B = \dfrac{130}{3}$

11. 290   **13.** 5   **15.** 28   **17.** $-kx^{-1} + 2k$ **19.** $a^{-5}b^{-3}c^{-1}$ **21.** $3xy - 5x$

23. 28   **25.** $-\dfrac{9}{8000}$ **27.** 6   **29.** $14\dfrac{8}{9}$

practice   **a.** 310

Before, 100%

After

c. $x = 11$; $A = B = 34$

problem set 7

1. 130

Before, 100%

After

3. 2300

Before, 100%

After

5. 20

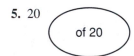

Before, 100%

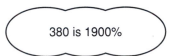

After

7. 2, 4, 6   **9.** 15   **11.** $m\angle a = 15°$; $m\angle b = 45°$; $m\angle c = 90°$; $m\angle d = 30°$

13. $-\dfrac{1}{18}$ **15.** $-\dfrac{47}{100}$ **17.** $2 - 6x^2yp^{-1}$ **19.** $2x^2y^{-10}$ **21.** $2x^3yp^{-1}$

23. $-\dfrac{5}{12}$ **25.** 16   **27.** -50 **29.** 38

practice

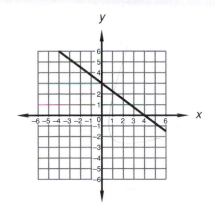

**problem set
8**

1. 2300   **3.** −13   **5.** 7, 9, 11

7. 190.4

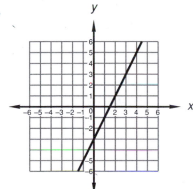

of 56

Before, 100%

190.4 is 340%

After

9.

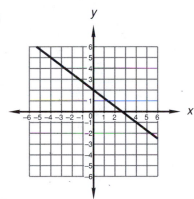

11. 5 m, 5 m, 5 m

13. $x = 20$; $y = 105$

15. $\dfrac{87}{35}$ **17.** $\dfrac{151}{56}$

19. $3x - 5xyp^2$ **21.** $20x^{-12}y^{-8}$

23. $4xy^{-3} - 7xy$

25. $\dfrac{1}{27}$ **27.** 24   **29.** $-1\dfrac{1}{2}$

practice   **a.** 500

**problem set
9**

1. 240   **3.** \$97,500   **5.** 250,000   **7.** $m = 50$; $p = 70$

9. $x = 10$; $y = 59$; $z = 120$

11.

13. 5   **15.** $\dfrac{6}{5}$ **17.** 0   **19.** $\dfrac{a^2}{8x^8y^2}$

21. $7x - 2xy$ **23.** $\dfrac{7}{16}$ **25.** -90

27. 14   **29.** 8

practice   **a.** $\sqrt{56}$

problem set 10

1. 1750   **3.** 5310   **5.** 4000 minas   **7.** 9π cm^2 **9.** $x = 5; P = Q = 105$

11.

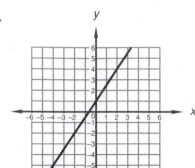

13. $\sqrt{33}$ **15.** 5   **17.** $-\dfrac{12}{5}$

19. $4 - 6xy^2p^{-2}$ **21.** $x^{-1}p^{-2}$ **23.** $9x^2yz^{-1}$

25. $-\dfrac{1}{3}$ **27.** $\dfrac{3}{32}$ **29.** 15

practice   **a.** $\dfrac{mz^4 + 3akz - 3y}{3bz^4}$ **c.** $x = 20; y = 80$

problem set 11

1. 780   **3.** \$3054   **5.** 8800   **7.** $z = 32; y = 116; x = 64; p = 64$ **9.** $2\pi^2$ cm

11. $\dfrac{kx^2 + abcx - m}{ax^3}$ **13.** $\dfrac{4m^2c^2x - 12cp^2 - 5p}{4c^2px}$ **15.** $\sqrt{89}$ **17.** $\dfrac{3}{28}$

19. $\dfrac{3}{7}$ **21.** $6 - 4y^2$ **23.** $8x^{10}y^3m^{-1}$ **25.** $-5ka$ **27.** $-\dfrac{1}{2}$ **29.** $5\dfrac{1}{4}$

practice   **a.** $y = -5$

problem set 12

1. 1600   **3.** 1200   **5.** -5 **7.** 30   **9.** $12\sqrt{6}$ cm   **11.** $\dfrac{mcx^2b + x^3b + c^2}{cx^2b}$

13. $\dfrac{b + a}{b}$ **15.** $3\sqrt{13}$ **17.** (a) $y = \dfrac{1}{3}x + 2$ (b) $y = -2$ **19.** 5.5

21. $-2a^{-3}y^3x$ **23.** $2x^{-4}y^8$ **25.** $2ax^{-1}$ **27.** $\dfrac{11}{32}$ **29.** 102

practice   **a.** $(5, -2)$

problem set 13

1. 450   **3.** 16, 18, 20, 22   **5.** 10 units^2 **7.** $AB = 3; DC = \sqrt{97}$ **9.** $(5, -7)$

11. $(13, 7)$ **13.** $\dfrac{4a^2 + 5k^2}{4k}$ **15.** $2\sqrt{13}$ **17.** (a) $x = 5$ (b) $y = -\dfrac{2}{3}x - 2$

19. -0.9 **21.** $3 - \dfrac{2x^2a^2}{3}$ **23.** $4x^{-3}a^{-6}$ **25.** 0   **27.** $\dfrac{9}{8}$ **29.** 11

practice   **a.** $y = -x + 2$

problem set 14

1. 2530   **3.** -10 **5.** $(5, -2)$ **7.** $(10, 18)$ **9.** $\dfrac{xcy^2 + x^2cy - 3x}{cy^2}$

11. $\dfrac{4x + c - cx^2y}{x}$ **13.** 8   **15.** (a) $y = \dfrac{2}{3}x + 4$ (b) $y = -4$

17. $y = -\dfrac{3}{4}x + 3$ **19.** $A = B = 15; C = 30; D = 30$ **21.** 110

23. $\dfrac{x}{y} + \dfrac{x^2}{3y^2}$ **25.** $1 - 4m^2$ **27.** $-\dfrac{11}{36}$ **29.** -5

practice $(2, 3)$

problem set 15

1. 7000   **3.** 4   **5.** $\left(\dfrac{8}{3}, 3\right)$ **7.** (4, 4)   **9.** $\dfrac{3a^2y^3 + 3xy^2 - mxa^2}{3a^2y^2}$

11. $\dfrac{cx + c^2 + ac^2x}{x}$ **13.** $\sqrt{85}$ **15.** (a) $y = \dfrac{9}{2}$ (b) $y = -\dfrac{2}{3}x - 1$

17. $y = -\dfrac{1}{7}x + \dfrac{33}{7}$ **19.** $A = B = 20; C = 40; D = 40$ **21.** $3 + 9xy^3p^2$

23. 6   **25.** y^{-5} **27.** $-\dfrac{1}{4}$ **29.** 11

practice $4x^2 + 9x + 34 + \dfrac{105}{x - 3}$

problem set 16

1. 420   **3.** 1050   **5.** (2, 3)   **7.** (1, −1)   **9.** $4x^3 - 2x^2 - 6x + 9$

11. $x^2 + 4x + 16 + \dfrac{56}{x - 4}$ **13.** $\dfrac{16c + 4c^3x - 3}{4c^2x}$ **15.** $\sqrt{53}$

17. (a) $y = -4$ (b) $y = -\dfrac{1}{2}x + 3$ **19.** $y = \dfrac{3}{5}x + \dfrac{17}{5}$

21. $A = 30; B = 120; C = 60; D = 60$ **23.** 11   **25.** $2 - \dfrac{y^2z}{3x^2}$ **27.** $-6x^4y^5$

29. −22

practice   **a.** $N_N = 58; N_O = 2$ **c.** $x = \dfrac{40}{9}; y = \dfrac{38}{9}$

problem set 17

1. 1000   **3.** 1440   **5.** $N_N = 210; N_D = -60$ **7.** $4x^3 + 10x^2 + 10x + 6$

9. $T_M = 5; T_W = 0$ **11.** $T_M = 1; T_R = 4$ **13.** $\dfrac{-2nxp^2 - cxynp^2 + 7x^2y^2}{ynp^2}$

15. $2\sqrt{13}$ **17.** (a) $y = -2x - 2$ (b) $x = 4$ **19.** $y = -\dfrac{2}{7}x + \dfrac{29}{7}$

21. $\dfrac{49}{5}$ cm   **23.** 11.09   **25.** $\dfrac{1}{y^2m} - \dfrac{3x^2m^2}{y}$ **27.** $-2xy^3 + 7y$ **29.** $-10\dfrac{7}{8}$

practice   **a.** 75   **c.** $a = \dfrac{45}{4}; b = \dfrac{27}{2}$

problem set 18

1. 40   **3.** 200,000   **5.** $N_D = 150; N_Q = 50$ **7.** $6x^3 + 8x^2 - 28x - 40$

9. $R_F = 96; R_S = 80$ **11.** $T_B = 8; T_G = 5$ **13.** $\dfrac{-3x^2y^2 - cxy^3 + 7c}{xy^3}$

15. $7\sqrt{2}$ **17.** (a) $y = -3$ (b) $y = -3x$ **19.** $y = \dfrac{5}{3}x - \dfrac{26}{3}$

21. $x = 6; P = 20$ **23.** $\dfrac{9}{20}$ **25.** $\dfrac{6x}{ypz^3} - \dfrac{27x^2}{z^2}$ **27.** 0   **29.** −10

practice   **a.** $N_N = 30; N_D = 50$ **c.** $x = \dfrac{25}{3}; y = \dfrac{20}{3}$

problem set 19

1. $N_N = 20; N_D = 40$ **3.** 198   **5.** −2   **7.** $x^5 - 2x^4 - 4x^3 + 8x^2 + 4x - 8$

9. $T_H = T_S = 2$ **11.** $T_R = 4; T_M = 1$ **13.** $\dfrac{ay - 2b - 2cx^2y}{2x^2y}$ **15.** $\sqrt{106}$

17. (a) $y = 2$ (b) $y = 2x$ **19.** $y = \dfrac{2}{7}x - \dfrac{29}{7}$

21. $A = 50$; $B = 130$; $C = D = 25$; 3.93 cm^2 **23.** -100 **25.** $-3x + 9x^{-1}y^3p$

27. $3x^2ay^{-1}$ **29.** 1

practice   **a.** $8\sqrt{10} - 6\sqrt{35}$ **c.** $y = \dfrac{1}{3}x + 2$

problem set 20   **1.** 560 **3.** $N_W = 10$; $N_E = 13$ **5.** $-4, -2, 0, 2$ **7.** $-2x^2 - 2x - 3 - \dfrac{1}{x-1}$

9. $72 - 50\sqrt{3}$ **11.** $50 - 75\sqrt{2}$ **13.** $\dfrac{2a^5 - 2a^5x^2 - 3x^3}{2x^2a^4}$

15. $z = \dfrac{24}{5}$; $A = \dfrac{28}{5}$ **17.** (a) $y = 1$ (b) $y = -x$ **19.** $y = -\dfrac{1}{12}x + \dfrac{31}{6}$

21. $\dfrac{25}{4}$ **23.** 0 **25.** $-\dfrac{y^8}{2x^2}$ **27.** $-\dfrac{129}{512}$ **29.** 3

practice   **a.** $N = 48$; $D = 60$

problem set 21   **1.** $N = 36$; $D = 60$ **3.** 2250 **5.** 25 **7.** $(6, 5)$ **9.** $T_K = 8$; $T_N = 16$

11. $144 - 24\sqrt{3}$ **13.** $\dfrac{m^2 - 3ax - ma^2x}{a^2x^2}$ **15.** 7

17. (a) $y = -2$ (b) $y = -2x$ **19.** $y = -\dfrac{3}{7}x + \dfrac{20}{7}$

21. $A = 50$; $B = C = 40$; $D = y = 50$ **23.** 24 **25.** $-10x^{-4} - 5x^{-3}$

27. $5xa$ **29.** 0

practice   **a.** $R_ET_E = R_DT_D$, $R_E = 14$, $R_D = 21$, $T_D = T_E - 3$; 126 miles

problem set 22   **1.** $R_ET_E = R_CT_C$, $R_E = 15$, $R_C = 30$, $T_C = T_E - 3$; 90 miles

3. $N = 560$; $D = 400$ **5.** 8 **7.** $(4, 4)$ **9.** $x = 18$; $y = \dfrac{33}{2}$

11. $48\sqrt{3} - 70$ **13.** $\dfrac{2x + 1}{x}$ **15.** 2×10^{-2}

17. (a) $x = -4$ (b) $y = -\dfrac{3}{2}x + 3$ **19.** $3\sqrt{10}$

21. $A = 40$; $B = C = 50$; $k = 40$; $m = 50$ **23.** 3.5 **25.** $1 - 3x^2y^{-7}p^{-2}$

27. $-6y$ **29.** 16

practice $(-1, -1)$

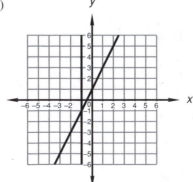

problem set 23

1. 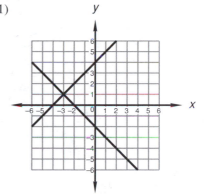 $R_M T_M = R_J T_J, R_M = 600, R_J = 800, T_M = T_J + 4$; 12 minutes

3. $\dfrac{110}{120}$ 5. $N_N = 50, N_Q = 10$ 7. $(10, 10)$ 9. $144\sqrt{3}$ 11. $50 - 30\sqrt{3}$

13. $\dfrac{5x^2 p - 4p^2 m + c}{p^2 m}$ 15. $(-3, 1)$

17. $y = \dfrac{1}{2}x$ 19. $y = -\dfrac{3}{8}x + \dfrac{11}{2}$

21. $x = \dfrac{28}{3}; \; y = \dfrac{18}{7}$ 23. 3

25. $1 - 3x^{-1}$ 27. $6mx^{-1}$ 29. 6

practice   a. $\dfrac{5}{3}$ c. $x = \dfrac{9}{2}; \; y = 10$

problem set 24

1. 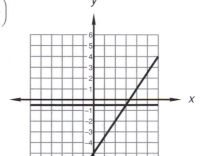 $R_F T_F = R_S T_S, T_S = 8, T_F = 6, R_F = 60$; 45 mph   3. $\dfrac{50}{70}$

5. 20 7. $(8, 7)$ 9. $720\sqrt{2}$ 11. $30 - 36\sqrt{6}$ 13. $\dfrac{4m^4 y^2 p + 6}{m^2 y}$

15. 1×10^{-19} 17. (a) $y = \dfrac{5}{6}x + 2$ (b) $y = -4$ 19. $2\sqrt{17}$ in.^{2}

21. $y = \dfrac{2}{9}x - \dfrac{37}{9}$ 23. $\dfrac{7}{16}$ 25. -250 27. $1 + 5x^3 y$

29. $-2p^2 xy^{-1} - 5p^2 x^7 y^{-1}$

practice   a. $2m^2 xy(4y^4 + 3mx - 1)$ c. $x = \dfrac{15}{4}; \; y = \dfrac{49}{4}$

problem set 25

1. 95 days   3. $N_B = 50; N_G = 10$ 5. $N_Q = 100; N_H = 100$

7. $x^2 + 5x + 25 + \dfrac{123}{x - 5}$ 9. $x^2 ym^2(y^2 m^3 + 12xm^2 - 3y)$

11. $x^2 yz(xyz^2 + z - 3x)$ 13. $180\sqrt{6}$ 15. $30\sqrt{3} - 20$

17. $\dfrac{ax^2 - cm^2 p + 2mp}{m^2 p}$ 19. $\left(3, -\dfrac{1}{2}\right)$

21. $x = 20; y = 10; P = \dfrac{99}{7}$

23. 7 25. $\dfrac{17}{7}$

27. $2x^2 + 3$ 29. 16

practice   a. $(x + 1)(x - 7)$ c. $(-3x^2)(x + 1)(x - 8)$

problem set 26

1.  $R_S T_S = R_O T_O, T_S = 20, T_O = 8, R_O = R_S + 60$; 800 miles

3. $N_F = 1000$; $N_T = 200$ **5.** 2080   **7.** $x^3 - x^2 + x - 1 - \dfrac{1}{x + 1}$

9. $2xym(3xm^4 - x + 2)$ **11.** $(x + 3)(x - 2)$ **13.** $(ab)(x + 2)(x - 1)$

15. $108\sqrt{2}$ **17.** $15\sqrt{6} - 12$ **19.** 6×10^{-22} **21.** $\sqrt{58}$

23. $x = 80$; $y = 30$; $m = 150$; $z = 20$ **25.** $BC = 12$ m; 452.16 m²

27. $-\dfrac{21}{2}$ **29.** $-8x^8 y$

practice   **a.** $\dfrac{6m^3 + 23m^2 + 2m + 8}{m^2(m + 4)}$

problem set 27

1. $R_H T_H = R_R T_R, R_H = 4, R_R = 20, T_H + T_R = 18$; 60 miles

3. $N_G = 26$; $N_B = 10$ **5.** 1800 grams   **7.** $x^2 + 2x + 4 + \dfrac{2}{x - 2}$

9. $mx^2 y(x^2 - y^2 - 4)$ **11.** $a(x + 5)(x - 1)$ **13.** $-ax(x + 8)(x - 3)$

15. $p(x - 8)(x - 7)$ **17.** $3\sqrt{2} - 36$ **19.** $6\sqrt{6} - 18$

21. $\dfrac{6x^3 + 33x^2 + 3x + 18}{x^2(x + 6)}$ **23.** $(2, 3)$

25. $A = \sqrt{65}$; $B = \dfrac{5\sqrt{65}}{7}$; $C = \dfrac{48}{7}$

27. $\dfrac{97}{3}$ **29.** $5a^{-2}b^{-1}y$

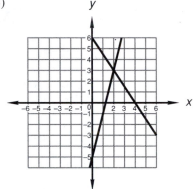

practice   **a.** $\dfrac{m}{z + x}$ **c.** $\dfrac{2\sqrt{2}}{3}$

problem set 28

1. $R_R T_R = R_P T_P, R_R = 60, R_P = 3, T_R + T_P = 21$; 60 miles

3. $N_G = 11$; $N_B = 4$ **5.** 142,500   **7.** $x^2 + 5x + 25 + \dfrac{118}{x - 5}$

9. $x^2 y^3 p(4p^2 - 16 - x^2 p^3)$ **11.** $a(x - 8)(x + 1)$ **13.** $-a^2(x + 1)^2$

15. $10\sqrt{3}$ **17.** 2×10^{-5} **19.** $\dfrac{\sqrt{15}}{20}$ **21.** $\dfrac{5x + 3}{(x + 1)^2}$

23. (a) $y = -4$ (b) $y = -\dfrac{1}{3}x + 2$ **25.** $x = 30$; 2.09 m²; 2.09 m

27. $\dfrac{139}{66}$ **29.** $1 - 12x^{-4}y^3 z^2$

practice $R_C T_C + R_M T_M = 66, T_C = 8, T_M = 7, R_M = 2R_C$; $R_C = 3$ mph, $R_M = 6$ mph

problem set 29

1. $\longmapsto\!\!\!\longrightarrow\!\!\!\longmapsto$ $R_W T_W + R_R T_R = 76, R_W = 4, R_R = 15, T_W + T_R = 8$; 4 hr

3. $\longmapsto\!\!\bullet\!\!\longmapsto$ $R_R T_R + R_M T_M = 7900, T_R = 6, T_M = 5, R_M = R_R - 400$;
$R_M = 500$ kph, $R_R = 900$ kph

5. 3,900,000   7. $x^2 + 2x + 5$ 9. $2x^2 y p^2 (p^2 - 3xp - 1)$ 11. $-m^2(x + 1)^2$

13. $1 + a$ 15. $18\sqrt{2} - 24$ 17. $\dfrac{m}{m + x}$ 19. $\dfrac{\sqrt{3}}{10}$ 21. $\dfrac{4x^2 + 14x + 24}{(x + 4)(x + 2)}$

23. $(1, 4)$ 25. 50 m; 31.4 m

27. $\dfrac{58}{49}$

29. $x^{-6} y^{15} z^{-9}$

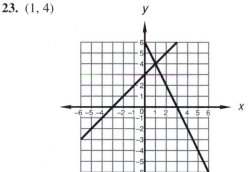

practice a. Valid; conclusion follows from premises

problem set 30

1. 480 grams

3. $\longmapsto\!\!\bullet\!\!\longmapsto$ $R_F T_F + R_B T_B = 68, T_F = 6, T_B = 4, R_F = 6$; 8 mph

5. $N_Q = 5; N_D = 10$ 7. $x^2 - x + 1 + \dfrac{1}{x + 1}$ 9. $3(x - 5)(x - 2)$

11. $a^2 b^2 (x + 7)(x + 2)$ 13. $-17\sqrt{2}$ 15. 7×10^{49} 17. $\dfrac{a}{p}$ 19. $\dfrac{\sqrt{2}}{15}$

21. $\dfrac{6x + 6}{(x + 2)(x + 3)}$ 23. $y = -\dfrac{1}{3}x + \dfrac{10}{3}$ 25. (C)   27. -18

29. $-a^2 x^2 y p^{-1} + 2x a^2 y p^{-1}$

practice $y = -4x + 6$

problem set 31

1. 40, 45, 50, 55

3. $\longleftarrow\!\!\!\longmapsto$ $R_J T_J = R_R T_R, R_J = 6, R_R = 30, T_J + T_R = 12$; 60 miles

5. 12,000 kg   7. $y = -3x + 3$ 9. $\dfrac{a}{a + b}$ 11. $\dfrac{3\sqrt{5}}{10}$ 13. $\sqrt{3}$

15. $1 + 4x$ 17. $\dfrac{x + 2}{x - 3}$ 19. $2ax(x - 5)(x - 4)$

21. $mc(x + 5)(x - 2)$ 23. $\dfrac{2}{25}$ 25. $x^3 - x^2 + x - 1 - \dfrac{1}{x + 1}$

27. $\dfrac{25}{4}$ 29. $A = 120; B = 150; C = 30; D = 30; E = 60; F = 30; P = 60$

practice a. $\dfrac{8\sqrt{15}}{15}$ c. $x = 1; p = 3$

problem set 32

1. $\longmapsto\!\!\longrightarrow\!\!\longmapsto$ $R_W T_W + R_J T_J = 56, R_W = 4, R_J = 8, T_W + T_J = 10;$
$D_W = 24$ miles; $D_J = 32$ miles

3. $N_N = 40; N_D = 70$ 5. 1200 7. $\dfrac{5\sqrt{6}}{6}$ 9. $-\dfrac{19\sqrt{15}}{15}$ 11. $A = 30°$

13. 1×10^{-10} 15. $1 + 4xy$ 17. $40\sqrt{2} - 60$ 19. $\dfrac{m^2 - 2}{m^2 - 5m}$

21. $x^3(x + 7)(x - 2)$ 23. $py(x + 6)(x - 2)$ 25. -30

27. $4x^4 + 2x^3 - 8x^2 + 12x + 8$ 29. $7\sqrt{2}$

practice

a. $\dfrac{mz + 3}{sz}$

problem set 33

1. 442 3. 2025 grams

5. $\longmapsto\!\!\longrightarrow\!\!\longmapsto$ $R_M T_M + R_C T_C = 540, R_M = 40, R_C = 60,$
$T_M + T_C = 11;$ 240 miles

7. $\dfrac{(bs + ax + bx)(s + x)}{ab(a + b)}$ 9. $x = 75; y = 70; z = 140$

11. $\dfrac{11\sqrt{10}}{10}$ 13. 23.28 cm   15. 1.5×10^{-22} 17. $12\sqrt{2}$

19. $\dfrac{ax + b(x + y) + x^2(cx + 4)}{x^2(x + y)}$ 21. $-x^2(x - 5)(x + 1)$

23. $ap(x - 5)(x + 4)$ 25. $\dfrac{36}{5}$ 27. $2x^3 + 4x^2 + 8x + 15 + \dfrac{30}{x - 2}$ 29. 12 ft

practice

$\longmapsto\!\!\longrightarrow\!\!\longmapsto$ 20 $R_T T_T + 20 = R_Z T_Z, R_Z = 30, R_T = 40, T_Z = T_T + 2;$ 11 a.m.

problem set 34

1. $\longmapsto\!\!\longrightarrow$ 60 $R_B T_B + 60 = R_E T_E, R_E = 40, R_B = 50, T_E = T_B + 3;$ 5 p.m.

3. $\longleftarrow\!\!\longmapsto$ $R_K T_K = R_Y T_Y, R_K = 10, R_Y = 3, T_K + T_Y = 13;$ 30 mi

5. 150 g   7. $\dfrac{y - a^2 b^2}{b - a^2}$ 9. 258.99 m^3 11. $y = \dfrac{1}{2}x$ 13. 0

15. 1.4×10^{-16} 17. $28\sqrt{3}$ 19. $\dfrac{ax^2 + bx^2 + cx(x + y)}{x^3(x + y)}$

21. $2x^2(x + 2)(x - 1)$ 23. $y(x - 2)^2$ 25. $-\dfrac{9}{2}$

27. $x^4 + 3x^3 + 5x^2 + 3x$ 29. $6\sqrt{2}$ cm^2

practice

a. $2160°$ c. $\dfrac{1}{8}$ e. $-\dfrac{1}{25}$

problem set 35

1. $1200 \;\;\bullet\!\!\longrightarrow$ $R_M T_M + 1200 = R_L T_L, R_L = 3R_M, T_L = T_M = 30;$
$R_M = 20$ yards per minute, $R_L = 60$ yards per minute

3. $\longmapsto\!\!\longrightarrow$ $R_D T_D = R_B T_B, T_D = 12, T_B = 4, R_B = R_D + 6;$ 36 miles

5. 100 tons   **7.** 325.6 m²   **9.** $\frac{1}{3}$ **11.** $-\frac{1}{16}$ **13.** $\frac{31\sqrt{35}}{35}$

15. $\frac{a-4b}{xy}$ **17.** $x = 35; y = 30; k = 150$ **19.** $x = 98; y = 93; p = 75$

21. $\frac{2x^2 - 2x + 4}{x^2(x+y)}$ **23.** $-a(x+7)(x-5)$ **25.** $36 - 14\sqrt{3}$

27. $-\frac{76}{5}$ **29.** $\frac{9}{4}$

practice   **a.** $\frac{x-9}{x-2}$

problem set 36

1. 1360   **3.** $\frac{15}{25}$

5. $R_J T_J + 80 = R_B T_B, T_B = T_J = 4, R_J = 30; R_B = 50$ mph

7. 3 in.   **9.** $\frac{(x-5)(x+3)}{(x-8)(x+4)}$ **11.** $-\frac{1}{9}$ **13.** $\frac{1}{9}$ **15.** $\frac{4-3x}{7+2x}$

17. 1×10^7 **19.** $24\sqrt{6} - 96$ **21.** $\frac{-3x^2 + 6x + 6}{x^2(x+2)(x+1)}$ **23.** $\frac{2}{3}$

25. (a) $y = 2$ (b) $y = \frac{1}{3}x - 2$ **27.** $x^3 - x^2 + x - 1 - \frac{1}{x+1}$ **29.** $3 - 3x^{-5}y^3$

practice   **a.** 120 grams   **c.** $x = 70; y = 5$

problem set 37

1. 4800 grams   **3.** 525   **5.** 200   **7.** $x - 2$ **9.** $\frac{1}{81}$ **11.** 81

13. $x = 100; y = 90; z = 80$ **15.** $\frac{ax+y}{ax-my}$ **17.** $\frac{7\sqrt{33}}{33}$

19. $x = 40; y = 110$ **21.** $\frac{x^2 - 4x}{(x+5)(x-1)}$ **23.** $\frac{41}{16}$ **25.** $30\sqrt{6} - 12$

27. $x^{-4}y^7$ **29.** 5×10^{20}

practice   **a.** $x^3 + 6x^2 + 12x + 8$

problem set 38

1. 15   **3.** 64 grams   **5.** 120 miles   **7.** $x^3 + 12x^2 + 48x + 64$ **9.** $0, 6, -8$

11. $\frac{x+4}{x-3}$ **13.** $-\frac{1}{8}$ **15.** -2 **17.** 111.64 m   **19.** $\frac{mp^2 - 20}{15p^2}$

21. $\frac{61\sqrt{22}}{22}$ **23.** $72 - 108\sqrt{2}$ **25.** $\frac{92}{17}$ **27.** $\sqrt{26}$ **29.** $p^{-20}y^{13}$

practice   **a.** $\frac{1}{2}, -\frac{1}{2}$ **c.** $x = 27; A = 63; z = 90$

problem set 39

1. 1820   **3.** 144 grams   **5.** 15 miles   **7.** $MP = 8$ ft; 200.96 ft²   **9.** $1, -1$

11. $x^3 - 3x^2 + 3x - 1$ **13.** -8 **15.** $\frac{a^2x^2 - 4}{x^2 + 6a}$ **17.** 1×10^{-45} **19.** $\frac{11\sqrt{6}}{6}$

21. $x = 30; y = 15$ **23.** 18   **25.** $\frac{55}{36}$ **27.** $\frac{2x^2 + 6xy + 2x + 5y}{y(x+1)^2}$

29. $3 + 8x^{-2}y^{-6}z^{-4}$

practice **a.** $\dfrac{5x}{p + sx + ayx}$

problem set 40

1. $1,968,000 **3.** 704 grams **5.** 150 miles **7.** $\dfrac{ym}{x + cm}$ **9.** $\dfrac{mc}{a - bc}$

11. $X = 21;\ Y = 69;\ Z = 90$ **13.** $0, 3, -6$ **15.** 8 **17.** $\dfrac{4x^2a^3 + y^2}{2y^2a^2 - 2}$

19. 2×10^8 **21.** $\dfrac{23\sqrt{21}}{21}$ **23.** 2 cm **25.** $-\dfrac{51}{2}$ **27.** $\dfrac{8x^2 + 11x + 6}{x(x + 2)(x + 1)}$

29. $1 - 3y^2p^{-2}$

practice **a.** $\dfrac{840}{12}$ ft **c.** $30(5280)(5280)(5280)(12)(12)(12)$ in.^{3}

problem set 41

1. $3, 5, 7, 9$ **3.** 400 **5.** $D_R = 64$ miles; $D_W = 12$ miles

7. $61(3)(3)(12)(12)$ in.^{2} **9.** $\dfrac{xm}{cm + k}$ **11.** $\dfrac{x^2\,km}{4\,pm - xy}$

13. $X = 70;\ Y = 60;\ K = 60$ **15.** $0, \dfrac{9}{4}$ **17.** 11.84 units^2 **19.** $\dfrac{1}{4}$

21. 7×10^{-22} **23.** $96 - 48\sqrt{3}$ **25.** $-\dfrac{17}{2}$

27. $\dfrac{-2x^3 - 2x^2 + 2x + 4}{x^2(x - 2)(x + 2)}$ **29.** $\dfrac{y^3}{x}$

practice **a.** 3×10^{10} **c.** 5×10^{38}

problem set 42

1. 396 **3.** 7200 grams **5.** $N_S = 2;\ N_L = 10$ **7.** 2×10^1

9. $40(3)(3)(3)(12)(12)(12)$ in.^{3} **11.** $\dfrac{axd}{pc - md}$ **13.** 8.37 cm^2 **15.** $\dfrac{5}{2}, -\dfrac{5}{2}$

17. $-\dfrac{1}{27}$ **19.** $-\dfrac{49\sqrt{33}}{33}$ **21.** $5\sqrt{5}$ **23.** $x = 26;\ y = 1$ **25.** $-\dfrac{21}{4}$

27. $-\dfrac{16}{(x - 2)(x + 2)}$ **29.** $1 - x^2$

practice **a.** 0.98 **c.** 0.48 **e.** 0.77 **g.** $52.56°$

problem set 43

1. 400,000 **3.** 184 grams **5.** 40 miles **7.** (a) 259.59 (b) 504.12

9. $4(12)(12)(12)$ in.^{3} **11.** $\dfrac{cxz}{m + ck}$ **13.** $a = \dfrac{\sqrt{113}}{2};\ b = \dfrac{21}{4};\ c = \dfrac{\sqrt{113}}{4}$

15. $\dfrac{9}{2}, -\dfrac{9}{2}$ **17.** 1 **19.** $\dfrac{4x^2 + 1}{ay^2 - 4x}$ **21.** $30\sqrt{6} - 6$ **23.** $y = -\dfrac{1}{7}x - \dfrac{37}{7}$

25. 6×10^{-7} **27.** $\dfrac{6x^3 - x^2 - 2x + 2}{x^2(x^2 - 1)}$ **29.** $\sqrt{37}$

practice **a.** $y = 2.46;\ H = 6.86$

problem set 44

1. 7   **3.** 1080   **5.** 50 miles   **7.** $B = 59$; $m = 12$; $x = 7.21$

9. $4(5280)(5280)$ ft^2 **11.** $\dfrac{3kpc}{lp + cd}$ **13.** $m\angle ABC = (180 - k)°$; $x = 1$, $y = 2$

15. $\dfrac{9}{5}, -\dfrac{9}{5}$ **17.** $\dfrac{x + 5}{x - 3}$ **19.** 1069.44 in.^{3} **21.** $-\dfrac{79\sqrt{34}}{34}$

23. (a) $y = 2$ (b) $y = \dfrac{1}{2}x - 4$ **25.** $-\dfrac{3}{32}$ **27.** $\dfrac{x^2 - 2x - 2}{x^2 - 4}$ **29.** $\dfrac{1}{8}$

practice   **a.** $\pm\sqrt{14}$ **c.** $-\dfrac{1}{7} \pm 2\sqrt{2}$

problem set 45

1. 400   **3.** 3024   **5.** $R_B = 40$ mph; $R_T = 70$ mph   **7.** $-7 \pm \sqrt{11}$

9. $-\dfrac{3}{4} \pm \sqrt{13}$ **11.** 6.50   **13.** $100{,}000(5280)(5280)(12)(12)$ in.^{2} **15.** $\dfrac{13}{3}$

17. $\dfrac{5cx}{m + ck}$ **19.** $\dfrac{5}{6}, -\dfrac{5}{6}$ **21.** $\dfrac{1}{243}$ **23.** $\dfrac{4x^2 p - 1}{6xp - p^2}$ **25.** $30\sqrt{3} - 20\sqrt{2}$

27. $\dfrac{43}{6}$ **29.** $\dfrac{-3x + 10}{x^2 - 4}$

practice   **a.** $\dfrac{-2\sqrt{21}}{7}$ **c.** $7^{3/4}$

problem set 46

1. 600   **3.** 212 grams   **5.** 20 mph   **7.** $\dfrac{38\sqrt{35}}{7}$ **9.** $6^{1/2}$ **11.** $x^{19/12}y^{29/12}$

13. $\dfrac{1}{4} \pm \sqrt{5}$ **15.** $A = 55.15$; $C = \sqrt{33}$ **17.** $\dfrac{bk}{ax - bc}$ **19.** $x = \dfrac{7}{3}$; $y = \dfrac{35}{12}$

21. $\dfrac{x - 2}{x - 7}$ **23.** $x = 170$; $y = 75$; $z = 115$ **25.** $42\sqrt{2} - 21$ **27.** $-\dfrac{29}{7}$

29. $\left(-\dfrac{8}{3}, \dfrac{1}{3}\right)$

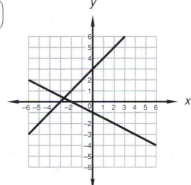

practice   **a.** $4^{7/6}$ **c.** $85(60)(60)\,\dfrac{\text{mi}}{\text{hr}}$

problem set 47

1. 4900   **3.** 2000   **5.** $D_W = 8$ km; $D_R = 40$ km   **7.** $\dfrac{805(5280)}{(60)(60)}\,\dfrac{\text{ft}}{\text{sec}}$

9. $5^{5/6}$ **11.** $x^{4/3}y^{7/6}$ **13.** $\dfrac{5\sqrt{6}}{2}$ **15.** 2.28   **17.** $3 \pm \sqrt{5}$ **19.** 5×10^3

21. $\dfrac{cx}{a - bx}$ **23.** $x = 30$; $y = 60$; $z = 60$; $s = 30$; $p = 60$; $m = 30$

25. $\dfrac{x - 2}{x - 5}$ **27.** $\left(\dfrac{12}{5}, -\dfrac{7}{5}\right)$ **29.** $0, -2, -10$

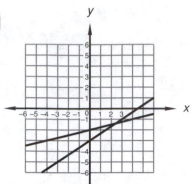

practice **a.** 105 **c.** 2

problem set 48

1. 1600 **3.** 252 grams **5.** $R_F = 30$ mph; $R_E = 60$ mph **7.** No solution

9. $\dfrac{60(5280)}{(60)(60)} \dfrac{\text{ft}}{\text{sec}}$ **11.** $m^{7/3}y^{5/6}$ **13.** $\dfrac{41\sqrt{10}}{10}$ **15.** $\dfrac{16}{7}, -\dfrac{12}{7}$

17. $C = 27$; $b = 4.46$ **19.** 1×10^{24} **21.** $20°$ **23.** $\sqrt{41}$ **25.** $-\dfrac{1}{8}$

27. $\sqrt{145}$ **29.** $y = -\dfrac{8}{5}x + \dfrac{41}{5}$

practice **a.** $y = \dfrac{5}{2}x - 11$

problem set 49

1. -3 **3.** 117.5 grams **5.** $220 \dfrac{\text{yd}}{\text{min}}$ **7.** 13.68 **9.** 1 **11.** $3^{5/6}$

13. $x^{3/2}y^{7/2}$ **15.** $\dfrac{-59\sqrt{14}}{14}$ **17.** $\dfrac{5}{3} \pm \sqrt{5}$ **19.** 2×10^{33} **21.** $\dfrac{mxy}{2y - 3x}$

23. $\dfrac{x - 3}{x - 2}$ **25.** $y = -\dfrac{1}{9}x + \dfrac{22}{9}$ **27.** $\dfrac{axy^2 - 3xy - 5a^2}{a^2y^4}$ **29.** 12

practice **a.** $\dfrac{9}{2} \pm \dfrac{\sqrt{53}}{2}$

problem set 50

1. $N_T = 20$; $N_W = 2$ **3.** 138 grams **5.** $180 \dfrac{\text{inches}}{\text{day}}$ **7.** $-6 \pm \sqrt{41}$

9. $y = 3x + 9$ **11.** 2 **13.** $\dfrac{400(3)(60)(60)}{5280} \dfrac{\text{mi}}{\text{hr}}$ **15.** $3^{5/4}$ **17.** $\dfrac{-62\sqrt{33}}{33}$

19. $\dfrac{2p + kpr}{mrx}$ **21.** $\dfrac{25}{4}$ **23.** $0, -4, 7$ **25.** -343

27. $y = -\dfrac{5}{3}x + \dfrac{5}{3}$ **29.** 12

practice **a.** $-6 + (9 - \sqrt{3})i$

problem set 51

1. 408 **3.** 355 grams **5.** 400 miles **7.** $-9 + (10 - \sqrt{5})i$ **9.** $-3i$

11. $1, -4$ **13.** 2.31 **15.** $\dfrac{20}{(12)(5280)(60)} \dfrac{\text{mi}}{\text{min}}$ **17.** $3^{7/6}$ **19.** $\dfrac{201\sqrt{26}}{26}$

21. 5×10^{22} **23.** $\dfrac{pR_1}{3R_1 - px}$ **25.** $y = \dfrac{2}{3}x + \dfrac{2}{3}$ **27.** $4p^2 - 2p$ **29.** $-\dfrac{50}{7}$

practice   800 ml 25%, 800 ml 5%

problem set 52

1. 50 ml 10%, 50 ml 40%   **3.** 460 grams   **5.** $D_L = 12$ km; $D_C = 16$ km

7. $5 - i$ **9.** $2 + 6i$ **11.** $\dfrac{5}{2} \pm \dfrac{\sqrt{33}}{2}$ **13.** 27.05   **15.** 67

17. $2^{13/8}$ **19.** $4x^{11/6}y^{19/6}$ **21.** 1×10^9 **23.** $\dfrac{x - 4}{x(x + 3)}$ **25.** $\dfrac{x^2 y - p^3}{mp - 1}$

27. $\dfrac{-x - 5}{x^2 - 9}$ **29.** $3x^2 + 12x + 47 + \dfrac{188}{x - 4}$

practice   **a.** 8%

problem set 53

1. 29%   **3.** 50 ml 40%, 200 ml 80%   **5.** 100 mph   **7.** $\dfrac{(32)(100)}{(2.54)(12)(3)}$ yd

9. $\dfrac{(0.063)(1000)(1000)(100)(100)}{(2.54)(2.54)(12)(12)(5280)(5280)}$ mi^2 **11.** $2 + 8i$ **13.** $-4 - 2i$

15. $3 \pm \sqrt{15}$ **17.** 60   **19.** $2^{4/15}$ **21.** $x^{11/15}y^{13/15}$ **23.** $-12\sqrt{2}$

25. $\dfrac{cmy}{cx + ky}$ **27.** $\sqrt{61}$ **29.** $x^2 - x - 1 + \dfrac{3}{x + 1}$

practice   **a.** $18.00R - 21.45U$

problem set 54

1. 100 liters 20%, 300 liters 70%   **3.** 32%   **5.** 300 miles   **7.** $-6.78R - 4.24U$

9. $100(1000)(1000)(100)(100)$ cm^2 **11.** $\dfrac{(60)(5280)(12)(2.54)}{(100)(1000)(60)(60)} \dfrac{\text{km}}{\text{sec}}$ **13.** $4 + 2i$

15. $-2 - 8i$ **17.** $\dfrac{7}{2} \pm \dfrac{\sqrt{61}}{2}$ **19.** $\dfrac{59}{3}$ **21.** 146   **23.** $\dfrac{cp}{py + px - a}$

25. $3^{7/15}$ **27.** -3 **29.** $\dfrac{xp^3 - 1}{xy - p^2}$

practice   **a.** $\dfrac{x(m + s)}{xt - a}$ **c.** 5, 6, 7 and $-2, -1, 0$

problem set 55

1. $-4, -3, -2$ and $-3, -2, -1$ **3.** 48 ml 5%, 12 ml 40%   **5.** 36, 39, 42

7. $\dfrac{cdx - mpx - cd}{c}$ **9.** $32.77R - 22.94U$

11. $\dfrac{4(3)(3)(3)(12)(12)(12)(2.54)(2.54)(2.54)}{(100)(100)(100)}$ m^3 **13.** $-2 - 2i$ **15.** $4 - 4i$

17. $\dfrac{5}{2} \pm \dfrac{3\sqrt{5}}{2}$ **19.** No solution   **21.** $2^{9/4}$ **23.** $x^3 y^{5/4}$ **25.** $\dfrac{-117\sqrt{22}}{11}$

27. 5   **29.** 1342.16 ft^3

practice   **a.** 70   **c.** 25

problem set 56

1. $-8, -6, -4$ and $-4, -2, 0$ **3.** 40 miles   **5.** 400   **7.** (a) 110   (b) 60

9. $\dfrac{bck + bmp - amk}{km}$ **11.** $-8.7R - 5U$ **13.** $\dfrac{60(1000)(100)}{(2.54)(60)(60)} \dfrac{\text{in.}}{\text{sec}}$

15. $3 + 13i$ **17.** $2^{11/30}$ **19.** $5^{7/6}$ **21.** $\dfrac{7}{2} \pm \dfrac{\sqrt{77}}{2}$ **23.** -1

25. $4x^3 + 12x^2 + 36x + 108 + \dfrac{323}{x - 3}$ **27.** $y = -8x - 45$ **29.** $\dfrac{17}{3}$

practice   **a.** 128 N/m^2

problem set 57

1. 25 N/m^2 **3.** 100 N/m^2 **5.** $-5, -3, -1, 1$ and $5, 7, 9, 11$

7. $\dfrac{mpx + cx - ap}{c + mp}$ **9.** $3.06R + 2.57U$ **11.** $\dfrac{40(60)(60)}{(2.54)(12)(5280)} \dfrac{\text{mi}}{\text{hr}}$

13. $-2 - 3i$ **15.** $6, -1$ **17.** $y = \dfrac{11}{5}x + \dfrac{41}{5}$ **19.** 4   **21.** $-\dfrac{14}{5}$

23. $x^{7/3} y^{1/6}$ **25.** $\dfrac{\sqrt{10}}{10}$ **27.** $\dfrac{x^2 y - p^5 z}{xz - 4p^5}$ **29.** $0, -4, 7$

practice   **a.** $-\dfrac{5}{6} \pm \dfrac{\sqrt{97}}{6}$

problem set 58

1. 600 K   **3.** 27%   **5.** 45 miles   **7.** $\dfrac{5}{4}, -1$ **9.** $x = 76$; $y = 92$; $z = 88$

11. $\dfrac{xy + ckx - ak}{ck + y}$ **13.** $7.07R - 7.07U$ **15.** $40(12)(12)(12)(2.54)(2.54)(2.54)$ cm^3

17. $2^{5/14}$ **19.** $-\dfrac{1}{8}$ **21.** $-2 + 5i$ **23.** $\dfrac{x + 2}{x - 5}$ **25.** $\dfrac{a^2 x^2 p^3 - 4m}{xa - 5mp^3}$

27. $4x^2 - 8x + 16 - \dfrac{37}{x + 2}$ **29.** $\dfrac{15}{7}$

practice   **a.** $N = 75F - 2200$ **c.** $\sqrt{34}\,\underline{/239°}$

problem set 59

1. $-7, -6, -5, -4$ and $-6, -5, -4, -3$ **3.** $D_W = 9$ miles; $D_R = 32$ miles

5. 7 liters   **7.** (a) 60   (b) 78   **9.** $2\sqrt{5}\,\underline{/207°}$ **11.** $\dfrac{1}{4} \pm \dfrac{\sqrt{41}}{4}$ **13.** $\dfrac{4}{3}, -1$

15. $\dfrac{4p - 4s}{c - 4x}$ **17.** $\dfrac{40(2.54)(60)(60)}{100} \dfrac{\text{m}}{\text{hr}}$ **19.** $6 + 5i$ **21.** 3   **23.** $3^{3/10}$

25. $-\dfrac{1}{8}$ **27.** $\dfrac{92\sqrt{26}}{13}$ **29.** $\dfrac{axy(a + y) + (x^2 + 2)(a + y) + ax^3 y}{ay^2(a + y)}$

practice   **a.** 24

problem set 60

1. 28   3. 16   5. $N_T = 40$; $N_P = 90$ 7. $2\sqrt{5}\,\underline{/117°}$ 9. $1, -\dfrac{3}{4}$

11. $Na = 0.037C + 3.9$ 13. $\dfrac{ad - cdy - my}{m + cd}$ 15. $-3 + 4i$

17. $\dfrac{(60)(60)(60)}{(2.54)(12)(5280)} \dfrac{\text{mi}}{\text{hr}}$ 19. $3^{7/6}$ 21. $\dfrac{1}{27}$ 23. $\dfrac{-23\sqrt{2}}{6}$

25. $\dfrac{32}{3}$ 27. $\left(\dfrac{21}{8}, \dfrac{11}{4}\right)$ 29. $y = \dfrac{2}{5}x + 18$

practice   250 gallons

problem set 61

1. 0.0003 kg/sec   3. 50 gallons   5. 100 kg   7. $2\sqrt{10}\,\underline{/342°}$

9. $\dfrac{2}{5} \pm \dfrac{2\sqrt{6}}{5}$ 11. $Pb = 42.6Sb - 31.9$ 13. $\dfrac{dkp - bdm - bdx - ak}{bd}$

15. $-2 - 7i$ 17. $\dfrac{40(60)(60)}{(2.54)(2.54)(2.54)} \dfrac{\text{in.}^3}{\text{hr}}$ 19. $2^{11/4}$ 21. $-\dfrac{1}{32}$ 23. $15\sqrt{6}$

25. $\dfrac{52}{5}$ 27. $\left(-\dfrac{8}{3}, \dfrac{5}{3}\right)$ 29. $y = \dfrac{3}{2}x + 7$

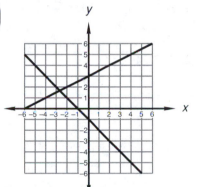

practice $\dfrac{1}{10} \pm \dfrac{\sqrt{59}}{10}i$

problem set 62

1. 2   3. 20 liters   5. 46.7%   7. $\sqrt{29}\,\underline{/111.8°}$ 9. $\dfrac{1}{4} \pm \dfrac{\sqrt{23}}{4}i$

11. $Bi = 16Hg - 72.6$ 13. $\dfrac{cp + ckr - amr}{mr}$ 15. $-2 - 5i$

17. $\dfrac{600}{(2.54)(2.54)(2.54)(12)(12)(12)(60)} \dfrac{\text{ft}^3}{\text{sec}}$ 19. $2^{9/4}$ 21. -32 23. $\dfrac{-79\sqrt{10}}{10}$

25. $\dfrac{25}{4}$ 27. $(-3, 1)$ 29. $y = -\dfrac{2}{7}x - \dfrac{59}{7}$

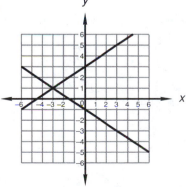

practice $-17.47R - 20.13U$

problem set 63

1. 3 **3.** 150 gallons **5.** 104 grams **7.** $11.82R + 24.63U$ **9.** (10, 20)

11. $\frac{1}{3} \pm \frac{\sqrt{11}}{3}i$ **13.** 3,440,000 cm² **15.** $\frac{bp + byz}{a + bc}$ **17.** $-i$

19. $\frac{4(5280)(5280)(5280)(12)(12)(12)(2.54)(2.54)(2.54)}{(100)(100)(100)(1000)(1000)(1000)}$ km³ **21.** $2^{9/20}$ **23.** $x^{7/4}y$

25. 8×10^{40} **27.** 80° **29.** 2

practice **a.** $\frac{b + 2bmz + z}{1 + 2mz}$ **c.** $1 + 3\sqrt{15} + 5i$

problem set 64

1. 1200 grams **3.** 40 liters 30%, 10 liters 80% **5.** 89.4%

7. $\frac{2am + ms + 3a^2}{2a^2 + as}$ **9.** $-5 - 5i$ **11.** $4 + 2\sqrt{6}$ **13.** $19 + 9i$ **15.** -10

17. $2\sqrt{13}\,\underline{/236.31°}$ **19.** $-\frac{1}{3} \pm \frac{\sqrt{14}}{3}i$ **21.** $W = -0.13Ir + 295$

23. $\frac{400(60)(60)}{(2.54)(2.54)(2.54)} \frac{\text{in.}^3}{\text{hr}}$ **25.** $\frac{139\sqrt{35}}{35}$

27. $\frac{dx + fmx - abf}{af}$ **29.** $0, -7, -8$

practice $R_C = 17; R_P = 51; T_C = \frac{108}{17}; T_P = \frac{96}{17}$

problem set 65

1. 48 kg **3.** 100 ml **5.** 80 kg **7.** $R_C = 15; R_P = 75; T_C = 7; T_P = 8$

9. $\frac{acx + bx + c}{ac + b}$ **11.** $\frac{ax + x + a^2}{a + 1}$ **13.** $-4 - 19i$ **15.** $5.48R + 9.14U$

17. $\left(3, -\frac{6}{7}\right)$ **19.** $-\frac{5}{4} \pm \frac{\sqrt{57}}{4}$ **21.** $m\widehat{AB} = (2y)°; x° = (2y)°$ **23.** $xy^{3/4}$

25. $\frac{-85\sqrt{2}}{6}$ **27.** $\frac{fx - acd - acfm}{ad + afm}$ **29.** 4×10^{10}

practice **a.** $\frac{1 + 3m}{z - 2}$ **c.** $a = \frac{7\sqrt{3}}{3}; b = \frac{14\sqrt{3}}{3}$

problem set 66

1. 50 grams **3.** 120 liters 20%, 80 liters 60% **5.** 800 grams **7.** $\frac{-x + 8}{x + 3}$

9. 4π in.² **11.** $R_P = 208; R_T = 52; T_P = 3; T_T = 7$ **13.** $\frac{a^2x^2 + ax + x}{ax + 1}$

15. $3 + \sqrt{6} + 4i$ **17.** $11.72R + 7.61U$ **19.** (8, 20) **21.** $-\frac{1}{2} \pm \frac{\sqrt{105}}{10}$

23. $Mg = 66.67Ca$ **25.** $x^{1/2}$ **27.** $\frac{-79\sqrt{10}}{10}$ **29.** $\frac{cdy + ky - dm}{cdp + kp}$

practice **a.** $\frac{-3 - \sqrt{7}}{2}$

problem set 67

1. 18,600   3. 100 pounds   5. 160 grams   7. $\dfrac{2\sqrt{2}-\sqrt{3}}{5}$ 9. $\dfrac{4x+5}{x-2}$

11. $R_P = 354;\ R_T = 59;\ T_P = 3;\ T_T = 5$ 13. $\dfrac{amx+a+x^2}{mx+1}$ 15. $4+i$

17. $5\underline{/216.87°}$ 19. $\dfrac{1}{8} \pm \dfrac{\sqrt{15}}{8}i$ 21. $x = \dfrac{5\sqrt{3}}{3};\ y = \dfrac{10\sqrt{3}}{3}$

23. $Ag = -0.2Au + 26$ 25. $x^{11/15}y^{13/15}p^{1/5}$

27. $\dfrac{aby+acy}{m+py}$ 29. $y = \dfrac{3}{7}x + \dfrac{34}{7}$

practice   a. 7.39×10^{-32} c. 2.39×10^{-10}

problem set 68

1. 2000   3. 200 pounds   5. 8400   7. (a) 3.99   (b) 0.14

9. All have equal areas.   11. $\dfrac{-2+\sqrt{2}}{2}$ 13. $3\sqrt{3}+5$ 15. $\dfrac{-2x-11}{x^2-16}$

17. $\dfrac{4rx^2+4mr+mx}{x^2+m}$ 19. $-23+14i$ 21. $37.42R - 24.22U$

23. $\dfrac{1}{4} \pm \dfrac{\sqrt{3}}{4}i$ 25. $\dfrac{(700)(60)}{(2.54)(2.54)(2.54)}\dfrac{\text{in.}^3}{\text{hr}}$ 27. $x^{17/6}y^{13/6}$

29. $\dfrac{my-apx-acxy}{px+cxy}$

practice 5.25×10^5 K

problem set 69

1. 2.08×10^{10} K   3. 80 tons   5. 5 p.m.   7. $\dfrac{6}{x-2}$ 9. $\dfrac{12+4\sqrt{2}}{7}$

11. $\dfrac{-6-8\sqrt{2}}{23}$ 13. $\dfrac{m^3+2m}{m^2+1}$ 15. $-20i$ 17. $-34.64R - 20U$

19. $(30, 20)$ 21. $-\dfrac{5}{6} \pm \dfrac{\sqrt{71}}{6}i$ 23. $4(12)(12)(12)(60)\dfrac{\text{in.}^3}{\text{hr}}$ 25. $3^{13/8}$

27. $\dfrac{-61\sqrt{2}}{3}$ 29. $-\dfrac{17}{11}$

practice $\dfrac{pyz+2psx-mx^2y}{mxy-2ps}$

problem set 70

1. 2×10^{13} K   3. $N_F = 40;\ N_T = 20$ 5. $-8, -6, -4$

7. $\dfrac{acrz-mxz-rtx}{tx-acz}$ 9. $\dfrac{4x^2+19x-26}{x^2+8x+12}$ 11. $\dfrac{3\sqrt{3}+1}{26}$

13. $R_B = 13;\ R_X = 26;\ T_B = 5;\ T_X = 4$ 15. $\dfrac{amy+m+xy^2}{ay^2+y}$ 17. $\sqrt{6}+i$

19. $\sqrt{41}\underline{/128.66°}$ 21. $\dfrac{1}{3}, 1$ 23. $a = 4\sqrt{3};\ b = 8$

25. (a) 5.12×10^{-9} (b) 6.33   27. $\dfrac{-73\sqrt{14}}{14}$ 29. $y = \dfrac{3}{4}x + \dfrac{13}{2}$

practice   a. $\dfrac{-b \pm \sqrt{b^2-4ac}}{2a}$

problem set 71

1. 20 N/m^2　　**3.** $130 \text{ ml } 10\%, 70 \text{ ml } 30\%$　　**5.** 3.2%　　**7.** $\frac{3}{2}, -1$

9. $x = 40; y = 50; z = 90$　　**11.** $\frac{4x + 9}{x - 2}$

13. $R_M = 40; R_P = 80; T_M = 4; T_P = 5$　　**15.** $\frac{a^2b + a^2 + b^2}{ab + a}$　　**17.** $-5 + i$

19. $4\sqrt{2}\underline{/315°}$　　**21.** $\frac{1}{3} \pm \frac{\sqrt{14}}{3}i$　　**23.** $Pb = 37B - 144$　　**25.** $x^{13/6}y^{7/2}$

27. $x^3 - 6x^2 + 12x - 8$　　**29.** (a) -6.56×10^{26}　(b) 6.86×10^{-2}

practice　　**a.** $-3.21R + 3.83U$

problem set 72

1. 100 ml　　**3.** 8200　　**5.** $2.5 \times 10^6 \text{ N/m}^2$　　**7.** $-1.03R + 2.82U$

9. $\frac{-b \pm \sqrt{b^2 - 4ac}}{2a}$　　**11.** $-\frac{7}{6} \pm \frac{\sqrt{13}}{6}$　　**13.** $\frac{3x^2 + 6x - 1}{x^2 - 2x - 8}$

15. $R_C = 33; R_P = 99; T_C = 5; T_P = 7$　　**17.** $\frac{x^3y^2 + x^2 + xy^2}{xy^2 + 1}$　　**19.** $6 + 7i$

21. $\frac{1}{4} \pm \frac{\sqrt{31}}{4}i$　　**23.** $2^{1/2}$　　**25.** $a^{7/4}y^{9/4}$　　**27.** $\frac{4\sqrt{13}}{3}$　　**29.** $y = -\frac{4}{3}x - \frac{28}{3}$

practice　　**a.** $\frac{5\sqrt{3} + 3}{6}$

problem set 73

1. 218 grams　　**3.** 96 miles　　**5.** $1.4 \times 10^{10} \text{ K}$　　**7.** 0　　**9.** $H = -9C + 1022$

11. $\sqrt{97}\underline{/66.04°}$　　**13.** $-\frac{5}{4} \pm \frac{\sqrt{15}}{4}i$　　**15.** $\frac{m_2 p}{ckm_2 + kp}$

17. $R_A = 40; R_B = 80; T_A = 4; T_B = 3$　　**19.** $-11 - 13i$　　**21.** $\frac{3}{2} \pm \frac{\sqrt{15}}{2}i$

23. $3^{9/4}$　　**25.** $ax^{1/4}y^{3/2}$　　**27.** $\frac{x + 5}{x + 3}$　　**29.** (a) 1.22×10^{34}　(b) 6.01×10^{-4}

practice　　$R_B = 30 \text{ mph}; R_D = 15 \text{ mph}; T_B = 4 \text{ hr}; T_D = 7 \text{ hr}$

problem set 74

1. $R_B = 7 \text{ mph}; R_G = 3.5 \text{ mph}; T_B = 4 \text{ hr}; T_G = 8 \text{ hr}$　　**3.** $1.5 \times 10^{14} \text{ K}$

5. $-1, 1, 3 \text{ and } 5, 7, 9$　　**7.** $\frac{6 - 5\sqrt{2}}{7}$　　**9.** $N = 70F - 2050$　　**11.** $5\sqrt{2}\underline{/225°}$

13. $-\frac{1}{6} \pm \frac{\sqrt{47}}{6}i$　　**15.** $\frac{ckmp}{m - acp}$　　**17.** $\frac{a^2x^2 - ax^2 - a^2}{ax - x}$

19. $\frac{2x - 4}{x^2 + 7x + 10}$　　**21.** $\frac{1}{6} \pm \frac{\sqrt{83}}{6}i$　　**23.** $2^{9/4}$　　**25.** $4x^{4/3}y^{25/6}$

27. $y = 4x + 8$　　**29.** 20

practice　　$\frac{-x^2 - 5x - 7}{x^2 - 3x - 4}$

problem set 75

1. $R_D = 60 \text{ mph}; R_E = 120 \text{ mph}; T_D = 2 \text{ hr}; T_E = 3 \text{ hr}$　　**3.** 25 days

5. $N_L = 18; N_P = 20$　　**7.** $\frac{x^2 - x - 11}{x^2 - x - 12}$　　**9.** $\frac{30 + \sqrt{3}}{39}$

11. $V = -1.25K + 146$ **13.** $18.06R + 1.58U$ **15.** $\frac{1}{5} \pm \frac{\sqrt{19}}{5}i$

17. $x = \frac{16\sqrt{3}}{3}$; $y = \frac{8\sqrt{3}}{3}$ **19.** $-9 - 6i$ **21.** $-\frac{1}{6} \pm \frac{\sqrt{23}}{6}i$

23. $2^{8/3}$ **25.** $m^{11/12}p^{31/12}$ **27.** (a) 2.53×10^{13} (b) 4.34 **29.** $\frac{86}{3}$

practice   **a.** $(33, 11, 12)$

problem set 76

1. $R_C = 4$ mph; $R_T = 12$ mph; $T_C = 8$ hr; $T_T = 6$ hr   **3.** 200 ml

5. $N_B = 5$; $N_R = 2$ **7.** $(15, 28, 5)$ **9.** $\sqrt{17}\underline{/194.04°}$ **11.** $-3 - \sqrt{2}$

13. $Al = 10.5B + 138$ **15.** $\frac{art - cmt - cmr}{mt + mr}$ **17.** $\frac{1}{14} \pm \frac{\sqrt{29}}{14}$ **19.** $7 - 8i$

21. 6 **23.** -32 **25.** $\frac{-19\sqrt{6}}{2}$ **27.** $y = \frac{11}{7}x + \frac{6}{7}$ **29.** $\frac{29}{54}$

practice   **a.** -3

problem set 77

1. $R_H = 40$ mph; $R_S = 60$ mph; $T_H = 8$ hr; $T_S = 4$ hr   **3.** 450 liters

5. 8 mph   **7.** 16   **9.** $(1, 2, 4)$ **11.** $-9.1R + 4.05U$

13. $\frac{9x + 8}{x^2 - 2x - 15}$ **15.** $\frac{-13 - 5\sqrt{5}}{4}$ **17.** $\frac{bmR_1 x}{aR_1 - amx}$ **19.** $(14, 10)$

21. $-2 \pm \frac{\sqrt{14}}{2}$ **23.** 2 cm, $2\sqrt{2}$ cm, 1 cm   **25.** $2^{7/12}$

27. $95 + 48\sqrt{5}$ **29.** $2\sqrt{41}$

practice $6.62R + 45.16U = 45.64\underline{/81.66°}$

problem set 78

1. $R_J = 135$ mph; $R_R = 45$ mph; $T_J = 7$ hr; $T_R = 3$ hr   **3.** 4000 grams

5. 20, 10 and $-20, -10$ **7.** 9   **9.** 25   **11.** $(1, 3, 4)$

13. $\frac{-3x^2 - 20x - 3}{x^2 + 5x - 14}$ **15.** $\frac{-6 - 4\sqrt{2}}{3}$ **17.** $-\frac{1}{2} \pm \frac{\sqrt{3}}{2}i$ **19.** $\frac{aR_1 + ax}{mxR_1}$

21. $-3 + 2i$ **23.** $400(2.54)(2.54)(2.54)(60) \frac{\text{cm}^3}{\text{min}}$ **25.** -2 **27.** $\frac{19\sqrt{10}}{5}$

29. 6.25×10^{-27}

practice   **a.** $(12{,}000)(1000)$ ml   **c.** $y = 3$; $z = 3\sqrt{2}$

problem set 79

1. \$7140   **3.** 27.3%   **5.** $N_{Swords} = 80$; $N_{Spears} = 280$

7. $\frac{20(12)(12)(12)(2.54)(2.54)(2.54)}{1000}$ liters   **9.** 1

11. $2.24R - 0.58U = 2.31\underline{/345.5°}$ **13.** $\frac{-1 - \sqrt{3}}{4}$ **15.** $\frac{-b \pm \sqrt{b^2 - 4ac}}{2a}$

17. $\frac{aR_1 R_2 - amR_2 x}{mR_1 x}$ **19.** $5 + 2i$ **21.** $1, -\frac{1}{3}$ **23.** $a = \sqrt{3}$; $b = \sqrt{6}$

25. $3^{13/4}$ **27.** $28 - \sqrt{2}$ **29.** $y = \frac{1}{5}x + 6$

practice   **a.** $80.36

problem set 80

1. $272   **3.** $N_B = 800$; $N_P = 1000$ **5.** $N_C = 10$; $N_D = 20$ **7.** 16

9. (2, 2, 4)   **11.** $2\sqrt{10}\underline{/108.43°}$ **13.** $7 - 4\sqrt{2}$ **15.** 5   **17.** $\dfrac{cmpx}{ac - mqx}$

19. (a) $m = \dfrac{5\sqrt{2}}{2}$; $n = \dfrac{5\sqrt{2}}{2}$ (b) $c = \dfrac{5}{2}$; $d = \dfrac{5\sqrt{3}}{2}$ **21.** (12, 8)

23. $\dfrac{600(60)}{(2.54)(2.54)(2.54)(12)(12)(12)} \dfrac{\text{ft}^3}{\text{hr}}$ **25.** $x^{7/4}y^{7/4}$

27. $\left(\dfrac{3}{7},\ -\dfrac{13}{7}\right)$ **29.** 0, -4, 8

practice   **a.** $-\dfrac{1}{6} + \dfrac{5}{6}i$

problem set 81

1. 625   **3.** $R_J = 25$ mph; $R_R = 75$ mph; $T_J = 15$ hr; $T_R = 5$ hr   **5.** 2800

7. $-\dfrac{4}{5} + \dfrac{1}{10}i$ **9.** 3   **11.** (2, 3, 1)   **13.** $\dfrac{-3x^2 - 2x - 2}{x^2 - 9}$

15. $\dfrac{-17 + \sqrt{3}}{26}$ **17.** $x = 45$; $y = 20$ **19.** $\dfrac{2a^4 - a^2}{a^2 + 1}$ **21.** (6, -6)

23. $4x^2 + 16x + 63 + \dfrac{254}{x - 4}$ **25.** $3^{17/8}$ **27.** $\dfrac{x^3y + 3x + 3y + 2x^2 + 2xy}{x^3y + x^2y^2}$

29. $\sqrt{170}$

practice   **a.** $\dfrac{3a + 3}{9a + ax + 9}$

problem set 82

1. 75   **3.** 4320 K   **5.** 276 grams, 32.4%   **7.** $\dfrac{p(my + 3)}{amy + 3a + by}$

9. $\dfrac{5}{17} - \dfrac{14}{17}i$ **11.** 2   **13.** (1, -6, -2)   **15.** $\sqrt{65}\underline{/119.74°}$ **17.** $7 - 4\sqrt{2}$

19. $\dfrac{16}{5}$ **21.** $\dfrac{az}{mx + amy}$ **23.** (5, 9)   **25.** $-\dfrac{2}{3}$, -1 **27.** $\dfrac{-29\sqrt{2}}{2}$

29. $\left(\dfrac{3}{8},\ \dfrac{3}{2}\right)$

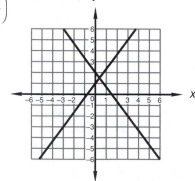

practice   **a.** $x^{6m/5}a^{7x/2}$ **c.** $m^{xz+x}z^{m+3}$

problem set 83

1. 400   **3.** 4 yd/sec   **5.** 640 ml 5%, 160 ml 20%   **7.** $a^{2x/3-3}m^{-2x+2}$

9. $\dfrac{a(ax + 1)}{ax^2 + 2x}$ **11.** $-1 - 3i$ **13.** (-2, 1, 2)   **15.** $-3.76R + 3.63U = 5.23\underline{/136°}$

17. $\dfrac{-2 - 3\sqrt{2}}{7}$ **19.** $\dfrac{acx}{a + cxy - px}$ **21.** $x = \dfrac{8\sqrt{3}}{3}$; $y = \dfrac{4\sqrt{3}}{3}$ **23.** $(7, -10)$

25. $\dfrac{-195\sqrt{7}}{14}$ **27.** $x^{8/3}y^{4/3}$ **29.** $\sqrt{145}$

practice   Inconsistent

problem set 84

1. 160   **3.** 824 grams, 6.8%   **5.** 350,000 tons   **7.** $y^{2b/3}x^{c+p+1}$ **9.** y^{a+2}

11. $\dfrac{bx - 1}{bpx - p - bx}$ **13.** $-\dfrac{1}{6} + \dfrac{7}{6}i$ **15.** $(1, 3, 3)$ **17.** $\sqrt{73}\,\underline{/69.44°}$

19. $\dfrac{34 - 23\sqrt{2}}{14}$ **21.** $\dfrac{mxy}{py + bpx - x^2 y}$ **23.** $-\dfrac{1}{4} \pm \dfrac{\sqrt{23}}{4}i$

25. $Y = -50B + 5050$ **27.** $\dfrac{-33\sqrt{15}}{5}$ **29.** 10

practice   **a.** $B = 27$; $T_D = 1$

problem set 85

1. 3   **3.** $N_D = 10$; $N_H = 20$ **5.** $-9, -7, -5$ and $-5, -3, -1$

7. $(0, -4)$ and $\left(\dfrac{16}{5}, \dfrac{12}{5}\right)$ **9.** $a^{x/2}b^{x/3-2}$ **11.** $\dfrac{a(c^2 + b)}{3c^2 + 2b}$ **13.** $\dfrac{10}{11} - \dfrac{7\sqrt{2}}{11}i$

15. $(2, 1, 1)$ **17.** $\dfrac{2x + 7}{x - a}$ **19.** $\dfrac{-7 - 5\sqrt{3}}{4}$ **21.** -2 **23.** $(9, 6)$

25. $Na = -5Mg + 360$ **27.** $\dfrac{rt}{s}$ **29.** $\dfrac{137\sqrt{15}}{10}$

practice   **a.**

$-8\ -7\ -6\ -5\ -4$

problem set 86

1. 714.22 liters   **3.** 900 liters   **5.** 16,800   **7.**

$-2\ -1\ \ 0\ \ 1\ \ 2$

9.

$-4\ -3\ -2\ -1\ \ 0$

11. $\left(-\dfrac{1}{2} + \dfrac{\sqrt{7}}{2}, \dfrac{1}{2} + \dfrac{\sqrt{7}}{2}\right)$ and $\left(-\dfrac{1}{2} - \dfrac{\sqrt{7}}{2}, \dfrac{1}{2} - \dfrac{\sqrt{7}}{2}\right)$

13. x^{3a+2} **15.** $-\dfrac{1}{2} + \dfrac{1}{2}i$ **17.** 23   **19.** $(-6, -3, -12)$ **21.** $\sqrt{109}\,\underline{/253.3°}$

23. $\dfrac{amy}{x + 2 - cy - mxy}$ **25.** $\dfrac{-25\sqrt{14}}{14}$ **27.** $\dfrac{(40)(60)}{(12)(12)(3)(3)}\dfrac{\text{yd}^2}{\text{hr}}$

29. $\left(-\dfrac{80}{23}, \dfrac{37}{23}\right)$

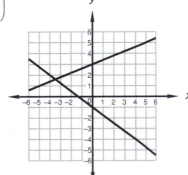

practice $\dfrac{47}{8}$

problem set 87

1. $R_B = 60$ mph; $R_J = 240$ mph; $T_B = 2$ hr; $T_J = 6$ hr

3. 480 in.³ 30%, 120 in.³ 80%   5. 39.34%, 1464 grams   7. $\dfrac{61}{19}$

9.

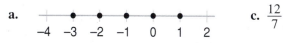

$$\begin{array}{cccccc} & 4 & 5 & 6 & 7 & 8 \end{array}$$

11. $\left(\dfrac{1}{5} + \dfrac{2\sqrt{19}}{5}, -\dfrac{2}{5} + \dfrac{\sqrt{19}}{5}\right)$ and $\left(\dfrac{1}{5} - \dfrac{2\sqrt{19}}{5}, -\dfrac{2}{5} - \dfrac{\sqrt{19}}{5}\right)$ 13. $m^{a-6}y^{-b}$

15. $-\dfrac{3}{2} - \dfrac{7}{2}i$ 17. (a) -1.92×10^{11} (b) 12.92   19. 49   21. $7 - 4\sqrt{3}$

23. $\dfrac{-74\sqrt{21}}{7}$ 25. $x = 105, y = 60$ 27. $\dfrac{(10)(1000)(100)}{(2.54)(60)(60)} \dfrac{\text{in.}}{\text{sec}}$ 29. $3^{17/8}$

practice a. $\sqrt{58}$

problem set 88

1. $-10, -8, -6$ and $0, 2, 4$ 3. $R_D = 50$ mph; $R_G = 70$ mph; $T_D = 8$ hr; $T_G = 16$ hr

5. $\dfrac{(1)(5)}{(0.0821)(251)}$ moles   7. $\sqrt{89}$ 9.

$$\begin{array}{ccccc} -2 & -1 & 0 & 1 & 2 \end{array}$$

11. $\left(\dfrac{1}{2} + \dfrac{\sqrt{3}}{2}, -\dfrac{1}{2} + \dfrac{\sqrt{3}}{2}\right)$ and $\left(\dfrac{1}{2} - \dfrac{\sqrt{3}}{2}, -\dfrac{1}{2} - \dfrac{\sqrt{3}}{2}\right)$ 13. $y^{7a/2+4}b^{1-2a}$

15. $\dfrac{c(cp^2 - 1)}{cp^2 - c^3 - 1}$ 17. (a) 1.37 (b) 2.14×10^{-4} 19. $(4, 2, -2)$

21. $4\sqrt{10}/288.43°$ 23. $-10\sqrt{3}$ 25. $\dfrac{-7 - 5\sqrt{3}}{2}$ 27. $-1 \pm i$

29. $x = \dfrac{49}{16}, y = 97$

practice a.

$$\begin{array}{ccccccc} -4 & -3 & -2 & -1 & 0 & 1 & 2 \end{array}$$

c. $\dfrac{12}{7}$

problem set 89

1. 2100   3. 560 liters   5. 80   7.

$$\begin{array}{cccccccc} -1 & 0 & 1 & 2 & 3 & 4 & 5 & 6 \end{array}$$

9. 4   11. $B = 14$; $T_D = 3$ 13. $x^{3a/2+1}y^{-b}$ 15. $\dfrac{k(am + 1)}{am^2 + m^2 + m}$

17. $-\dfrac{14}{17} + \dfrac{5}{17}i$ 19. $(1, 4, -2)$ 21. $3.13R - 7.47U = 8.10/-67.27°$

23. $\dfrac{amx + cmx}{x - acm - c^2m}$ 25. 0   27. $\dfrac{1 \pm \sqrt{141}}{14}$

29. $x = 15$; $A = 70$; $B = 110$

practice $\left(\dfrac{8}{3}, \dfrac{5}{3}, \dfrac{4}{3}\right)$

problem set 90

1. $R_P = 30$ mph; $R_R = 80$ mph; $T_P = 10$ hr; $T_R = 2$ hr   3. $N_B = 5$; $N_R = 10$

5. 1536   7. $(3, 2, -2)$ 9.

$$\begin{array}{ccccc} 1 & 2 & 3 & 4 & 5 \end{array}$$

11. $(-2 + \sqrt{5}, 2 + \sqrt{5})$ and $(-2 - \sqrt{5}, 2 - \sqrt{5})$ 13. $y^c m^{-3b/2}$

15. $\dfrac{p(x - p)}{x^2 - px - px^2}$ 17. $\dfrac{20}{3}\pi$ cm   19. $6.06R + 7.77U$

21. $\dfrac{5 + 3\sqrt{3}}{4}$ **23.** $5 + 2i$ **25.** $3^{3/4}$ **27.** $-\dfrac{1}{12} \pm \dfrac{\sqrt{119}}{12}i$ **29.** $\dfrac{95}{13}$

practice   **a.**

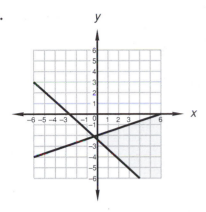

problem set 91

1. 2960 mm Hg   **3.** 75.8%   **5.** 49   **7.**

9.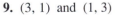
−1 0 1 2 3

11. $B = 15;\ T_D = 3$ **13.** $x^{2b-ab/2}$

15. $\dfrac{m(a^2 + m)}{a^3 + am + a^2m}$ **17.** $-\dfrac{12}{5} - \dfrac{9}{5}i$

19. $-2\sqrt{2} - 3$ **21.** $\dfrac{dmry}{x - bdmy}$

23. $O_P = 7I_P - 100$ **25.** $2 - 6i$

27. $\dfrac{-b \pm \sqrt{b^2 - 4ac}}{2a}$ **29.** $3x^2 - 6x + 10 - \dfrac{18}{x + 2}$

practice   **a.** $W = 1$ mph; $T = 10$ hr

problem set 92

1. $W = 4$ mph; $T = 5$ hr   **3.** 28 kph   **5.** 25 s   **7.**

9. $(3, 1)$ and $(1, 3)$

11.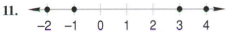
−2 −1 0 1 2 3 4

13. $\dfrac{x(xy + 1)}{y + xy^2 + xy}$ **15.** $-\dfrac{18}{25} + \dfrac{1}{25}i$

17. 36   **19.** $5\sqrt{17}\,\underline{/104.04°}$

21. $\dfrac{7 + 3\sqrt{2}}{2}$ **23.** $\dfrac{bR_2 xy}{aR_2 - bxy}$

25. $2i$ **27.** $-\dfrac{1}{6} \pm \dfrac{\sqrt{59}}{6}i$ **29.** $\dfrac{-3x^2 - 9x + 4}{x^2 - 9}$

practice   **a.** Two real number solutions

problem set 93

1. 10 mph   **3.** 2 mph   **5.** 800 K   **7.** Two complex number solutions

9.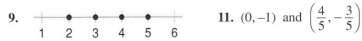
1 2 3 4 5 6

11. $(0, -1)$ and $\left(\dfrac{4}{5}, -\dfrac{3}{5}\right)$

13. $\dfrac{x(bc + 1)}{bc + 1 + ac}$ **15.** $\dfrac{19}{25} - \dfrac{8}{25}i$ **17.** 36   **19.** $\dfrac{36 + 25\sqrt{2}}{2}$

21. $2^{11/15}$ **23.** $\dfrac{m^2 d + bm^2 x - cdx}{dx}$ **25.** $-4i$ **27.** $-\dfrac{1}{6} \pm \dfrac{\sqrt{23}}{6} i$

29. $x^2 - 6x + 18 - \dfrac{56}{x + 3}$

practice   **a.** Function   **c.** Function   **e.** 50

problem set 94

1. 0.515 mole   **3.** 2 mph   **5.** 16   **7.** 8

9.

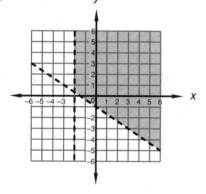

11. $(1, 1, 1)$ **13.** $x^{-7a/6} y^{2-2a}$

15. $\dfrac{x(ab^2 - 1)}{a^2 b^2 - a + b^2}$ **17.** $-i$

19. $0R + 0U$ **21.** $\dfrac{-13 - 2\sqrt{2}}{7}$

23. $\dfrac{aR_1 R_2 + bR_1 R_2 - R_2 x}{R_1 x}$

25. $-3 + i$ **27.** $-\dfrac{1}{4} \pm \dfrac{\sqrt{57}}{4}$

29. $N = -1.2R + 210$

practice   **a.** $(-1, -7), \left(\dfrac{7}{3}, 3\right)$

problem set 95

1. 48 km   **3.** 738.27 ml   **5.** $B = 12$ mph; $W = 3$ mph

7. $(\sqrt{5}, \pm\sqrt{11}), (-\sqrt{5}, \pm\sqrt{11})$ **9.** a, d   **11.**

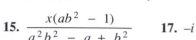

13. $a^{-x/2+1} x^a$ **15.** $\dfrac{5(2x + 2)}{5x + 4}$

17. $-1 + 4i$

19. $23.83R - 6.88U = 24.80 \underline{/-16.10°}$

21. $\dfrac{2 - \sqrt{2}}{2}$ **23.** $\dfrac{qx}{amq + bmq - rx}$

25. $\dfrac{-79\sqrt{21}}{21}$ **27.** 13 liters

29. (a) 11   (b) 5

practice   **a.** $\left(\dfrac{5}{2} + \dfrac{\sqrt{61}}{2}, -\dfrac{5}{6} + \dfrac{\sqrt{61}}{6}\right)$ and $\left(\dfrac{5}{2} - \dfrac{\sqrt{61}}{2}, -\dfrac{5}{6} - \dfrac{\sqrt{61}}{6}\right)$

problem set 96

1. 260   **3.** 10 kph   **5.** 350 ml 20%, 150 ml 60%   **7.** $(0, 2), (0, -2)$ **9.** a, b

11.

13. $a^{x/2+1/2} b^x$

15. $2^{8/3}$ **17.** $1 - i$

19. $2.34R + 3.52U = 4.23 \underline{/56.39°}$

21. $\dfrac{-3 + \sqrt{7}}{2}$

23. $\dfrac{cp + mpr - dm}{m}$

25. 1.81 moles   **27.** $\left(\dfrac{10}{9}, \dfrac{22}{9}\right)$ **29.** $3\sqrt{7}$

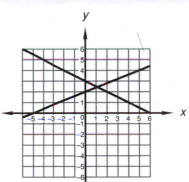

practice   (37, 20)

problem set 97

1. $N_T = 96, N_U = 60$ **3.** 1500 K   **5.** 2500   **7.** (12, 20)

9. $(\sqrt{3}, \pm 2\sqrt{2}), (-\sqrt{3}, \pm 2\sqrt{2})$ **11.** 17   **13.**

```
+---•---•---•---+
0   1   2   3   4
```

15. $\dfrac{p(mx^2 + 1)}{m^2 x^3 + mx - m^2 x}$ **17.** $2^{7/12}$ **19.** 49   **21.** $\dfrac{10 - 7\sqrt{2}}{4}$

23. $\dfrac{-40\sqrt{3}}{3}$ **25.** $\dfrac{ar^2 - x - cr^2 x}{1 + cr^2}$ **27.** 98 cm^2 **29.** $\dfrac{(400)(60)(60)}{(2.54)(2.54)(2.54)} \dfrac{\text{in.}^3}{\text{hr}}$

practice $\dfrac{89}{24}$

problem set 98

1. $D_R = 48$ miles; $D_W = 48$ miles

3. $T_C = 24$ hr; $T_M = 6$ hr; $R_C = 50$ mph; $R_M = 60$ mph   **5.** 2   **7.** $\dfrac{139}{75}$

9. $\left(-\dfrac{5}{6} + \dfrac{\sqrt{97}}{6}, \dfrac{5}{2} + \dfrac{\sqrt{97}}{2}\right)$ and $\left(-\dfrac{5}{6} - \dfrac{\sqrt{97}}{6}, \dfrac{5}{2} - \dfrac{\sqrt{97}}{2}\right)$ **11.** a, b, c

13.

```
◄──•───•───+───+───•───•──►
  -6  -5  -4  -3  -2  -1
```

15. $\dfrac{y(xy + 1)}{xy + 1 + y^2}$

17. $-2 + 3i$ **19.** $S = -50P + 380$ **21.** $-\dfrac{\sqrt{2}}{2}$ **23.** $\dfrac{nxz + nyz}{an - xz - yz}$

25. $3 - i$ **27.** $-\dfrac{1}{10} \pm \dfrac{3\sqrt{11}}{10}i$ **29.** $\dfrac{ay - x^2}{4y^2 - 3}$

practice

```
◄──•───•──┼─┼─┼──•───•──►
  -7  -6  -5  5  6   7
```

problem set 99

1. 87.1%   **3.** 20 mph   **5.** 480 ml 5%, 720 ml 10%

7.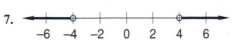

9. $\dfrac{43}{42}$

11. $\left(6, \dfrac{4}{3}\right), (-4, -2)$ **13.** $(2, -2, 1)$ **15.**

17. $x^{4-7a/4}$ **19.** $2^{28/15}$ **21.** -3

23. $1.69R + 9.63U = 9.78\underline{/80.05°}$

25. $\dfrac{6 - 4\sqrt{3}}{3}$ **27.** $4i$ **29.** 0.116 liter

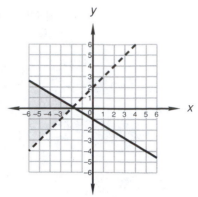

practice   **a.** $y = (x - 1)^2 + 2$

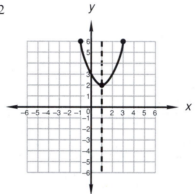

problem set 100

1. $R_B = 80$ mph; $R_S = 40$ mph; $T_B = 4$ hr; $T_S = 6$ hr

3. $N_C = 60$; $N_S = 40$ **5.** $-4, -3, -2$ and $10, 11, 12$

7. $y = -(x - 2)^2 + 8$

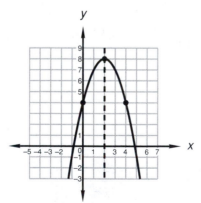

9.

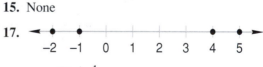

11. $\left(\dfrac{1}{2}, 1\right)$ **13.** $(\sqrt{3}, \pm\sqrt{2}), (-\sqrt{3}, \pm\sqrt{2})$

15. None

17.

19. $\dfrac{xy + 1}{x^2 y^2 + xy - y}$ **21.** $-1 + 2i$

23. $-4.25R + 9.05U = 10\underline{/115.16°}$

25. $\dfrac{-17 - \sqrt{5}}{4}$ **27.** $5 + 5i$

29. $\dfrac{(400)(12)(12)(12)}{60} \dfrac{\text{in.}^3}{\text{sec}}$

practice $P_P = \$13,036$

problem set 101

1. $P_P = \$60$ **3.** $\$13,228$ **5.** 6 mph

7. $y = (x + 1)^2 + 1$ **9.**

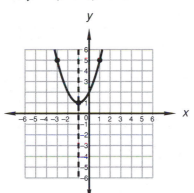

$-2 \quad -1 \quad 0 \quad 1 \quad 2$

11. $\left(-\dfrac{1}{8}, \dfrac{7}{4}\right)$

13. $(\sqrt{6},\ \pm\sqrt{10}), (-\sqrt{6},\ \pm\sqrt{10})$ **15.** $-\dfrac{15}{4}$

17.

$-14 \quad -12 \quad -10 \quad -8 \quad -6 \quad -4$

19. $\dfrac{my - 1}{my^2 - 2y}$ **21.** $-\dfrac{12}{13} - \dfrac{5}{13}i$

23. $9.26R + 7.52U = 11.93\underline{/39.08°}$

25. $\dfrac{8 - 5\sqrt{2}}{2}$ **27.** $7 + 4i$

29. $-\dfrac{1}{4} \pm \dfrac{\sqrt{31}}{4}i$

practice

a. 30 **c.** $fg(x) = x^2 + 2x - 24;\ D = \{\text{Positive integers}\}$

problem set 102

1. 460 miles   **3.** 9 **5.** $\$1200$ **7.** -76

9. $y = (x + 2)^2 - 2$

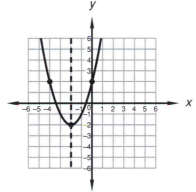

11.

$-4 \quad -3 \quad -2 \quad 2 \quad 3 \quad 4$

13. $\left(\dfrac{3}{2} + \dfrac{\sqrt{17}}{2},\ -3 + \sqrt{17}\right),$

$\left(\dfrac{3}{2} - \dfrac{\sqrt{17}}{2},\ -3 - \sqrt{17}\right)$

15. $(1, 3, 4)$

17.

$-4 \quad -3 \quad -2 \quad -1 \quad 0 \quad 1 \quad 2$

19. $x^{-5a/2}y^{8a/3}$ **21.** $x^{23/12}y^{5/6}$

23. $-2 + i$ **25.** $\dfrac{bx - amx^2 + am}{ax^2 - a}$ **27.** $2 + 3i$ **29.** $\dfrac{-b \pm \sqrt{b^2 - 4ac}}{2a}$

practice

a. $4x^2 - 8xy + 16y^2$

problem set 103

1. $\$890$ **3.** $T_D = 8$ hr; $T_M = 4$ hr; $R_D = 90$ mph; $R_M = 50$ mph

5. $22, 33, 44$ **7.** $x = 5, y = 12, P = 80$ **9.** $y = (x + 2)^2 + 2$

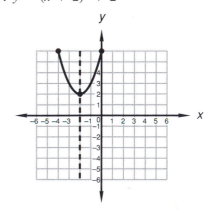

11.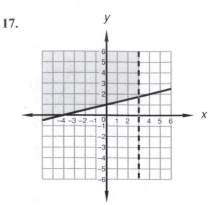
(number line with point at 2, arrow, marks 1 2 3 4 5)

13. $(2, 3)$

15. $\left(\dfrac{5}{2}, \pm\dfrac{\sqrt{15}}{2}\right), \left(-\dfrac{5}{2}, \pm\dfrac{\sqrt{15}}{2}\right)$

17.

19. $x^{2ab-5b/2}$ **21.** $x^{13/20}y^{17/20}$

23. $\dfrac{21}{29} - \dfrac{20}{29}i$

25. $-5.25R + 1.09U = 5.36\underline{/168.27°}$

27. 16 **29.** $9i$

practice   **a.** $\dfrac{513}{100,000,000}$

problem set 104

1. $N_D = 45$; $N_G = 4$ **3.** 360 ml 10%, 40 ml 20%

5. 500% of cost, 83.3% of price   **7.** $\dfrac{1418}{99,000}$ **9.** (a) $m = 3\sqrt{2}$, $n = 3$;
(b) $x = 4\sqrt{3}$, $y = 8$

11. a, c **13.** (number line, filled points from -3 to 5, marks -5 -3 -1 1 3 5)   **15.** $(9, 20)$ **17.** $(2, 4, 6)$

19. (number line, open circle at -2 to filled at 2, marks -3 -2 -1 0 1 2 3)   **21.** $\dfrac{k^2x - 1}{k^3x^2 - kx - k}$

23. $\dfrac{9}{10} + \dfrac{17}{10}i$ **25.** $2 - 6i$ **27.** $40.01R - 28.05U = 48.86\underline{/-35.03°}$

29. $1, -\dfrac{5}{4}$

practice   **a.** $-\dfrac{3}{2}, 4$

problem set 105

1. 1284 grams   **3.** 60 ml   **5.** 25% of selling price; 33.3% of cost

7. $x^3 - 5x^2 + x - 5$ **9.** $\dfrac{165}{9990}$ **11.** $x^2 - xy + y^2$

13. (number line, filled point at 0, marks -2 -1 0 1 2)   **15.** $(14, 40)$

17. $(1 + \sqrt{3}, -1 + \sqrt{3}), (1 - \sqrt{3}, -1 - \sqrt{3})$ **19.** $(-2, 2)$

21. $0, 2, -\dfrac{1}{3}$ **23.** $-\dfrac{2}{3}, -2$ **25.** $\dfrac{5}{2}, -1$ **27.** $x^{11a/6+8}y^{2b/3}$ **29.** $\dfrac{-25 - 11\sqrt{5}}{10}$

practice $\left(\dfrac{16}{13}, \dfrac{15}{13}, -\dfrac{85}{13}\right)$

**problem set
106**

1. $550   **3.** 12 mph   **5.** $(2, -1, -3)$ **7.** $4\sqrt{33}\,\pi^2\ \text{cm}^2$ **9.** $\dfrac{7006}{9,990,000}$

11. $y = -(x + 1)^2 - 2$ **13.** $\dfrac{73}{20}$

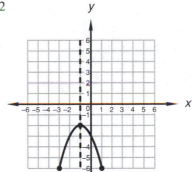

15. $\left(\dfrac{3}{10} + \dfrac{\sqrt{89}}{10}, -\dfrac{3}{2} + \dfrac{\sqrt{89}}{2}\right), \left(\dfrac{3}{10} - \dfrac{\sqrt{89}}{10}, -\dfrac{3}{2} - \dfrac{\sqrt{89}}{2}\right)$

17. $(3, -4)$ **19.** $\dfrac{5}{2} - \dfrac{1}{2}i$ **21.** $8.17R - 6.29U = 10.31\underline{/-37.6°}$ **23.** $\varnothing$

25. $-\dfrac{1}{3}, -2$ **27.** $-\dfrac{3}{2}, -5$ **29.** $5, -\dfrac{2}{3}$

practice   52

**problem set
107**

1. 320 ml 30%, 80 ml 60%   **3.** 78   **5.** $P_P = \$420, M = \$980$ **7.** $(1, 1, 2)$

9. $x^2 - xy + y^2$ **11.**

13. $(12, 9)$ **15.** $\left(\dfrac{1}{5} + \dfrac{2\sqrt{19}}{5}, -\dfrac{2}{5} + \dfrac{\sqrt{19}}{5}\right), \left(\dfrac{1}{5} - \dfrac{2\sqrt{19}}{5}, -\dfrac{2}{5} - \dfrac{\sqrt{19}}{5}\right)$

17.   **19.** $x^{23/12}y^{17/12}$ **21.** $4\sqrt{5}\underline{/116.57°}$

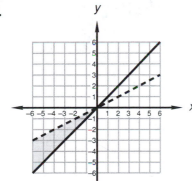

23. $\dfrac{p}{m^2\,px - 1}$ **25.** $10\sqrt{2}\ \text{m}$

27. $0, \dfrac{5}{2}, -3$ **29.** $\dfrac{74,139}{999}$

practice $(4p^2a^3 - xy^4)(16p^4a^6 + 4p^2a^3xy^4 + x^2y^8)$

**problem set
108**

1. $3840   **3.** 96   **5.** $B = 20\ \text{mph}; T = 4\ \text{hr}$

7. $(2x^4z^2 - my^3)(4x^8z^4 + 2x^4z^2my^3 + m^2y^6)$ **9.** 3

11. $y = (x - 1)^2 - 2$ **13.** $(10, 8)$ **15.** $(1, 2, 4)$

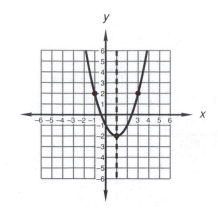

17.

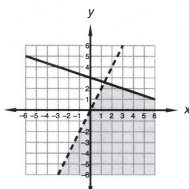

19. $7 - 2\sqrt{5}$ **21.** $-\dfrac{\sqrt{15}}{3}$

23. $\dfrac{(40)(60)(60)}{5280} \dfrac{\text{mi}}{\text{hr}}$

25. $W = 15E - 500$

27. $\dfrac{-b \pm \sqrt{b^2 - 4ac}}{2a}$

29. $0, -\dfrac{1}{3}, -3$

practice   **a.** $m + 2m^{1/2}a^{1/2} + a$

problem set 109

1. 72 **3.** $R_C = 6$ mph; $R_L = 3$ mph; $T_C = 10$ hr; $T_L = 21$ hr   **5.** 60 ml

7. $x^{1/2} + 2x^{1/4}y^{1/4} + y^{1/2}$ **9.** $(xy^2 - 3m)(x^2y^4 + 3xy^2m + 9m^2)$ **11.** $\dfrac{101,319}{99,000}$

13. ⊶━━● $\begin{array}{ccccc} -7 & -6 & -5 & -4 & -3 \end{array}$ **15.** $\left(6, \dfrac{1}{2}\right), (-1, -3)$ **17.** $(3, 3, 3)$

19. $\dfrac{6}{13} - \dfrac{9}{13}i$ **21.** $3^{4/3}$ **23.** $x^{-5a/3-2/3}y^{3b/2}$ **25.** $\dfrac{ka(k^2a - 1)}{k^3a - k - a^2}$

27. $\dfrac{(1000)(1000)}{60} \dfrac{\text{ml}}{\text{sec}}$ **29.** $0, 4, -\dfrac{1}{2}$

practice   **a.** ←━⊶━━⊶━→ $\begin{array}{cccccc} -2 & -1 & 0 & 1 & 2 & 3 \end{array}$

problem set 110

1. 63 **3.** 765 grams   **5.** 33.3% of S_P; 50% of cost

7. ←●━●━┊┊━●━●→ $\begin{array}{cccccc} -2 & -1 & 0 & 4 & 5 & 6 \end{array}$ **9.** $x - 2x^{1/2}y^{1/4} + y^{1/2}$

11. $(2x^3 - y^2p)(4x^6 + 2x^3y^2p + y^4p^2)$ **13.** $\dfrac{1361}{99,900}$

15. ←━⊶━●━→ $\begin{array}{ccccc} 0 & 1 & 2 & 3 & 4 \end{array}$ **17.** $(2, 3, -4)$ **19.** $(1, 2, -2)$ **21.** $-\dfrac{1}{3} - \dfrac{1}{3}i$

23. $a^{4-5b/2}x^{2-b/2}$ **25.** $\dfrac{23\sqrt{6}}{6}$ **27.** $\sqrt{241}\underline{/-75.07°}$ **29.** $0, \dfrac{1}{4}, -1$

practice $N_N = 6;\ N_D = 17;\ N_Q = 12$

problem set 111

1. $N_N = 15;\ N_D = 10;\ N_Q = 3$ **3.** 12   **5.** 400

7.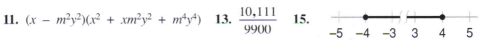
     9. $x - 2x^{1/2}y^{-1/2} + y^{-1}$

11. $(x - m^2y^2)(x^2 + xm^2y^2 + m^4y^4)$ **13.** $\dfrac{10,111}{9900}$ **15.**

17. $(-10, 10)$ **19.** $\left(\dfrac{1}{2} + \dfrac{\sqrt{7}}{2}, -\dfrac{1}{2} + \dfrac{\sqrt{7}}{2}\right), \left(\dfrac{1}{2} - \dfrac{\sqrt{7}}{2}, -\dfrac{1}{2} - \dfrac{\sqrt{7}}{2}\right)$

21. $\dfrac{40(2.54)(60)(60)}{100}\ \dfrac{\text{m}}{\text{hr}}$ **23.** $\dfrac{6 + 4\sqrt{3}}{3}$ **25.** $x^{3/2}y$

27.

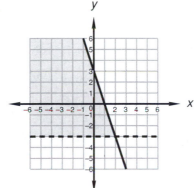

29. $\dfrac{5}{2}, -2$

practice   **a.**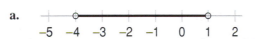

problem set 112

1. $N_N = 5;\ N_D = 5;\ N_Q = 10$ **3.** $1440

5. $R_F = 200$ mph; $R_H = 400$ mph; $T_F = 9$ hr; $T_H = 3$ hr

7.   **9.** $x + x^{1/2}y^{1/2} - x^{1/2}y^{-1/4} - y^{1/4}$

11. $\dfrac{39,742}{9900}$ **13.**   **15.** a, b   **17.** $-\dfrac{1}{10} \pm \dfrac{\sqrt{79}}{10}i$

19. $(8, 12)$ **21.** $\left(-\dfrac{1}{2}, -8\right), (4, 1)$ **23.** $-\dfrac{5}{6} + \dfrac{1}{6}i$ **25.** $x^{25/6}y^{4/3}$

27. $\dfrac{36 - 5\sqrt{2}}{89}$ **29.** $-\dfrac{1}{2}, 2$

practice   **a.** -5.26 **c.** 1.56×10^{-2} **e.** 6.92×10^{14}

problem set 113

1. $N_N = 10;\ N_D = 5;\ N_Q = 4$ **3.** 140 ml 60%, 60 ml 70%   **5.** 3400 K

7. (a) e^{5163} (b) 136.77   **9.**

11.   **13.**

15. $(2p^2k^5 - xm^2)(4p^4k^{10} + 2p^2k^5xm^2 + x^2m^4)$ **17.** $y = -(x - 2)^2 + 3$

19. $\dfrac{1}{6} \pm \dfrac{\sqrt{85}}{6}$ **21.** $(0, -2), \left(\dfrac{6}{5}, \dfrac{8}{5}\right)$

23. $(2, 3, 4)$ **25.** $\dfrac{8 + 3\sqrt{2}}{4}$

27. $-\dfrac{31\sqrt{3}}{3}$ **29.** $1, -\dfrac{2}{3}$

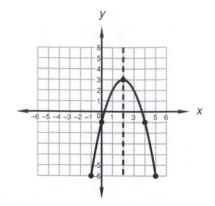

practice *A*

problem set 114

1. $-4, -2, 0$ and $8, 10, 12$

3. $R_F = 800$ mph; $R_S = 400$ mph; $T_F = 6$ hr; $T_S = 5$ hr **5.** 17 **7.** A

9. (a) -4.77 (b) 4.47 **11.** $(3 + 2\sqrt{10})$ cm **13.** \$112

15.

```
  +---•---•---•---•---•---•---+
 -2  -1   0   1   2   3   4   5
```

17.

```
  +---+---•---•---+
      8   9  10  11  12
```

19. $y = (x - 2)^2 + 3$ **21.** $(3, 15)$ **23.** $(3 + \sqrt{10}, 3 - \sqrt{10}), (3 - \sqrt{10}, 3 + \sqrt{10})$

25. $O = 22.5I - 1512$

27. $\dfrac{-99\sqrt{10}}{20}$ **29.** $0, -\dfrac{2}{5}, -1$

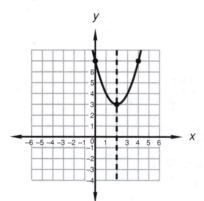

practice **a.** -3.57 **c.** $40,637$

problem set 115

1. 320 grams **3.** $R_B = 120$ mph; $R_H = 240$ mph; $T_B = 4$ hr; $T_H = 5$ hr **5.** 24

7. 4.53 **9.** $10^{3.412}$ **11.** \$225 **13.** $\dfrac{S\sqrt{2}}{2}; \dfrac{S^2}{2}$

15.

```
  ←---○---++-++---○---→
 -6  -5  -4   1   2   3
```

17.

```
  +---•---•---•---•---•---+
 -3  -2  -1   0   1   2   3
```

19. $x^{1/2} - 2x^{1/4}y^{-1/4} + y^{-1/2}$ **21.** $\dfrac{1037}{990}$ **23.** $\dfrac{23}{4}$ **25.** $(2, 4, -4)$

27. $-8 - 3i$ **29.** $\dfrac{1}{6} \pm \dfrac{\sqrt{73}}{6}$

practice **a.** 210

problem set 116

1. $-48, -42, -36$ **3.** $B = 10$ mph; $W = 3$ mph   **5.** 29   **7.** 6.89×10^{14}

9. \$3024   **11.** 15,120   **13.** $215°$ **15.**

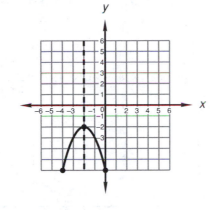

17.

19. $y = -(x + 2)^2 - 2$

21. $(3m^3p - x^4y)(9m^6p^2 + 3m^3px^4y + x^8y^2)$

23. $(15, 21)$

25. $(-3, -3, -3)$

27. $\dfrac{6 + \sqrt{6}}{2}$

29. 49

practice

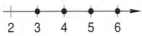

problem set 117

1. $N_N = 8$; $N_D = 6$; $N_Q = 2$

3. $R_G = 40$ mph; $R_M = 65$ mph; $T_G = 5$ hr; $T_M = 10$ hr

5. $M = \$3456$; $P_P = \$864$ **7.** $\dfrac{13}{162}$ **9.** 1.79   **11.** \$3038   **13.** E

15. $\dfrac{4\sqrt{2}}{3}$ **17.**   **19.**

21. $y = -(x - 2)^2 + 2$

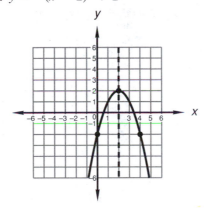

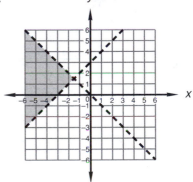

23. $(m - 2p^2k^3)(m^2 + 2mp^2k^3 + 4p^4k^6)$

25. $(16, 9)$ **27.** $-\dfrac{1}{5} + \dfrac{8}{5}i$

29. $x^{29/12}y^{1/4}$

practice   **a.** 12   **c.** 4

problem set 118

1. 4977   **3.** $N_N = 9$; $N_G = 9$ **5.** 41   **7.** 332,640   **9.** 1.01

11. 1,562,500   **13.** $9.08 \times 10^{-6} \dfrac{\text{mole}}{\text{liter}}$ **15.** 15   **17.** $\dfrac{162}{99,000}$

19. All integers   **21.** 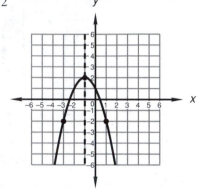   **23.** $-3 \pm i$

25. $(x^2y - 3p^2m^3)(x^4y^2 + 3x^2yp^2m^3 + 9p^4m^6)$ **27.** $(3, 3, -3)$ **29.** $3^{7/10}$

practice   **a.**

problem set 119

1. 1500 ml   **3.** 1000   **5.** 41   **7.**   **9.** 7

11. $\dfrac{547}{50}$ **13.** $(xy^3 - 4p^4m^3)(x^2y^6 + 4xy^3p^4m^3 + 16p^8m^6)$

15. $e^{40.57} = 4.15 \times 10^{17}$ **17.** B **19.** $5.89 \times 10^{-3} \dfrac{\text{mole}}{\text{liter}}$ **21.** $(-3 - \sqrt{2}, 4)$

23. $(1, 0, -6)$ **25.** $y = -(x + 1)^2 + 2$

27. $\dfrac{281}{75}$ **29.** $a^{b/2}x^{b/2}$

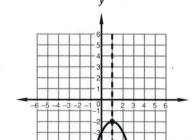

practice $B_N = 15;\ K_N = 20$

problem set 120

1. 210 liters   **3.** $M_N = 36;\ S_N = 2$ **5.** $L_N = 30;\ M_N = 24$

7.   **9.** 11,232,000   **11.** 4.31×10^4

13. 2.52×10^{-1} **15.** $7.41 \times 10^{-4} \dfrac{\text{mole}}{\text{liter}}$ **17.** $\dfrac{2161}{99,900}$ **19.** $y = -(x - 1)^2 - 2$

21. $(6, 4)$ **23.** $7.62R + 5.96U = 9.67 \underline{/38.03°}$

25. 220   **27.** $x^3 - 2x^{3/2}y^{3/2} + y^3$

29. $OB = 6;\ AB = 15;\ OP = \dfrac{18}{5}$

practice

problem set 121

1. 20 liters   **3.** $B = 10$ mph; $W = 4$ mph   **5.** $R_N = 10;\ Y_N = 5$ **7.** $\dfrac{13}{216}$

9. 490   **11.**   **13.** 3   **15.** 6.29×10^{-1}

17. $100°;\ 50°;\ 130°$ **19.** $0.99 \dfrac{\text{mole}}{\text{liter}}$ **21.** $16\sqrt{3}$ cm^2; $96\sqrt{3}$ cm^2

23. $y = -(x - 2)^2 - 3$

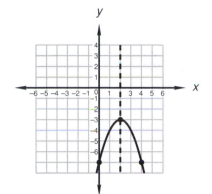

25. $(2, 3, -2)$

27. $(1, 2, 3)$

29. $0, \dfrac{7}{3}, -1$

practice   **a.** 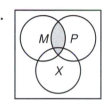   **c.**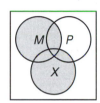

problem set 122

1. $N_N = 47; N_D = 22$ **3.** 28   **5.** $D_N = 5; P_N = 10$ **7.** 840

9. (a) $\log MN = \log M + \log N$ (b) $\log \dfrac{M}{N} = \log M - \log N$

(c) $\log M^N = N \log M$

11. ![number line] $-3\ -2\ -1\ \ 0\ \ 1\ \ 2\ \ 3$ **13.** ![number line] $-9\ -7\ -5\ -3\ -1\ \ 1\ \ 3$

15. 3.74×10^1 **17.** 8.49   **19.** 2.79   **21.** $5.88 \times 10^{-9} \dfrac{\text{mole}}{\text{liter}}$

23. $(226.08)(3)(3)(12)(12)$ in.^{2} **25.** $\dfrac{167}{99{,}000}$ **27.** $-1 - \dfrac{1}{2}i$ **29.** $x^{3/4}y^{17/12}$

practice   **a.** Two lines that are parallel to and 3 feet from the given line

problem set 123

1. See Lesson 123   **3.** See Lesson 123   **5.** See Lesson 123

7. $-1, 1, 3$ and $7, 9, 11$ **9.** 80 miles   **11.** $M_N = 24; S_N = 4$

13. $\{1, 3, 5, 7, 8, 10\}$ **15.** ![number line] $-1\ \ 0\ \ 1\ \ 2\ \ 3$

17. ![number line] $-8\ -6\ -4\ -2\ \ 0\ \ 2$ **19.** 1.07×10^{-4} **21.** C   **23.** $\dfrac{63}{2}$

25. $(am^9 - p^2y^{12})(a^2m^{18} + am^9p^2y^{12} + p^4y^{24})$

27. $y = -(x - 2)^2 + 3$ **29.** $3 - 4i$

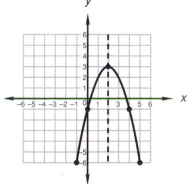

practice   **a.** SSS; AAAS; SAS; HL

problem set   **1.** See Section 124.C   **3.** See Section 124.C   **5.** See Section 124.C
124   **7.** See Lesson 123   **9.**

2   3   4   5   6

11.

-7 -5 -3 -1   1   3

13. 11   **15.** (a) $\log MN = \log M + \log N$

(b) $\log \dfrac{M}{N} = \log M - \log N$ (c) $\log M^N = N\log M$ **17.** 625   **19.** $\dfrac{251}{280}$

practice   **a.** See Section 125.D   **c.** See Section 125.D

problem set   **1.** See Section 125.D   **3.** See Section 124.C   **5.** See Section 125.D
125   **7.** See Lesson 123   **9.** See Lesson 123   **11.** 4791   **13.** 0.181

15.

1   2   3   4   5

17. 4096   **19.** 6.22×10^{11}

problem set   **1.** See Section 124.C   **3.** See Section 126.D   **5.** See Section 124.C
126   **7.** See Lesson 123   **9.** See Lesson 123   **11.** 6561   **13.** -3.73   **15.** 2.28

17.

1   2   3   4   5   6   7

19. 59

practice   **a.** See Lesson 127

problem set   **1.** See Lesson 127   **3.** See Section 124.C   **5.** See Section 124.C
127

7. See Lesson 123   **9.** See Lesson 123   **11.** 109,760   **13.** $7.76 \times 10^{-6} \dfrac{\text{mole}}{\text{liter}}$

15. 2.43×10^{-15} **17.** $y = -(x + 1)^2 + 5$ **19.** 840

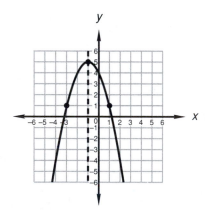

practice   **a.** See Lesson 128

problem set   **1.** See Section 124.C   **3.** See Section 124.C   **5.** See Section 124.C
128   **7.** See Lesson 127   **9.** See Section 125.D   **11.** 36   **13.** 124   **15.** $2\sqrt{\pi}$ cm

17. $C = 10, D = 5\sqrt{3}$ **19.** 60   **21.** $\dfrac{20}{7}$ **23.** $x = \dfrac{21}{2}, y = \dfrac{45}{2}$ **25.** 31.4 cm

practice   **a.** See Section 129.A

problem set 129

1. 4.43 cm^2 **3.** 65   **5.** 50   **7.** $P = \dfrac{16}{5}$; $Q = 10$ **9.** $x = \sqrt{3}$; $y = \sqrt{3}$

11. $\dfrac{3\sqrt{\pi}}{\pi}$ **13.** See Lesson 123   **15.** See Section 124.C   **17.** $10\sqrt{\pi}$ cm   **19.** 40

21. 204,800   **23.** 4.63×10^{-17} **25.** $y = (x - 1)^2 - 3$

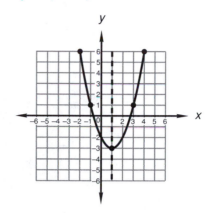

Index

Absolute value
  and conjunctions, 392–393
  definition of, 7
  and disjunctions, 393, 465
  inequalities, 392–394, 464–466
  and negative numbers, 393–394
  of numbers, 7–8
  of numerals, 7–8
  of slope, 70
  of zero, 8
Abstract equations, 231–232, 284
Abstract fractional equations, 177–179
Acute angles, 4
Addition. *See also* Elimination
  commutative property of, 8
  distributive property of, 36–37
  of fractions, 64–66
  and functions, 405–406
  of like terms, 34–35
  rule for equations, 37–39, 80–81
  of vectors, 259–260
Additive property of equality, 37
Age word problems, 468–469
Algebra, rules of, 149–150, 325
Algebraic expressions
  definition of, 34
  use of denominator-numerator
    same-quantity rule in, 149,
    325–326
Alternate angles. *See* Angles
Altitude, 15, 270. *See also*
  Pythagorean theorem
Angles
  bisecting of, 481
  copying of, 481
  definition in, 4
  degrees of, 4–6
  equality of in triangles, 95–96
  equidistance from sides of, 491
  formed with transversals, 25
  initial position of, 5
  inscribed, 66–68, 234–235,
    507–509
  kinds of, acute, 4
    adjacent, 5
    alternate, 136

Angles, kinds of (*Cont.*)
    central, 18–19
    complementary, 5, 6
    corresponding, 136
    dihedral, 501
    exterior, 136–137, 141
    interior, 136–137, 141
    lower base, 497
    negative, 291–292
    obtuse, 4
    reflex, 4
    remote interior, 141
    right, 4
    straight, 4
    supplementary, 5, 6
    upper base, 497
    vertical, 5, 135–136
  measurement of, 4–6
    in circles, 18–19, 234–235
    in inscribed quadrilaterals,
      157–158
    and parallel lines, 118
    in parallelograms, 166–167
    in polygons, 156–157
    in a rhombus, 175–176
    in right triangles, 190–193
    in trapezoids, 497
    in triangles, 23–24
  in overlapping right triangles,
    121–122
  in overlapping triangles, 113–114
  proofs of, 235–236
  in similar triangles, 92
  terminal position of, 5
  use of to solve equations, 88–89
Antilogarithms, 439–440
Approximately equal to, 10–11
Archimedes, 15
Arcs
  definition of, 18
  intercepted, 234–235
  major, 234
  measure of, 234–235
  measuring angles with, 497–499
  minor, 234
  theorems of, 497–499

Area
  calculating, 11–13
  of circles, 11–12
  of isosceles triangles, 75–76
  of rectangles, 11
  of surface (*see* Surface area)
  of triangles, 12
Atoms, 164–166
Axioms
  definition of, 134–135
  substitution, 73–74
  transitive, 341–342
  trichotomy, 341–342
Axis
  definition of, 14
  of symmetry, 397

Bars, and repeating digits, 412–414
Binomials
  and completing the square,
    212–213
  definition of, 53
Boat-in-the-river problems, 366–368
Braces, and set-builder notation, 458

Calculators, use of
  with antilogarithms, 439–440
  with exponential equations,
    448–449
  with logarithms, 438–439
  with pi, 11
  with powers, 277–279
  with right triangles, 190–193
  with roots, 277–279
  with scientific notation, 275–277
  with trigonometric functions,
    187–188
Cancellation, 117
Change sides—change signs. *See*
  Transposition
Chemical compounds, 164–166
Chemical mixture problems, 220–222,
  252–254

Abbreviations

U.S. CUSTOMARY		METRIC	
UNIT	ABBREVIATION	UNIT	ABBREVIATION
inch	in.	meter	m
foot	ft	centimeter	cm
yard	yd	millimeter	mm
mile	mi	kilometer	km
ounce	oz	gram	g
pound	lb	kilogram	kg
degree Fahrenheit	°F	degree Celsius	°C
pint	pt	cubic centimeter	cc
quart	qt	liter	L
gallon	gal	milliliter	mL

OTHER ABBREVIATIONS	
second	s
hour	hr
square	sq.
square mile	sq. mi
square centimeter	sq. cm

Equivalence Table for Units

LENGTH	
U.S. CUSTOMARY	METRIC
12 in. = 1 ft	10 mm = 1 cm
3 ft = 1 yd	1000 mm = 1 m
5280 ft = 1 mi	100 cm = 1 m
1760 yd = 1 mi	1000 m = 1 km

WEIGHT	MASS
U.S. CUSTOMARY	METRIC
16 oz = 1 lb	1000 g = 1 kg
2000 lb = 1 ton	

LIQUID MEASURE	
U.S. CUSTOMARY	METRIC
16 oz = 1 pt	1000 mL = 1 L
2 pt = 1 qt	1 cc = 1 mL
4 qt = 1 gal	

CONVERSION
2.54 cm = 1 inch
3600 s = 1 hr